T0292888

STATISTICAL ANALYSIS OF CONTINGENCY TABLES

STATISTICAL ANALYSIS OF CONTINGENCY TABLES

MORTEN W. FAGERLAND
STIAN LYDERSEN
PETTER LAAKE

CRC Press
Taylor & Francis Group
Boca Raton   London   New York

CRC Press is an imprint of the
Taylor & Francis Group, an **informa** business

A CHAPMAN & HALL BOOK

CRC Press
Taylor & Francis Group
6000 Broken Sound Parkway NW, Suite 300
Boca Raton, FL 33487-2742

Library of Congress Cataloging-in-Publication Data

Names: Fagerland, Morten W., 1974- | Lydersen, Stian. | Laake, Petter, 1947-
Title: Statistical analysis of contingency tables / Morten W. Fagerland,
Stian Lydersen, Petter Laake.
Description: Boca Raton, Florida : CRC Press, [2017] | Includes
bibliographical references and index.
Identifiers: LCCN 2017009369| ISBN 9781466588172 (hardback) | ISBN
9781315374116 (e-book) | ISBN 9781466588189 (adobe reader) | ISBN
9781315356556 (epub) | ISBN 9781315337494 (mobi/kindle)
Subjects: LCSH: Contingency tables. | Mathematical statistics.
Classification: LCC QA278.75 .F34 2017 | DDC 519.5--dc23
LC record available at https://lccn.loc.gov/2017009369

Visit the Taylor & Francis Web site at
http://www.taylorandfrancis.com

and the CRC Press Web site at
http://www.crcpress.com

Printed and bound in the United States of America by
Edwards Brothers Malloy on sustainably sourced paper

Contents

Preface

Researchers in many fields, such as biology, medicine, social and behavioral sciences, law, and economics regularly encounter data presented in contingency tables. The history of analysis of contingency tables goes back to the highly influential article by Karl Pearson in 1900. Since then, an almost inexhaustible amount of research has been done, and a vast number of different methods for the analysis of contingency tables is described in the literature. This is even the case for seemingly simple contingency tables such as the ubiquitous 2×2 table.

For the last fifteen years, we have taught courses and done research on statistical methods for the analysis of contingency tables. During this work, the need for an *up-to-date book* on this topic became increasingly clear to us. Beyond describing how to derive and calculate *effect size estimates, confidence intervals*, and *hypothesis tests* for contingency tables, we have emphasized *evaluations* and comparisons of methods, illustrations of the methods on *real-life data*, and *recommendations for practical use*. Large parts of the material in this book have until now been available only in specialized journals, and some parts of it have not been previously published at all.

This book should be accessible to anyone who has taken a basis course in statistics, but as a textbook in contingency tables, it aims primarily at the master's or PhD level.

This book covers contingency tables with unordered and ordered categories, with paired or unpaired data. It covers one- and two-dimensional tables, as well as three-dimensional tables in the form of stratified two-dimensional tables. Tables of higher dimensions are outside the scope of this book. *Logistic regression* is studied in several chapters; however, loglinear models are only briefly considered. *Meta-analysis* is covered in depth as part of Chapter 10 on stratified 2×2 tables. Although we have aimed at giving a comprehensive presentation of the analysis of contingency tables, several topics are not covered. Multivariate analyses, such as factor analysis of categorical data, latent class analysis, correspondence analysis, and item response theory are, for instance, not covered.

Chapter 1 deals with various introductory material, including a general introduction to categorical data and contingency tables in Sections 1.1–1.3. Sections 1.5–1.8 and Section 1.11 cover general theory that may be skipped by readers who use this book as a *reference book* or as a *guide for practical applications*.

Chapters 2–11 are organized by table size and type. Each chapter deals with one particular type of table, characterized by size (the number of rows and the number of columns), whether the variables are ordinal or nominal, whether the tables are stratified or not, and whether the observations are paired or unpaired. The start of each chapter describes relevant study and sampling designs, followed by real-life data examples and relevant statistical models. The main part of each chapter contains descriptions of the most important and interesting statistical methods. The methods are divided into confidence intervals for relevant effect measures—such as the difference between probabilities, the ratio of probabilities, and the odds ratio—and hypothesis tests. Within each section, the methods are illustrated with data examples, and their properties are evaluated according to criteria described in Section 1.4. Each chapter ends with recommendations for which methods to use and when to use them. These chapters are to a large extent *self-standing*, such that a reader will find all that he or she needs to analyze a particular table type in the corresponding chapter, without having to read the preceding chapters.

To ease guidance for the reader, we also provide an appendix with a *list of the 250 methods* described in these chapters, with recommendations marked, and with page references to where the methods are defined. Also included in the appendix is a list of examples with page references to where the examples are introduced and analyzed.

Chapter 12 deals with *sample size and power calculations*, and includes descriptions of the superiority, equivalence, and non-inferiority study designs.

The last chapter in this book, Chapter 13, contains miscellaneous topics: *diagnostic accuracy, inter-rater agreement, missing data, structural zeros, categorization of continuous variable, and ecological inference*.

At the book's website `http://contingencytables.com`, you will find *Matlab* and *R-code* for almost all the methods described in this book. The website also includes a list of errata and contact information for the authors. Please contact us if you discover any errors or have comments or questions about the book.

Morten W. Fagerland, Stian Lydersen, Petter Laake
Oslo/Trondheim, June 2017

1

Introduction

1.1 Categorical Variables

A *categorical variable* is a variable that can take on one of a countable, but usually fixed, number of values. The term *discrete variable* is synonymous with categorical variable. Continuous—or scale—variables, on the other hand, such as blood pressure or a person's height, can attain any real value within their range.

Some examples of categorical variables are:

- The sex of an offspring, with two possible outcomes: female or male.

- In an individual, a certain disease may be either present or not present.

- The alleles of the apolipoprotein E gene (APOE) may be of the genotype $\epsilon2$, $\epsilon3$, or $\epsilon4$. Several studies have shown that the presence of the APOE $\epsilon4$ genotype increases the risk of Alzheimer's disease. The number of APOE $\epsilon4$ alleles in a human is a categorical variable, with values 0, 1, or 2.

- Self-rated degree of depression. One of the questions on the EORTC QLQ-30 quality of life questionnaire from the European Organization for Research and Treatment of Cancer is: During the past week did you feel depressed? The response categories are 1 (not at all), 2 (a little), 3 (quite a bit), and 4 (very much).

- Blood types are categorized according to several systems. One such system is the ABO system, in which the blood type is classified as O, A, B, or AB.

Some authors use the term discrete numerical data or count data if the categorical data reflects the number of events, such as the number of offspring, or the number of visits to a general practitioner during one year. The number of APOE $\epsilon4$ alleles may also be regarded as a count variable. For categorical variables, it may be useful or even necessary to assign numbers to the categories before recording or analyzing them, as for self-rated degree of depression above.

A categorical variable is called *binary* or *dichotomous* if there are only two categories, such as sex of an offspring, or presence of a disease. When there are

1

more than two categories, we can distinguish between an *ordinal* or *nominal* variable. The variable is referred to as ordinal when there is a well-defined quantitative ordering between the categories, such as the number of APOE $\epsilon 4$ alleles, or degree of depression in the EORTC QLQ-30 questionnaire. The "distance" between these categories need not be equal, nor even well-defined, such as for degree of depression above. Categorical variables with no ordering, such as blood type, are referred to as nominal categorical variables. The distinction between ordinal and nominal variables is important for choosing the appropriate method of analysis.

In this book, we only consider categorical variables with a fixed or limited number of possible categories, such as in the examples above. In other situations, it may be appropriate to use a count variable with no limit to the number of possible outcomes. Some places in this book, we briefly mention these situations, such as in Section 4.3, where we encounter the Poisson distribution, which is an appropriate probability model for cross-sectional and cohort studies that use the sampling model called the no sums fixed model (see Section 4.2).

The remaining sections of this chapter give background material relevant for several of the subsequent chapters. Section 1.2 introduces the term contingency table and shows basic notation common for different table types. In Section 1.3, we describe important concepts of statistical estimation and inference, including effect measures, effect estimates, confidence intervals, and hypothesis tests. In later chapters, we present different methods for performing statistical inference, and we shall evaluate and compare the properties of most of the methods we consider. Section 1.4 introduces some common properties of confidence intervals and tests, and the criteria for evaluating them. Generalized linear models (GLMs) are introduced in Section 1.5, and the exponential family is briefly described in Section 1.6. The next five sections present several important, general concepts and approaches to constructing statistical methods for contingency tables: the Wald, likelihood ratio, and score principles (Section 1.7); deviance, likelihood ratio, and Pearson chi-squared tests for GLMs (Section 1.8); exact methods (Section 1.9); the mid-P approach (Section 1.10); and Bayes inference (Section 1.11).

1.2   Contingency Tables

A *contingency table*, also called a *cross-tabulation*, is a table in matrix format that displays the observed counts of categorical variables. It is most often used to describe and analyze the relationship between two or more categorical variables; however, in this book, we shall also treat the situation with only one categorical variable—either with two categories (Chapter 2) or more than two categories (Chapter 3)—as a contingency table problem.

We classify contingency tables according to the number of categorical variables they display. A tabulation of the observed counts of a single categorical variable is a one-way contingency table, a cross-tabulation of two categorical variables is a two-way contingency table, and so on. Three-way contingency tables are quite common—they occur naturally in meta-analyses and in stratified analyses—but four-way or higher-order contingency tables are not. In this book, we restrict our attention to one-, two-, and three-way contingency tables.

Table 1.1 shows how we form a one-way contingency table of a single categorical variable with c outcome categories. We put the categories in the columns, and we often refer to this as a $1 \times c$ table. The observed counts are denoted by n_i, where $i = 1, \ldots, c$ indicates the outcome categories. The sum of the observed counts are denoted by $N = \sum_i n_i$.

TABLE 1.1

The general setup for the observed counts of a one-way $1 \times c$ contingency table, where $c \geq 2$

	Variable 1			
	Category 1	Category 2	...	Category c	Total
Sample	n_1	n_2	...	n_c	N

The general setup for the two-way contingency table is shown in Table 1.2. We now assume that we have two categorical variables: Variable 1 with r outcome categories, and Variable 2 with c outcome categories. We let Variable 1 form the rows and Variable 2 the columns, such that we have a $r \times c$ contingency table. The observed counts are denoted by n_{ij}, where $i = 1, \ldots, r$ indexes the outcomes of Variable 1, and $j = 1, \ldots, c$ indexes the outcomes of Variable 2. The sum of all the n_{ij}s is given by $N = \sum_{i,j} n_{ij}$. An additional feature of the two-way contingency table is the marginal sums. These are given as n_{i+} (row margins) and n_{+j} (column margins). The row margins sum to N, and the column margins sum to N.

TABLE 1.2

The general setup for the observed counts of a two-way $r \times c$ contingency table, where $r \geq 2$ and $c \geq 2$

	Variable 2				
Variable 1	Category 1	Category 2	...	Category c	Total
Category 1	n_{11}	n_{12}	...	n_{1c}	n_{1+}
Category 2	n_{21}	n_{22}	...	n_{2c}	n_{2+}
$\vdots$	$\vdots$	$\vdots$	$\ddots$	$\vdots$	$\vdots$
Category r	n_{r1}	n_{r2}	...	n_{rc}	n_{r+}
Total	n_{+1}	n_{+2}	...	n_{+c}	N

The three-way contingency table is slightly more elaborate to construct, because we now have three categorical variables to tabulate. Table 1.3 shows a simple—but quite common—situation, where Variables 1 and 2 each have two outcome categories and Variable 3 has $K \geq 2$ outcome categories. Table 1.3 is thus a $2 \times 2 \times K$ contingency table. From such a table, we may calculate a large number of marginal and total sums. Table 1.3 may also be considered as a stratified 2×2 table, with Variable 3 as a stratification variable. Then, the marginals and total sums are reduced to those in each of the K tables, and our interest is on the association between Variable 1 and Variable 2 in each table and their variation over the strata.

TABLE 1.3

The general setup for the observed counts of a three-way $2 \times 2 \times K$ contingency table, where $K \geq 2$

	Variable 2		
Variable 1	**Category 1**	**Category 2**	**Variable 3**
Category 1	n_{111}	n_{121}	1
Category 2	n_{211}	n_{221}	
Category 1	n_{112}	n_{122}	2
Category 2	n_{212}	n_{222}	
$\vdots$	$\vdots$	$\vdots$	$\vdots$
Category 1	n_{11K}	n_{12K}	K
Category 2	n_{21K}	n_{22K}	

If we allow Variables 1 and 2 to have more than two outcome categories, we obtain the general $r \times c \times K$ three-way contingency table, where $r \geq 2$, $c \geq 2$, and $K \geq 2$. The notation for this table, and the manner of displaying it, is a straightforward extension of Table 1.3.

1.3 Statistical Inference for Contingency Tables

In this section, we first describe the concepts and the notation for contingency tables. Then, we describe four main concepts for statistical estimation and inference: effect measures, effect estimates, confidence intervals, and hypothesis testing.

1.3.1 Notation and Concepts

First, we define the general parameter structure for contingency tables, and we start with the two-way $r \times c$ table. We denote Variable 1 by Y_1 and Variable 2 by Y_2. Often, we are interested in the association between an outcome and an explanatory variable. The explanatory variable may be a grouping variable. Then, Variable 2 (Y_2) will denote the outcome variable, and Variable 1 (Y_1) will denote the explanatory (or grouping) variable. Let π_{ij} denote the joint probability that Y_1 equals category i and Y_2 equals category j:

$$\pi_{ij} = \Pr(Y_1 = i, Y_2 = j),$$

where $i = 1, 2, \ldots, r$ and $j = 1, 2, \ldots, c$. In the case of a binary outcome variable, we usually denote the two possible outcomes as success and failure, where success is the outcome of interest, such as the occurrence of an event. In that case, let $Y_2 = 1$ if the outcome is a success, and let $Y_2 = 0$ if the outcome is a failure.

It is also useful to define marginal sums of probabilities, as with observed counts, and the notation is shown in Table 1.4. In specific applications, the observed data arise from different sampling models, and each sampling model defines which of the parameters are fixed (by design), which parameters are unknown, and how the parameters are connected through a probability model. The probability model also accounts for possible dependencies in the data—as determined by the sampling model—for instance, if observations are matched in pairs. The specifics will be treated in more detail in each of the chapters in this book.

TABLE 1.4
The joint probabilities of an $r \times c$ contingency table, where $r \geq 2$ and $c \geq 2$

Variable 1	Variable 2 Category 1	Category 2	...	Category c	Total
Category 1	π_{11}	π_{12}	$\cdots$	π_{1c}	π_{1+}
Category 2	π_{21}	π_{22}	$\cdots$	π_{2c}	π_{2+}
$\vdots$	$\vdots$	$\vdots$	$\ddots$	$\vdots$	$\vdots$
Category r	π_{r1}	π_{r2}	$\cdots$	π_{rc}	π_{r+}
Total	π_{+1}	π_{+2}	$\cdots$	π_{+c}	1

Because we often regard Y_2 as the outcome variable and Y_1 as the explanatory variable, we will consider the association between Y_2 and Y_1 via the conditional probabilities of Y_2 given Y_1. Let $\pi_{j|i}$ denote this conditional probability, given by

$$\pi_{j|i} = \Pr(Y_2 = j \mid Y_1 = i) = \pi_{ij}/\pi_{i+}.$$

TABLE 1.5

Conditional probabilities within the rows of an $r \times c$ table

Variable 1	Variable 2				Total
	Category 1	Category 2	...	Category c	
Group 1	$\pi_{1\|1}$	$\pi_{2\|1}$	$\cdots$	$\pi_{c\|1}$	1
Group 2	$\pi_{1\|2}$	$\pi_{2\|2}$	$\cdots$	$\pi_{c\|2}$	1
$\vdots$	$\vdots$	$\vdots$	$\ddots$	$\vdots$	$\vdots$
Group r	$\pi_{1\|r}$	$\pi_{2\|r}$	$\cdots$	$\pi_{c\|r}$	1

The notation is shown in Table 1.5.

The general one-way $(1 \times c)$ contingency table has parameters $\pi_1, \pi_2, \ldots, \pi_c$, which sum to one: $\sum_i \pi_i = 1$. The interpretation of the parameters is straightforward: $\pi_i = \Pr(Y_1 = i)$ is the probability that Variable 1 (the only variable) equals category i.

The notation for the parameters of the three-way $(r \times c \times K)$ contingency table follows as a straightforward extension of the two-way case. The connection between the parameters, however, depends on the specific application and may not be stated as a general case. We return to the details in Chapter 10, which deals with three-way—or stratified two-way—tables.

Note that for the 2×2 table in Chapter 4, for the $r \times 2$ table in Chapter 5, and for stratified 2×2 tables in Chapter 10, we use the notation π_i and $1 - \pi_i$ for $i = 1, 2, \ldots, r$ instead of $\pi_{1|i}$ and $\pi_{2|i}$ for the conditional probabilities.

1.3.2   Effect Measures and Effect Estimates

An *effect measure* is a measure of the strength of the relationship between two—or sometimes more than two—variables. We often call this relationship an association, or the effect of the explanatory variable on the outcome variable. We denote an arbitrary effect measure by θ, which is a function of one or more parameters: $\theta = f(\boldsymbol{\pi})$, where $\boldsymbol{\pi}$ is a vector of conditional probabilities.

One example of an effect measure is the difference between two conditional probabilities. The data in this case can be summarized in a 2×2 table, see Table 1.2 and the conditional probabilities in Table 1.5, both tables with $r = c = 2$. We denote the difference between the conditional probabilities by Δ, and it is defined as $\Delta = \pi_{1|1} - \pi_{1|2}$.

From a sample of observations, we can calculate an estimate of the effect measure, which we refer to as the *effect estimate*. We distinguish the effect estimate from the effect measure by putting a "hat" on top of the symbol for the effect measure, such as $\hat{\theta}$. The effect measure is the unknown relationship in the population, and the effect estimate is its estimate based on the available data.

Some textbooks distinguish between an *estimator*, which is a function of the random variable(s), and an *estimate*, which is the numerical result when values are plugged into this function. In this book, we will, in general, use the term *estimate* in both senses.

1.3.3 Confidence Intervals

The effect estimate is always associated with uncertainty. A *confidence interval* quantifies this uncertainty by giving a range of plausible values of the effect measure given the data. The *nominal coverage*, also called the *confidence level* or the *confidence coefficient*, is usually set to 95%, which we interpret as the amount of confidence we have that the true but unknown effect measure is contained in the interval.

If we could observe many data sets of the same type and size, and compute a confidence interval for each of them, the long-term proportion of the confidence intervals that contain the true value is called the *actual coverage* for the confidence interval. We prefer this coverage to equal the nominal coverage. This ideal is possible to fulfill in many situations with continuous variables, such as the confidence interval for normally distributed variables; however, because of their discrete nature, this is generally not possible for categorical variables. The difference between the actual and nominal coverage of different confidence interval methods will be an important topic in this book.

1.3.4 Hypothesis Tests

In most settings, the *null hypothesis* states that the effect measure of interest is equal to the *null value*, which usually indicates no effect or no association. Exceptions include non-inferiority and equivalence studies, see Section 12.2.2. Difference-based effect measures have null value equal to zero: $H_0 : \theta_{\text{diff}} = 0$; ratio-based effect measures have null value equal to one: $H_0 : \theta_{\text{ratio}} = 1$.

The purpose of *hypothesis testing* is to use the observed data to test the null hypothesis against the *alternative hypothesis* (H_A), which usually equals the research hypothesis and states how the effect measure differs from the null value. For most problems, we can formulate three different alternative hypotheses. The most common one is a *two-sided* alternative hypothesis, in which the effect measure under the alternative hypothesis may take on any possible value except the null. For the difference between two binomial probabilities, the two-sided hypothesis setup is

$$H_0 : \Delta = 0 \quad \text{versus} \quad H_A : \Delta \neq 0.$$

The second and third options are to define the values of the effect measure under the alternative hypothesis in one direction only. We thus have two different *one-sided* alternative hypotheses:

$$H_0 : \Delta = 0 \quad \text{versus} \quad H_A : \Delta > 0$$

and

$$H_0 : \Delta = 0 \quad \text{versus} \quad H_A : \Delta < 0.$$

This book emphasizes two-sided alternative hypotheses, because these are the most relevant hypotheses for life science research. In cases where we do not specify whether we consider a one- or two-sided alternative hypothesis, a two-sided alternative is assumed.

A *test statistic* is a function of the observed data and measures the extent to which the data comply with the null hypothesis. It is usually defined such that large values of the test statistic indicate less agreement with the null hypothesis. The test statistic is used to calculate a *P-value*, which is defined as the probability of obtaining the observed data or data that agree equally or less with the null hypothesis than the observed, if we assume the null hypothesis to be true:

$$P\text{-value} = \Pr\left[T(\mathbf{x}) \geq T(\mathbf{n}) \,|\, H_0\right].$$

Here, $T()$ is the test statistic, $\mathbf{n}$ denotes the observed table, and $\mathbf{x}$ denotes any possible table under the assumed statistical model. It is common to define a *nominal significance level* (α), and say that we reject the null hypothesis in favor of the alternative hypothesis if P-value $< \alpha$. The nominal significance level is usually set to 5%, that is $\alpha = 0.05$.

For hypothesis tests, the *actual significance level* is the probability of rejecting the null hypothesis, when the null hypothesis is true. We prefer that the actual significance level is equal to the nominal significance level. As for confidence intervals, this is generally not possible to achieve for categorical data. An important topic in this book will be to investigate how the actual significance levels of different hypothesis tests vary and how well they agree with the nominal significance level.

Twice the Smallest Tail

One general approach for computing a two-sided P-value is the *twice the smallest tail* principle: the two-sided P-value is set to twice the smallest of the two corresponding one-sided P-values. If this results in a P-value exceeding 1, set it equal to 1. An attractive property of the twice the smallest tail principle is that rejecting the two-sided alternative hypothesis at level α implies that both the null hypotheses with one-sided alternatives should be rejected at level $\alpha/2$.

1.4 Properties of Confidence Intervals and Hypothesis Tests

1.4.1 Basic Properties

Statistical methods ought to be based on a theoretical framework; we like to avoid methods that are purely pragmatic, without any justification for their expressions. The theoretical framework does not have to be complex and difficult. On the contrary, we prefer methods that are consistent with theory that is as general, and as simple, as possible. Examples of general and simple statistical theory are the Wald, likelihood ratio, and score principles (see the upcoming Section 1.7).

The practical utility of a method is largely dependent on its *ease of computation*. Methods of *closed-form expression* are usually easy to compute. The expressions may be simple or quite elaborate; however, as long as the calculations only involve evaluations of standard mathematical functions, they may be performed with a desktop calculator or in a spreadsheet. Methods that are not of closed-form expression must use an *iterative algorithm*, such as the Newton-Raphson or secant method. An iterative algorithm is an algorithm without a fixed number of steps, where each step (iteration) produces an improved approximate solution to the problem at hand, for instance that of finding the lower confidence limit of a confidence interval. The algorithm stops when one or more of a predefined set of criteria—such as the required precision of the approximation—are fulfilled. Some amount of computer language programming skills is required to do this. If the expressions to be calculated within each iteration are fairly simple, a dedicated researcher should be able to code such methods without too much trouble. Some exact methods, however, have an extra layer of complexity to their calculations. They require one or more algorithms, such as maximization of a non-linear, non-monotone function, within each iteration. These methods are downright difficult to compute. If a method is widely available in statistical software packages, it is not that important whether the method is easy or difficult to compute; however, most statistical software packages include only the standard—and most simple—methods and do not provide a wide range of alternatives. In this book, we shall make note of the methods that are of closed-form expression, and give weight to ease of computation as a beneficial property when we compare and recommend methods.

Another basic property of statistical methods is that they give sensible results in nontrivial situations. This is not the case with many commonly used methods for contingency tables, which often fail to handle tables with one or more zero cell counts. Then, several things can happen: (i) no result is possible, for instance, if the method tries to divide by zero; (ii) the result is meaningless, such as a P-value below 0 or above 1, or a *degenerate interval*, which is a confidence interval of zero width; (iii) *overshoot*: a confidence interval with limits outside the permissible range, such as a confidence interval for the difference between two probabilities that has a lower limit below -1 or an upper limit above 1. Fortunately, problems with zero cell counts are usually restricted to the most simple statistical methods, and small modifications to these methods can rectify the situation. Unfortunately, the simple methods with problems are the ones in most widespread use.

Finally, we prefer to use a confidence interval and a hypothesis test that give consistent results for the same data. Here, *consistency* is taken to mean that both the interval and the test indicate the same strength of the relationship under study. By this, we do not mean that they have to provide 100% consistent results in every conceivable situation, but that they, in general, tend to agree well. This is somewhat of a subjective judgement, and we could instead define a more formal requirement of consistency: that the interval and test derive from the same statistical framework, so that they are, in fact, always in agreement. We do, however, think this is too restrictive. It would, in practice, disqualify many excellent intervals and tests, and we consider a small discrepancy between the results of intervals and tests to be a small price to pay for having a wide range of methods from which to choose. In this case—as in other cases in this book—we take the pragmatic view, and we shall not put much emphasis on consistency between intervals and tests when we decide on which methods to recommend.

1.4.2 Evaluation Criteria for Confidence Intervals

Coverage Probability

Coverage probability is the most important attribute of a confidence interval. The coverage probability of a confidence interval method is defined as the probability that the interval will cover the true value. We prefer this probability to be close to the nominal coverage level $1 - \alpha$, usually set to 95%. The *exact criterion* states that the minimum attainable coverage probability should be no less than the nominal level. Exact intervals comply with this criterion, whereas most other intervals do not. There are two views on how to align coverage with the nominal $1 - \alpha$. One view states that the exact criterion is the gold standard and that intervals with coverage below the nominal level should be avoided. One disadvantage with this criterion is that the typical coverage probability of exact intervals is usually much larger than the nominal level, often 97% or 98% for a 95% nominal level. We refer to such intervals as *conservative intervals*, whereas intervals with coverage below the nominal level are sometimes referred to as *liberal intervals*.

The other view is that it is more appropriate to align the mean instead of the minimum coverage with $1 - \alpha$, at least as long as the violations of the nominal level are small, and not too frequent. One argument in favor of the second view is that coverage should be considered as a *moving average*, thus smoothing out occasional dips in coverage (Newcombe and Nurminen, 2011). On a more pragmatic note, aligning mean instead of minimum coverage with the nominal level leads to narrower intervals. In this book, we take the pragmatic view; however, we do not recommend intervals that violate the nominal level more often than not, nor intervals with a non-negligible probability of very low coverage. Furthermore, we prefer an interval with coverage slightly above the nominal level over an interval with coverage slightly below the nominal level.

The coverage probability of a confidence interval depends on the nominal level and the parameters of the statistical model that describe the table type and size. When we evaluate coverage probabilities, we can calculate them exactly with *complete enumeration*: for fixed values of the parameters, the coverage probability for a given interval is the sum of the probabilities of all possible tables with confidence limits that enclose the true value of the effect measure:

$$\text{CP}(\text{parameters}) = \sum_{\text{all tables}} I(L \leq \theta \leq U) \cdot \text{Pr}(\text{table}).$$

Here, $I()$ is the indicator function, L and U are the lower and upper confidence limits, and θ is the effect measure of interest (which depends on the parameters). The *indicator function* is a function that equals one if the expression evaluated as its argument is true and zero otherwise. The statistical model for the table provides an expression for $\text{Pr}(\text{table})$. For each type of table and for each effect measure we consider in this book, the details needed to perform the calculations will be presented.

Interval Width

If two or more intervals have similar coverage probabilities, we prefer the narrowest one. We expect an interval with low coverage to be shorter than an interval with greater coverage, so it does not make sense to compare interval widths without also considering coverage. Interval width is thus an evaluation index of secondary interest.

As with the coverage probability, we can calculate the expected width of an interval exactly with complete enumeration:

$$\text{Width}(\text{parameters}) = \sum_{\text{all tables}} (U - L) \cdot \text{Pr}(\text{table}).$$

When we evaluate the widths of intervals for ratio-based effect measures, such as the ratio of probabilities or the odds ratio, we use $\log(U) - \log(L)$ instead of $U - L$.

Interval Location

Another index of secondary interest is the location of the interval. We say that an interval is located too *distally* if it is located too far out from the centre of symmetry of the possible values for the effect measure (the midpoint). If the interval is located too close to the midpoint, we say that the interval is too *mesially* located. A $1 - \alpha$ confidence interval has ideal location if both the left and right non-coverage are equal to $\alpha/2$.

To quantify the extent of non-symmetry, we use the index MNCP/NCP defined by Newcombe (2011). MNCP/NCP, which takes on values in the range $[0, 1]$, measures the proportion of mesial non-coverage probability (MNCP) to the total non-coverage probability (NCP). The limits 0.4 and 0.6 are often used to define satisfactory location. Values of MNCP/NCP below 0.4 indicate that the interval is too mesially located, and values of MNCP/NCP above 0.6 indicate that the interval is too distally located. Mesial and distal directions are defined in relation to the true value of the effect measure of interest.

As with coverage probability and width, we use complete enumeration to calculate MNCP/NCP exactly. NCP is obtained from the coverage probability as $\text{NCP} = 1 - \text{CP}$, and MNCP is calculated as

$$\text{MNCP(parameters)} = \sum_{\text{all tables}} I(L > \theta \geq m \text{ or } U < \theta \leq m) \cdot \text{Pr(table)},$$

where m is the midpoint. For ratio-based effect measures, we use $\log(L)$, $\log(\theta)$, $\log(m)$, and $\log(U)$ instead of L, θ, m, and U.

1.4.3 Evaluation Criteria for Hypothesis Tests

Actual Significance Level

The actual significance level of a hypothesis test is the probability of rejecting the null hypothesis when the null hypothesis is true. We prefer this probability to be close to the nominal significance level (α), which is usually 5%. As with confidence intervals and coverage probability, we have a corresponding exact criterion for tests, which states that the actual significance level should never exceed the nominal level. The main disadvantage with tests that comply with the exact criterion is that the actual significance level usually is quite a bit lower than the nominal level, maybe 2% or 3% for a 5% nominal level. A test with actual significance levels lower than the nominal level is called a *conservative test*, and the P-values for these tests are higher than necessary. Tests with actual significance levels above the nominal level are sometimes referred to as *liberal tests*. In this book, we take a pragmatic view and prefer to use tests that have actual significance levels close to the nominal level, even if the actual significance level sometimes exceeds the nominal level. To be able to recommend tests with this property, we add the requirement that the infringements on the nominal level are small and not too frequent. If faced with a choice between a test that has actual significance levels slightly above the nominal level and a test that has actual significance levels slightly below the nominal level, we prefer the latter.

The actual significance level of a test depends on the nominal level and the parameters under the null hypothesis. We can use complete enumeration to calculate the actual significance level exactly: for fixed values of the parameters, the actual significance level of a test is the sum of the probability of all possible tables that lead to rejection of the null hypothesis at level α:

$$\text{ASL(parameters)} = \sum_{\text{all tables}} I(P\text{-value} \leq \alpha) \cdot \text{Pr(table} \mid H_0).$$

Here, $I()$ is the indicator function and $\text{Pr(table} \mid H_0)$ is an expression for the probability of observing any possible table (under the null hypothesis). Details for the specific table types and sizes throughout this book will be provided in their respective chapters.

Power

Power is the probability to obtain P-value $\leq \alpha$ under the alternative hypothesis. We compare the power between tests that have acceptable actual significance levels, and we prefer methods with high power. Power is calculated in a similar manner as the actual significance levels:

$$\text{Power(parameters)} = \sum_{\text{all tables}} I(P\text{-value} \leq \alpha) \cdot \Pr(\text{table} \mid H_A);$$

however, the expression for the probability of observing any possible table will be different under the null and alternative hypotheses.

1.5 Generalized Linear Models

1.5.1 Introduction

Generalized linear models (GLMs) were introduced by Nelder and Wedderburn (1972) as a class of regression models, see also McCullagh and Nelder (1989, Chapter 2). Let Y denote the outcome variable, and let $\mathbf{x}^{\mathrm{T}} = \{x_1, x_2, \ldots, x_p\}$ be a set of explanatory variables. Note the difference in notation from Section 1.3, in which Y_2 is the outcome variable and Y_1 is the (only) explanatory variable. Then, let

$$\mu(\mathbf{x}) = \mathrm{E}(Y \mid \mathbf{x})$$

be the conditional expectation of Y given $x_1, x_2, \ldots, x_p$. The *link function*, $g()$, given by

$$g\left[\mu(\mathbf{x})\right] = \alpha + \beta_1 x_1 + \beta_2 x_2 + \ldots + \beta_p x_p, \tag{1.1}$$

defines the link between the conditional expectation and the linear combination of the explanatory variables. From Equation 1.1, it follows that

$$\mu(\mathbf{x}) = g^{-1}\left(\alpha + \beta_1 x_1 + \beta_2 x_2 + \ldots + \beta_p x_p\right).$$

The term $\mu(\mathbf{x})$ expresses the *systematic part* of a GLM. The *random part* is described by the conditional distribution of Y given $x_1, x_2, \ldots, x_p$.

The *null model* has only one parameter, α, which is common to all the observations. The *saturated model* is a full model in the sense that it gives a perfect fit to the data. As such, it serves as a baseline for measuring the fit of alternative models, see Section 1.8 and McCullagh and Nelder (1989, p. 33).

Three link functions are particularly useful: the *linear link*, the *log link*, and the *logit link*, see Section 1.5.2. When the link function makes the parameters of the linear function $g()$ equal to the natural parameters of the density of the exponential family distribution, this link is the *canonical link function*, see Section 1.6. The canonical links for the normal, the Poisson, and the binomial distributions are given in Table 1.6.

TABLE 1.6
Canonical link functions

Distribution	Canonical link function
Normal	Identity
Poisson	Log
Binomial	Logit

Because of the properties of the distributions of the exponential family, the canonical links are especially attractive. Nevertheless, all three link functions are of interest for the binomial distribution, because they are directly connected to the effect measures of interest. In the next section, we therefore give a brief overview of the three link functions for the binomial distribution.

1.5.2 Link Functions for the Binomial Distribution

Let $Y = 1$ if the outcome is a success and $Y = 0$ if the outcome is a failure. The probability of success, given the explanatory variables $x_1, x_2, \ldots, x_p$, is denoted by

$$\pi(\mathbf{x}) = \Pr(Y = 1 \mid x_1, x_2, \ldots, x_p).$$

The linear link is given by

$$\pi(\mathbf{x}) = \alpha_{\text{linear}} + \beta_{1,\text{linear}} x_1 + \beta_{2,\text{linear}} x_2 + \ldots + \beta_{p,\text{linear}} x_p;$$

the log link is given by

$$\log\left[\pi(\mathbf{x})\right] = \alpha_{\log} + \beta_{1,\log} x_1 + \beta_{2,\log} x_2 + \ldots + \beta_{p,\log} x_p; \tag{1.2}$$

and the logit link is given by

$$\log\left[\frac{\pi(\mathbf{x})}{1 - \pi(\mathbf{x})}\right] = \alpha_{\text{logit}} + \beta_{1,\text{logit}} x_1 + \beta_{2,\text{logit}} x_2 + \ldots + \beta_{p,\text{logit}} x_p.$$

The parameterization of the link function is the same for all three link functions, and we may use a general description of the parameterization, often formulated as

$$\text{link}\left[\pi(\mathbf{x})\right] = \alpha + \beta_1 x_1 + \beta_2 x_2 + \ldots + \beta_p x_p.$$

The Linear Link for the 2 × 2 Table

For a 2×2 table (Chapter 4), we denote the explanatory variable by Y_1, which we usually take to be a grouping of the data. The two possible values of Y_1 are $Y_1 = 1$ (Group 1) and $Y_1 = 2$ (Group 2). The outcome is denoted by Y_2, with $Y_2 = 1$ for a success and $Y_2 = 0$ for a failure. We now have two parameters: $\pi_{1|1} = \Pr(Y_2 = 1 \,|\, Y_1 = 1)$ and $\pi_{1|2} = \Pr(Y_2 = 1 \,|\, Y_1 = 2)$, and the linear link function is now

$$\pi_{1|i} = \alpha_{\text{linear}} + \beta_{i,\text{linear}},$$

with $\beta_{2,\text{linear}} = 0$. It follows that

$$\beta_{1,\text{linear}} = \pi_{1|1} - \pi_{1|2},$$

which is the *difference between probabilities*. Note that Group 2 is the reference group.

The Log Link for the r × 2 Table

The log link will be described for ordered $r \times 2$ tables, which is the topic of Chapter 5. We now have r groups, which we regard as ordered, and a dichotomous outcome (success and failure). The general model for the probability of a success is given in Equation 1.2. Now, let $Y_1 = 1, 2, \ldots, r$ be the possible values of the grouping variable. As for the 2×2 table, the outcome is denoted by Y_2, with $Y_2 = 1$ for a success and $Y_2 = 0$ for a failure. Also, let $\beta_{1,\text{log}} = 0$, which defines Group 1 as the reference group. We can now simplify Equation 1.2 to

$$\log(\pi_{1|i}) = \alpha_{\text{log}} + \beta_{i,\text{log}},$$

or equivalently,

$$\beta_{i,\text{log}} = \log(\pi_{1|i}) - \log(\pi_{1|1}).$$

This can be written as

$$\exp(\beta_{i,\text{log}}) = \pi_{1|i}/\pi_{1|1},$$

which is the *ratio of probabilities* of success for Group i relative to Group 1.

The Logit Link for Stratified 2 × 2 Tables

We use stratified 2×2 tables (Chapter 10) to introduce the logit link. We have K strata that defines a set of K 2×2 tables. As usual, Y_1 is a grouping of the data with $Y_1 = 1$ for Group 1 and $Y_1 = 2$ for Group 2, and Y_2 is the outcome variable with $Y_2 = 1$ for a success and $Y_2 = 0$ for a failure. Let $Y_3 = 1, 2, \ldots, K$ denote the stratification variable. The conditional probability of success, given Group i and stratum k is denoted by

$$\pi_{1|ik} = \Pr(Y_2 = 1 \,|\, Y_1 = i, Y_3 = k).$$

The logit link with only main effects is given by

$$\text{logit}(\pi_{1|ik}) = \alpha_{\text{logit}} + \beta_{i,\text{logit}} + \gamma_{k,\text{logit}}, \tag{1.3}$$

for $i = 1, 2$ and $k = 1, 2, \ldots K$, where we assume that $\beta_{2,\text{logit}} = \gamma_{1,\text{logit}} = 0$, which defines Group 2 and stratum 1 as reference categories. Here, $\beta_{1,\text{logit}}$ is the effect on the logit of success of being in Group 1 relative to being in Group 2, conditional on the strata. The parameter vector $\{\gamma_{2,\text{logit}}, \gamma_{3,\text{logit}}, \ldots, \gamma_{K,\text{logit}}\}$ is the stratum effects, relative to stratum 1. With the same argument as for the linear and log links, we now get that

$$\beta_{1,\text{logit}} = \text{logit}(\pi_{1|1k}) - \text{logit}(\pi_{1|2k}),$$

or equivalently,

$$\exp(\beta_{1,\text{logit}}) = \frac{\pi_{1|1k}/(1 - \pi_{1|1k})}{\pi_{1|2k}/(1 - \pi_{1|2k})},$$

for $k = 2, 3, \ldots, K$. Thus, $\exp(\beta_{1,\text{logit}})$ is the odds ratio of success in Group 1 relative to Group 2, which is the same for all strata. This is an important characteristic of a model with logit link and only main effects.

When the effect of the grouping (the row variable) varies with the strata, we reformulate Equation 1.3 to include an *interaction* term:

$$\text{logit}(\pi_{1|ik}) = \alpha_{\text{logit}} + \beta_{i,\text{logit}} + \gamma_{k,\text{logit}} + \delta_{ik,\text{logit}}, \tag{1.4}$$

for $i = 1, 2$, and $k = 1, 2, \ldots, K$, where $\beta_{2,\text{logit}} = \gamma_{1,\text{logit}} = \delta_{11,\text{logit}} = 0$, and $\delta_{2k,\text{logit}} = 0$ for $k = 1, 2, \ldots, K$. In Equation 1.4, $\beta_{1,\text{logit}}$ is the overall effect of being in Group 1, $\gamma_{k,\text{logit}}$ is the effect of being in stratum k, and the $\delta_{1k,\text{logit}}$ for $k = 2, 3, \ldots, K$ are the interaction effects between the grouping and the stratification variables. Note that there are $2K$ parameters in Equation 1.4. Thus, the model with interactions is a saturated model.

From Equation 1.4, we now have that

$$\exp(\delta_{1k,\text{logit}}) = \frac{\pi_{1|1k}/(1 - \pi_{1|1k})}{\pi_{1|2k}/(1 - \pi_{1|2k})}, \tag{1.5}$$

for $k = 2, 3, \ldots, K$. Here, $\exp(\delta_{1k,\text{logit}})$ varies over the strata, which is a characteristic of the linear model with interaction. The assumption of no interaction between the grouping and the stratification variables means that we impose the $K - 1$ restrictions that $\exp(\delta_{1k,\text{logit}}) = 1$ for $k = 2, 3, \ldots, K$.

Because $\pi_{1|1k} = \pi_{11k}/\pi_{1+k}$ and $\pi_{1|2k} = \pi_{12k}/\pi_{2+k}$, Equation 1.5 can be written as

$$\exp(\delta_{1k,\text{logit}}) = \frac{\pi_{11k}/(1 - \pi_{11k})}{\pi_{12k}/(1 - \pi_{12k})}, \tag{1.6}$$

which means that the effect measure with the logit link is the same, regardless of whether it is expressed via the conditional (Equation 1.5) or joint (Equation 1.6) probabilities. This is a fundamental property of the logit link that will be used many times throughout this book.

1.5.3 Loglinear Models versus Logit Link

The generalized linear models are of interest when we study the conditional probabilities of success, given a set of explanatory variables. The logit link is of particular interest, because it is the canonical link function for the binomial distribution, see Table 1.6. An alternative to a linear logit model is a loglinear model, in which we study the unconditional probabilities, rather than the conditional probabilities. For an introduction to loglinear models for contingency tables, see Agresti (2013, Chapter 9).

For an $r \times c$ contingency table, let π_{ij} denote the probability of outcome j in Group i. Further, let $\mu_{ij} = N\pi_{ij}$ be the expected cell count in cell (i, j), where N is the total number of observations. As usual, Y_1 is the row variable, and Y_2 is the column variable. A saturated loglinear model for the $r \times c$ contingency table is then

$$\log(\mu_{ij}) = \lambda + \lambda_i^{Y_1} + \lambda_j^{Y_2} + \lambda_{ij}^{Y_1 Y_2}, \tag{1.7}$$

where $\lambda_1^{Y_1} = \lambda_1^{Y_2} = \lambda_{i1}^{Y_1 Y_2} = \lambda_{1j}^{Y_1 Y_2} = 0$ for $i = 1, 2, \ldots, r$ and $j = 1, 2, \ldots, c$. Then,

$$\lambda_{ij}^{Y_1 Y_2} = \log\left(\mu_{11}\mu_{ij}/\mu_{i1}\mu_{1j}\right),$$

see, for instance, Agresti (2013, p. 373). $\lambda_{ij}^{Y_1 Y_2} = 0$ means independence between Y_1 and Y_2. For an $r \times 2$ table, we have that

$$
\begin{aligned}
\mathrm{logit}\left[\Pr(Y_2 = 1 \mid Y_1 = i)\right] &= \log(\mu_{i1}) - \log(\mu_{i2}) \\
&= \left(\lambda_1^{Y_2} - \lambda_2^{Y_2}\right) + \left(\lambda_{i1}^{Y_1 Y_2} - \lambda_{i2}^{Y_1 Y_2}\right) \\
&= \alpha + \beta_i,
\end{aligned}
$$

with $\beta_1 = 0$. Thus, β_i is equal to $\beta_{i,\mathrm{logit}}$ in a logit formulation for an $r \times 2$ table.

For stratified 2×2 tables, a saturated loglinear model is given by

$$\log(\mu_{ijk}) = \lambda + \lambda_i^{Y_1} + \lambda_j^{Y_2} + \lambda_k^{Y_3} + \lambda_{ij}^{Y_1 Y_2} + \lambda_{ik}^{Y_1 Y_3} + \lambda_{jk}^{Y_2 Y_3},$$

where the first category is the reference category for all variables. Then, we have that

$$\log\left[\frac{\mu_{11k}\mu_{22k}/\mu_{12k}\mu_{21k}}{\mu_{111}\mu_{221}/\mu_{12k}\mu_{21k}}\right] = \lambda_{222}^{Y_1 Y_2 Y_3},$$

for $k = 2, 3, \ldots, K$. For the stratified 2×2 table, $\lambda_{222}^{Y_1 Y_2 Y_3}$ is the only free parameter of the third order interaction. Thus, there is no third order interaction if $\lambda_{222}^{Y_1 Y_2 Y_3} = 0$. Then $\mu_{11k}\mu_{22k}/\mu_{12k}\mu_{21k} = \mu_{111}\mu_{221}/\mu_{12k}\mu_{21k}$, that is, the odds ratios are the same for all strata. This is equivalent to no second order interaction in the logit formulation.

A loglinear model of the outcome, Y_2, with explanatory variables Y_1 and Y_3 describes a symmetric relationship between the variables. There are several loglinear parameterizations that have an equivalent logit parameterization, see Table 1.7 and Agresti (2013, p. 355). Note that the second order interaction terms in the loglinear parameterization are linear effects in the logit parameterizations, and that third order interaction terms in the loglinear parameterizations are second order interaction in the logit parameterizations.

TABLE 1.7
Loglinear and logit parameterizations of a three-way contingency table

Loglinear parameterization	Logit parameterization
$\lambda + \lambda_{ij}^{Y_1 Y_2} + \lambda_{ik}^{Y_1 Y_3}$	$\alpha + \beta_i$
$\lambda + \lambda_{ik}^{Y_1 Y_3} + \lambda_{ik}^{Y_1 Y_3}$	$\alpha + \gamma_k$
$\lambda + \lambda_{ij}^{Y_1 Y_2} + \lambda_{jk}^{Y_2 Y_3} + \lambda_{ik}^{Y_1 Y_3}$	$\alpha + \beta_i + \gamma_k$
$\lambda + \lambda_{ij}^{Y_1 Y_2} + \lambda_{jk}^{Y_2 Y_3} + \lambda_{ik}^{Y_1 Y_3} + \lambda_{ijk}^{Y_1 Y_2 Y_3}$	$\alpha + \beta_i + \gamma_k + \delta_{ik}$

1.6   Exponential Family

The exponential family of distributions is important in the analysis of contingency tables. The logit link, see Section 1.5, is crucial, and the concept of *sufficiency* is important in the construction of uniformly most powerful unbiased tests. Let the distribution of X be $f(x, \theta)$. Then, the statistic T is sufficient for θ if it is as informative to observe T as X. More formally, the conditional distribution of X given T does not depend on θ. These tests play an important part, particularly for 2×2 tables paired 2×2 tables, and stratified 2×2 tables.

The following brief introduction to the exponential family is based on the seminal textbook by Lehmann and Romano (2005) and the review article by Gart (1970).

1.6.1   Definitions

A probability distribution is in the natural exponential family if its distribution can be written on the form

$$f(\mathbf{x}, \boldsymbol{\theta}) = C(\boldsymbol{\theta}) \exp \left[\sum_{j=1}^{p} \theta_j T_j(\mathbf{x}) \right] h(\mathbf{x}), \qquad (1.8)$$

where $\boldsymbol{\theta} = \{\theta_1, \theta_2, \ldots, \theta_p\}$ is a p-dimensional vector of parameters, and $\mathbf{x} = \{x_1, x_2, \ldots, x_p\}^{\mathrm{T}}$ is a vector of observations, see Lehmann and Romano (2005, p. 46). The functions $T_j(\mathbf{x})$ comprise a set of sufficient statistics. Let now the distribution be of the form

$$f(\mathbf{x}, \boldsymbol{\theta}, \boldsymbol{\lambda}) = C(\boldsymbol{\theta}, \boldsymbol{\lambda}) \exp\left[\sum_{j=1}^{p} \theta_j T_j(\mathbf{x}) + \sum_{j=1}^{q} \lambda_j U_j(\mathbf{x})\right].$$

Assume that our interest is in the parameter vector $\boldsymbol{\theta}$, and let $\boldsymbol{\lambda}$ be a vector of nuisance parameters. For a distribution in the exponential family, the conditional distribution of T given U is then independent of $\boldsymbol{\lambda}$, see lemma 2.7.2 in Lehmann and Romano (2005, p. 48). As we will see below, conditioning on the margins in a 2×2 table, or in stratified 2×2 tables, is appropriate for testing independence in a 2×2 table, and conditional independence in stratified 2×2 tables.

The parameters of Equation 1.8 are called *natural parameters*. A link function that makes the parameters of the linear function g of Equation 1.1 equal to the natural parameters is a canonical link function, see Section 1.5. The canonical link functions for the normal, Poisson, and binomial distributions are given in Table 1.6.

As an example, let X be binomially distributed with parameter π and sample size n. This can be considered a 1×2 table, with X successes and $n - X$ failures. The probability distribution can be written as

$$\begin{aligned}
f(x, \pi) &= \binom{n}{x} \pi^x (1 - \pi)^{n-x} \\
&= \left[\frac{1}{1 + \exp(\theta)}\right]^n \exp(\theta x) \binom{n}{x},
\end{aligned}$$

where $\theta = \log\left[\pi/(1 - \pi)\right]$, which is the natural parameter, with

$$C(\theta) = \left[\frac{1}{1 + \exp(\theta)}\right]^n$$

and

$$h(x) = \binom{n}{x}.$$

Thus, the logit link is the canonical link function for the binomial distribution. As a consequence, the logit link will be of particular interest and importance in this book.

1.6.2 UMP Unbiased Tests

The natural exponential family is important in the construction of uniformly most powerful (UMP) tests and UMP unbiased tests. Unbiased means that the probability of rejection of an alternative is always above the nominal significance level. For the binomial distribution, we use the logit link, which in this case is the canonical link. We assume that X is binomially distributed with parameter π and sample size n. The one-sided null hypothesis $H_0 : \pi \geq \pi_0$ versus $H_A : \pi < \pi_0$ is equivalent to $H_0 : \theta \geq \theta_0$ versus $H_A : \theta < \theta_0$, where $\theta = \log\left[\pi/(1-\pi)\right]$. According to Corollary 3.4.1 in Lehmann and Romano (2005, p. 67), the test that rejects H_0 when X is small is the UMP test. The exact binomial test for the one-sided hypothesis in Section 2.5.5 is a UMP test.

Most often, we will use two-sided alternatives, which for the example above, means testing the null hypothesis $H_0 : \pi = \pi_0$ against $H_A : \pi \neq \pi_0$. We will consider unbiased tests. In this case, a UMP test will not exist; however, with the condition of unbiasedness, a UMP unbiased test exists. The UMP unbiased test rejects H_0 versus the two-sided alternative when X is small or X is large. The P-value is found by calculating the cumulative binomial distribution under the null hypothesis, see Lehmann and Romano (2005, p. 113).

The UMP Unbiased Test for the 2 × 2 Table

The 2 × 2 table is the topic of Chapter 4, and the notation for the observed counts is given in Table 4.1. There are different probability models for the 2 × 2 table, depending on the study design, see Sections 4.2 and 4.3. Here, we consider the row margins fixed model, where the sample sizes (the row margins) are fixed by design. Let now X_1 and X_2 be the random variables for the number of successes in Group 1 and Group 2, respectively. The observed numbers of successes are n_{11} and n_{21}, see Table 4.1. We assume that X_1 and X_2 are binomially distributed with parameters $\pi_{1|1}$ and $\pi_{1|2}$ and sample sizes n_{1+} and n_{2+}. The distribution function is given by

$$f(x_1, x_2 \,|\, \pi_{1|1}, \pi_{1|2}, n_{1+}, n_{2+}) =$$

$$\binom{n_{1+}}{x_1} \pi_{1|1}^{x_1}(1-\pi_{1|1})^{n_{1+}-x_1} \cdot \binom{n_{2+}}{x_2} \pi_{1|2}^{x_2}(1-\pi_{1|2})^{n_{2+}-x_2}.$$

As in Section 1.5.2, for a 2 × 2 table with the logit link, we have that

$$\log\left(\frac{\pi_{1|1}}{1-\pi_{1|1}}\right) = \alpha + \beta \quad \text{and} \quad \log\left(\frac{\pi_{1|2}}{1-\pi_{1|2}}\right) = \alpha.$$

Then,

$$\beta = \log\left[\frac{\pi_{1|1}}{1-\pi_{1|1}}\right] - \log\left[\frac{\pi_{1|2}}{1-\pi_{1|2}}\right]$$

and

$$\theta = \exp(\beta) = \frac{\pi_{1|1}/(1-\pi_{1|1})}{\pi_{1|2}/(1-\pi_{1|2})},$$

which is the odds ratio. If we rewrite the distribution function in the form of a distribution in the exponential family, we obtain

$$f(x_1, x_2 \,|\, \pi_1, \pi_2, n_{1+}, n_{2+}) = C(\beta, \alpha)\exp\left[\beta x_1 + \alpha(x_1 + x_2)\right]h(x_1, x_2).$$

Then, X_1 and $X_1 + X_2$ are sufficient for β and α, respectively. With results for the exponential family, see Lehmann and Romano (2005, p. 48), the conditional distribution of X_1 given $X_1 + X_2$ can be derived. The null hypothesis is now $H_0 : \theta = 1$. Under the null hypothesis, the conditional distribution simplifies to

$$f(x_1 \mid x_1 + x_2 = n_{+1}) = \frac{\dbinom{n_{1+}}{x_1}\dbinom{n_{2+}}{n_{1+} - x_1}}{\dbinom{N}{n_{1+}}}. \tag{1.9}$$

Thus, the UMP unbiased test of independence is the Fisher exact test in Section 4.4.5, although with a slightly different notation here. For details, see Gart (1970) or Lehmann and Romano (2005, p. 126).

The UMP Unbiased Test for Stratified 2×2 Tables

The 2×2 table is generalized to stratified 2×2 tables in Chapter 10. The observed counts are shown in Table 10.1, with the parameters of each table in Table 10.4. In Section 1.5.2, we exemplified the logit link for stratified 2×2 tables. We simplify the notation in Section 1.5.2 by letting

$$\text{logit}(\pi_{1|ik}) = \alpha_k + \beta_i,$$

for $i = 1, 2$ and $k = 1, 2, \ldots, K$, where $\beta_2 = 0$. Then,

$$\log\left(\frac{\pi_{1|1k}}{1 - \pi_{1|1k}}\right) = \alpha_k + \beta_1 \quad \text{and} \quad \log\left(\frac{\pi_{1|2k}}{1 - \pi_{1|2k}}\right) = \alpha_k,$$

for each stratum $k = 1, 2, \ldots, K$. Then,

$$\theta = \exp(\beta_1) = \frac{\pi_{1|1k}/(1 - \pi_{1|1k})}{\pi_{1|2k}/(1 - \pi_{1|2k})}.$$

The odds ratios are the same for each stratum, because there is no interaction terms in the model. The null hypothesis $H_0 : \theta = 1$ is conditional independence, under the assumption of no interaction. In this case, $X_{1k} + X_{2k}$ are sufficient statistics for the α_k. With the same argument as for the 2×2 table, we obtain the conditional distribution under the null hypothesis as the product of hypergeometric distributions given in Equation 1.9 over the K strata. This is the conditional distribution in Equation 10.12, inserted $\theta = 1$. Thus, the Gart exact test in Equation 10.33 is the UMP unbiased test for the null hypothesis: $H_0 : \theta = 1$.

The UMP Unbiased Test for the Paired 2×2 Table

The paired 2×2 table is covered in Chapter 8. Two events, A and B are observed, and each observation of an event A is paired with an event B.

Table 8.1 shows the observed counts of a paired 2×2 table, and Table 8.5 gives the joint probabilities. The probability distribution is given by

$$f(\mathbf{x} \mid \pi_{11}, \pi_{12}, \pi_{21}, \pi_{22}, N) = \frac{N!}{x_{11}! x_{12}! x_{21}! x_{22}!} \pi_{11}^{x_{11}} \pi_{12}^{x_{12}} \pi_{21}^{x_{21}} \pi_{22}^{x_{22}}.$$

The null hypothesis is $H_0 : \pi_{1+} = \pi_{+1}$, which is equivalent to $H_0 : \pi_{12} = \pi_{21}$, see Section 8.5.1. The joint distribution can be rewritten as

$$f(\mathbf{x} \mid \pi_{11}, \pi_{12}, \pi_{21}, \pi_{22}, N) = \pi_{22}^{N} \exp \left[x_{21} \log \left(\frac{\pi_{21}}{\pi_{12}} \right) + \right.$$

$$\left. (x_{12} + x_{21}) \log \left(\frac{\pi_{21}}{\pi_{22}} \right) + x_{11} \log \left(\frac{\pi_{11}}{\pi_{22}} \right) \right] \frac{N!}{x_{11}! x_{12}! x_{21}! x_{22}!}.$$

The UMP unbiased test is obtained by conditioning on $n_d = n_{12} + n_{21}$ and n_{11}. Only x_{12} and $n_d - x_{12}$ remain as variables, and the conditional distribution is

$$f(x_{12} \mid \mu, n_d, n_{11}) = \binom{n_d}{x_{12}} \mu^{x_{12}} (1 - \mu)^{n_d - x_{12}},$$

where $\mu = \pi_{12}/(\pi_{12} + \pi_{21})$, see Section 8.5.3 and Lehmann and Romano (2005, p. 138). Thus, the McNemar exact conditional test is the UMP unbiased test.

Concluding Remarks

So far in this section, we have assumed that the row margins are fixed in the sampling. But, as discussed in Section 4.3 for the 2×2 table and in Section 10.3 for the stratified 2×2 table, there are three sampling models of general interest: the row margins fixed model, the case-control design (fixed column margins), and the total sum fixed model (random sampling, no fixed margins). If we condition on both the row and column margins, the same test will be derived, and this test is the UMP unbiased test, regardless of the sampling model (see, for instance, Lehmann and Romano (2005, pp. 132–135)).

The problem of testing the null hypothesis of independence in $r \times c$ tables is much more complicated than for 2×2 or stratified 2×2 tables. Under the null hypothesis of independence, we have $(r - 1)(c - 1)$ restrictions on the parameters. Aaberge (2000) studied four interesting test problems, for which UMP unbiased tests can be derived. We briefly mention two of those here: the test for Hardy-Weinberg equilibrium, proposed by Haldane (1954), and the Fisher-Freeman-Halton test in Section 7.5.3.

1.7 Wald, Likelihood Ratio, and Score Inference

In this section, we introduce the Wald, likelihood ratio, and score methods for statistical inference. All three methods are based on the *maximum likelihood principle*. Here, the Wald, likelihood ratio, and score methods will be explained for a single parameter. For the general expressions for several parameters, we refer the reader to Lehmann and Casella (1998, Section 6.5) or Agresti (2013, Section 1.3).

In a general notation, let θ denote the parameter of interest. The *likelihood function*, denoted by $l(\theta)$ is the probability of the observed data, given a particular statistical model for θ. It is expressed as a function of θ and the observed data. The *maximum likelihood* (ML) estimate $\hat{\theta}$ is the value of θ that maximizes $l(\theta)$. A formula for the ML estimate is called a *maximum likelihood estimator* (MLE). The MLE has desirable properties: it has a large-sample normal distribution, and it converges to the parameter of interest as the number of observations increases. Furthermore, the MLE is asymptotically efficient with large sample standard errors no greater than those obtained from other estimation methods.

Rather than work with the likelihood function, which consists of a product of terms, it is usually mathematically simpler to work with the *log-likelihood function* $L(\theta)$, which is a sum of terms. Because the logarithm is a monotonous increasing function, the log-likelihood function is maximized for the same value of θ as is the likelihood function. Because $L(\theta)$ is usually concave, the ML estimate is the value of θ for which the derivative of $L(\theta)$ with respect to θ is zero. The second derivative of the log likelihood gives the curvature near θ. In matrix form, these second derivatives are known as the *Hessian matrix*. For the one parameter case, the Hessian matrix is defined as $H(\theta) = \partial^2 L(\theta)/\partial^2\theta$. The *observed information* is given by $I(\theta) = -H(\theta)$. The *Fisher information* is given by the expected value of the observed information: $F(\theta) = -\mathrm{E}\left[\partial^2 L(\theta)/\partial^2\theta\right]$, which is also called the *expected information*. There are discussions on whether one should use the observed or the expected information matrix for variance estimation of the maximum likelihood estimate. Efron and Hinkley (1978) and McCullagh and Nelder (1989, p. 342) argue for the use of observed information; however, these are "fine-tuned" differences, because there are no first-order differences in the asymptotic distributions. Note, however, that for small sample sizes, the variance estimates may differ, depending on whether it is estimated as observed or expected information.

In the following, we explain the Wald, likelihood ratio, and score methods in terms of hypothesis testing; however, there is a close connection between tests and confidence intervals for all three methods. A confidence interval can be formed by *inverting hypothesis tests*: a $1 - \alpha$ confidence interval for θ is the set of θ_0 for which the test of $H_0 : \theta = \theta_0$ has a P-value exceeding α.

We first explain the Wald method. We have a single parameter θ with

ML estimate $\hat{\theta}$. Let the null hypothesis be H_0: $\theta = \theta_0$, and let $\widehat{\text{SE}}(\hat{\theta})$ denote the standard error estimate of $\hat{\theta}$. $\widehat{\text{SE}}(\hat{\theta})$ may be expressed as the expected information $F(\theta)$ evaluated at $\hat{\theta}$. The *Wald statistic* is defined as

$$Z_{\text{Wald}} = (\hat{\theta} - \theta_0)/\widehat{\text{SE}}(\hat{\theta}), \qquad (1.10)$$

and it has an approximate standard normal distribution under H_0. The Wald statistic may also be expressed in chi-squared form as $T_{\text{Wald}} = Z_{\text{Wald}}^2$. T_{Wald} is asymptotically chi-squared distributed with one degrees of freedom. In the multiparameter case, T_{Wald} is asymptotically chi-squared distributed with degrees of freedom equal to the difference in the number of free parameters under the null and alternative hypotheses.

The second method uses the *likelihood ratio statistic*, which is defined as follows: Let l_0 and l_1 denote the maximized likelihood under the null hypothesis, $H_0 : \theta = \theta_0$, and the alternative hypothesis, respectively. The likelihood ratio is defined as $\Lambda = l_0/l_1$. The likelihood ratio statistic is given by

$$T_{\text{LR}} = -2\log\Lambda = -2\log(l_0/l_1) = -2(L_0 - L_1),$$

where L_0 and L_1 denote the maximized log-likelihood functions. This statistic is asymptotically chi-squared distributed under the null hypothesis. Under the null hypothesis, only one parameter is specified, and the likelihood ratio test has one degree of freedom.

The third method uses the *score statistic*. It is based on the score function

$$u(\theta) = \partial L(\theta)/\partial\theta$$

evaluated at θ_0. The null standard error of the score function for a single parameter θ equals the expected information $F(\theta)$ evaluated at θ_0. The chi-squared form of the score statistic is

$$T_{\text{score}} = \frac{\left[\partial L(\theta)/\partial\theta\right]^2}{-\text{E}\left[\partial^2 L(\theta)/\partial\theta^2\right]},$$

evaluated at θ_0. In the multiparameter case, the degrees of freedom equal the difference in the number of free parameters under the null and alternative hypotheses. In the single parameter setting, the score statistic in standard normal form is given by

$$Z_{\text{score}} = \frac{\partial L(\theta)/\partial\theta}{\sqrt{-\text{E}\left[\partial^2 L(\theta)/\partial\theta^2\right]}},$$

evaluated at θ_0, which is asymptotically normally distributed. Note that the score statistic uses the standard error under the null hypothesis, whereas the Wald statistic in Equation 1.10 uses the standard error of the ML estimate.

For linear regression with a normally distributed dependent variable, the

Wald, likelihood ratio, and score methods give identical results. In other settings, there can be marked differences. Wald methods are more commonly used than score methods. This is in large part because Wald methods are readily available in statistical software packages, and they are often expressed in simple closed-form mathematical formulas. The latter property is seldom the case with methods based on the score and likelihood ratio statistics; an iterative algorithm is often required to compute these methods.

The Wald, likelihood ratio, and score methods are asymptotic, in the sense that the derived test statistics have large-sample normal or chi-squared distributions. This implies that the nominal coverage and nominal significance level will equal the actual coverage and actual significance level only approximately, with better approximations in large than small samples. Asymptotic tests produce P-values by referring the observed value of the test statistic to the reference distribution; here, the standard normal distribution:

$$\text{asymptotic } P\text{-value} = \Pr\left[Z \geq \left| Z_{\text{test statistic}}(\mathbf{n}) \right| \right],$$

where Z is a standard normal variable.

1.8 Deviance, Likelihood Ratio Tests, and Pearson Chi-Squared Tests for Generalized Linear Models

The generalized linear models were introduced in Section 1.5 as a family of regression models. The link function specified the expectation of the response variable as a linear function of the explanatory variables. The random part of the model specifies the distribution of the response variable. Cross-classified data may be considered as grouped or ungrouped. Count data presented in a contingency table are considered as grouped, whereas data presented per individual are ungrouped, see Agresti (2013, p.138). For ungrouped data, the response variable follows a Bernoulli distribution. The binomial, Poisson, and Bernoulli distributions are all members of the exponential family, see Section 1.6.

For generalized linear models with binary data, the parameter estimates and their asymptotic covariance matrix are derived in McCullagh and Nelder (1989, p. 114–117). Note that for the exponential families with canonical links, the observed and expected information are identical (McCullagh and Nelder, 1989, p. 43).

Model checking is an important part of statistical analysis. In this section, we introduce the likelihood ratio test and the Pearson chi-squared test for testing goodness of fit and model comparisons. Saturated and nested models are important concepts. A *saturated model* has a maximum number of parameters with a one-to-one relationship between the parameters and the cells in

the contingency tables. The maximum likelihood estimates of the cell probabilities of a success are simply the cell proportions. Such a model is never of
interest in itself; however, it is used as a baseline for comparing models. For
such comparisons, *nested models* are important.

A generalized linear model M_0 is nested within M_1 if M_1 contains at
least one regression term more than M_0. Let the saturated model be M_s.
The *deviance* for model M_1 is a measure of discrepancy between M_1 and the
saturated model M_s. The deviance is defined from the likelihood ratio test by
$D_1 = -2\big[\log(l_1) - \log(l_s)\big]$. For model M_0, the deviance is $D_0 = -2\big[\log(l_0) -
\log(l_s)\big]$. Then, the difference in deviances $D_0 - D_1$ is equal to $-2\big[\log(l_0) -
\log(l_1)\big]$, which is the likelihood ratio statistic described in Section 1.7. Because
M_0 is nested within M_1, $D_0 - D_1 \geq 0$, and the larger the difference, the poorer
the model fit for M_0 relative to M_1.

Suppose that cell i in a contingency table has n_i observations. The fitted
cell counts m_{i1} under model M_1 is the total number of observations times the
estimated cell probabilities under M_1. When M_s is a saturated model, the
deviance is simply a comparison of the fitted and the observed cell counts,
expressed as

$$D_1 = T_{\mathrm{LR}} = 2 \sum_i n_i \log\left(\frac{n_i}{m_{i1}}\right),$$

which is chi-squared distributed with degrees of freedom equal to the number
of restrictions put on the parameters in model M_1, see, for instance, Rao
(1965, p. 349).

Let M_0 be nested within M_1. The deviance for model M_0 versus the saturated model M_s is obtained by inserting the fitted cell counts under M_0 in
the denominator. More interesting, however, is the difference between the deviances of M_0 and M_1. Denote the fitted cell counts under M_0 by m_{i0}. The
deviance and the likelihood ratio test statistic then takes the form

$$D_0 - D_1 = T_{\mathrm{LR}} = 2 \sum_i m_{i1} \log\left(\frac{m_{i1}}{m_{i0}}\right).$$

This test statistic is chi-squared distributed with degrees of freedom equal to
the number of restrictions put on the parameters in model M_0, relative to
that of model M_1.

The Pearson chi-squared statistic for comparing model M_1 with the saturated model M_s is given by

$$T_{\mathrm{Pearson}} = \sum_i \frac{(n_i - m_{i1})^2}{m_{i1}},$$

and for comparing model M_0 with model M_1, the Pearson chi-squared statistic
is

$$T_{\mathrm{Pearson}} = \sum_i \frac{(m_{i1} - m_{i0})^2}{m_{i0}},$$

see Rao (1965, p. 329). The Pearson chi-squared statistics have the same asymptotic distributions as the likelihood ratio statistics.

The likelihood ratio (deviance) and Pearson chi-squared statistics are often referred to as goodness-of-fit tests when they are used to compare a model M_1 with the saturated model M_s. Because the fitted cell counts of the saturated model are equal to the observed cell counts, the deviance is a measure of the goodness of fit of model M_0, relative to the observed cell counts.

Any contingency table can be organized as grouped data or as ungrouped data. Ungrouped data are recorded per subject and may accordingly be analyzed per subject, rather than as a contingency table. Logistic regression with categorical explanatory variables is a method of analyzing ungrouped data that is relevant for many types of contingency tables. When data are analyzed per subject, the saturated model and the deviance will be different than those of an analysis of grouped data. Thus, the deviance cannot be used as a goodness-of-fit test. The difference between the deviances, expressed as a likelihood ratio test, on the other hand, is the same for ungrouped and grouped analyses, see Hosmer et al. (2013, p. 12).

1.9 Exact Methods

The asymptotic methods for statistical inference described in the previous section rely on large-sample approximations to their test statistics. These approximations may be quite poor in small samples. As alternatives to asymptotic methods, exact methods guarantee the preservation of coverage and significance level. That is, the actual coverage is at least as high as the nominal coverage, and the actual significance level does not exceed the nominal significance level. Exact tests produce P-values by summing the (exact) probabilities of all possible tables ($\mathbf{x}$) that agree less than or equally with the null hypothesis than does the observed table ($\mathbf{n}$):

$$\text{exact } P\text{-value} = \sum_{\text{all tables}} I\big[T(\mathbf{x}) \geq T(\mathbf{n})\big] \cdot f(\mathbf{x} \,|\, H_0). \qquad (1.11)$$

Here, $I()$ is the indicator function, $T()$ is defined such that large values indicate less agreement with the null hypothesis than do small values, and $f()$ is an expression for the probability distribution of possible tables under the null hypothesis. This approach may seem straightforward; however, in many situations, $f()$ contains one or more unknown parameters (nuisance parameters) that need to be managed. This will be further explained in Section 4.4.7, where we consider exact unconditional tests for the 2×2 table.

Exact confidence intervals can be obtained by inverting exact tests, as explained (for asymptotic tests and intervals) in the previous section.

1.10   The Mid-P Approach

In principle, the exact approach described in the previous section seems ideal. But there is a serious complication. For any possible combination of parameter values, the actual coverage of an exact interval is at least as large as the nominal coverage, but it can be much larger. The resulting confidence interval is typically wider than with other (asymptotic) methods. Similarly, because there is a finite number of possible outcomes, there is also a finite number of possible P-values, of which, for instance, 0.05 may not be one. This implies that exact methods for categorical data are conservative.

One possible remedy is to use the mid-P approach. The mid-P approach was originally proposed by Lancaster (1961) and can be used for both hypothesis tests and confidence intervals. It is useful across many different situations; see the review by Berry and Armitage (1995) and the several applications in this book.

The mid-P value is the sum of the probabilities of all possible tables that agree less with the null hypothesis than does the observed table, plus half the point probability of the tables that agree equally with the null hypothesis as the observed table:

$$\text{mid-}P \text{ value} = \sum_{\text{all tables}} I\big[T(\mathbf{x}) > T(\mathbf{n})\big] \cdot f(\mathbf{x}|H_0) +$$

$$0.5 \cdot \sum_{\text{all tables}} I\big[T(\mathbf{x}) = T(\mathbf{n})\big] \cdot f(\mathbf{x}|H_0). \qquad (1.12)$$

This expression is a small modification of the expression for the exact P-value in Equation 1.11. A mid-P test is thus based on a particular exact test, and a mid-P interval is based on a particular exact interval. Although the mid-P approach may seem like an ad hoc solution, it has solid theoretical justifications, as shown in Hwang and Yang (2001). Because they are based on exact methods, mid-P tests and mid-P intervals are sometimes called quasi-exact.

Of course, there is no free lunch. The mid-P approach reduces the conservatism of exact methods, but at the price of not preserving the nominal coverage (confidence intervals) and nominal significance level (tests). The good news, however, is that in a wide variety of designs and models, the violations of the nominal coverage and significance level are not of a serious magnitude. There are many good reasons for using the mid-P approach; see, for instance, the excellent summary in Hirji (2006, pp. 218–219). An ideal P-value is uniformly distributed on $[0, 1]$ under the null hypothesis, whereas an exact P-value for categorical data is skewed to the right. The mid-P value, on the other hand, has an expectation equal to 0.5 and is distributed approximately uniformly on $[0, 1]$. Furthermore, for analysis of categorical data, confidence intervals and tests based on the mid-P approach have, in most cases, better properties than those based on asymptotic normal theory. A notable exception is testing for equality of paired binomial distributions (Chapter 8), where the McNemar asymptotic test is better than the McNemar mid-P test.

1.11 Bayes Inference

Most of the methods covered in this book are derived from the frequentist paradigm. This implies that the unknown parameter θ is regarded as an unknown constant, and not as a random variable. In a Bayesian paradigm, on the other hand, the parameter is regarded as a random variable Θ with some probability distribution. The *prior distribution* reflects our knowledge about the parameter before any data is taken into account. This can be based on communication with experts within the field of interest, or from other data sets that contain information relevant to the problem at hand. The prior distribution is updated with respect to the given data, yielding the *posterior distribution*. This posterior distribution can be expressed using *Bayes Theorem* from probability theory, given by the expression

$$f_{\Theta|X}(\theta\,|\,x) = \frac{f_\Theta(\theta)f_{X|\Theta}(x\,|\,\theta)}{f_X(x)}.$$

Here, $f_\Theta(\theta)$ is the prior distribution for Θ, and $f_{X|\Theta}(x\,|\,\theta)$ is the probability distribution for our data given θ, which is identical to the likelihood function. The denominator $f_X(x)$ describes the marginal distribution of X, taking the random variation of Θ into account. This is not a function of θ, so the posterior distribution is proportional to the product of the prior distribution and the likelihood. If the posterior distribution belongs to the same class of distributions as the prior, it is said to be conjugate for the likelihood at hand. For the binomial likelihood, the beta distribution has this property.

The Bayesian parallel to the confidence interval is the *credible interval* (also called the posterior interval). A $1 - \alpha$ credible interval is usually constructed as the region of the posterior distribution lying between the $\alpha/2$ and $1 - \alpha/2$ percentiles. As explained in Agresti and Min (2005a), Bayesian intervals can infer some performance benefits—in the frequentist sense—relative to ordinary frequentist intervals, particularly for small sample sizes. Agresti and Min recommend using non-informative priors and show that this leads to good coverage performance for credible intervals for the difference between probabilities, ratio of probabilities, and odds ratio for the 2×2 table. Brown et al. (2001) note similar good results for Bayesian intervals for the binomial probability.

In Section 2.4.5, we shall use the beta distribution to construct a Bayesian interval for the binomial probability. Another Bayesian interval will be considered for the ratio of proportions in the 2×2 table (see Section 4.7.3). A more thorough treatment of Bayesian analysis of contingency tables can be found in Agresti (2013, Section 3.6).

2

The 1×2 Table and the Binomial Distribution

2.1 Introduction

We consider the situation where we sample n observations of a dichotomous variable and count the number of a certain outcome of interest. The outcome of interest is usually called "success", not necessarily being the most favorable outcome, but because it is the outcome we count. If the probability of success is the same in each of the n observations, and if the outcomes are independent, the number X of successes is binomially distributed with parameters n and π, where n is known, and π is the probability of success. Note that we use a different notation in this chapter—for reasons of simplicity and tradition—than the general notation defined in Chapter 1.

One example for which the binomial distribution is a realistic model is the number of males among n singleton births. If we include twins in the sample, the sex of the offspring would not be independent, since monozygous twins are the same sex. Then, the assumptions for the binomial distribution would not hold. Another example is a randomized cross-over clinical trial where n patients receive treatment A and treatment B in a randomized order. If X and $n - X$ of the patients report to prefer treatment A and B, respectively, then X may be regarded as binomially distributed.

Table 2.1 shows the contingency table setup for the one-sample dichotomous variable. It is usually superfluous to display the results in this way; however, we do it to tie this case to the general contingency table situation and to illustrate the notation. Usually, it would be sufficient to give the results simply as X/n, perhaps with a percentage, as in 12/48 (25%).

TABLE 2.1
The observed counts of a 1×2 table

	Success	Failure	Total
Sample	X	$n - X$	n

This chapter is organized as follows. Section 2.2 introduces two examples where the binomial distribution is appropriate; the first one is an observational study with moderate to large sample sizes, and the other one is a randomized cross-over trial with a small sample size. Section 2.3 shows how to estimate the binomial parameter π and its standard error. Confidence intervals for π are the topic of Section 2.4, and tests for the null hypothesis that π equals a given probability are considered in Section 2.5. In both these sections, the methods are: (i) presented in detail; (ii) applied to the two data examples; and (iii) evaluated and compared according to strict criteria. Finally, Section 2.6 gives recommendations for which confidence intervals and tests to use.

2.2   Examples

2.2.1   The Probability of Male Births to Indian Immigrants in Norway

In Norway, a large study showed a proportion of 51.3% male offsprings in the years 1967–2003 (Lippert et al., 2005). A similar proportion of males is seen in many countries around the world, and proportions in the range 50.7%–51.7% are considered normal (Dai and Li, 2015). In some districts in China and India, however, substantially higher proportions of males have been reported (Jha et al., 2006; Dai and Li, 2015). Singh et al. (2010) studied sex distributions in children born to Indian and Pakistani immigrants in Norway. The numbers of males per births among Indian immigrants in the years 1987–1996 for birth order 1 to 4 are shown in Table 2.2. For the purposes of this chapter, we regard each row (each birth order) as one separate sample of dichotomous outcomes, to be analyzed apart. The four samples are not independent of one another, although the outcomes within each sample are independent. We may thus regard the number of males in each row as binomially distributed. The observed proportions of males are 46.9% (1st birth), 49.5% (2nd birth), 61.7% (3rd birth), and 73.3% (4th birth).

TABLE 2.2
Male and female births among Indian immigrants in Norway, 1987–1996 (Singh et al., 2010)

Birth order	Males	Females	Total
1st	250	283	533
2nd	204	208	412
3rd	103	64	167
4th	33	12	45

The data shown in Table 2.2 are a subset of the data material presented in Singh et al. (2010). They conclude that the male proportions in higher birth order children seem to have increased among Indian immigrants, but not among Pakistani immigrants, after the introduction of ultrasound scanning technology in Norway in 1987. We return to this example in Section 2.4.9, where we estimate 95% confidence intervals for the probability of a male offspring, and in Section 2.5.7, where we test whether the probability of a male offspring for each birth order is equal to the proportion in the general Norwegian population (51.3%).

2.2.2 A Randomized Cross-Over Trial of Probiotics Versus Placebo for Patients with Irritable Bowel Syndrome

Ligaarden et al. (2010) report a double blind randomized cross-over trial of 16 patients with irritable bowel syndrome who underwent two three-week periods of daily intake of the probiotic L. plantarum MF 1298 or placebo separated by a four-week washout period. The primary outcome measure of the trial was treatment preference (the period with least symptoms). Thirteen (81%) of the 16 patients preferred the placebo period to the period with L. plantarum MF 1298. Sections 2.4.9 and 2.5.7 present confidence intervals and tests, respectively, for the probability that a patient prefers placebo.

2.3 Estimation of the Binomial Parameter π

Let X be binomially distributed with parameters n and π. The natural estimate of π is the sample proportion

$$\hat{\pi} = \frac{X}{n}. \tag{2.1}$$

Some authors use the term proportion also for the (usually unknown) binomial probability π. We reserve the term proportion for the sample proportion.

The sample proportion in Equation 2.1 turns out to be the maximum likelihood estimate of π, and we derive it as follows: the binomial probability distribution is given by

$$f(x \mid \pi, n) = \binom{n}{x} \pi^x (1 - \pi)^{n-x}, \tag{2.2}$$

for $x = 0, 1, \ldots, n$. The expression $\binom{n}{x}$ is termed the *binomial coefficient*, defined as

$$\binom{n}{x} = \frac{n!}{x!(n - x)!}.$$

The likelihood function (see Section 1.7) equals the probability in Equation 2.2, regarded as a function of the parameter π:

$$l(\pi) = \binom{n}{x} \pi^x (1 - \pi)^{n-x}.$$

The MLE $\hat{\pi}$ for π is the value of π that maximizes $l(\pi)$. As noted in Section 1.7, it is easier to maximize the log-likelihood function. For the binomial distribution, the log-likelihood becomes

$$L(\pi) = \log\big[l(\pi)\big] = \log\binom{n}{x} + x\log(\pi) + (n - x)\log(1 - \pi).$$

This function is maximized by equating its derivative to zero:

$$\frac{\partial}{\partial \pi} L(\pi) = \frac{x}{\pi} - \frac{n - x}{1 - \pi} = 0.$$

Solving this equation for π gives $\pi = x/n$, so the MLE for π is the sample proportion $\hat{\pi} = X/n$.

The standard deviation of an estimate is commonly termed its *standard error* (*SE*). The standard error of $\hat{\pi}$ is

$$SE(\hat{\pi}) = \sqrt{\frac{\pi(1 - \pi)}{n}},$$

which is estimated by

$$\widehat{SE}(\hat{\pi}) = \sqrt{\frac{\hat{\pi}(1 - \hat{\pi})}{n}}.$$

The estimates $\hat{\pi}$ and $\widehat{SE}(\hat{\pi})$ can be used as the basis for construction of confidence intervals (Section 2.4) and hypothesis tests (Section 2.5).

2.4   Confidence Intervals for the Binomial Parameter

2.4.1   The Wald Interval

The Wald confidence interval for π is based on the fact that the Wald statistic

$$\frac{\hat{\pi} - \pi}{\widehat{SE}(\hat{\pi})} = \frac{\hat{\pi} - \pi}{\sqrt{\dfrac{\hat{\pi}(1 - \hat{\pi})}{n}}} \tag{2.3}$$

is approximately standard normally distributed. The confidence limits for the Wald interval is given by

$$\hat{\pi} \pm z_{\alpha/2} \sqrt{\frac{\hat{\pi}(1 - \hat{\pi})}{n}}.$$

Here, $z_{\alpha/2}$ denotes the upper $\alpha/2$ percentile of the standard normal distribution. For a 95% confidence interval, $\alpha = 0.05$ and $z_{\alpha/2} \approx 1.96$. The Wald interval produces the zero-width (degenerate) intervals $(0,0)$ when $X = 0$ and $(1,1)$ when $X = n$. Overshoot (interval limits outside $[0,1]$, see Section 1.4) may happen for values of X close to 0 or n. For example, when $X = 1$ and $n = 10$, the Wald interval is $(-0.086, 0.286)$.

For small sample sizes, a modification called a *continuity correction* may be applied to the Wald statistic to bring standard normal tail probabilities in closer agreement with binomial tail probabilities (which are used to construct exact intervals and exact tests). The *Wald interval with continuity correction* is given by

$$\hat{\pi} \pm \left[z_{\alpha/2} \sqrt{\frac{\hat{\pi}(1 - \hat{\pi})}{n}} + \frac{1}{2n} \right].$$

The Wald interval with continuity correction does not produce zero-width intervals. It gives, however, wider intervals than the Wald interval without continuity correction—due to the additional $1/2n$ term—and thereby overshoots more often. For example, when $n = 10$, only $X = 4$, $X = 5$, and $X = 6$ gives interval limits bounded by 0 and 1.

2.4.2 The Likelihood Ratio Interval

The likelihood ratio principle (Section 1.7) can be used to construct a confidence interval for π. Let π_0 denote an arbitrary value between 0 and 1, and define the following likelihood ratio statistic:

$$T_{\mathrm{LR}}(\hat{\pi} \mid \pi_0) = 2 \left[x \log \left(\frac{\pi_0}{\hat{\pi}} \right) + (n - x) \log \left(\frac{1 - \pi_0}{1 - \hat{\pi}} \right) \right].$$

The likelihood ratio interval is defined as those values of π_0 that satisfy $T_{\mathrm{LR}}(\hat{\pi} \mid \pi_0) \geq z_{\alpha/2}^2$. This interval is more difficult to calculate than the Wald intervals because the confidence limits do not have closed-form expressions. We must use an iterative algorithm to find the lower (L) and upper(U) confidence limits such that $L < U$,

$$T_{\mathrm{LR}}(\hat{\pi} \mid L) = z_{\alpha/2}^2,$$

and

$$T_{\mathrm{LR}}(\hat{\pi} \mid U) = z_{\alpha/2}^2.$$

2.4.3 The Wilson Score Interval

A score interval for the binomial parameter was first proposed by Wilson (1927), and we shall denote it as the Wilson score interval. It is obtained by equating the score statistic,

$$\frac{\hat{\pi} - \pi_0}{\sqrt{\dfrac{\pi_0(1 - \pi_0)}{n}}}, \tag{2.4}$$

to $\pm z_{\alpha/2}$ and solving for π_0. Note that the score statistic uses the standard error under the null hypothesis of $\pi = \pi_0$, whereas the Wald statistic in Equation 2.3 uses the standard error of the sample proportion. In contrast to the likelihood ratio interval, the Wilson score interval has confidence limits with closed-form expressions:

$$\frac{2n\hat{\pi} + z_{\alpha/2}^2 \pm z_{\alpha/2}\sqrt{z_{\alpha/2}^2 + 4n\hat{\pi}(1 - \hat{\pi})}}{2\left(n + z_{\alpha/2}^2\right)}. \tag{2.5}$$

The Wilson score interval plays an important role in the analysis of contingency tables, not only because it has favorable properties as a confidence interval for the binomial parameter, but also because it is explicitly used in the calculations of MOVER (method of variance estimates recovery) and hybrid score intervals for the 2×2 table (Chapter 4) and the paired 2×2 table (Chapter 8).

A *score interval with continuity correction* is also possible. The lower (L) and upper (U) confidence limits of this interval are expressed in closed form as

$$L = \frac{2n\hat{\pi} + z_{\alpha/2}^2 - 1 - z_{\alpha/2}\sqrt{z_{\alpha/2}^2 - 2 - (1/n) + 4n\hat{\pi}(1 - \hat{\pi}) + 4\hat{\pi}}}{2\left(n + z_{\alpha/2}^2\right)}$$

and

$$U = \frac{2n\hat{\pi} + z_{\alpha/2}^2 + 1 + z_{\alpha/2}\sqrt{z_{\alpha/2}^2 + 2 - (1/n) + 4n\hat{\pi}(1 - \hat{\pi}) - 4\hat{\pi}}}{2\left(n + z_{\alpha/2}^2\right)}.$$

The score interval with continuity correction does not contain the maximum likelihood estimate ($\hat{\pi}$) when $X = 0$ (for which $\hat{\pi} = 0$ and $L > 0$) or $X = n$ (for which $\hat{\pi} = 1$ and $U < 1$).

2.4.4 The Agresti-Coull Interval

Agresti and Coull (1998a) noted that while the Wilson score interval is of closed-form expression (Equation 2.5), it may be too elaborate to use when teaching elementary statistics courses. They show that the midpoint of the Wilson score interval is a weighted average of $\hat{\pi}$ and $1/2$:

$$\text{midpoint} = \hat{\pi}\left(\frac{n}{n + z_{\alpha/2}^2}\right) + \frac{1}{2}\left(\frac{z_{\alpha/2}^2}{n + z_{\alpha/2}^2}\right).$$

The midpoint shrinks the Wilson score interval toward $1/2$, with less shrinkage as n increases. For the purpose of constructing a 95% confidence interval, $z_{\alpha/2}^2 = 1.960^2 \approx 4$, and

$$\text{midpoint} \approx \frac{X+2}{n+4}.$$

Based on this observation, Agresti and Coull suggested that adding two successes and two failures to the data, and computing the Wald interval as if the data were $\tilde{X} = X + 2$ successes in $\tilde{n} = n + 4$ trials, is a good and simple approximation to the score interval. Thus, the Agresti-Coull 95% confidence interval is given by

$$\hat{\pi} \pm z_{\alpha/2} \sqrt{\frac{\tilde{\pi}(1-\tilde{\pi})}{\tilde{n}}},$$

where $\tilde{\pi} = \tilde{X}/\tilde{n}$.

The added success and failure counts are often denoted by *pseudo-counts* or *pseudo-frequencies*.

2.4.5 The Jeffreys Interval

Section 1.11 introduced the Bayesian paradigm to statistical inference, including the credible interval (see Section 1.11), which is the Bayesian equivalent to the confidence interval. If the prior distribution is the beta distribution with parameters a and b, denoted by $B(a, b)$, the posterior distribution is $B(a + x, b + n - x)$, with expectation $(a + x)/(a + b + n)$. A *Bayesian credible interval* is given by the lower (L) and upper (U) $\alpha/2$ percentiles in the posterior distribution:

$$L = B(\alpha/2; x + a, n - x + b)$$

and

$$U = B(1 - \alpha/2; x + a, n - x + b),$$

where $B(z; a, b)$ is the lower z-quantile of the beta distribution with parameters a and b. Note that the interval (L, U) equals the Clopper-Pearson exact interval (see Section 2.4.7) if $x + a$ successes were observed among $n + a + b$ observations.

If we use Jeffreys non-informative prior, $B(0.5, 0.5)$, we obtain the Jeffreys interval:

$$L = B(\alpha/2; x + 0.5, n - x + 0.5)$$

and

$$U = B(1 - \alpha/2; x + 0.5, n - x + 0.5).$$

2.4.6 The Arcsine Interval

The variance of the binomial distribution varies considerably across the range of π, which makes it difficult to construct a simple confidence interval on the form $\hat{\pi} \pm z_{\alpha/2}\widehat{SE}$ with satisfactory properties for all values of π. Several variance stabilizing transformations have been suggested to remove this obstacle. Here, we consider an approach based on two variance stabilizing steps. The first step, due to Anscombe (1948), consists of replacing $\hat{\pi}$ with $\tilde{\pi} = (X+3/8)/(n+3/4)$. The second step is to use an inverse sine transformation. We refer to the resulting interval (L, U) as the arcsine interval:

$$L = \sin^2 \left[\arcsin\left(\sqrt{\tilde{\pi}}\right) - \frac{z_{\alpha/2}}{2\sqrt{n}} \right]$$

and

$$U = \sin^2 \left[\arcsin\left(\sqrt{\tilde{\pi}}\right) + \frac{z_{\alpha/2}}{2\sqrt{n}} \right].$$

The arcsine interval, like the score interval with continuity correction, does not contain the maximum likelihood estimate $(\hat{\pi})$ when $X = 0$ (for which $\hat{\pi} = 0$ and $L > 0$) or $X = n$ (for which $\hat{\pi} = 1$ and $U < 1$).

2.4.7   Exact Intervals

An exact confidence interval for π can be constructed by inverting two one-sided exact binomial tests (see the upcoming Section 2.5.5). This method of constructing a confidence interval is called the *tail method* (Agresti, 2003). To obtain the lower (L) and upper (U) confidence limits of this interval, we solve the two equations

$$\sum_{i=x}^{n} \binom{n}{i} L^i (1 - L)^{n-i} = \alpha/2$$

and

$$\sum_{i=0}^{x} \binom{n}{i} U^i (1 - U)^{n-i} = \alpha/2,$$

with respect to L and U. An iterative algorithm is needed to solve the equations. The resulting interval is called the *Clopper-Pearson exact interval*, which is the shortest interval that guarantees that both the left and right non-coverage will be at most $\alpha/2$ for all values of π (Newcombe, 2013, p. 69). If we use the connection between the binomial sum and the cumulative beta distribution, we arrive at a simple non-iterative expression for the Clopper-Pearson interval (Brown et al., 2001):

$$L = B(\alpha/2; x, n - x + 1)$$

and

$$U = B(1 - \alpha/2; x + 1, n - x),$$

where $B(z; a, b)$ is the lower z-quantile of the beta distribution with parameters a and b.

Several attempts have been made to construct exact intervals that are less conservative than the Clopper-Pearson interval (Sterne, 1954; Crow, 1956; Blyth and Still, 1983; Casella, 1986). These intervals are, however, quite complicated to calculate or contain irregularities (Casella, 1986). An elegant solution was proposed by Blaker (2000), who sought to improve exact inference in discrete distributions, including the binomial, negative binomial, hypergeometric, and Poisson distributions. The *Blaker exact interval* inverts one two-sided exact test of size α. It thereby does not guarantee that the non-coverage in each tail is limited by $\alpha/2$, but only that the total non-coverage is limited by α. For $k = 0, 1, \ldots, n$, define the function

$$\gamma(k, \pi_*) = \min\left[\sum_{i=k}^{n} \binom{n}{i} \pi_*^i (1 - \pi_*)^{n-i}, \sum_{i=0}^{k} \binom{n}{i} \pi_*^i (1 - \pi_*)^{n-i} \right], \quad (2.6)$$

where π_* denotes an arbitrary confidence limit. Let $\gamma(x, \pi_*)$ denote the value of γ for the observed data. The confidence limits of the Blaker exact interval are the two solutions of π_* that satisfy the equation

$$\sum_{k=0}^{n} I\left[\gamma(k, \pi_*) \leq \gamma(x, \pi_*)\right] \cdot \binom{n}{k} \pi_*^k (1 - \pi_*)^{n-k} = \alpha,$$

where $I()$ is the indicator function.

2.4.8 Mid-*P* Intervals

A mid-*P* (Section 1.10) version of the Clopper-Pearson exact confidence interval is obtained as the solutions L and U to the equations

$$\sum_{i=x}^{n} \binom{n}{i} L^i (1 - L)^{n-i} - \frac{1}{2} \binom{n}{x} L^x (1 - L)^{n-x} = \alpha/2 \quad (2.7)$$

and

$$\sum_{i=0}^{x} \binom{n}{i} U^i (1 - U)^{n-i} - \frac{1}{2} \binom{n}{x} U^x (1 - U)^{n-x} = \alpha/2. \quad (2.8)$$

We call this interval the *Clopper-Pearson mid-P interval*. No simplification with the beta distribution is applicable here, and Equations 2.7 and 2.8 must be solved iteratively. If $x = 0$, no solution to Equation 2.7 is possible, and we set $L = 0$. Similarly, if $x = n$, we set $U = 1$.

We may also apply the mid-*P* approach to the Blaker exact interval. The confidence limits of the resulting *Blaker mid-P interval* are the two solutions of π_* that satisfy the equation

$$\sum_{k=0}^{n} I\left[\gamma(k, \pi_*) < \gamma(x, \pi_*)\right] \cdot \binom{n}{k} \pi_*^k (1 - \pi_*)^{n-k}$$

$$+ \quad 0.5 \cdot \sum_{k=0}^{n} I\left[\gamma(k, \pi_*) = \gamma(x, \pi_*)\right] \cdot \binom{n}{k} \pi_*^k (1 - \pi_*)^{n-k} = \alpha,$$

where γ is defined in Equation 2.6.

2.4.9   Examples

The Probability of Male Births to Indian Immigrants in Norway (Table 2.2)

For the 1st order births ($\hat{\pi} = 250/533 = 0.469$), the sample size is large and
the observed proportion is not close to either end of the parameter space (0
or 1). As expected, all the confidence interval methods in Sections 2.4.1–2.4.8
produce approximately the same interval. The lower 95% confidence limit is
$L = 0.426$ (or 0.427), and the upper 95% confidence limit is $U = 0.511$ (or
0.512). Similar confidence limits across the different methods are also obtained
for the 2nd order births ($\hat{\pi} = 204/412 = 0.495$), for which the 95% Wilson
score interval is (0.447 to 0.543); and the 3rd order births ($\hat{\pi} = 103/167 = 0.617$), for which the 95% Wilson score interval is (0.541 to 0.687). Note that
the proportion of male births in the general Norwegian population, $\hat{\pi} = 0.513$,
is contained in the 95% confidence interval for the 2nd order births, but it is
higher than the upper 95% confidence limit for the 1st order births and lower
than the lower 95% confidence limit for the 3rd order births (see Figure 2.1).

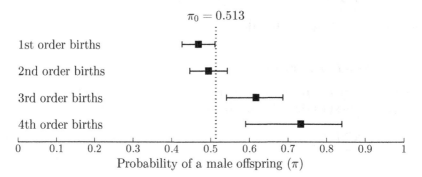

FIGURE 2.1
95% Wilson score confidence intervals based on the data in Table 2.2

For the 4th order births, the estimate of the probability of a male offspring
is $\hat{\pi} = 33/45 = 0.733$, which is considerably higher than the observed proportions for the 1st, 2nd, and 3rd order births. It is also considerably higher than
the observed proportion in the general Norwegian population ($\hat{\pi} = 0.513$);
however, the sample size is quite small ($n = 45$), so we expect wider confidence intervals than those for the 1st, 2nd, and 3rd order births. Table 2.3
summarizes the results. The Wilson score, Blaker mid-P, Agresti-Coull, and
Jeffreys intervals are the shortest ones. The two intervals with continuity correction and the Clopper-Pearson exact interval are the widest intervals. Although there are some differences between the intervals in Table 2.3, the sizes
of these differences are small, and it is unlikely that the choice of interval will
affect how the results are interpreted. For all 12 intervals, the lower confi-

dence limit ($\approx$ 59%) for the probability of a male offspring for the 4th order birth is noticeably higher than the proportion of male offsprings in the general Norwegian population ($\approx$ 51%).

TABLE 2.3
95% Confidence intervals for the probability of a male offspring for the 4th order birth ($\hat{\pi} = 0.733$) based on the data in Table 2.2

Interval	Confidence limits		Width
	Lower	Upper	
Wald	0.604	0.863	0.258
Wald with continuity correction	0.593	0.874	0.281
Likelihood ratio	0.594	0.847	0.254
Wilson score	0.590	0.840	0.251
Score with continuity correction	0.578	0.849	0.271
Agresti-Coull	0.588	0.841	0.253
Jeffreys	0.593	0.845	0.253
Arcsine	0.592	0.848	0.256
Clopper-Pearson exact	0.581	0.854	0.273
Blaker exact	0.589	0.852	0.263
Clopper-Pearson mid-P	0.591	0.847	0.256
Blaker mid-P	0.590	0.843	0.252

A Randomized Cross-Over Trial of Probiotics Versus Placebo for Patients with Irritable Bowel Syndrome

The probability that a patient with irritable bowel syndrome prefers placebo to the probiotic L. plantarum MF is estimated to be $\hat{\pi} = 13/16 = 0.813$. Table 2.4 gives 95% confidence intervals for π. The Wald interval and the Wald interval with continuity correction overshoots the upper boundary of π; both have upper confidence limits greater than 1. A convenient way to fix this problem is to set the upper confidence limit to 1; however, this is not a satisfactory solution, because the resulting confidence interval will underestimate the uncertainty in the data. This problem is, by itself, a good reason for abandoning the two Wald intervals entirely. As we shall see in the next section, another good reason for avoiding the Wald intervals is that they have poor coverage probabilities.

Among the other confidence intervals in Table 2.4, two intervals stand out as being wider than the others: the score interval with continuity correction and the Clopper-Pearson exact interval. The Wilson score interval is again the shortest interval, and when we evaluate the intervals in the next section, we shall see that it also has the best average coverage probability. For the probability that a patient with irritable bowel syndrome prefers placebo to the probiotic L. plantarum MF, the 95% Wilson score interval is (0.570 to

0.934). This interval does not contain $\pi = 0.5$, which is the probability that reflects no preference for either placebo or L. plantarum MF.

TABLE 2.4
95% Confidence intervals for the probability that a patient prefers placebo to probiotics ($\hat{\pi} = 0.813$) based on data from Ligaarden et al. (2010)

	Confidence limits		
Interval	**Lower**	**Upper**	**Width**
Wald	0.621	1.004	0.382
Wald with continuity correction	0.590	1.035	0.445
Likelihood ratio	0.583	0.950	0.367
Wilson score	0.570	0.934	0.364
Score with continuity correction	0.537	0.950	0.413
Agresti-Coull	0.560	0.940	0.380
Jeffreys	0.579	0.944	0.365
Arcsine	0.575	0.952	0.378
Clopper-Pearson exact	0.544	0.960	0.416
Blaker exact	0.566	0.947	0.381
Clopper-Pearson mid-*P*	0.570	0.950	0.380
Blaker mid-*P*	0.566	0.935	0.369

2.4.10 Evaluation of Intervals

Evaluation Criteria

We evaluate confidence intervals for the binomial parameter by calculating coverage probabilities, expected width, and location (see Section 1.4). All three items depend on π, n, and α, and the calculations can be performed exactly with complete enumeration. The coverage probability is defined as

$$\mathrm{CP}(\pi, n, \alpha) = \sum_{x=0}^{n} I(L \leq \pi \leq U) \cdot f(x \,|\, \pi, n),$$

where $I()$ is the indicator function, $L = L(x, \alpha)$ and $U = U(x, \alpha)$ are the lower and upper $100(1-\alpha)\%$ confidence limits of an interval for the data with x successes in n observations, and $f()$ is the binomial probability distribution (Equation 2.2). The expected interval width is

$$\mathrm{Width}(\pi, n, \alpha) = \sum_{x=0}^{n} (U - L) \cdot f(x \,|\, \pi, n),$$

and the mesial non-coverage probability—which is needed to calculate the MNCP/NCP location index (see Section 1.4)—is defined as

$$\mathrm{MNCP}(\pi, n, \alpha) = \sum_{x=0}^{n} I(L > \pi \geq 0.5 \text{ or } U < \pi \leq 0.5) \cdot f(x \,|\, \pi, n).$$

The total non-coverage probability (NCP) is obtained from the coverage probability as $NCP = 1 - CP$.

Evaluation of Coverage Probability

One informative way to assess coverage probabilities is to fix all parameters except one, and plot the coverage probability as a function of the remaining parameter. Figure 2.2 shows the coverage probabilities of the Wald, Wilson score, and Clopper-Pearson exact intervals as functions of the binomial probability (π), when we fix $\alpha = 0.05$ and $n = 25$. We have chosen $n = 25$ because the general coverage properties of the different interval methods for the binomial probability are well illustrated with this sample size. We shall use $n = 25$ throughout this section and note in the text when interesting coverage properties for other sample sizes are not sufficiently illustrated by the plots.

The coverage probability is a highly discontinuous function of π and it may change quite drastically, particularly for the asymptotic (non-exact) intervals, with small changes in π. Although the differences in coverage probabilities between the three intervals in Figure 2.2 are quite clear, it is often difficult to assess and compare the typical performance of the intervals when faced with such discontinuities. Newcombe and Nurminen (2011) make the argument that in these cases, a smoothed curve, obtained as the moving average of the coverage probabilities, provides a more realistic assessment of the coverage achieved in practice. Thus, for the remainder of this section, we plot moving averages superimposed on plots of coverage probabilities.

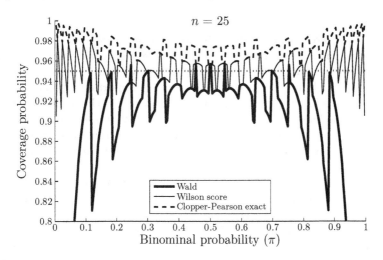

FIGURE 2.2
Coverage probabilities of three confidence intervals for the binomial probability

Figure 2.3 shows the results of the Wald interval and the Wald interval with continuity correction for $n = 25$. The Wald interval has unacceptably low average coverage probabilities for all values of π. The Wald interval with continuity correction is a considerable improvement; however, when π is close to either 0 or 1, the coverage probability quickly drops to unacceptable low levels. As we shall soon see, several readily available intervals perform better.

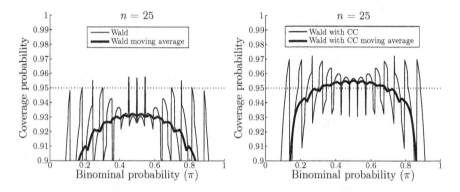

FIGURE 2.3
Coverage probabilities (with moving averages over the range $[\pi - 0.1, \pi + 0.1]$) of the Wald interval and the Wald interval with continuity correction (CC) for the binomial probability

The Wald interval performs quite poorly when $n = 25$. Figure 2.4 shows how it performs for $n = 100$ and $n = 200$. The average coverage probability is still markedly below the nominal level, and we clearly need a very large sample size to obtain good results. This is an important observation. The Wald interval is in widespread use, partly because it is the default confidence interval for the binomial probability in most statistical software packages. Its poor performance in small samples may be well known; however, it may come as a surprise that the coverage probability often is quite low when $n \geq 100$ as well.

Figure 2.5 shows the coverage probabilities of the Wilson score interval and the score interval with continuity correction. The average coverage probability of the Wilson score interval (left panel) is excellent; it is always close to—and slightly above—the nominal level. The score interval with continuity correction (right panel) is strictly conservative for the parameters shown in Figure 2.5; however, it is not an exact interval, and no guarantee can be made that the coverage will be at least $1 - \alpha$. It turns out that for 1 000 evenly distributed π-values between 0 and 1—but not including the endpoints—and for all n in the range 5–100, the coverage probability of the score interval with continuity correction is below 95% in 6 (0.006%) out of 96 000 cases.

Figure 2.5 is a good illustration of the kind of performance we look for

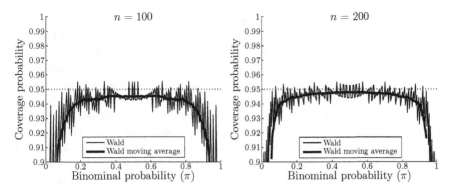

FIGURE 2.4
Coverage probabilities (with moving averages over the range $[\pi - 0.05, \pi + 0.05]$) of the Wald interval for the binomial probability

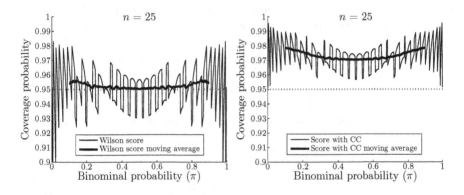

FIGURE 2.5
Coverage probabilities (with moving averages over the range $[\pi - 0.1, \pi + 0.1]$) of the Wilson score interval and the score interval with continuity correction (CC) for the binomial probability

in a confidence interval. We may take one of two views. The first view is that the coverage probability should never be below the nominal level. A 95% confidence interval thus should always (no matter the parameter values) cover the true effect measure in no less than 95% of cases. Exact intervals adhere to this criterion, whereas asymptotic intervals do not. Even though the score interval with continuity correction is not an exact interval, the right panel of Figure 2.5 illustrates that the consequence of the exact criterion can be severe conservatism: the average coverage probability is in the range 97–98%, which indicates that the score interval with continuity correction produces wide intervals.

The second view is more pragmatic. It states that it is better to align the mean (and not the minimum) coverage probability with $1 - \alpha$. Thus, we permit the coverage probability to dip below the nominal level as long as the average coverage probability is close to, and preferably slightly above, the nominal level. We also add a condition stating that the fluctuations around the average value should not be unreasonably large. The Wilson score interval is the archetypical interval that exemplifies this view, as shown in the left panel of Figure 2.5. Aligning the mean coverage probability with the nominal level is appropriate if we expect a 95% confidence interval to have 95% actual coverage. With an exact interval, on the other hand, we typically cannot expect an actual coverage of more than, say, 97%. For most practical applications, we recommend the pragmatic view, a recommendation that we share with many other researchers in the field, for instance Agresti and Coull (1998a) and Newcombe (1998c).

An example of the coverage probabilities of the likelihood ratio and Agresti-Coull intervals is shown in Figure 2.6. The likelihood ratio interval has average coverage probabilities slightly below the nominal level and a tendency toward severe dips in coverage for small and large values of π. This trait is not unique for $n = 25$ and extends to many other values of n. The Agresti-Coull interval performs much better. Its average coverage probability is a smoother function of π than that of the likelihood ratio interval, which makes for a more predictable performance. Coverage is a bit conservative for small and large values of π—at least compared with the Wilson score interval—although not at the same level as the score interval with continuity correction.

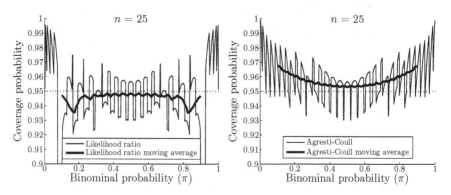

FIGURE 2.6
Coverage probabilities (with moving averages over the range $[\pi - 0.1, \pi + 0.1]$) of the likelihood ratio interval and the Agresti-Coull interval for the binomial probability

The performances of the Bayesian Jeffreys interval and the arcsine interval are illustrated in Figure 2.7. These two intervals often perform very similarly.

The average coverage probability of the Jeffreys interval is usually slightly below the nominal level, whereas the average coverage probability of the arcsine interval is usually slightly above the nominal level; however, the difference between the average coverage probabilities is quite small. As with the likelihood ratio interval, a disadvantage with both the Jeffreys interval and the arcsine interval is that coverage may dip to unacceptable low levels for small and large values of π.

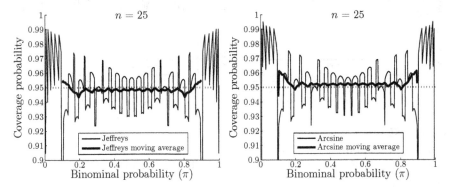

FIGURE 2.7
Coverage probabilities (with moving averages over the range $[\pi - 0.1, \pi + 0.1]$) of the Jeffreys intervall and the arcsine interval for the binomial probability

Figure 2.8 illustrates the coverage probabilities of the Clopper-Pearson exact interval and the Blaker exact interval. The Clopper-Pearson interval is very conservative, and its average coverage probabilities are very similar to those of the Wilson score interval with continuity correction. The Blaker interval is considerably less conservative. For $n < 50$, it has average coverage probabilities about one percentage point closer to the nominal level than does the Clopper-Pearson interval. For these small sample sizes (and $\alpha = 0.05$), the coverage probability is typically 97–98% with the Clopper-Pearson interval and 96–97% with the Blaker interval. Even for $n = 100$, there is a considerable difference in average coverage probabilities at about 0.6 percentage points in favor of the Blaker interval.

Finally, we show how the two mid-P intervals perform in Figure 2.9. Compared with Figure 2.8, we see that the mid-P intervals have coverage probabilities much closer to the nominal level than their exact counterparts. The price to pay for the mid-P intervals is occasional dips in coverage below the nominal level. The Clopper-Pearson mid-P interval is, on average, slightly more conservative than the Blaker mid-P interval. This is to be expected because the Clopper-Pearson exact interval is more conservative than the Blaker exact interval. What Figure 2.9 does not show is that the Blaker mid-P interval sometimes has coverage probabilities markedly below the nominal level for π in the range 0.4–0.6. A similar plot as in Figure 2.9 but with $n = 40$ would reveal this anomaly. The Clopper-Pearson mid-P interval is thereby a "safer" method, which we shall prefer ahead of the Blaker mid-P interval.

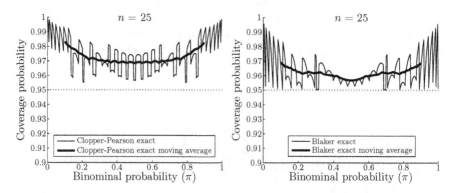

FIGURE 2.8

Coverage probabilities (with moving averages over the range $[\pi - 0.1, \pi + 0.1]$) of the Clopper-Pearson exact interval and the Blaker exact interval for the binomial probability

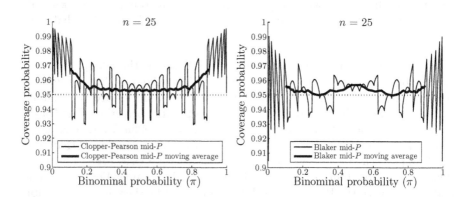

FIGURE 2.9

Coverage probabilities (with moving averages over the range $[\pi - 0.1, \pi + 0.1]$) of the Clopper-Pearson mid-P interval and the Blaker mid-P interval for the binomial probability

If we consider that aligning average coverage probabilities with the nominal level is an appropriate guide to choosing a confidence interval method, we may be interested in how the average coverage probability changes with the number of observations. Figure 2.10 shows this for a selection of seven intervals. For each value of $n = 5, 6, \ldots, 100$, average coverage probabilities are computed over 1 000 evenly distributed π-values in the range 0 to 1 (but not including the endpoints). A clear advantage with the Wilson score interval can be seen, which holds for all sample sizes from 5 to 100. Figure 2.10 should discourage use of the Wald interval for all but the largest sample sizes.

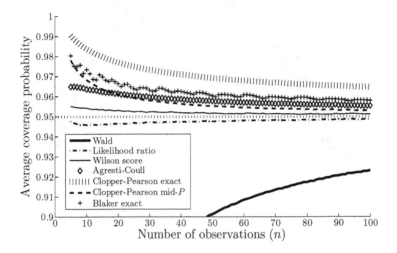

FIGURE 2.10
Average coverage probabilities of seven confidence intervals for the binomial probability as functions of the number of observations

Evaluation of Width

As with the coverage probability, we may plot the expected interval width as a function of the binomial probability. Figure 2.11 shows the results for six key intervals when $n = 25$. If we restrict our attention to the centermost half of the π-range, the Wilson score interval and Agresti-Coull interval are the two narrowest intervals, followed by the Clopper-Pearson mid-P interval. The Clopper-Pearson exact interval is by far the widest of the six intervals. For the smallest and largest values of π, the differences between the intervals are much smaller, and the ranking of the intervals according to width is less clear.

The most striking observation from Figure 2.11 is that the expected width of the Wald interval is not smaller. The Wald interval has about the same expected width as the Blaker exact interval, in spite of the fact that the coverage probability of the Wald interval is that much lower than any of the other intervals. Usually, a low coverage probability is accompanied by a narrow interval; but with the Wald interval, you do not even get a tight interval as compensation for the poor coverage probability.

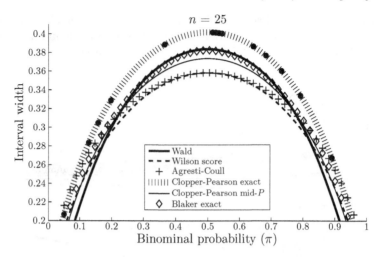

FIGURE 2.11
Expected width of six confidence intervals for the binomial probability

Evaluation of Location

Figure 2.12 illustrates the location properties of the Wald, Wilson score, and Blaker exact intervals. The Wald interval has MNCP/NCP > 0.6 for the majority of the range of π. This means that mesial non-coverage predominates over distal non-coverage, and we say that the Wald interval is too distally located with respect to 0.5, the center of symmetry for π. Another way of putting it is to say that Wald intervals tend to be located too far from 0.5. The greater the MNCP/NCP index, the greater the amount of asymmetry. Asymmetry can also be the result of too mesial location (MNCP/NCP < 0.4, where smaller MNCP/NCP values indicate greater amount of asymmetry). Too mesial location means that confidence intervals tend to be located too close to 0.5. The Wilson score interval has this property, although the amount of asymmetry of the Wilson score interval is not as great as that of the Wald interval. The Agresti-Coull interval (not shown in Figure 2.12) has similar MNCP/NCP values as the Wilson score interval. Also shown in Figure 2.12 is the location of the Blaker exact interval. It is slightly too mesially located but not by much. The Clopper-Pearson exact interval and the two mid-P intervals all have similar location as the Blaker exact interval.

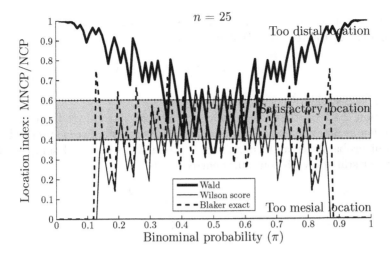

FIGURE 2.12
Location, as measured by the MNCP/NCP index, of three confidence intervals for the binomial probability

2.5 Tests for the Binomial Parameter

2.5.1 The Null and Alternative Hypotheses

We consider the null hypothesis that the probability of success (π) is equal to a given probability (π_0):

$$H_0 : \pi = \pi_0 \quad \text{versus} \quad H_A : \pi \neq \pi_0.$$

Here, the alternative hypothesis is two-sided: we do not specify whether $\pi > \pi_0$ or $\pi < \pi_0$. A one-sided alternative hypothesis would be either $H_A : \pi > \pi_0$ or $H_A : \pi < \pi_0$. This book mainly considers two-sided alternative hypotheses, and if not otherwise stated, a two-sided alternative hypothesis is always implied.

2.5.2 The Wald Test

The Wald test statistic is obtained by substituting π_0 for π in Equation 2.3:

$$Z_{\text{Wald}}(\hat{\pi}) = \frac{\hat{\pi} - \pi_0}{\sqrt{\dfrac{\hat{\pi}(1 - \hat{\pi})}{n}}}. \tag{2.9}$$

The Wald test statistic is asymptotically standard normally distributed, and we can express the (two-sided) Wald P-value as

$$P\text{-value} = \Pr\left[Z \geq \left|Z_{\text{Wald}}(\hat{\pi})\right|\right], \tag{2.10}$$

where Z is a standard normal variable.

Section 2.4.1 on the Wald confidence interval introduced the continuity correction as an approach to achieve closer agreement between standard normal tail probabilities and binomial tail probabilities in small samples. The Wald test statistic with continuity correction is

$$Z_{\text{WaldCC}}(\hat{\pi}) = \frac{|\hat{\pi} - \pi_0| - \dfrac{1}{2n}}{\sqrt{\dfrac{\hat{\pi}(1 - \hat{\pi})}{n}}},$$

and P-values for the *Wald test with continuity correction* is obtained from Equation 2.10, with Z_{WaldCC} instead of Z_{Wald}.

For both the Wald test and the Wald test with continuity correction, the estimated standard error will be zero—and the test will not be possible to compute—if $X = 0$ or $X = n$.

2.5.3   The Likelihood Ratio Test

The likelihood ratio test statistic is given by

$$T_{\text{LR}}(\hat{\pi}) = 2\left[x \log\left(\frac{\hat{\pi}}{\pi_0}\right) + (n - x) \log\left(\frac{1 - \hat{\pi}}{1 - \pi_0}\right)\right].$$

We obtain a P-value for the likelihood ratio test by comparing the observed value of the test statistic with the chi-squared distribution with one degree of freedom:

$$P\text{-value} = \Pr\left[\chi_1^2 \geq T_{\text{LR}}(\hat{\pi})\right].$$

As with the two Wald tests, the likelihood ratio test will not be possible to compute when $X = 0$ or $X = n$.

2.5.4   The Score Test

The test statistic for the score test is obtained by substituting π_0 for π in Equation 2.4:

$$Z_{\text{score}}(\hat{\pi}) = \frac{\hat{\pi} - \pi_0}{\sqrt{\dfrac{\pi_0(1 - \pi_0)}{n}}}. \tag{2.11}$$

The score test statistic is approximately standard normally distributed, and P-values can be calculated as in Equation 2.10, with Z_{score} instead of Z_{Wald}.

A continuity corrected version of the score test statistic, which is also approximately standard normally distributed, can be expressed as

$$Z_{\text{scoreCC}}(\hat{\pi}) = \frac{|\hat{\pi} - \pi_0| - \dfrac{1}{2n}}{\sqrt{\dfrac{\pi_0(1 - \pi_0)}{n}}}. \tag{2.12}$$

Both the score test and the *score test with continuity correction* handle the situations where $X = 0$ or $X = n$.

2.5.5 Exact Tests

An exact test of the null hypothesis $H_0 : \pi = \pi_0$ is constructed as follows. Let X be binomially distributed with parameters n and π, and let x be the observed value. The exact right tail and left tail P-values are given as

$$P_{\text{right}} = \Pr(X \geq x) = \sum_{i=x}^{n} \binom{n}{i} \pi_0^i (1 - \pi_0)^{n-i} \tag{2.13}$$

and

$$P_{\text{left}} = \Pr(X \leq x) = \sum_{i=0}^{x} \binom{n}{i} \pi_0^i (1 - \pi_0)^{n-i}. \tag{2.14}$$

The one-sided P-values in Equations 2.13 and 2.14 are for the one-sided alternative hypotheses $H_A : \pi > \pi_0$ and $H_A : \pi < \pi_0$, respectively. A two-sided twice the smallest tail (see Section 1.3) P-value for the alternative hypothesis $H_A : \pi \neq \pi_0$ is

$$P\text{-value} = 2 \cdot \min(P_{\text{right}}, P_{\text{left}}).$$

We call this test the *exact binomial test*.

Section 2.4.7 presented the Blaker exact interval as an approach to reduce the conservatism of the Clopper-Pearson exact interval. Blaker (2000) focused on confidence intervals but formulated the problems in terms of preference functions and confidence curves such that both hypothesis tests and confidence intervals can be derived within the same framework. The *Blaker exact test* is defined as

$$P\text{-value} = \sum_{k=0}^{n} I\big[\gamma(k, \pi_0) \leq \gamma(x, \pi_0)\big] \cdot \binom{n}{k} \pi_0^k (1 - \pi_0)^{n-k}, \tag{2.15}$$

where $I()$ is the indicator function and γ is defined in Equation 2.6.

2.5.6 Mid-P Tests

Exact tests are well known to be quite conservative, and the power of these tests can be low. As discussed in Section 1.10, one approach to reduce conservatism is to calculate mid-P values. We can construct mid-P tests based on both the exact binomial test and the Blaker exact test. We start with the exact binomial test.

If we include only half of the point probability of the observed outcome in Equations 2.13 and 2.14, we obtain the one-sided mid-P values:

$$
\begin{aligned}
\text{mid-}P_{\text{right}} \;=\;& \Pr(X \geq x) - \frac{1}{2}\Pr(X = x) \\
=\;& \sum_{i=x}^{n} \binom{n}{i} \pi_0^i (1 - \pi_0)^{n-i} - \frac{1}{2}\binom{n}{x} \pi_0^x (1 - \pi_0)^{n-x}
\end{aligned}
$$

and

$$
\begin{aligned}
\text{mid-}P_{\text{left}} \;=\;& \Pr(X \leq x) - \frac{1}{2}\Pr(X = x) \\
=\;& \sum_{i=0}^{x} \binom{n}{i} \pi_0^i (1 - \pi_0)^{n-i} - \frac{1}{2}\binom{n}{x} \pi_0^x (1 - \pi_0)^{n-x}.
\end{aligned}
$$

The two-sided twice the smallest tail mid-P value for the *mid-P binomial test* is given as:

$$
\text{mid-}P \text{ value} = 2 \cdot \min(\text{mid-}P_{\text{right}}, \text{mid-}P_{\text{left}}).
$$

A similar mid-P adjustment can also be made to Equation 2.15. We shall refer to the resulting test as the *Blaker mid-P test*:

$$
\begin{aligned}
\text{mid-}P \text{ value} \;=\;& \sum_{k=0}^{n} I\big[\gamma(k, \pi_0) < \gamma(x, \pi_0)\big] \cdot \binom{n}{k} \pi_0^k (1 - \pi_0)^{n-k} \\
+\;& 0.5 \cdot \sum_{k=0}^{n} I\big[\gamma(k, \pi_0) = \gamma(x, \pi_0)\big] \cdot \binom{n}{k} \pi_0^k (1 - \pi_0)^{n-k}.
\end{aligned}
$$

2.5.7   Examples

The Probability of Male Births to Indian Immigrants in Norway (Table 2.2)

One interesting research question in this study is whether the probability of male births to Indian immigrants in Norway is equal to the probability of male births in the general Norwegian population ($\pi_0 = 0.513$). The relevant null hypothesis is $H_0 : \pi = 0.513$, which we shall test against the two-sided alternative $H_A : \pi \neq 0.513$.

For the 1st order births, we have the observed proportion $\hat{\pi} = 250/533 = 0.469$, and the P-values for the nine tests in Sections 2.5.2–2.5.6 range from 0.0420 (Wald) to 0.0469 (score with continuity correction and exact binomial). All the tests are significant at the 5% level and suggest (although mildly) that the probability of a male for the 1st order birth is slightly less among Indian immigrants in Norway than in the general Norwegian population.

For the 2nd order births, the observed proportion of males ($\hat{\pi} = 204/412 = 0.495$) is very similar to $\pi_0 = 0.513$, and none of the tests indicates a true difference (P-values from 0.47 to 0.50).

The results for the 3rd and 4th order births are shown in Table 2.5. The observed proportions of male births are now considerably higher than in the general Norwegian population, and even though the sample sizes are quite small—particularly for the 4th order births—all nine tests have P-values < 0.01 (3rd order births) and ≤ 0.005 (4th order births). We note that the P-value for the Wald test is somewhat lower than the other P-values in Table 2.5, especially for the 4th order births. As we shall see in the next section, the actual significance level of the Wald test is considerably higher than the nominal level. This results in small P-values and increased chance of false positive findings.

TABLE 2.5

Two-sided P-values for $H_0 : \pi = 0.513$, based on data from Singh et al. (2010)

Test	3rd birth 103/167 $\hat{\pi} = 0.617$	4th birth 33/45 $\hat{\pi} = 0.733$
Wald	0.0058	0.0008
Wald with continuity correction	0.0074	0.0015
Likelihood ratio	0.0070	0.0025
Score	0.0073	0.0031
Score with continuity correction	0.0092	0.0050
Exact binomial	0.0089	0.0043
Blaker exact	0.0083	0.0041
Mid-P binomial	0.0072	0.0029
Blaker mid-P	0.0075	0.0034

A Randomized Cross-Over Trial of Probiotics Versus Placebo for Patients with Irritable Bowel Syndrome

If the patients were indifferent to whether they were treated with placebo or probiotics, the probability of a patient preferring placebo would be $\pi_0 = 0.5$. To test whether probiotics may have a beneficial (or harmful) effect, we formulate the null hypothesis $H_0 : \pi = 0.5$, which we test against the two-sided alternative $H_A : \pi \neq 0.5$. The observed proportion of patients who prefer placebo is $\hat{\pi} = 13/16 = 0.813$. Table 2.6 gives the results for nine tests for the binomial probability. The P-values differ more in this example than in the previous one, mainly because we now have a very small sample size. The asymptotic (non-exact) tests are based on large-sample approximations and may perform poorly for $n < 20$, although some of the tests, such as the score, have good average properties even in quite small samples. The exact tests are safe choices, although they are slightly conservative. For this example, even the most conservative tests (score with continuity correction and exact binomial) suggest, with a reasonable amount of confidence, that patients prefer placebo to the probiotic L. plantarum MF 1298.

TABLE 2.6

Two-sided P-values for $H_0 : \pi = 0.5$, based on data from Ligaarden et al. (2010)

Test	P-value
Wald	0.0014
Wald with continuity correction	0.0039
Likelihood ratio	0.0094
Score	0.0124
Score with continuity correction	0.0244
Exact binomial	0.0213
Blaker exact	0.0127
Mid-P binomial	0.0127
Blaker mid-P	0.0085

2.5.8   Evaluation of Tests

Evaluation Criteria

We evaluate tests for the binomial parameter by calculating their actual significance levels and power. The actual significance level and power depends on π, n, and the nominal significance level α. For any given combination of $\{\pi, n, \alpha\}$, we can calculate the actual significance level if $\pi = \pi_0$, or power if $\pi \neq \pi_0$, exactly with complete enumeration: we sum the probabilities of all possible outcomes with P-values less than the nominal significance level (α). The actual significance level is given by

$$\mathrm{ASL}(\pi_0, n, \alpha) = \sum_{x=0}^{n} I\big[P(x) \leq \alpha\big] \cdot f(x \,|\, \pi_0, n),$$

and the power is given by

$$\mathrm{Power}(\pi, n, \alpha) = \sum_{x=0}^{n} I\big[P(x) \leq \alpha\big] \cdot f(x \,|\, \pi, n),$$

where $I()$ is the indicator function, $P(x)$ is the P-value for a test on the data with x successes in n observations, and $f()$ is the binomial probability distribution (Equation 2.2).

Evaluation of Actual Significance Level

For a given sample size (n), we can make a plot of the actual significance level as a function of the binomial probability (π). Figure 2.13 shows actual significance levels of the Wald, likelihood ratio, and score tests for $n = 20$. The actual significance level is a highly discontinuous function of π with spikes spanning a wide range of values. Although it is quite obvious from Figure 2.13 that the actual significance levels of the Wald test, for most values of π, are considerably higher than the nominal level, it may be difficult to assess and compare the typical behavior of different tests when faced with such discontinuities. Similar discontinuous curves arose when we calculated coverage probabilities for different confidence interval methods for π in Section 2.4.10. Newcombe and Nurminen (2011) make the argument that when assessing confidence intervals for π, a smoothed curve, obtained as the moving average of the coverage probabilities, provides a more realistic assessment of the coverage achieved in practice. This argument can also be made for plots of actual significance levels. Thus, for the remainder of this section, we shall plot moving averages superimposed on plots of actual significance levels.

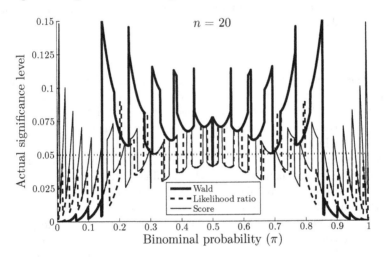

FIGURE 2.13
Actual significance levels of three tests for the binomial probability

The Wald test did not perform well in Figure 2.13; but, with only $n = 20$ observations, we may suspect that this is simply due to the small sample size, and that the actual significance levels of the Wald test will be close to the nominal level for larger sample sizes. As Figure 2.14 illustrates, however, we must increase the sample size by a factor of more than 10 to obtain satisfactory results. For $n = 200$, the actual significance levels of the Wald test are still considerably larger than the nominal level, particularly for values of π close to 0 or 1. Even for $n = 500$, the average actual significance level is about 5.5–6.0% when $\pi < 0.1$ or $\pi > 0.9$.

A continuity correction to the Wald test is a great improvement over the standard Wald test. Figure 2.15 shows the actual significance levels of the Wald test with continuity correction for $n = 30$. We have chosen $n = 30$

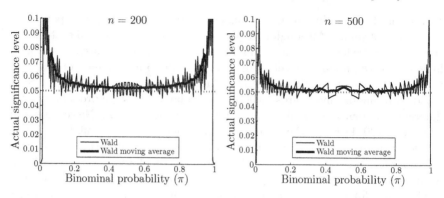

FIGURE 2.14
Actual significance levels (with moving averages over the range $[\pi - 0.05, \pi + 0.05]$) of the Wald test

because the general behavior of the actual significance levels of the different tests for the binomial probability is well illustrated with this sample size. We shall use $n = 30$ throughout most of this section and note in the text when interesting actual significance levels for other sample sizes are not sufficiently illustrated by the plots.

The moving average curve in Figure 2.15 is well within 0.04 and 0.06 for most values of π, and usually below the nominal level. In general, we prefer significance levels that are slightly below the nominal level over significance levels that are slightly above the nominal level. As we shall see later in this book, such beneficial effects of a continuity correction are the exception, not the rule. In most cases, continuity corrections make both tests and confidence intervals overly conservative: tests have unnecessary low power and intervals are unnecessarily wide.

Also shown in Figure 2.15 are the actual significance levels of the likelihood ratio test for $n = 30$. These are closer to the nominal level than the Wald test with continuity correction; however, for small and large values of π, the actual significance levels can be considerably lower than the nominal level.

Figure 2.16 illustrates the actual significance levels of the score test (left panel) and the score test with continuity correction (right panel). The performance of the score test (without continuity correction) is excellent. The moving average curve is slightly below the nominal level for almost the entire range of π. With continuity correction, the moving average curve drops down to between 0.02 and 0.03 (for $n = 30$), and illustrates the typical effect of the continuity correction. Note that the performance of the score test with continuity correction is very similar to the performance of the exact binomial test.

Exact tests are guaranteed to have actual significance levels bounded by

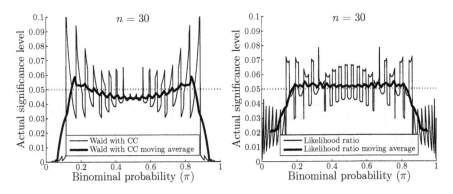

FIGURE 2.15
Actual significance levels (with moving averages over the range $[\pi - 0.05, \pi + 0.05]$) of the Wald test with continuity correction and the likelihood ratio test

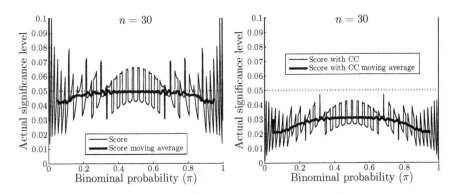

FIGURE 2.16
Actual significance levels (with moving averages over the range $[\pi - 0.05, \pi + 0.05]$) of the score test and the score test with continuity correction

the nominal level for all values of π (and all values of n). As we can see in Figure 2.17, which shows results for the exact binomial (left panel) and Blaker exact (right panel) tests, a disadvantage with exact tests is that the actual significance level often is considerably lower than the nominal level. The average significance level of the exact binomial test in the left panel of Figure 2.17 is mostly between 2% and 3%, which is quite different from the nominal 5%. The Blaker exact test is a considerable improvement; however, its average actual significance level is seldom above 4%.

One common approach to reduce the conservatism of exact tests is to calculate mid-P values instead of P-values. The mid-P binomial and Blaker mid-P tests shown in Figure 2.18 have significance levels considerably closer

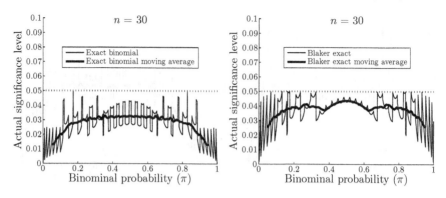

FIGURE 2.17
Actual significance levels (with moving averages over the range $[\pi - 0.05, \pi + 0.05]$) of the exact binomial and Blaker exact tests

to the nominal level than do the exact tests; however, the tests are no longer exact, and they frequently violate the nominal level. Nevertheless, the average performance of the mid-P tests is almost as good as that of the score test in Figure 2.16. Note that the actual significance level of the Blaker mid-P test has greater oscillations and a propensity toward more frequent and larger violations of the nominal level than that of the mid-P binomial test.

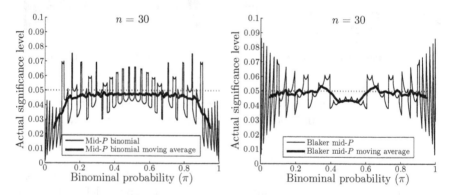

FIGURE 2.18
Actual significance levels (with moving averages over the range $[\pi - 0.05, \pi + 0.05]$) of the mid-P binomial and Blaker mid-P tests

If we prefer to align average instead of maximum actual significance levels with the nominal level, the two best-performing tests are the score test and the mid-P binomial test. This holds not only for $n = 30$, as shown in Figures 2.15–2.18, but also for other sample sizes as well. To further compare the score and the mid-P binomial tests, we show their actual significance levels for $n = 15$ observations in Figure 2.19. With such a small sample size, we clearly see that the score test maintains a reasonable significance level throughout all relevant values of π, whereas the mid-P binomial test is more conservative. The score test performs, in general, slightly better than the mid-P binomial test for small sample sizes; however, for $n \geq 50$, the performance of the two tests is practically equal.

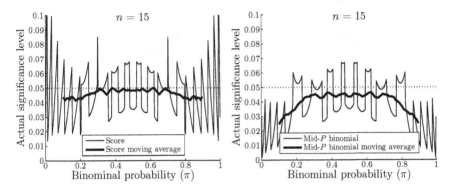

FIGURE 2.19
Actual significance levels (with moving averages over the range $[\pi - 0.1, \pi + 0.1]$) of the score and mid-P binomial tests

Evaluation of Power

In Figures 2.13–2.19, we plotted the actual significance level as a function of the binomial probability (π) for fixed values of the number of observations (n). To illustrate power, it is more informative to fix π and plot power as a function of n. One example is given in Figure 2.20, where we plot the power of the Wald, score and exact binomial tests for the null hypothesis $H_0 : \pi = 0.4$, when the true value of the binomial parameter is $\pi = 0.2$. As usual, the alternative hypothesis is two-sided ($H_A : \pi \neq 0.4$). The Wald test has the greatest power, followed by the score test. The exact binomial test has power below that of the other two tests. This ordering of power reflects the differences in actual significance levels: the Wald test has actual significance levels above the nominal level, the score test has actual significance levels close to the nominal level, whereas the exact binomial test has actual significance levels below the nominal level.

The score and the mid-P binomial tests had similar actual significance levels. They also have similar power. Sometimes the score test is slightly better than the mid-P binomial test, and sometimes the mid-P binomial test is slightly better than the score test; it depends on the choices of π and π_0. Based on power, we cannot rank one test ahead of the other.

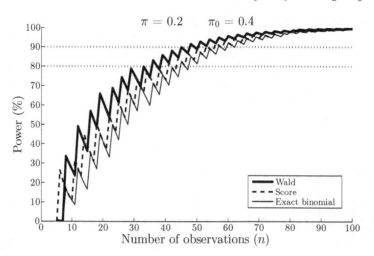

FIGURE 2.20
Power of three tests for the binomial probability

Figure 2.17 showed that the actual significance levels of the Blaker exact test were considerably closer to the nominal level than those of the exact binomial test. This advantage extends to power, as seen in Figure 2.21, where we have restricted the range of the y-axis to cover only the most interesting range of power values. Figure 2.21 illustrates that for studies designed to have 80% or 90% power, it is possible to reduce the sample size by planning to use the Blaker exact test instead of the exact binomial test.

2.6   Recommendations

2.6.1   Summary

The ideal method is simple to describe, simple to calculate or available in every standard software package, has strong theoretical grounds, preferably based on general statistical principles, and excellent properties. Unfortunately, the ideal method does not exist, and to arrive at general recommendations, we must compare methods with different strengths and weaknesses. This requires a balancing act, where simplicity and availability are weighted against performance properties such as coverage probabilities (confidence intervals) and actual significance levels (tests). Before we go into the details of our recommendations, we make the following general remarks, germane to both confidence intervals and tests for the binomial parameter:

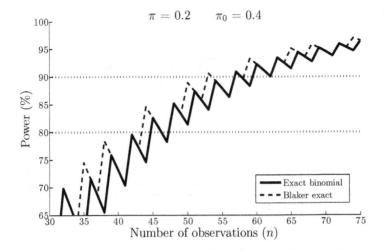

FIGURE 2.21
Power of two exact tests for the binomial probability

- Wald methods do not perform well, even for quite large sample sizes

- Score methods have excellent properties and are clearly superior to methods based on the Wald and likelihood ratio approaches

- Exact methods can be very conservative; however, the Blaker approach can reduce conservatism considerably

- Another approach to reduce the conservatism of exact methods is to use a mid-P modification

Table 2.7 provides a summary of the recommended confidence intervals and tests, and gives the sample sizes for which the recommended methods are appropriate. The small, medium, and large labels cannot be given precise definitions, they will vary from one analysis to the other, and some subjectivity needs to be applied. As a rule of thumb, small may be taken as less than 50 observations, medium as between 50 and 100 observations, and large as more than 100 observations. Sections 2.6.2–2.6.3 discuss the recommendations in more detail and summarize the merits of the different methods.

2.6.2 Confidence Intervals for the Binomial Parameter

The traditional Wald interval does not perform well. It has coverage below the nominal level, and often by a considerable margin. The practical problem with too low coverage is that the Wald interval will include the true binomial probability less often than the nominal level specifies. The Wald interval is

TABLE 2.7

Recommended confidence intervals (CIs) and tests for the binomial parameter

Analysis	Recommended methods	Sample sizes
CIs for the binomial parameter	Wilson score*	all
	Blaker exact	small/medium
	Clopper-Pearson mid-P	medium
	Agresti-Coull*	medium/large
Tests for the binomial parameter	Score*	all
	Blaker exact	small/medium
	Mid-P binomial	medium

*These methods have closed-form expression

also too distally located, which means that it tends to be located too close to either 0 or 1, as compared with the true binomial probability. And considering that the expected width of the Wald interval is not particularly narrow, it has nothing but simplicity and ubiquitousness on its side. We do not recommend it, unless the sample size is very large, say 500 or more observations.

The Wilson score interval, on the other hand, can be recommended for any sample size. It has excellent average coverage probability even for the smallest of sample sizes ($n = 5$), and the expected width is small compared with other interval methods. The Wilson score interval can be slightly too mesially located, which means that the interval tends to be located closer to 0.5 than the true binomial probability; however, we consider such small amounts of asymmetry to be of little practical concern. Although the expression for the Wilson score interval (see Equation 2.5) is more elaborate than that of the Wald interval (or the Agresti-Coull interval), it is a closed-form expression, and may be calculated by a desktop calculator or in a spreadsheet.

If an exact interval is required, we recommend the Blaker exact interval rather than the Clopper-Pearson exact interval. The Blaker exact interval has coverage probabilities considerably closer to the nominal level, and shorter expected interval width, than does the Clopper-Pearson exact interval. Unfortunately, the Blaker exact interval is not well supported by statistical software packages. Moreover, it does not have a simple formulation via the beta distribution, as does the Clopper-Pearson exact interval, and must be computed with an iterative algorithm. We still make it our recommended exact interval for the binomial parameter, and urge software developers to include it in future editions of their applications.

Two other intervals can be recommended. Despite its simplicity and close relation to the Wald interval, the Agresti-Coull interval has vastly improved coverage probabilities compared with the Wald interval. It does not perform quite as good as the Wilson score interval; however, for moderate and large sample sizes, the Agresti-Coull interval is quite satisfactory. The Clopper-

Pearson mid-P interval is more complex to calculate than the Agresti-Coull or the Wilson score intervals. It is based on the Clopper-Pearson exact interval; however, it does not have a simple beta distribution formula, and requires an iterative algorithm to compute its confidence limits. The performance of the Clopper-Pearson mid-P interval is very good, though, and we include it in our list of recommended intervals for the binomial parameter.

2.6.3 Tests for the Binomial Parameter

The performance of the Wald test mimics the performance of the Wald interval. The actual significance levels of the Wald test can be much larger than the nominal significance level, and this problem can be observed for quite large sample sizes. The probability that the Wald test indicates a false positive finding is thereby increased, as compared with the nominal level, and often beyond acceptable levels. As with the Wald interval, we do not recommend the Wald test unless the sample size is very large, say 500 or more observations.

Among the remaining tests, we find three methods to recommend, each corresponding to a confidence interval method that was recommended in the previous section. The rationales for recommending these tests are, in principle, the same as those for the corresponding confidence intervals. Here, we repeat them only briefly: the score test has excellent average actual significance levels, including the case when $n = 5$; the Blaker exact test is superior to the exact binomial test (which might also be called the Clopper-Pearson exact test); the mid-P binomial test (corresponding to the Clopper-Pearson mid-P interval) performs almost as good as the score test; however, it is more complex to calculate.

3

The 1 × c Table and the Multinomial Distribution

3.1 Introduction

In this chapter, we consider a simple extension of the situation in Chapter 2 (the 1 × 2 table), where the data consisted of a sample from a dichotomous variable. Now, we sample N observations of a categorical variable with c possible outcomes, and count the number of occurrences of each of the c outcomes. For each observation, let π_i denote the probability of outcome i, and let n_i be the number of observations with outcome i, with $i = 1, 2, \ldots, c$. Naturally, $\sum_{i=1}^{N} \pi_i = 1$ and $\sum_{i=1}^{N} n_i = N$. If the N observations are independent, $n_1, n_2, \ldots, n_c$ are said to be multinomially distributed with parameters N and $\pi_1, \pi_2, \ldots, \pi_c$. For example, blood samples from N individuals can be classified into the $c = 4$ blood types: O, A, B, or AB. If these individuals are unrelated, the number of each blood type may be regarded as multinomially distributed. Blood types for family members, on the other hand, are dependent and cannot appropriately be described by the multinomial distribution.

Let $x_1, x_2, \ldots, x_c$ denote any possible table such that $\sum_{i=1}^{N} x_i = N$. The multinomial probability distribution is given by

$$f(x_1, \ldots, x_c \mid \pi_1, \ldots, \pi_c; N) = \begin{pmatrix} N \\ x_1 \ x_2 \ \ldots \ x_c \end{pmatrix} \pi_1^{x_1} \pi_2^{x_2} \cdots \pi_c^{x_c}, \qquad (3.1)$$

where

$$\begin{pmatrix} N \\ x_1 \ x_2 \ \ldots \ x_c \end{pmatrix} = \frac{N!}{x_1! x_2! \cdots x_c!}.$$

If $c = 2$, Equation 3.1 reduces to a binomial distribution for x_1, see Equation 2.2. In fact, the marginal distribution for any x_i is binomial with parameters N and π_i.

Table 3.1 shows the contingency table setup for the one-sample multinomial variable, where the observed counts are summarized in a 1 × c table.

This chapter—the shortest in this book—is organized as follows. Section 3.2 introduces an example that will be used to illustrate the methods later in the chapter. Tests for hypotheses about the multinomial parameters $\pi_1, \pi_2, \ldots, \pi_c$ are presented, applied to the data example, and evaluated in

TABLE 3.1

The observed counts of a $1 \times c$ table

	Category 1	Category 2	...	Category c	Total
Sample	n_1	n_2	...	n_c	N

Additional notation:

$\mathbf{n} = \{n_1, n_2, \ldots, n_c\}$: the observed table

$\mathbf{x} = \{x_1, x_2, \ldots, x_c\}$: any possible table

Section 3.3. Sections 3.4 and 3.5 consider two alternative hypothesis setups for $\pi_1, \pi_2, \ldots, \pi_c$. Simultaneous confidence intervals for the multinomial parameters and the differences between them are derived in Section 3.6. Finally, Section 3.7 provides recommendations.

3.2 Example: Distribution of Genotype Counts of SNP rs6498169 in Patients with Rheumatoid Arthritis

Table 3.2 shows the observed genotype counts for SNP rs 6498169 in a sample of rheumatoid arthritis (RA) patients from a study by Skinningsrud et al. (2010). The observed proportions of genotypes AA, AG, and GG are 35.7%, 49.1%, and 15.2%, respectively. The study also included a control group of 2110 healthy individuals. For simplicity, we shall assume that the genotype probabilities for AA, AG, and GG in the general population are equal to the observed proportions in the control group, namely 40.2%, 47.9%, and 11.9%, respectively. The research question is: does the genotype distribution in the RA sample deviate significantly from that of the general population?

TABLE 3.2

Genotype counts for SNP rs 6498169 in RA patients

	AA	AG	GG	Total
RA patients	276 (36%)	380 (49%)	118 (15%)	774

3.3 Tests for the Multinomial Parameters

3.3.1 Estimation of the Multinomial Parameters

Maximum likelihood estimates of $\pi_1, \pi_2, \ldots, \pi_c$ may be derived in a similar manner as we derived the maximum likelihood estimate of the binomial parameter π in Section 2.3. We do not provide the details here, but simply state that the maximum likelihood estimates are equal to the sample proportions:

$$\hat{\pi}_i = \frac{n_i}{N},$$

for $i = 1, 2, \ldots, c$.

3.3.2 The Null and Alternative Hypotheses

A fully specified null hypothesis is

$$H_0 : \pi_i = \pi_{i,0}, \tag{3.2}$$

for all $i = 1, 2, \ldots, c$, where $\pi_{i,0}$ are specified constants that sum to one: $\sum_{i=1}^{c} \pi_{i,0} = 1$. The two-sided alternative hypothesis is

$$H_A : \pi_i \neq \pi_{i,0},$$

for one or more i. Note that the null hypothesis

$$H_0 : \pi_1 = \pi_2 = \ldots = \pi_c$$

is a special case of Equation 3.2, obtained by letting $\pi_{i,0} = 1/c$.

In the following, we describe four tests: two asymptotic tests, one exact test, and one mid-P test. The two asymptotic tests rely on large-sample approximations to the null distribution of the test statistics. In samples with small expected counts (the next section shows how to calculate them), the null distribution of the test statistic can deviate from the purported chi-squared distribution. One rule of thumb is to use the asymptotic tests only if all expected counts are above 1 and at most 20% of the expected counts are below 5 (Siegel and Castellan, 1988, p. 49); however, as we will see in our evaluations in Section 3.3.8, the Pearson chi-squared test performs well also for quite small sample sizes.

3.3.3 The Pearson Chi-Squared Test

The *Pearson chi-squared statistic* has the general form

$$T_{\text{Pearson}}(n_1, n_2, \ldots, n_c) = \sum_{i=1}^{c} \frac{(O_i - E_i)^2}{E_i},$$

where O_i and E_i are the observed and expected counts, respectively, in category i. For the null hypothesis in Equation 3.2, the Pearson chi-squared statistic is

$$T_{\text{Pearson}}(n_1, n_2, \ldots, n_c) = \sum_{i=1}^{c} \frac{(n_i - N\pi_{i,0})^2}{N\pi_{i,0}}, \tag{3.3}$$

where higher values indicate increasing deviations from the null hypothesis. Under the null hypothesis, T_{Pearson} is approximately chi-squared distributed with $c - 1$ degrees of freedom, thus

$$P\text{-value} = \Pr\left[\chi_{c-1}^2 \geq T_{\text{Pearson}}(n_1, n_2, \ldots, n_c)\right].$$

The Pearson chi-squared test for the multinomial parameters is also a score test. When $c = 2$, the test statistic in Equation 3.3 is equal to the square of the score test statistic for the binomial parameter (Equation 2.11).

To *identify cells with deviating counts*, we suggest that simultaneous confidence intervals for the multinomial parameters are estimated, as described in Section 3.6. We declare cell i to have deviating counts if the null value $\pi_{i,0}$ is not contained in the simultaneous confidence interval for π_i. With this approach, the probability of making at least one incorrect statement is less than or equal to the significance level used for the test.

3.3.4   The Likelihood Ratio Test

The likelihood ratio test for the multinomial parameters is a straightforward extension of the likelihood ratio test for the binomial parameter in Section 2.5.3. The test statistic is

$$T_{\text{LR}}(n_1, n_2, \ldots, n_c) = 2 \sum_{i=1}^c n_i \log\left(\frac{n_i}{N\pi_{i,0}}\right),$$

which is chi-squared distributed with $c - 1$ degrees of freedom under the null hypothesis. By convention, if any $n_i = 0$, let $n_i \log(n_i/N\pi_{i,0}) = 0$.

3.3.5   The Exact Multinomial Test

Recall from Section 1.9 that the general expression for an exact P-value is

$$\text{exact } P\text{-value} = \sum_{\text{all tables}} I\left[T(\mathbf{x}) \geq T(\mathbf{n})\right] \cdot f(\mathbf{x}\,|\,H_0), \tag{3.4}$$

where $I()$ is the indicator function, $T()$ is a test statistic, $\mathbf{n}$ denotes the observed table, $\mathbf{x}$ denotes any possible table that might be observed, and $f()$ is the probability of observing $\mathbf{x}$ under the null hypothesis. For the $1 \times c$ table, $\mathbf{n} = \{n_1, n_2, \ldots, n_c\}$, $\mathbf{x} = \{x_1, x_2, \ldots, x_c\}$, and $f()$ is the multinomial probability distribution in Equation 3.1. Any valid test statistic can be used as $T()$. Here, we shall use the Pearson chi-squared statistic in Equation 3.3.

The sums over all possible tables in Equation 3.4 can be expressed explicitly only for specific values of c. For $c = 3$, for example, we have that

$$\text{exact } P\text{-value} = \sum_{x_1=0}^{N} \sum_{x_2=0}^{N-x_1} I\left[T_{\text{Pearson}}(\mathbf{x}) \geq T_{\text{Pearson}}(\mathbf{n})\right] \cdot f(\mathbf{x}\,|\,\boldsymbol{\pi}_0, N),$$

where $\mathbf{x} = \{x_1, x_2, N - x_1 - x_2\}$, $\mathbf{n} = \{n_1, n_2, n_3\}$, and $\boldsymbol{\pi}_0 = \{\pi_{1,0}, \pi_{2,0}, \pi_{3,0}\}$ are the probabilities under the null hypothesis.

3.3.6   The Mid-*P* Multinomial Test

The general expression for a mid-*P* value was given in Equation 1.12. The mid-*P* value is computed in the same way as the exact *P*-value, except that we only include half the point probability of tables that agree equally with the null hypothesis as the observed table. A mid-*P* test based on the exact multinomial test from the preceding section can thus be expressed (for $c = 3$) as

$$
\text{mid-}P \text{ value} = \sum_{x_1=0}^{N} \sum_{x_2=0}^{N-x_1} I\big[T_{\text{Pearson}}(\mathbf{x}) > T_{\text{Pearson}}(\mathbf{n})\big] \cdot f(\mathbf{x}\,|\,\boldsymbol{\pi}_0, N) +
$$

$$
0.5 \cdot \sum_{x_1=0}^{N} \sum_{x_2=0}^{N-x_1} I\big[T_{\text{Pearson}}(\mathbf{x}) = T_{\text{Pearson}}(\mathbf{n})\big] \cdot f(\mathbf{x}\,|\,\boldsymbol{\pi}_0, N).
$$

3.3.7   Example

Distribution of Genotype Counts of SNP rs6498169 in Patients with Rheumatoid Arthritis (Table 3.2)

The purpose of this example is to investigate whether the probabilities of genotypes AA, AG, and GG in RA patients differ from those of the general population. We have the following null hypothesis:

$$
H_0: \ \pi_1 = 0.402, \quad \pi_2 = 0.479, \quad \pi_3 = 0.119, \tag{3.5}
$$

where π_1 is the probability of genotype AA, π_2 is the probability of genotype AG, and π_3 is the probability of genotype GG. The alternative hypothesis is that at least one of the equalities in Equation 3.5 does not hold.

The Pearson chi-squared statistic is

$$
\begin{aligned}
T_{\text{Pearson}} &= \frac{(276 - 774 \cdot 0.402)^2}{774 \cdot 0.402} + \frac{(380 - 774 \cdot 0.479)^2}{774 \cdot 0.479} \\
&\quad + \frac{(118 - 774 \cdot 0.119)^2}{774 \cdot 0.119} \\
&= 3.97 + 0.23 + 7.28 \\
&= 11.48.
\end{aligned}
$$

The degrees of freedom are $c - 1 = 2$. The *P*-value for the Pearson chi-squared test is $\Pr(\chi_2^2 \geq 11.48) = 0.0032$, so there is evidence that the probability distribution in RA patients deviates from the general population.

Next, we calculate the likelihood ratio test. The likelihood ratio test statistic is

$$T_{LR} = 2 \cdot \left(276 \cdot \log \frac{276}{774 \cdot 0.402} + 380 \cdot \log \frac{380}{774 \cdot 0.479} \right.$$

$$\left. + 118 \cdot \log \frac{118}{774 \cdot 0.119} \right)$$

$$= 2 \cdot (-33.08 + 9.37 + 29.23)$$

$$= 11.04.$$

With $c - 1 = 2$ degrees of freedom, the P-value for the likelihood ratio test is $\Pr(\chi^2_2 \geq 11.04) = 0.0040$, which is similar to that of the Pearson chi-squared test.

As we will see in the upcoming Section 3.6.2, where we calculate the simultaneous confidence intervals for π_1, π_2, and π_3, there are deviating cell counts for AA and GG.

The number of observations in this example is $N = 774$, which is sufficiently large that we expect the chi-squared approximation to the distributions of the T_{Pearson} and T_{LR} statistics to be more than satisfactory. We therefore do not consider the exact multinomial and mid-P multinomial tests for the data in Table 3.2. For small sample sizes, however, we may be worried about the accuracy of the chi-squared approximation.

Suppose that we select 10 of the 774 RA patients, and that the observed counts of their genotypes are as shown in Table 3.3. The expected counts for $N = 10$ and the probabilities given in Equation 3.5 are 4.02, 4.79, and 1.19, which fail the rule of thumb (see p. 69) that no more than 20% of the expected counts should be below 5. The P-values for the Pearson chi-squared and likelihood ratio tests are $P = 0.035$ and $P = 0.027$, respectively; however, we are disinclined to trust them, so we turn to the exact and mid-P tests.

TABLE 3.3
Genotype counts for SNP rs 6498169 in a subset of 10
RA patients

	AA	**AG**	**GG**	**Total**
RA patients	6 (60%)	1 (10%)	3 (30%)	10

Out of a total number of 66 possible ways to distribute the 10 observations to the three categories, 40 agree less with the null hypothesis than the observed data, as measured by the Pearson chi-squared statistic. The exact P-value is thus the sum of the probabilities of these 40 tables, where the probability of any one table is given by the multinomial probability distribution in Equation 3.1. The result is $P = 0.0479$, which is a little bit higher than the P-values for the two asymptotic tests.

To calculate the mid-P test, we identify the tables that have an equal value of T_{Pearson} as the observed table. In this case, the observed table is the only table with that particular value ($T_{\text{Pearson}} = 6.727$). The point probability of the observed table is 0.0029, and we obtain the mid-P value by subtracting half the point probability from the exact P-value: mid-$P = 0.0479 - 0.5 \cdot 0.0029 = 0.0465$.

We remark that the above calculations are for illustration purposes only; the probabilities for the general population are based on a finite control sample (with 2110 individuals). A more appropriate treatment of this problem would be to add a row to Table 3.5 with the observed counts of AA, AG, and GG for the control sample, and analyze the data as a 2×3 table, possibly with ordered categories, as described in Chapter 6.

3.3.8   Evaluation of Tests

Evaluation Criteria

The evaluation of tests for the multinomial parameters are based on calculations of actual significance levels and power (see Section 1.4 for general descriptions). Both quantities depend on π_i, N, and the nominal significance level α. For any combination of parameters, we can calculate the actual significance level (if $\pi_i = \pi_{i,0}$ for all $i = 1, 2, \ldots, c$) or power (if $\pi_i \neq \pi_{i,0}$ for at least one i) exactly with complete enumeration: we sum the probabilities of all possible tables with N observations that lead to P-values less than the nominal level (α). An expression for "all possible tables" requires that we specify c, the number of categories. Here, we give formulas for the actual significance level and power for $c = 3$ only. The formulas for higher values of c are straightforward extensions of those below. The actual significance level for $c = 3$ is given by

$$\text{ASL}(\pi_{1,0}, \pi_{2,0}, \pi_{3,0}, N, \alpha) = \sum_{x_1=0}^{N} \sum_{x_2=0}^{N-x_1} I\big[P(\mathbf{x}) \leq \alpha\big] \cdot f(\mathbf{x} \,|\, \pi_{1,0}, \pi_{2,0}, \pi_{3,0}; N),$$

and the power is given by

$$\text{Power}(\pi_1, \pi_2, \pi_3, N, \alpha) = \sum_{x_1=0}^{N} \sum_{x_2=0}^{N-x_1} I\big[P(\mathbf{x}) \leq \alpha\big] \cdot f(\mathbf{x} \,|\, \pi_1, \pi_2, \pi_3; N),$$

where $\mathbf{x} = \{x_1, x_2, N - x_1 - x_2\}$, $I()$ is the indicator function, $P(\mathbf{x})$ is the P-value for a test on table $\mathbf{x}$, and $f()$ is the multinomial probability distribution (Equation 3.1).

Evaluation of Actual Significance Level

For most of the table types in this book, we have one or two unknown probabilities to estimate and test hypotheses about. In this chapter, we have $c \geq 3$ unknown probabilities, of which $c-1$ are independent (the c probabilities sum to one). Instead of plotting the actual significance level as a function of one of these probabilities—such as we usually do when we have only one independent probability—we here select a set of probabilities and plot the actual significance level as a function of the number of observations.

Figures 3.1 and 3.2 show the results for a three-level multinomial variable, when the probability of each category is $1/3$. The Pearson chi-squared test has, in general, actual significance levels slightly below the nominal level, with small oscillations for varying values of N. For a method this simple, the Pearson chi-squared test performs excellently. The likelihood ratio test violates the nominal level for almost all sample sizes. Its actual significance levels approach the nominal level with increasing N, although more slowly than the Pearson chi-squared test. The actual significance levels of the exact multinomial and the mid-P multinomial tests are quite close to the nominal level, even more so than the Pearson chi-squared test. The conservatism of the exact test is very modest.

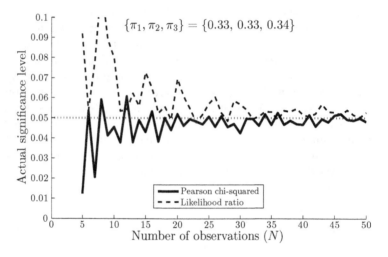

FIGURE 3.1
Actual significance levels of the Pearson chi-squared and likelihood ratio tests for the multinomial parameters

Figures 3.3 and 3.4 illustrate the effect of unevenly distributed probabilities on the actual significance levels. The first category is associated with a 60% probability, whereas category two and three have probabilities 30% and 10%, respectively. The Pearson chi-squared test still performs well, although its actual significance levels vary slightly more than in Figure 3.1. The disadvantage with the likelihood ratio test is even more pronounced than before, with large violations of the nominal level for N in the range 20–40. The exact multinomial test is slightly more conservative than in Figure 3.2, whereas the mid-P multinomial test still has actual significance levels very close to the nominal level.

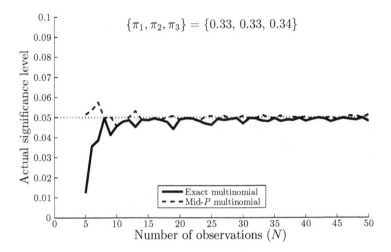

FIGURE 3.2
Actual significance levels of the exact and mid-P multinomial tests for the
multinomial parameters

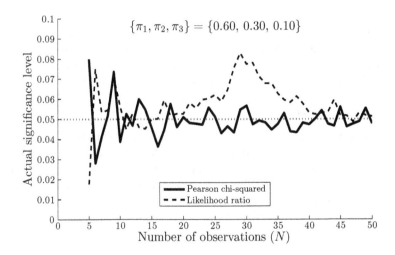

FIGURE 3.3
Actual significance levels of the Pearson chi-squared and likelihood ratio tests
for the multinomial parameters

So far, we only have considered a three-level multinomial variable. Figure 3.5 shows an example of the actual significance levels of the two asymptotic
tests when $c = 4$. As before, the Pearson chi-squared test performs surpris-

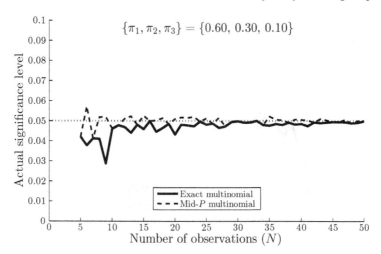

FIGURE 3.4
Actual significance levels of the exact and mid-P multinomial tests for the multinomial parameters

ingly well, while the likelihood ratio test fails to produce appropriate actual significance levels. Both the exact multinomial and mid-P multinomial tests— not shown here—perform excellently: they have actual significance levels that are very close to the nominal level for sample sizes as low as $N = 10$.

Evaluation of Power

The power of the four tests are, in general, quite similar to each other across different scenarios. Figure 3.6 shows an example of the power of the Pearson chi-squared test and the exact multinomial test to reject the null hypothesis $H_0 : \pi_1 = \pi_2 = \pi_3 = 1/3$ when the correct probabilities are $\pi_1 = 0.6$, $\pi_2 = 0.2$, and $\pi_3 = 0.2$. The results for the likelihood ratio and mid-P multinomial tests are not notably different from the Pearson chi-squared and exact multinomial tests.

3.4   An Incompletely Specified Null Hypothesis

Sometimes, we are interested in a null hypothesis that is not fully specified like Equation 3.2. Rather, the null hypothesis may specify the probability distribution in terms of one or more parameters that need to be estimated from the data. To test such a hypothesis, we may use a simple modification of the

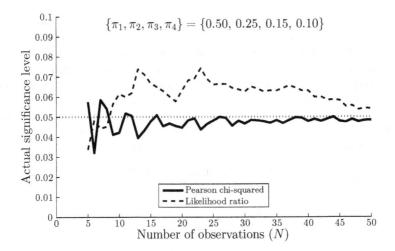

FIGURE 3.5
Actual significance levels of the Pearson chi-squared and likelihood ratio tests
for the multinomial parameters

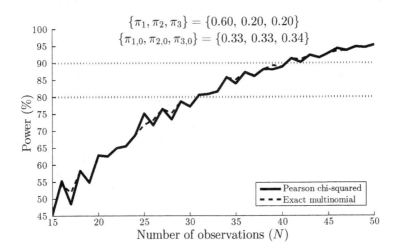

FIGURE 3.6
Power of the Pearson chi-squared and exact multinomial tests for the multinomial parameters

Pearson chi-squared test: if s parameters are estimated, the null distribution of
the Pearson chi-squared statistic has $c - 1 - s$ degrees of freedom. We illustrate
the details with an example based on the data in Table 3.2.

The genotype probabilities are said to be in Hardy-Weinberg equilibrium if $\pi_1 = \pi_{AA} = \pi_A^2$, $\pi_2 = \pi_{AG} = 2\pi_A\pi_G$, and $\pi_3 = \pi_{GG} = \pi_G^2$, where $\pi_G = 1 - \pi_A$. The three probabilities are thus modeled in terms of one parameter π_A. The maximum likelihood estimate for π_A turns out to be $\hat{\pi}_A = (2n_1 + n_2)/2N$. This corresponds to the observed proportion of allele A among the $2N$ allele observations. The estimate for the RA sample in Table 3.2 is $\hat{\pi}_A = (2 \cdot 276 + 380)/2 \cdot 774 = 0.602$. The estimated group probabilities under Hardy-Weinberg equilibrium are $\hat{\pi}_1 = 0.602^2 = 0.362$, $\hat{\pi}_2 = 2 \cdot 0.602 \cdot (1 - 0.602) = 0.479$, and $\hat{\pi}_3 = (1 - 0.602)^2 = 0.158$. The Pearson chi-squared statistic is

$$
\begin{aligned}
T_{\text{Pearson}} &= \frac{(276 - 774 \cdot 0.362)^2}{774 \cdot 0.362} + \frac{(380 - 774 \cdot 0.479)^2}{774 \cdot 0.479} \\
&\quad + \frac{(118 - 774 \cdot 0.158)^2}{774 \cdot 0.158} \\
&= 0.062 + 0.231 + 0.151 \\
&= 0.444,
\end{aligned}
$$

with degrees of freedom given by $3 - 1 - 1 = 1$. We calculate the P-value to be $\Pr(\chi_1^2 \geq 0.444) = 0.51$, so there is no evidence to suggest that the distribution of genotype probabilities in the RA patients deviates from Hardy-Weinberg equilibrium.

3.5   Ordered-Restricted Trend Tests

Suppose that an experiment is performed on N subjects, and that a multinomial variable with c possible outcomes is measured. If the categories of the multinomial variable have a natural ordering, one may expect that the experiment would have either no effect (represented by the null hypothesis) or a gradual effect (represented by the alternative hypothesis) on the outcome. The null hypothesis in this case is

$$
H_0 : \quad \pi_1 = \pi_2 = \ldots = \pi_c,
$$

which is equal to the null hypothesis in Equation 3.2 with $\pi_{i,0} = 1/c$ for all i. The alternative hypothesis is

$$
H_A : \quad \pi_1 \leq \pi_2 \leq \ldots \leq \pi_c, \tag{3.6}
$$

with at least one strict inequality.

Chacko (1966) derived a test for this hypothesis setup based on the Pearson chi-squared statistic (Equation 3.3). The *Chacko test* statistic is

$$
T_{\text{Chacko}}(n_1, n_2, \ldots, n_c) = \frac{c}{N} \sum_{i=1}^{m} t_i \left[\bar{n}_{(t_i)} - \frac{N}{c} \right]^2.
$$

Under the null hypothesis, T_{Chacko} is asymptotically chi-squared distributed with $m - 1$ degrees of freedom. The quantities m, t_i, and $\bar{n}_{(t_i)}$ are obtained through a process called the *ordering process* (Chacko, 1963):

1. Arrange the observed counts as $n_1, n_2, \ldots, n_c$.

2. If for some i, we have that $n_i > n_{i+1}$, substitute n_i and n_{i+1} with the average: $(n_i + n_{i+1})/2$. We now have a set of $c - 1$ quantities, one of which has weight 2.

3. Repeat step 2 until we obtain an ordered set of monotone, non-decreasing quantities.

4. Denote the number of quantities in the final set by m.

5. Denote the quantities by $\bar{n}_{(t_i)}$, where $i = 1, 2, \ldots, m$.

6. Denote the weight of quantity number i by t_i. t_i is the number of original counts that were pooled to obtain the quantity.

We illustrate the Chacko test with a simple, hypothetical example. Suppose that the observed counts of a five-level multinomial variable are as shown in Table 3.4. The ordering process substitutes $n_2 = 4$ and $n_3 = 3$ with the average 3.5, and $n_4 = 11$ and $n_5 = 9$ with the average 10. The final set of monotone, non-decreasing quantities is then $\{\bar{n}_{(t_1)}, \bar{n}_{(t_2)}, \bar{n}_{(t_3)}\} = \{1, 3.5, 10\}$. The weights are $\{t_1, t_2, t_3\} = \{1, 2, 2\}$, and $m = 3$. The Chacko test statistic is $T_{\text{Chacko}} = 12.27$, and the P-value is $\Pr(\chi_2^2 \geq 12.27) = 0.0022$.

TABLE 3.4

Observed counts of a hypothetical experiment

	Cat 1	Cat 2	Cat 3	Cat 4	Cat 4	Total
Sample	1 (3.6%)	4 (14%)	3 (11%)	11 (39%)	9 (32%)	28

Cat = Category

Equation 3.6 is just one of many possible ordered-restricted alternative hypotheses. We refer to Robertson (1978) for extensions of the Chacko test to alternative hypotheses such as $\pi_1 \leq \pi_2 = \pi_3 \leq \pi_4$.

3.6 Confidence Intervals for the Multinomial Parameters

Although there are c parameters $\pi_1, \pi_2, \ldots, \pi_c$ in the multinomial distribution, in some instances, one might be interested in confidence interval estimation for one specific parameter, say, π_i. Because n_i is binomially distributed with parameters N and π_i, we can use the confidence interval methods for the binomial parameter (Section 2.4) also for π_i. The Wilson score interval is recommended for all sample sizes.

Most often, however, we are interested in estimating simultaneous confidence intervals for the entire set of parameters $\pi_1, \pi_2, \ldots, \pi_c$, and we want the overall confidence level to be $1 - \alpha$, as opposed to having a confidence level of $1 - \alpha$ for each confidence interval. A simultaneous confidence interval is derived such that for a given confidence level $1 - \alpha$, all the c confidence intervals will include the population value with a probability of at least $1 - \alpha$. For an introduction to simultaneous confidence intervals for multinomial parameters, see Miller (1981, Section 2.1).

3.6.1 Simultaneous Confidence Intervals for the Multinomial Parameters

To derive simultaneous confidence intervals for the multinomial parameters, we use the maximum likelihood estimates of $\pi_1, \pi_2, \ldots, \pi_c$, which equal the sample proportions $\hat{\pi}_i = n_i/N$, for $i = 1, 2, \ldots, c$. Asymptotically, the vector $\hat{\pi} = \{\hat{\pi}_1, \hat{\pi}_2, \ldots, \hat{\pi}_c\}$ is normally distributed with mean $\pi = \{\pi_1, \pi_2, \ldots, \pi_c\}$ and covariance matrix Σ, which has elements

$$\sigma_{ii} = \pi_i(1 - \pi_i), \quad i = 1, 2, \ldots, c, \tag{3.7}$$

and

$$\sigma_{ij} = -\pi_i\pi_j, \quad i \neq j. \tag{3.8}$$

Then, $(\hat{\pi} - \pi)^{\mathrm{T}}\Sigma^{-1}(\hat{\pi} - \pi)$ is chi-squared distributed with $c - 1$ degrees of freedom. A confidence region in the $c - 1$ dimensional space is given as

$$\left\{ \pi : (\hat{\pi} - \pi)^{\mathrm{T}}\Sigma^{-1}(\hat{\pi} - \pi) \leq \chi^2_{c-1}(\alpha) \right\}, \tag{3.9}$$

where $\chi^2_{c-1}(\alpha)$ is the upper α percentile of the chi-squared distribution with $c - 1$ degrees of freedom. Simultaneous confidence intervals based on this confidence region are Wilson score-type intervals, and they generalize the Wilson score interval for the binomial parameter in Section 2.4.3.

Because $\hat{\pi}$ is a consistent estimate of π, $\hat{\Sigma}$ with $\hat{\pi}$ inserted into Equations 3.7 and 3.8 is consistent for Σ, and $(\hat{\pi} - \pi)^{\mathrm{T}}\hat{\Sigma}^{-1}(\hat{\pi} - \pi)$ is chi-squared distributed with $c - 1$ degrees of freedom. We thus obtain the confidence region given by

$$\left\{ \pi : (\hat{\pi} - \pi)^{\mathrm{T}}\hat{\Sigma}^{-1}(\hat{\pi} - \pi) \leq \chi^2_{c-1}(\alpha) \right\}. \tag{3.10}$$

Note that the quadratic form in Equation 3.10 is a modified Pearson chi-squared statistic with n_i instead of $N\pi_{i,0}$ in the denominator in Equation 3.3. Under the null hypothesis $H_0 : \pi_1 = \pi_2 = \ldots = \pi_c$, the modified Pearson statistic and the Pearson statistic are asymptotically equivalent.

Simultaneous confidence intervals based on the confidence region in Equation 3.10 are Wald-type intervals that generalize the Wald confidence interval for the binomial parameter in Section 2.4.1.

In the following, we describe four sets of simultaneous confidence intervals for the multinomial parameters. First, we describe two Wald-type intervals based on Equation 3.10, then we describe two Wilson score-type intervals based on Equation 3.9

Gold (1963) derived simultaneous confidence intervals that use the *Scheffé adjustment*. The *Gold Wald intervals* are given by

$$(L_1, U_1)_i = \hat{\pi}_i \mp \sqrt{\chi^2_{c-1}(\alpha) \frac{\hat{\pi}_i(1 - \hat{\pi}_i)}{N}}.$$

These confidence intervals adjust the confidence level for all linear combinations of the multinomial parameters. The Scheffé adjustment is thus overly strict for the purpose of estimating the c confidence intervals of interest here.

The *Bonferroni adjustment*, however, is less conservative and more appropriate in this setting (see also Hjort (1988) and Sverdrup (1990) and the note on page 394). The intervals derived by Goodman (1965) are similar to the Gold Wald intervals, except that they use the Bonferroni adjustment instead of the Scheffé adjustment. The limits of the *Goodman Wald intervals* are given by

$$(L_2, U_2)_i = \hat{\pi}_i \mp \sqrt{\chi^2_1(\alpha/c) \frac{\hat{\pi}_i(1 - \hat{\pi}_i)}{N}}.$$

The Wald interval for the binomial parameter is not recommended unless the sample size is very large (see Section 2.6 and Table 2.7). The Wilson score interval, however, is recommended for all sample sizes, and we now present two generalizations of it to simultaneous confidence intervals for the multinomial parameters.

Quesenberry and Hurst (1964) derived Wilson score-type intervals based on the Scheffé adjustment, which we shall refer to as the *Quesenberry-Hurst Wilson score intervals*:

$$(L_3, U_3)_i = \frac{\chi^2_{c-1}(\alpha) + 2N\hat{\pi}_i}{2\chi^2_{c-1}(\alpha) + 2N} \mp \frac{\sqrt{\left[\chi^2_{c-1}(\alpha)\right]^2 + 4N\chi^2_{c-1}(\alpha)\hat{\pi}_i(1 - \hat{\pi}_i)}}{2\chi^2_{c-1}(\alpha) + 2N}.$$

Wilson score-type intervals with Bonferroni adjustment were derived by Goodman (1965), and we shall refer to these as the *Goodman Wilson score intervals*:

$$(L_4, U_4)_i = \frac{\chi^2_1(\alpha/c) + 2N\hat{\pi}_i}{2\chi^2_1(\alpha/c) + 2N} \mp \frac{\sqrt{\left[\chi^2_1(\alpha/c)\right]^2 + 4N\chi^2_1(\alpha/c)\hat{\pi}_i(1 - \hat{\pi}_i)}}{2\chi^2_1(\alpha/c) + 2N}.$$

It is also possible, as shown by Goodman (1965), to derive simultaneous confidence intervals for the differences $\pi_i - \pi_j$. The *Goodman Wald intervals for differences* is given by

$$(L_5, U_5)_i = \hat{\pi}_i - \hat{\pi}_j \mp \sqrt{\chi^2_{c-1}(\alpha) \frac{\hat{\pi}_i + \hat{\pi}_j - (\hat{\pi}_i - \hat{\pi}_j)^2}{N}}.$$

These intervals are Scheffé-adjusted; however, Bonferroni versions are also possible:

$$(L_6, U_6)_i = \hat{\pi}_i - \hat{\pi}_j \mp \sqrt{\chi_1^2(\alpha/c)\frac{\hat{\pi}_i + \hat{\pi}_j - (\hat{\pi}_i - \hat{\pi}_j)^2}{N}}.$$

Finally, we mention that Sison and Glaz (1995) derived simultaneous confidence intervals for the multinomial parameters based on the doubly truncated Poisson distribution; these intervals work well when the number of multinomial parameters is high.

3.6.2   Example

Distribution of Genotype Counts of SNP rs6498169 in Patients with Rheumatoid Arthritis (Table 3.2)

Table 3.5 presents simultaneous confidence intervals for the data in Table 3.2. Also included in the table are the results for the (unadjusted) Wilson score intervals for reference. We observe only small differences between the different methods, which is expected due to the large sample size. We note, however, that there is a tendency for the Wald intervals to be wider than the Wilson score intervals. Furthermore, the Bonferroni intervals (Goodman Wald and Goodman Wilson score) tend to be shorter than the Scheffé intervals (Gold Wald and Quesenberry-Hurst Wilson score). This is because the Scheffé intervals adjust for all linear combinations of the multinomial parameters, whereas the Bonferroni intervals only adjust for the c different parameters.

TABLE 3.5
Simultaneous confidence intervals for the probabilities of the genotypes for SNP rs 6498169, based on data from 774 RA patients (Skinningsrud et al., 2010)

Intervals	AA $\hat{\pi}_1 = 0.357$	AG $\hat{\pi}_2 = 0.491$	GG $\hat{\pi}_3 = 0.153$
Wilson score*	0.324 to 0.391	0.456 to 0.526	0.129 to 0.180
Gold Wald (L_1, U_1)	0.314 to 0.399	0.447 to 0.535	0.121 to 0.184
Goodman Wald (L_2, U_2)	0.315 to 0.398	0.448 to 0.534	0.122 to 0.183
Quesenberry-Hurst Wilson score (L_3, U_3)	0.316 to 0.400	0.447 to 0.535	0.124 to 0.187
Goodman Wilson score (L_4, U_4)	0.317 to 0.399	0.448 to 0.534	0.124 to 0.186

*Unadjusted (not simultaneous) intervals

The null values for the multinomial parameters are $\pi_{1,0} = 0.402$, $\pi_{2,0} = 0.479$, and $\pi_{3,0} = 0.119$. We note that the simultaneous confidence intervals for AA and GG do not contain their respective null values. Thus, we declare that these cells have deviating cell counts, with correction for multiplicity.

We may also estimate the differences between the three pairs of probabilities with the Goodman Wald intervals for differences, here with Bonferroni adjustment (L_6, U_6):

$$\hat{\pi}_1 - \hat{\pi}_2 = -0.134 \ (\text{-}0.213 \text{ to } \text{-}0.056)$$
$$\hat{\pi}_1 - \hat{\pi}_3 = 0.204 \ (0.145 \text{ to } 0.263)$$
$$\hat{\pi}_2 - \hat{\pi}_3 = 0.339 \ (0.276 \text{ to } 0.401).$$

We find that there is no difference between the probabilities of genotypes AA and AG; however, the probability of genotype GG is different from those of AA and AG.

3.7 Recommendations

3.7.1 Summary

Table 3.6 provides a summary of the recommended tests and confidence intervals for the multinomial parameters, and gives the sample sizes—in broad qualitative categories—for which the recommended methods are appropriate. In the following, we discuss the recommendations in more detail and summarize the merits of the different methods.

TABLE 3.6

Recommended tests and confidence intervals (CIs) for the multinomial parameters

Analysis	Recommended methods*	Sample sizes
Tests for the fully specified null hypothesis in Equation 3.2	Pearson chi-squared	all
	Exact multinomial	small/medium
	Mid-P multinomial	small/medium
Tests for an incompletely specified null hypothesis	Pearson chi-squared[†]	all
Ordered-restricted trend tests	Chacko	all
Simultaneous CIs	Quesenberry-Hurst Wilson score	all
	Goodman Wilson score	all
	Goodman Wald for differences	all

*All recommended methods have closed-form expression
[†]With reduced degrees of freedom

3.7.2   Tests for the Multinomial Parameters

The Pearson chi-squared test is a remarkably good test even for small sample
sizes. Because of its computational simplicity—it has an uncomplicated closed-
form expression—the Pearson chi-squared test is a very attractive choice of
test for the multinomial parameters. It sometimes violates the nominal significance level; however, the infringements are neither frequent nor great. The
actual significance level of the exact multinomial test, on the other hand, has
the nominal level as an upper bound. An interesting quality of the exact multinomial test is that its level of conservatism, in contrast with most other exact
methods for categorical data, is very small. The mid-P version of the exact
multinomial test is also an excellent test, although the benefit of the mid-P
approach is less evident for the $1 \times c$ table than, for instance, the 2×2 table.
The mid-P multinomial test is at its most beneficial when the probabilities
are unequally distributed across the categories, such as in the example in Figure 3.4, where $\pi_1 = 0.6$, $\pi_2 = 0.3$, and $\pi_3 = 0.1$. The likelihood ratio test
has, in general, actual significance levels above the nominal level and is not
recommended.

3.7.3   Confidence Intervals for the Multinomial Parameters

As for the binomial parameter (see Section 2.6), we do not recommend Wald-
type intervals, unless the sample size is very large. The two Wilson score-type
intervals by Quesenberry and Hurst (1964) and Goodman (1965) are our primary recommendations for simultaneous confidence interval estimation. Because the Quesenberry-Hurt Wilson score intervals are adjusted for all linear
combinations of the multinomial parameters (Scheffé adjustment), the width
of these intervals will, in general, be wider than the Goodman Wilson score
intervals, which use the less conservative Bonferroni adjustment. This difference in widths will be small for small values of c, but it will increase with
increasing c. When c is large and the expected number of observations are
low, the intervals presented in Sison and Glaz (1995) can be recommended.

Simultaneous confidence intervals for the differences between the probabilities $\pi_i - \pi_j$ can be estimated with the Goodman Wald intervals for differences,
which come with either the Scheffé adjustment or the Bonferroni adjustment.

4

The 2×2 Table

4.1 Introduction

This chapter considers tests for association, effect measures, and confidence intervals for the 2×2 table of unpaired data. Several study designs lead to data that can be summarized in an (unpaired) 2×2 table, including randomized controlled trials, cohort studies, unmatched case-control studies, and cross-sectional studies. Binary matched-pairs data (Chapter 8) can also be presented in a 2×2 table; however, the rows and columns are defined differently in the two cases, and the statistical analyses are not the same. We shall refer to the 2×2 table of paired data in Chapter 8 as the paired 2×2 table.

Table 4.1 shows the general notation for the observed counts of a 2×2 table. It also shows the usual designation of rows and columns, which is appropriate for most but not all study designs. The rows comprise a dichotomized grouping of the data, for instance, into treatment or exposure groups. The columns represent a dichotomous outcome, where the two possible outcomes are referred to as success and failure. Success does not necessarily indicate a favorable outcome but rather the outcome of interest, for instance, the presence of a certain disease. The convention of letting the rows represent groups and the column outcomes is not universal. It is, in fact, arbitrary, and we might as well reverse rows and columns without other consequences than a slight change of notation.

TABLE 4.1
The observed counts of a 2×2 table

	Success	Failure	Total
Group 1	n_{11}	n_{12}	n_{1+}
Group 2	n_{21}	n_{22}	n_{2+}
Total	n_{+1}	n_{+2}	N

Additional notation:
$\mathbf{n} = \{n_{11}, n_{12}, n_{21}, n_{22}\}$: the observed table
$\mathbf{x} = \{x_{11}, x_{12}, x_{21}, x_{22}\}$: any possible table

Section 4.2 discusses five different sampling models and their associated study designs and presents data examples that illustrate different ways in which a 2×2 table might arise. Parameters and probability models are described in Section 4.3. Section 4.4 presents tests of association, and Sections 4.5–4.8 present confidence intervals for the difference between probabilities, the number needed to treat, the ratio of probabilities, and the odds ratio, respectively. Section 4.9 gives recommendations for the practical use of the methods in Sections 4.4–4.8. This chapter is partly based on Lydersen et al. (2009) and Fagerland et al. (2015).

4.2   Sampling Models, Study Designs, and Examples

A 2×2 table may be the result of different sampling models. The sampling model represents the way the data were collected. It is closely related to the study design, in that a particular study design makes use of a particular sampling model. Five such models are described below. The most common model is the row margins fixed model (Section 4.2.2). If not otherwise stated, we shall assume this model for the methods and examples in this chapter.

4.2.1   Both Margins Fixed Model

Under the both margins fixed model, both the row and column sums (n_{1+}, n_{2+}, n_{+1}, n_{+2}) are fixed beforehand. This sampling model is hardly ever used in practice. Nevertheless, it is important because it forms the basis for the derivation of the famous Fisher exact test (see Section 4.4.5). The classic example is "a lady tasting a cup of tea" (Fisher, 1937): Fisher's colleague Muriel Bristol claims she can taste whether milk or tea was added first to her cup. Four tea-first and four milk-first cups are presented to her in randomized order. She is told that there are four of each kind, and is asked to identify which is which. A possible result is given in Table 4.2. The feature of this model that makes it irrelevant in most practical cases is the pre-specification of the numbers of cups of each kind. That is akin to a clinical trial where the number of patients that will benefit from each treatment is decided upon before treatments start.

4.2.2   Row Margins Fixed Model

In clinical trials, one set of marginal sums (the row sums n_{1+} and n_{2+}) usually is fixed beforehand, typically the number of patients in each treatment group. One example is shown in Table 4.3, which summarizes the results of a double-blind trial of high dose versus standard dose of epinephrine (adrenaline) in children with cardiac arrest (Perondi et al., 2004). Seven (21%) of the 34 children in the standard dose group and one (2.9%) of the 34 children in the high dose group survived 24 hours.

TABLE 4.2
A possible result of Muriel Bristol's blind
taste test

| | Guessed | | |
Poured	Milk first	Tea first	Total
Milk first	3	1	4*
Tea first	1	3	4*
Total	4*	4*	8*

*Fixed by design

TABLE 4.3
Treatment of epinephrine in children with
cardiac arrest (Perondi et al., 2004)

| | Survival at 24h | | |
Treatment	Yes	No	Total
Standard dose	7 (21%)	27 (79%)	34*
High dose	1 (2.9%)	33 (97%)	34*
Total	8 (12%)	60 (88%)	68*

*Fixed by design

4.2.3 Column Margins Fixed Model

The column marginals are fixed when, for instance, sampling is done separately for cases and controls. A case-control study starts by identifying the cases, defined by the presence of a certain disease or condition. A control group is then selected, consisting of individuals with similar characteristics as the cases but without the disease or condition. The purpose of a case-control study is not to compare an outcome but rather an exposure between the cases and controls. The number of cases and the number of controls are fixed before collection of the exposure status. Table 4.4 shows the association between GADA positivity (high levels of glutamic acid decarboxylase) and immune dysregulation, polyendocrinopathy, enteropathy, X-linked (IPEX) syndrome in children (Lampasona et al., 2013). GADA is an autoantibody marker of type I diabetes and eczema. Nine (69%) of 13 IPEX patients (the cases) and four (29%) of 14 IPEX-like patients (the controls) were positive for GADA.

4.2.4 Total Sum Fixed Model

In cross-sectional studies of association, usually only the total sum N is fixed beforehand. Consider the data in Table 4.5, which is a cross-classification of CHRNA4 genotypes and presence of exfoliation syndrome (XFS) in the eyes of healthy adults (Ritland et al., 2007). Here, only the total number

TABLE 4.4
Exposure to GADA for children with IPEX
versus IPEX-like syndromes (Lampasona et al.,
2013)

	Cases/controls		
GADA	**IPEX**	**IPEX-like**	**Total**
Positive	9 (69%)	4 (29%)	13 (48%)
Negative	4 (31%)	10 (71%)	14 (52%)
Total	13*	14*	27*

*Fixed by design

of participants ($N = 88$) was fixed before genotype determination and eye examination. No participants with genotype CHRNA4-CC showed presence of XFS, whereas 15 (21%) of 72 participants with genotype CHRNA4-TC/TT showed presence of XFS. With an alternative but equally correct way to look at, all of the 15 participants with presence of XFS had genotype CHRNA4-TC/TT. For the remaining 73 participants without presence of XFS, 57 (78%) had genotype CHRNA4-TC/TT and 16 (22%) had genotype CHRNA4-CC.

TABLE 4.5
Genotype and presence of XFS in the eyes (Ritland et al.,
2007)

	XFS		
Genotype	**Yes**	**No**	**Total**
CHRNA4-CC	0	16 (18%)	16 (18%)
CHRNA4-TC/TT	15 (17%)	57 (65%)	72 (82%)
Total	15 (17%)	73 (83%)	88* (100%)

*Fixed by design

4.2.5 No Sums Fixed Model

In cohort and cross-sectional studies, it is also possible that no sums are fixed by design. This might happen if the participants of the study consist of those who are present at a specific place during a specific time interval, and the exact number of participants is not known. One example is a study including all patients visiting a specific clinic during one month. In zoology, the study sample might comprise the animals that appear in a particular forest region during 24 hours. In neither of these examples, the number of study individuals is known at the design phase of the study.

4.2.6 Inverse Sampling

Inverse sampling, also called negative binomial sampling, is a useful technique in studies of rare events. If conventional sampling is used, that is if the group sizes are determined ahead of the actual sampling, we run the risk of observing zero events (successes) in one or both groups. In studies using inverse sampling, the numbers of events in both groups (n_{11} and n_{21}) are fixed beforehand, and inclusion of participants in the study continues until these numbers are reached. Although inverse sampling is seldom used, an extensive literature exists on the statistical analysis of inverse sampling studies. A review is provided by Lui (2004), and some recent developments can be found in Zou (2010). In this book, the inverse sampling model will not be considered further.

4.3 Probabilities, Probability Models, and Link Functions

4.3.1 Probabilities and Probability Models

Let π_{ij} denote the joint probability that Variable 1 equals category i and Variable 2 equals category j (Table 4.6). Depending on the sampling model, some of these probabilities are unknown and some are fixed by design.

TABLE 4.6
The joint probabilities of a general 2 × 2 table, with unspecific row and column designations

	Variable 2		
Variable 1	**Category 1**	**Category 2**	**Total**
Category 1	π_{11}	π_{12}	π_{1+}
Category 2	π_{21}	π_{22}	π_{2+}
Total	π_{+1}	π_{+2}	1

In many situations, Variable 1 (Y_1) will be a grouping variable, and Variable 2 (Y_2) is an outcome variable. Then, we consider the association between the two variables via the conditional probabilities of Y_2 given Y_1. We denote the conditional probabilities by $\pi_{j|i}$, and we have that $\pi_{1|i} + \pi_{2|i} = 1$. We can thus simplify the notation and define $\pi_i = \pi_{1|i}$, such that we only have two probabilities to estimate, see Table 4.7.

TABLE 4.7

Conditional probabilities within the rows
of a 2 × 2 table

	Success	Failure	Total
Group 1	π_1	$1 - \pi_1$	1
Group 2	π_2	$1 - \pi_2$	1

Both Margins Fixed Model

Because all marginal sums are fixed, we only have one unknown parameter. We may take any one of the π_{ij} in Table 4.6 as this unknown parameter; however, none of them will, by themselves, measure association between rows and columns. Instead, we use $\theta = \pi_{11}\pi_{22}/\pi_{12}\pi_{21}$ as the unknown parameter for the both margins fixed model. θ is the odds ratio, to which we will return in Section 4.8. Any realization of a 2 × 2 table with both margins fixed can be determined from one cell count, say x_{11}. The sampling distribution of x_{11} is the Fisher non-central hypergeometric distribution:

$$f(x_{11} \mid \theta, n_{1+}, n_{2+}, n_{+1}, n_{+2}) = \frac{\binom{n_{1+}}{x_{11}} \binom{n_{2+}}{n_{+1} - x_{11}} \theta^{x_{11}}}{\sum_{t=n_0}^{n_1} \binom{n_{1+}}{t} \binom{n_{2+}}{n_{+1} - t} \theta^t}, \qquad (4.1)$$

where $n_0 = \max(0, n_{+1} - n_{2+})$ and $n_1 = \min(n_{1+}, n_{+1})$. If $\theta = 1$, Equation 4.1 simplifies to the hypergeometric distribution:

$$f(x_{11} \mid n_{1+}, n_{2+}, n_{+1}, n_{+2}) = \frac{\binom{n_{1+}}{x_{11}} \binom{n_{2+}}{n_{+1} - x_{11}}}{\binom{N}{n_{+1}}}. \qquad (4.2)$$

Row Margins Fixed Model

We assume that the row sums (n_{1+} and n_{2+}) are fixed and that our interest is on the two conditional probabilities in Table 4.7. The sampling distribution for each row is the binomial distribution. The joint probability distribution for an arbitrary table—now completely defined by x_{11} and x_{21}—is thus the product of two independent binomial distributions:

$$f(x_{11}, x_{21} \mid \pi_1, \pi_2, n_{1+}, n_{2+}) =$$
$$\binom{n_{1+}}{x_{11}} \pi_1^{x_{11}}(1 - \pi_1)^{n_{1+}-x_{11}} \binom{n_{2+}}{x_{21}} \pi_2^{x_{21}}(1 - \pi_2)^{n_{2+}-x_{21}}. \qquad (4.3)$$

If we condition on the column sums n_{+1} and n_{+2}, the resulting distribution is the Fisher non-central hypergeometric distribution in Equation 4.1 with $\theta = \dfrac{\pi_1(1 - \pi_2)}{\pi_2(1 - \pi_1)}$.

Column Margins Fixed Model

Suppose that the column sums (n_{+1} and n_{+2}) are fixed. This is the situation when cases and controls are sampled, see the example in Table 4.4. The sampling distribution in each column is the binomial distribution with parameters γ_1 and γ_2, given in Table 4.8.

TABLE 4.8
Conditional probabilities within
the columns of a 2 × 2 table

Exposure	Cases	Controls
Positive	γ_1	γ_2
Negative	$1 - \gamma_1$	$1 - \gamma_2$
Total	1	1

Sampling is done independently among the cases and the controls, and the joint probability distribution—now completely defined by x_{11} and x_{12}—is given by

$$f(x_{11}, x_{12} \mid \gamma_1, \gamma_2, n_{+1}, n_{+2}) =$$

$$\binom{n_{+1}}{x_{11}} \gamma_1^{x_{11}} (1 - \gamma_1)^{n_{+1} - x_{11}} \binom{n_{+2}}{x_{12}} \gamma_2^{x_{12}} (1 - \gamma_2)^{n_{+2} - x_{12}}.$$

It is possible to estimate the probabilities γ_1 and γ_2; however, our interest is not in the γ_j but rather in the π_i. To find π given γ, we use Bayes theorem and obtain

$$\begin{aligned}
\pi_1 &= \gamma_1 \pi_{+1} / \pi_{1+}, \\
1 - \pi_1 &= \gamma_2 \pi_{+2} / \pi_{1+}, \\
\pi_2 &= (1 - \gamma_1) \pi_{+1} / \pi_{2+}, \\
1 - \pi_2 &= (1 - \gamma_2) \pi_{+2} / \pi_{2+}.
\end{aligned} \tag{4.4}$$

As we see, given γ_1 and γ_2, it is not possible to derive π_1 and π_2 unless the marginal probabilities π_{1+} and π_{+1} are known.

Total Sum Fixed Model

Now, none of the marginal sums are fixed, and we have four unknown parameters: π_{11}, π_{12}, π_{21}, and π_{22}. (Only three parameters are independent because $\pi_{11} + \pi_{12} + \pi_{21} + \pi_{22} = 1$.) The sampling distribution of an arbitrary table $\mathbf{x}$ is the multinomial distribution:

$$f(\mathbf{x} \mid \pi_{11}, \pi_{12}, \pi_{21}, \pi_{22}, N) = \frac{N!}{x_{11}! x_{12}! x_{21}! x_{22}!} \pi_{11}^{x_{11}} \pi_{12}^{x_{12}} \pi_{21}^{x_{21}} \pi_{22}^{x_{22}}.$$

No Sums Fixed Model

Under the no sums fixed model, there is no upper limit to the number of successes and failures, and we need an open-ended distribution to describe the sampling distribution of the cell counts. Let μ_{ij} denote the parameter in a Poisson distribution for the cell count x_{ij}. The probability distribution for $\mathbf{x}$ equals the product of the Poisson probabilities for each cell count:

$$f(\mathbf{x} \mid \mu_{11}, \mu_{12}, \mu_{21}, \mu_{22}) = \prod_{i=1}^{2} \prod_{j=1}^{2} \frac{\mu_{ij}^{x_{ij}}}{x_{ij}!} \exp(-\mu_{ij}).$$

Table 4.9 shows a summary of the unknown parameters and probability models associated with each sampling model.

TABLE 4.9
Unknown parameters and probability models

Sampling model	Unknown parameters	Probability model
Both margins fixed	θ^*	Hypergeometric
Row margin fixed	π_1, π_2	Two independent binomials
Column margin fixed	γ_1, γ_2	Two independent binomials
Total sum fixed	$\pi_{11}, \pi_{12}, \pi_{21}, \pi_{22}$ †	Multinomial
No sums fixed	$\mu_{11}, \mu_{12}, \mu_{21}, \mu_{22}$	Multivariate Poisson

$^*\theta = \pi_{11}\pi_{22}/\pi_{12}\pi_{21}$
$^\dagger\pi_{11} + \pi_{12} + \pi_{21} + \pi_{22} = 1$

4.3.2   Link Functions for the 2 × 2 Table

In Section 1.5.2, we introduced the link functions for the binomial distribution, with row margins fixed, and gave an example with the linear link. With the notation in this chapter, the linear link has the form

$$\pi_j = \alpha_{\text{linear}} + \beta_{j,\text{linear}}.$$

The log link has the form

$$\log(\pi_j) = \alpha_{\text{log}} + \beta_{j,\text{log}},$$

and the logit link has the form

$$\log\left(\frac{\pi_j}{1 - \pi_j}\right) = \alpha_{\text{logit}} + \beta_{j,\text{logit}},$$

For all the link functions, we set $\beta_2 = 0$. For the three link functions, we get that

$$\beta_{1,\text{linear}} = \pi_1 - \pi_2,$$

$$\exp(\beta_{1,\text{log}}) = \pi_1/\pi_2,$$

$$\exp(\beta_{1,\text{logit}}) = \frac{\pi_1/(1-\pi_1)}{\pi_2/(1-\pi_2)}.$$

When the row margins are fixed, it is possible to estimate all parameters. For the total or no sum fixed models, we can estimate the parameters after conditioning on the row margins. In a case-control study, however, we are not able to estimate $\pi_1 - \pi_2$ or π_1/π_2, see Equation 4.4. The odds ratio, on the other hand, can be expressed in terms of either the π_j or the γ_i:

$$\theta = \frac{\pi_{11}/\pi_{21}}{\pi_{12}/\pi_{22}} = \frac{\pi_1/(1-\pi_1)}{\pi_2/(1-\pi_2)} = \frac{\gamma_1/(1-\gamma_1)}{\gamma_2/(1-\gamma_2)}.$$

Regardless of which margins are fixed, the odds ratio can be used, which makes the odds ratio a versatile effect measure. Note that $\gamma_1 = \gamma_2$ if and only if $\pi_1 = \pi_2$, in which case, $\theta = 1$ and $\pi_{ij} = \pi_{i+}\pi_{+j}$ for $i,j = 1,2$.

4.4 Tests for Association

4.4.1 The Null and Alternative Hypotheses

We consider the null hypothesis of no association between the variables defining the rows and columns. Table 4.10 shows the relevant two-sided null and alternative hypotheses for each of the relevant sampling models. In clinical trials and cohort studies, this means that the probability of success is independent of treatment (clinical trials) or exposure (cohort studies). In a case-control study, it means that the probability of being exposed is equal for cases and controls. In cross-sectional studies, independence means that the conditional probabilities given the row variable equals the column margins, and, likewise, that the conditional probabilities given the column variable equals the row margins. Unless otherwise stated, we assume sampling under the row margins fixed model. For the total sum fixed model, we condition on the row sums and proceed as for the row margins fixed model.

4.4.2 The Pearson Chi-Squared Test

Karl Pearson introduced the (now famous) chi-squared statistic in 1900 as the sum of the squared relative differences between observed and expected

TABLE 4.10

Null and two-sided alternative hypotheses by sampling model

Sampling Model	Null hypothesis	Alternative hypothesis
Both margins fixed	$\theta = 1$	$\theta \neq 1$
Row margins fixed	$\pi_1 = \pi_2$	$\pi_1 \neq \pi_2$
Column margins fixed	$\gamma_1 = \gamma_2$	$\gamma_1 \neq \gamma_2$
Total sum fixed	$\pi_{ij} = \pi_{i+}\pi_{+j}$ *	$\pi_{ij} \neq \pi_{i+}\pi_{+j}$ *
No sums fixed	$\mu_{ij} = \mu_{i+}\mu_{+j}$ *	$\mu_{ij} \neq \mu_{i+}\mu_{+j}$ *

*$i,j = 1,2$

frequencies (Pearson, 1900). The estimated expected cell counts of the 2×2 table under the null hypothesis are

$$m_{ij} = n_{i+}n_{+j}/N, \quad i,j = 1,2, \tag{4.5}$$

and the Pearson chi-squared test statistic is

$$T_{\text{Pearson}}(\mathbf{n}) = \sum_{i,j} \frac{(n_{ij} - m_{ij})^2}{m_{ij}} = \frac{N(n_{11}n_{22} - n_{12}n_{21})^2}{n_{1+}n_{2+}n_{+1}n_{+2}}. \tag{4.6}$$

The asymptotic P-value for the Pearson chi-squared test is obtained as

$$P\text{-value} = \Pr\left[\chi_1^2 \geq T_{\text{Pearson}}(\mathbf{n})\right],$$

where χ_1^2 is chi-squared distributed with one degree of freedom.

Yates (1934) proposed the following modification of Equation 4.6:

$$T_{\text{YatesCC}}(\mathbf{n}) = \sum_{i,j} \frac{(|n_{ij} - m_{ij}| - 1/2)^2}{m_{ij}}$$

$$= \frac{N(|n_{11}n_{22} - n_{12}n_{21}| - N/2)^2}{n_{1+}n_{2+}n_{+1}n_{+2}}. \tag{4.7}$$

The modification in Equation 4.7 is called a continuity correction, and the purpose of the correction is to obtain asymptotic P-values closer to exact P-values. The use of Yates's continuity correction—and other continuity corrections suggested in the literature—has been widely debated; see, for instance, the historical review in Hitchcock (2009). Many now consider such corrections to be no more than "interesting historic curiosities" (Hirji, 2006, p. 149).

4.4.3   The Likelihood Ratio Test

Section 1.7 introduced the general likelihood ratio statistic on the form $-2\log(l_0/l_1)$, where l_0 and l_1 denote the maximum likelihood values under the null and alternative hypotheses, respectively. In the case of the 2×2 table, the likelihood ratio test statistic is

$$T_{\mathrm{LR}}(\mathbf{n}) = 2 \sum_{i,j} n_{ij} \log \frac{n_{ij}}{m_{ij}}, \quad \text{where a term is 0 if } n_{ij} = 0, \quad (4.8)$$

and a P-value is obtained by comparing the observed value of the test statistic with the chi-squared distribution with one degree of freedom.

4.4.4 Tests Based on the Difference between Proportions

The (unstandardized) difference between the observed proportions of success was suggested as a test statistic for confidence interval estimation of the difference between probabilities and the ratio of probabilities by Santner and Snell (1980):

$$Z_{\mathrm{SS}}(\mathbf{n}) = \frac{n_{11}}{n_{1+}} - \frac{n_{21}}{n_{2+}}.$$

The Santner-Snell statistic can be highly discrete with small sample sizes. A better option is to divide the difference by its estimated standard error under the null hypothesis. This is a score statistic (see Section 1.7) and can be expressed as

$$Z_{\mathrm{score}}(\mathbf{n}) = \frac{\dfrac{n_{11}}{n_{1+}} - \dfrac{n_{21}}{n_{2+}}}{\sqrt{\dfrac{n_{+1}}{N}\dfrac{n_{+2}}{N}\left(\dfrac{1}{n_{1+}} + \dfrac{1}{n_{2+}}\right)}}.$$

A test based on the Z_{score} statistic is equivalent to the Pearson chi-squared test because $T_{\mathrm{Pearson}} = Z_{\mathrm{score}}^2$.

If we divide the difference by its standard error under the alternative hypothesis instead of the null hypothesis, we obtain the unpooled Z statistic:

$$Z_{\mathrm{unpooled}}(\mathbf{n}) = \frac{\dfrac{n_{11}}{n_{1+}} - \dfrac{n_{21}}{n_{2+}}}{\sqrt{\dfrac{n_{11}n_{12}}{n_{1+}^3} + \dfrac{n_{21}n_{22}}{n_{2+}^3}}}. \quad (4.9)$$

The three statistics Z_{SS}, Z_{score}, and Z_{unpooled} have asymptotic standard normal distributions. In the following, we refer to the test based on Equation 4.9 as the Z-*unpooled test*. P-values for the Z-unpooled test are obtained as

$$P\text{-value} = \Pr\left[Z \geq \left|Z_{\mathrm{unpooled}}(\mathbf{n})\right|\right],$$

where Z is a standard normal variable.

4.4.5   The Fisher Exact Test

Section 1.9 introduced exact tests and showed how they differ from asymptotic tests in their approach to computing P-values. Recall that if $T()$ is a test statistic, $\mathbf{n}$ denotes the observed table, and $\mathbf{x}$ denotes any possible table under the prevailing sampling model, the general expression for an exact P-value is

$$\text{exact } P\text{-value} = \Pr\big[T(\mathbf{x}) \geq T(\mathbf{n}) \,|\, H_0\big].$$

In words, the exact P-value is the sum of the probabilities of all possible tables that agree less than or equally with the null hypothesis than does the observed table. Section 4.3 showed how the probability distribution of possible tables depends on the sampling model. If we condition on all the marginal totals of the table (the both margins fixed model), the distribution of possible tables can be described by the hypergeometric probability distribution (Equation 4.2). The resulting test is called the *Fisher exact test*. Sometimes, since we condition on marginal sums that are not fixed by design (n_{+1} and n_{+2}), we shall refer to this as an *exact conditional test*. The exact conditional P-value is given by

$$\text{exact cond. } P\text{-value} = \Pr\big[T(\mathbf{x}) \geq T(\mathbf{n}) \,|\, H_0, \mathbf{n}_{++}\big], \qquad (4.10)$$

where $\mathbf{n}_{++} = \{n_{1+}, n_{2+}, n_{+1}, n_{+2}\}$ denotes the fixed row and column sums. Because we know that the hypergeometric distribution is the probability distribution of possible tables under the both margins fixed model, we can formulate the expression for the exact conditional P-value explicitly:

$$P\text{-value} = \sum_{x_{11}=n_0}^{n_1} I\big[T(\mathbf{x}) \geq T(\mathbf{n})\big] \cdot f(x_{11} \,|\, \mathbf{n}_{++}). \qquad (4.11)$$

where $n_0 = \max(0, n_{+1} - n_{2+})$ and $n_1 = \min(n_{1+}, n_{+1})$, $I()$ is the indicator function, and $f(x_{11} \,|\, \mathbf{n}_{++})$ is the probability of the table with x_{11} successes in Group 1, given by Equation 4.2. Exact conditional tests are guaranteed to have significance levels below the nominal level, see, for instance, Lydersen et al. (2009, p. 1165).

So far, we have not specified which test statistic to use in Equation 4.11. Many different statistics have been suggested, and each gives rise to a different exact test (although they often give identical P-values). The test commonly referred to as the Fisher exact test is due to Irwin (1935)—it is sometimes called the *Fisher-Irwin test*—and uses the hypergeometric probability distribution in Equation 4.2 as a test statistic. Because this statistic gives larger values for tables that agree more with the null hypothesis than do tables that agree less with the null hypothesis, we reverse the inequality in Equation 4.11 and obtain the following expression for the Fisher-Irwin version of the Fisher exact test:

$$P\text{-value} = \sum_{x_{11}=n_0}^{n_1} I\big[f(x_{11} \,|\, \mathbf{n}_{++}) \leq f(n_{11} \,|\, \mathbf{n}_{++})\big] \cdot f(x_{11} \,|\, \mathbf{n}_{++}). \qquad (4.12)$$

We consider two other versions of the Fisher exact test: the one obtained with the Pearson chi-squared statistic,

$$P\text{-value} = \sum_{x_{11}=n_0}^{n_1} I\big[T_{\text{Pearson}}(\mathbf{x}) \geq T_{\text{Pearson}}(\mathbf{n})\big] \cdot f(x_{11} \mid \mathbf{n}_{++}),$$

where T_{Pearson} denotes the Pearson chi-squared statistic in Equation 4.6, and the one obtained with the likelihood ratio statistic,

$$P\text{-value} = \sum_{x_{11}=n_0}^{n_1} I\big[T_{\text{LR}}(\mathbf{x}) \geq T_{\text{LR}}(\mathbf{n})\big] \cdot f(x_{11} \mid \mathbf{n}_{++}),$$

where T_{LR} denotes the likelihood ratio statistic in Equation 4.8.

4.4.6 The Fisher Mid-P Test

By conditioning on all marginal sums, an exact conditional test, such as the Fisher exact test, greatly reduces the parameter space. When x_{11} can attain only a few values, the hypergeometric distribution is highly discrete, as in the forthcoming example in Table 4.11. The P-value in turn is restricted to a few possible values. This discreteness leads to unnecessary conservative testing, with actual significance levels notably less than the nominal level. One approach to reduce conservatism is to calculate a mid-P value (see Section 1.10) instead of a P-value:

$$\begin{aligned} \text{mid-}P\text{ value} \;=\;& \Pr\big[T(\mathbf{x}) > T(\mathbf{n}) \mid H_0, \mathbf{n}_{++}\big] \\ +\;& 0.5 \cdot \Pr\big[T(\mathbf{x}) = T(\mathbf{n}) \mid H_0, \mathbf{n}_{++}\big]. \end{aligned}$$

Contrast this definition with that of the exact conditional P-value in Equation 4.10, and observe that they are equal except that the mid-P value includes only half the probability of the observed table.

A mid-P value may, in principle, be based on any exact P-value. If we use the Fisher exact test (Fisher-Irwin version) as our starting point, we define the Fisher mid-P test as

$$\text{mid-}P\text{ value} = \sum_{x_{11}=n_0}^{n_1} I\big[f(x_{11} \mid \mathbf{n}_{++}) < f(n_{11} \mid \mathbf{n}_{++})\big] \cdot f(x_{11} \mid \mathbf{n}_{++})$$

$$+ \; 0.5 \cdot \sum_{x_{11}=n_0}^{n_1} I\big[f(x_{11} \mid \mathbf{n}_{++}) = f(n_{11} \mid \mathbf{n}_{++})\big] \cdot f(x_{11} \mid \mathbf{n}_{++}).$$

This mid-P value is quite easy to calculate; it only involves quantities that are included in the calculations of the exact P-value upon which it is based.

4.4.7   Exact Unconditional Tests

Although the mid-P test in Section 4.4.6 is an excellent approach to reduce conservatism, it is not an exact test, and it frequently violates the nominal significance level. This section presents exact unconditional tests that do not suffer from unnecessary conservatism or violations of the nominal significance level. By "unconditional test", we usually mean a test that does not assume any fixed marginal sums, save those fixed by design. Unless the sampling model happens to be the both margins fixed model—we have yet to experience this in practice—a complicating feature of exact unconditional tests is the presence of one or more unknown parameters (under the null hypothesis), known as *nuisance parameters*. Nuisance parameters are parameters that we have no interest in estimating, but they need to be handled. One solution, originally proposed by Barnard (1945a,b, 1947), is to define the exact unconditional P-value as the maximum of all possible P-values across the nuisance parameter space.

If we assume the row margins fixed model, we have one nuisance parameter: the common success probability $\pi = \pi_1 = \pi_2$ under the null hypothesis. The exact unconditional P-value is hence

$$\max_{0 \leq \pi \leq 1} \left\{ \Pr\big[T(\mathbf{x}) \geq T(\mathbf{n}); \pi \mid H_0\big] \right\}. \tag{4.13}$$

Under the total sum fixed model, two nuisance parameters arise: π_{1+} and π_{+1}. The maximization must now be performed over a two-dimensional area $[0, 1] \times [0, 1]$. We shall restrict our treatment of exact unconditional tests to those defined for the row margins fixed model.

The probability of observing an arbitrary table $\mathbf{x}$ under H_0 is then given by Equation 4.3 with $\pi = \pi_1 = \pi_2$, which we rewrite slightly to

$$f(\mathbf{x} \mid \pi, \mathbf{n}_+) = \binom{n_{1+}}{x_{11}} \binom{n_{2+}}{x_{21}} \pi^{x_{11}+x_{21}} (1 - \pi)^{N - x_{11} - x_{21}},$$

where $\mathbf{n}_+ = \{n_{1+}, n_{2+}\}$ denotes the fixed row sums. An explicit expression for the exact unconditional P-value is

$$P\text{-value} = \max_{0 \leq \pi \leq 1} \left\{ \sum_{\Omega(\mathbf{x}|\mathbf{n}_+)} I\big[T(\mathbf{x}) \geq T(\mathbf{n})\big] \cdot f(\mathbf{x} \mid \pi, \mathbf{n}_+) \right\}, \tag{4.14}$$

where $\Omega(\mathbf{x}|\mathbf{n}_+)$ denotes the set of all tables with row sums equal to $\mathbf{n}_+$.

Different test statistics can be used in Equation 4.14. The test suggested by Suissa and Shuster (1985), which is available with the StatXact software (Cytel Inc., Cambridge, MA), uses the Pearson chi-squared statistic (Equation 4.6). We shall refer to this test as the *Suissa-Shuster exact unconditional test*; however, it is sometimes referred to as the unconditional Z pooled test.

The likelihood ratio statistic (Equation 4.8) and the unpooled Z statistic (Equation 4.9) may also be used to derive exact unconditional tests. In the latter case, we use $T_{\text{unpooled}} = Z^2_{\text{unpooled}}$ so that the test statistic is consistent with the notation in Equations 4.13 and 4.14.

Another exact unconditional test is obtained by using the *P*-value from the Fisher exact test as test statistic. Because small *P*-values provide evidence against the null hypothesis, we reverse the inequality in Equation 4.14 when using this test statistic. The resulting test is called the *Fisher-Boschloo test* owing to an equivalent test suggested by Boschloo (1970).

The Berger and Boos Procedure

It may be argued that maximizing π over the entire nuisance parameter space, $0 \leq \pi \leq 1$, is unreasonable because the interval contains values that are highly unlikely in light of the observed data. This argument was the crux of Fisher's criticism (Fisher, 1945) of Barnard's proposition of the exact unconditional test. The Berger and Boos procedure is a remedy (Berger and Boos, 1994). It restricts the nuisance parameter space to C_γ: a $100(1-\gamma)\%$ exact confidence interval for π, where γ is taken to be very small. To make sure that the actual significance level is bounded by the nominal level, the value of γ is added to the *P*-value:

$$P\text{-value} = \max_{\pi \in C_\gamma} \left\{ \sum_{\Omega(\mathbf{x}|\mathbf{n}_+)} I\big[T(\mathbf{x}) \geq T(\mathbf{n})\big] \cdot f(\mathbf{x} \mid \pi, \mathbf{n}_+) \right\} + \gamma.$$

For the Suissa-Shuster exact unconditional test, Lydersen et al. (2012b) found $\gamma = 0.0001$ to be approximately optimal under rather general conditions. In addition to avoiding computation over unrealistic values of the nuisance parameter, Berger (1996) states two other advantages of using the Berger and Boos procedure: (i) maximization over C_γ is computationally easier than over $0 \leq \pi \leq 1$; and (ii) the resulting test can have higher power than the ordinary exact unconditional test. The user manual of the software package StatXact also notes that using the Berger and Boos procedure provides greater computational stability (StatXact 11, 2015, p. 528). A common method to form the confidence interval C_γ is to use the Clopper-Pearson exact interval (see Section 2.4.7).

4.4.8 Examples

A Lady Tasting a Cup of Tea (Table 4.2)

Consider the hypothetical data from the experiment of the tea tasting ability of Muriel Bristol in Table 4.2. Since all marginal sums are fixed by design, the Fisher exact test and the Fisher mid-*P* test are the natural choices as tests for association. There are only five possible tables, defined by $x_{11} = 0, 1, 2, 3, 4$. Table 4.11 shows the probability of each possible table and how the two-sided *P*-value for the Fisher exact test is calculated: $P = 0.0143 + 0.2286 + 0.2286 + 0.0143 = 0.4857$. A one-sided *P*-value for the alternative hypothesis

$H_A : \theta > 1$ is obtained by only considering tables that agree equally or more with H_A than the observed table. That leaves two tables, $x_{11} = 3$ and $x_{11} = 4$, and the one-sided P-value is $P = 0.2286 + 0.0143 = 0.2429$.

TABLE 4.11

Quantities involved in the calculation of the Fisher exact test (Fisher-Irwin) on the tea tasting data in Table 4.2

x_{11}	$f(x_{11})$	$f(x_{11}) \leq f(n_{11})$	Cumulative probability
0	0.0143	Yes	0.0143
1	0.2286	Yes	0.2429
2	0.5143	No	0.2429
3*	0.2286	Yes	0.4714
4	0.0143	Yes	0.4857

*The observed table

To calculate the (two-sided) Fisher mid-P test, we include only half the probability of the tables for which the test statistic equals the observed value: mid-$P = 0.0143 + 0.5 \cdot 0.2286 + 0.5 \cdot 0.2286 + 0.0143 = 0.2571$.

An asymptotic test, such as the Pearson chi-squared test, can also be used to test association under both margins fixed sampling. Since asymptotic tests rely on a large sample approximation to the distribution of the test statistic, the accuracy of this approximation is poor when the sample size is small or unevenly distributed across the table cells. With a total sample size of $N = 8$ in Table 4.2, it should be quite obvious that asymptotic tests are not appropriate in this case. For slightly larger sample sizes, we may use the expected cell counts in Equation 4.5 to decide whether or not the approximation holds. The much-used *Cochran's criterion* (Cochran, 1954b) states that the Pearson chi-squared statistic should not be used if any of the expected counts are less than five ($m_{ij} < 5$; $i, j = 1, 2$). In Table 4.2, all expected counts equal two.

A Randomized Clinical Trial of High Versus Standard Dose of Epinephrine (Table 4.3)

The expected cell counts for the data in Table 4.3 are 4, 4, 30, and 30. Since two of the counts are less than five, the accuracy of asymptotic tests is in doubt. Instead, we turn to exact and mid-P tests. The data in Table 4.3 stems from a randomized clinical trial (Perondi et al., 2004), where the numbers of patients in both treatment groups were fixed at 34. The use of the Fisher exact test for this example imposes an unnecessary level of discreteness of the sample space. If the number of successes is fixed at $n_{+1} = 8$, nine tables are possible, $x_{11} = 0, 1, \ldots, 8$, and only five of those have different values of the test statistic. An exact unconditional test, on the other hand, does not assume that n_{+1} and n_{+2} are fixed. Hence, the sample space consists of $(n_{1+} + 1)(n_{2+} + 1) = 35 \times 35 = 1225$ possible tables.

Table 4.12 shows the results of applying exact conditional (Fisher exact) tests, the Fisher mid-P test, and exact unconditional tests to the data in

Table 4.3. All the exact unconditional tests have lower P-values than the exact conditional tests. The lowest P-value ($P = 0.0281$) is obtained from the exact unconditional tests with the Pearson chi-squared and the unpooled Z statistics. Also note that the Fisher mid-P value is quite close to the lowest P-values of the exact unconditional tests. Figure 4.1 illustrates how the P-values of the four exact unconditional tests are determined by maximization over the nuisance parameter space.

TABLE 4.12

Results of exact and mid-P tests on the epinephrine trial data in Table 4.3

Test	Statistic	P-value	π^*_{max}
Fisher exact (Fisher-Irwin)	Hypergm[†] distribution	0.0544	n/a
Fisher exact	Pearson chi-squared	0.0544	n/a
Fisher exact	Likelihood ratio	0.0544	n/a
Fisher mid-P	Hypergm[†] distribution	0.0297	n/a
Suissa-Shuster exact uncond.	Pearson chi-squared	0.0281	0.132
Exact unconditional	Likelihood ratio	0.0402	0.076
Exact unconditional	Unpooled Z	0.0281	0.132
Fisher-Boschloo exact uncond.	Fisher exact test	0.0385	0.500

[*]Value of the common success probability at which the maximum P-value occurred
[†]Hypergm = Hypergeometric

The Berger and Boos procedure may be applied to all the exact unconditional tests in Table 4.12. Following the recommendation in Lydersen et al. (2012b), we use $\gamma = 0.0001$. Under the null hypothesis, an estimate of the common success probability is $\hat{\pi} = n_{+1}/N = 8/68 = 0.118$. A 99.99% Clopper-Pearson exact confidence interval for π is $C_\gamma = (0.0193, 0.330)$. The resulting exact unconditional P-values that use C_γ as the parameter space for π are shown in Table 4.13. Compared with the results in Table 4.12, the P-value of the Fisher-Boschloo test changes noticeably with use of the Berger and Boos procedure, whereas the P-values of the other tests change marginally. Figure 4.1 shows that the value of π that maximizes the P-value is included in C_γ for all the tests except for the Fisher-Boschloo test (lower right panel).

TABLE 4.13

Results of exact unconditional tests with Berger and Boos procedure ($\gamma = 0.0001$) on the epinephrine trial data in Table 4.3

Test	Statistic	P-value	π^*_{max}
Suissa-Shuster exact uncond.	Pearson chi-squared	0.0282	0.132
Exact unconditional	Likelihood ratio	0.0403	0.076
Exact unconditional	Unpooled Z	0.0282	0.132
Fisher-Boschloo exact uncond.	Fisher exact test	0.0313	0.330

[*]Value of the common success probability at which the maximum P-value occurred

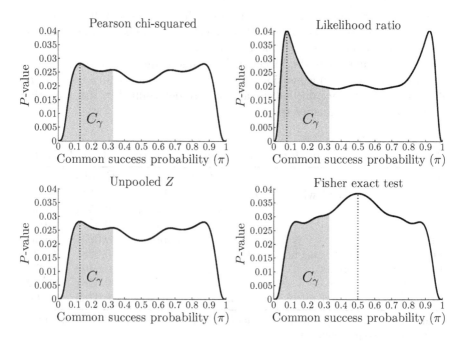

FIGURE 4.1

P-value as a function of the common success probability (π) for four exact unconditional tests on the epinephrine trial data in Table 4.3. The dotted vertical lines show the maximum P-values and their corresponding values of π. The shaded area indicates the C_γ interval.

A Case-Control Study of the Effect of GADA Exposure on IPEX Syndrome (Table 4.4)

For the case-control data in Table 4.4, the smallest expected cell count is 6.26, and by Cochran's criterion (see page 100), we may safely use asymptotic tests. Using Equation 4.6, we find that the observed value of the Pearson chi-squared statistic is

$$T_{\text{Pearson}}(\mathbf{n}) = \frac{27 \cdot (9 \cdot 10 - 4 \cdot 4)^2}{13 \cdot 14 \cdot 13 \cdot 14} = 4.464,$$

and the P-value for the Pearson chi-squared test is $\Pr(\chi_1^2 \geq 4.464) = 0.0346$. The results for this and three other asymptotic tests are summarized in Table 4.14. There is considerable variation in P-values also for this example, most notably that the Pearson chi-squared test with continuity correction has a much higher P-value than the other tests. The purpose of the correction is to mimic the results of the Fisher exact test, which for these data, has $P = 0.0570$. The Fisher mid-P test and the Suissa-Shuster exact unconditional tests give $P = 0.0391$ and $P = 0.0523$, respectively.

TABLE 4.14
Results of asymptotic tests on the case-control
data in Table 4.4

Test	$T(n)$	P-value
Pearson chi-squared	4.464	0.0346
Pearson chi-squared with CC*	2.984	0.0841
Likelihood ratio	4.593	0.0321
Z-unpooled	5.339[†]	0.0209

*CC = continuity correction
[†]$T_{\text{unpooled}}(\mathbf{n}) = Z^2_{\text{unpooled}}(\mathbf{n})$

A Cross-Sectional Study of the Association between CHRNA4 Genotypes and XFS (Table 4.5)

Only the total sum $N = 88$ is fixed by design. We calculate the expected cell counts and observe that $m_{11} = 2.73$, whereas the other counts are greater than ten. Hence, exact and mid-P tests are in order, and we show the results for these tests in Table 4.15. Since only the total sum is fixed by design, it is possible to construct an unconditional exact test by maximizing over two nuisance parameters, see Lydersen et al. (2009). Here, we condition on the row sums, and proceed as if the data were sampled under the row margins fixed model. The exact unconditional P-values are shown in Table 4.15 with ($\gamma = 0.0001$) and without ($\gamma = 0$) the Berger and Boos procedure.

TABLE 4.15
Results of exact and mid-P tests on the cross-sectional data in Table 4.5

		P-value	
Test	Statistic	$\gamma = 0$	$\gamma = 10^{-4}$
Fisher exact (Fisher-Irwin)	Hypergm dist*	0.0629	n/a
Fisher exact	Pearson chi-squared	0.0629	n/a
Fisher exact	Likelihood ratio	0.0629	n/a
Fisher mid-P	Hypergm dist*	0.0447	n/a
Suissa-Shuster exact uncond.	Pearson chi-squared	0.0815	0.0499
Exact unconditional	Likelihood ratio	0.0181	0.0182
Exact unconditional	Unpooled Z	0.0135	0.0136
Fisher-Boschloo exact uncond.	Fisher exact test	0.0523	0.0478

*Hypergm dist = Hypergeometric distribution

There is considerably less agreement between the exact unconditional tests in Table 4.15 than was the case for the epinephrine data in Table 4.12, although use of the Berger and Boos procedure evens out some of the inconsistencies. We may speculate that the zero in the upper left cell of Table 4.5 is the reason for the discrepancies between the tests; zero cell counts have the tendency to add a layer of complexity to the analysis of contingency tables, although exact tests and confidence intervals usually are quite adept at handling such cases. Two plots that look behind the scenes at the computations of the tests may shed some light on why the P-values of the tests differ. Figure 4.2 illustrates that the sets (black regions) of possible tables that agree equally or less with the null hypothesis than do the observed table, as defined by the four test statistics used by the four exact unconditional tests, are quite different. This means that the tests add up different sets of probabilities to calculate the P-values. The distributions of P-values across the common success probability (π) are shown in Figure 4.3, along with shaded areas that represent the 99.99% confidence interval (C_γ) for π. For the tests that use the Pearson chi-squared statistic and the Fisher exact test as test statistics, the maximum P-value occurs outside the C_γ interval, thus making the exact unconditional P-value smaller with than without the Berger and Boos procedure.

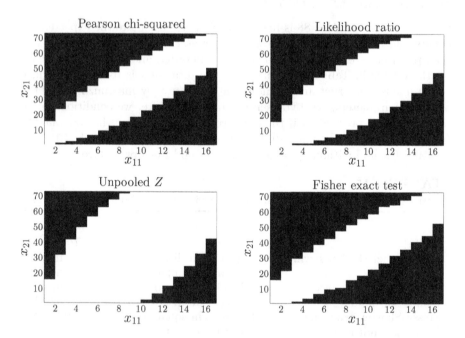

FIGURE 4.2
Black regions correspond to tables that agree less than or equally with the null hypothesis than do the observed table, according to four test statistics on the cross-sectional data in Table 4.5

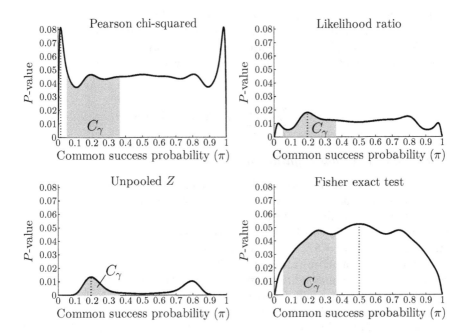

FIGURE 4.3

P-value as a function of the common success probability (π) for four exact unconditional tests on the cross-sectional data in Table 4.5. The dotted vertical lines show the maximum *P*-values and their corresponding values of π. The shaded area indicates the C_γ interval.

4.4.9 Evaluation of Tests

Evaluation Criteria

We evaluate tests for association by calculating their actual significance levels and power. We shall assume sampling under the row margins fixed model, such that the probability of an arbitrary 2 × 2 table can be described by the product of two independent binomial probabilities (Equation 4.3). Under the null hypothesis (H_0: $\pi_1 = \pi_2$), the actual significance level depends on the common success probability $\pi = \pi_1 = \pi_2$, the marginal row sums n_{1+} and n_{2+}, and the nominal significance level α. A set of particular values of these quantities is called a *parameter space point*. For a given parameter space point, we can calculate the actual significance level exactly with complete enumeration; that is, we sum the probabilities of all possible tables that lead to rejection of the null hypothesis at level α:

$$\text{ASL}(\pi, n_{1+}, n_{2+}, \alpha) = \sum_{x_{11}=0}^{n_{1+}} \sum_{x_{21}=0}^{n_{2+}} I\big[P(\mathbf{x}) \leq \alpha\big] \cdot f(\mathbf{x} \,|\, \pi, n_{1+}, n_{2+}),$$

where $I()$ is the indicator function, $P(\mathbf{x})$ is the P-value for a test on $\mathbf{x} = \{x_{11}, n_{1+} - x_{11}, x_{21}, n_{2+} - x_{21}\}$, and $f()$ is the probability of table $\mathbf{x}$ under the null hypothesis, given in Equation 4.3 with $\pi = \pi_1 = \pi_2$.

Power is calculated in a similar manner; however, we now assume that $\pi_1 \neq \pi_2$:

$$\text{Power}(\pi_1, \pi_2, n_{1+}, n_{2+}, \alpha) = \sum_{x_{11}=0}^{n_{1+}} \sum_{x_{21}=0}^{n_{2+}} I\big[P(\mathbf{x}) \leq \alpha\big] \cdot f(\mathbf{x} \,|\, \pi_1, \pi_2, n_{1+}, n_{2+}).$$

Evaluation of Actual Significance Level

By fixing the row margins (n_{1+} and n_{2+}), we can plot the actual significance level as a function of π, the common success probability. Figure 4.4 shows the results of five commonly used tests for the fixed row margins $n_{1+} = n_{2+} = 25$. Here, and elsewhere in this book—unless otherwise stated—we use a nominal significance level of $\alpha = 5\%$. The Pearson chi-squared test violates the nominal significance level for most values of π, with a maximum level of 6.49%. The Suissa-Shuster exact unconditional test performs best of the five tests in this example, and the Fisher mid-P test performs similarly but with slightly lower actual significance levels. The actual significance levels of the Fisher exact test and the Pearson chi-squared test with continuity correction are low, most often about half the nominal level.

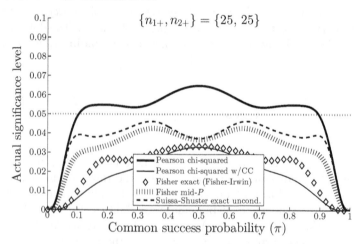

FIGURE 4.4
Actual significance levels of five commonly used tests

The Pearson chi-squared test is somewhat unfairly included in Figure 4.4, because the actual significance levels are calculated over many scenarios with small cell counts, for which we do not expect an asymptotic test to perform well. The test is included to illustrate the limited utility of asymptotic tests for the analysis of small-sample data. The main problem with the asymptotic tests is that the actual significance level is not bounded by—and can sometimes be considerably higher than—the nominal level. A notable exception is the Pearson chi-squared test with continuity correction, which performs similarly to an exact conditional test. The practical consequence of an inflated actual significance level is an (unwanted) increased probability of false positive findings. We shall consider the asymptotic tests for larger sample sizes later in this section.

The Fisher exact test is very conservative in Figure 4.4, with a maximum actual significance level of 3.33%. The situation is similar with the other two versions of the Fisher exact test—those using the Pearson chi-squared and likelihood ratio statistics. Figure 4.5 shows the actual significance levels of the three versions of the Fisher exact tests and the Fisher mid-P test for a combination of small and unequal row margins. There is little that separates the three exact conditional tests; however, the easy-to-implement mid-P adjustment makes a considerable difference. Figure 4.5 also illustrates that the Fisher mid-P test may violate the nominal significance level. This behavior is typical for mid-P tests (and mid-P intervals); however, the degree of infringements of the nominal level is usually low (Mehta and Walsh, 1992; Lydersen and Laake, 2003).

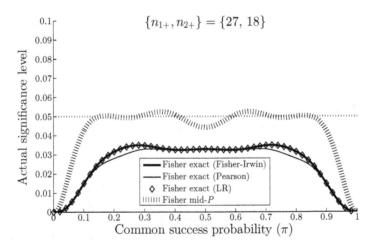

FIGURE 4.5
Actual significance levels of three versions of the Fisher exact test and the Fisher mid-P test

The Fisher mid-P test is a better choice than the Fisher exact test for small-sample data if small violations of the nominal significance level are acceptable. Another option, which guarantees that the nominal level is maintained, is to use an exact unconditional test. Figure 4.6 shows a comparison of four exact unconditional tests (without the Berger and Boos procedure) for the row margins $n_{1+} = 24$ and $n_{2+} = 14$. The tests perform similarly, although the Suissa-Shuster and Fisher-Boschloo tests are slightly better than the other two tests. The difference between the exact unconditional tests with and without the Berger and Boos procedure is not easy to capture visually; the curves are most often identical or close to one another. An example of a clear difference is shown in Figure 4.7, where the tests with the Berger and Boos procedure has actual significance levels closer to the nominal level than do the tests without the Berger and Boos procedure.

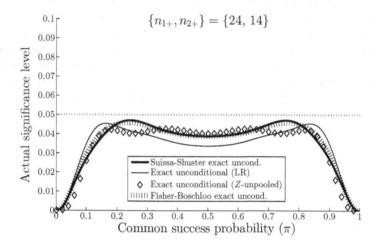

FIGURE 4.6
Actual significance levels of four exact unconditional tests (without the Berger and Boos procedure)

As discussed in Section 4.4.8, the asymptotic tests need a larger sample size than we have considered so far to perform well. Figure 4.8 shows the actual significance levels of four asymptotic tests for a combination of medium-sized row margins, $n_{1+} = n_{2+} = 75$. The Pearson chi-squared test performs much better than in Figure 4.4, with only a few moderately sized violations of the nominal level (maximum level: 5.65%). The Z-unpooled test is slightly worse than the Pearson chi-squared test, whereas the likelihood ratio test has severely high actual significance levels for values of π close to zero and one. The Pearson chi-squared test with continuity correction still has very low actual significance levels. To get an idea of the extent of this behavior, we calculated the actual significance levels of the asymptotic tests with both row margins equal to 500 (Figure 4.9). The actual significance levels of the likelihood ratio test are now almost equal to those of the Pearson chi-squared test for all parameters value except for $\pi = 0.01$ and $\pi = 0.99$. The Pearson chi-squared

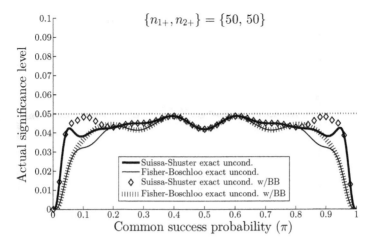

FIGURE 4.7
Actual significance levels of the Suissa-Shuster and Fisher-Boschloo exact unconditional tests with and without the Berger and Boos (BB) procedure

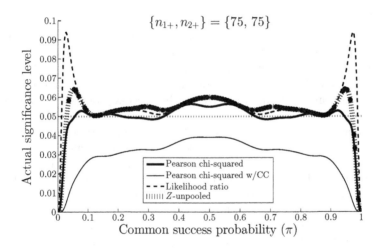

FIGURE 4.8
Actual significance levels of four asymptotic tests for a pair of medium-sized row margins

test with continuity correction has actual significance levels around 4% for most of the parameter space.

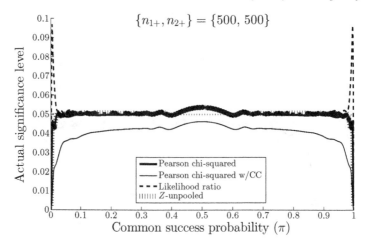

FIGURE 4.9

Actual significance levels of four asymptotic tests for a pair of large row margins

Evaluation of Power

To illustrate power, we fix the success probabilities and plot power as a function of the row margins. We start by looking at the five commonly used tests considered in Figure 4.4: the Pearson chi-squared test with and without continuity correction, the Fisher exact test, the Fisher mid-P test, and the Suissa-Shuster exact unconditional test. We fix the success probabilities in Group 1 and Group 2 at $\pi_1 = 0.2$ and $\pi_2 = 0.5$, respectively, and we keep the row margins equal ($n_{1+} = n_{2+}$). Figure 4.10 shows the power of the tests for row margins ranging from 30 to 70. The between-tests differences in power correspond to the between-tests differences in actual significance level observed in Figure 4.4, although to a lesser extent: the Pearson chi-squared has the greatest power, followed by the Fisher mid-P and Suissa-Shuster exact unconditional tests. The power of the Fisher exact test and the Pearson chi-squared test with continuity correction is notably lower than the power of the other tests. We do not illustrate the power of the two versions of the Fisher exact test that uses the Pearson chi-squared and likelihood ratio statistics, because the results for these tests are very similar to the shown Fisher-Irwin test.

The power of the four exact unconditional tests (without the Berger and Boos procedure) is shown in Figure 4.11. The tests have similar power for most of the marginal row sums. For the shown combination of parameter values, the test using the likelihood ratio statistic has slightly lower power than the other tests, and the Fisher-Boschloo test sometimes has power slightly above that of the other tests. This result may not hold for other combinations of π_1

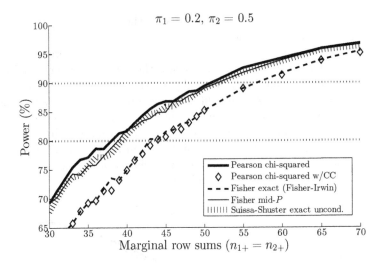

FIGURE 4.10
Power of five commonly used tests

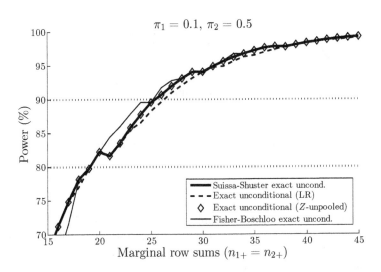

FIGURE 4.11
Power of four exact unconditional tests (without the Berger and Boos procedure)

and π_2. If we were to target 80% or 90% power, the needed sample size would be the same or almost the same for all four tests.

Figure 4.12 compares the power with and without the Berger and Boos procedure for two exact unconditional tests: the Suissa-Shuster and Fisher-Boschloo tests. There seems to be a small power advantage of using the Berger and Boos procedure; however, this difference is almost exclusively present for small success probabilities. Plots of, for instance, $\pi_1 = 0.2$ vs $\pi_2 = 0.4$ or $\pi_1 = 0.3$ vs $\pi_2 = 0.5$, do not display a clear power advantage with the Berger and Boos procedure. Nevertheless, as discussed in Section 4.4.7, the Berger and Boos procedure may elicit other benefits than a power increase, such as increased computational speed and stability.

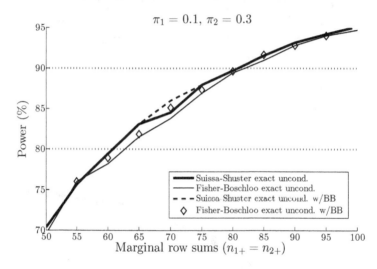

FIGURE 4.12

Power of two exact unconditional tests with and without the Berger and Boos procedure

For the asymptotic tests, we make three observations: (i) the power of the Pearson chi-squared test with continuity correction is considerably lower than the power of the other three tests (Figures 4.13 and 4.14); (ii) if one of the success probabilities is small, the power of the likelihood ratio and Z-unpooled tests is slightly greater than the power of the Pearson chi-squared test (Figure 4.13); and (iii) the power of the Pearson chi-squared, likelihood ratio, and Z-unpooled tests are nearly identical for non-small success probabilities (Figure 4.14). An explanation for the observed increase in power of the likelihood ratio and Z-unpooled tests can be seen in Figures 4.8 and 4.9, where the actual significance levels of the two tests are larger than the nominal level for values of π close to zero or one.

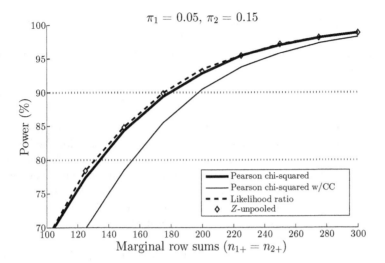

FIGURE 4.13
Power of four asymptotic tests for a low value of π_{11}

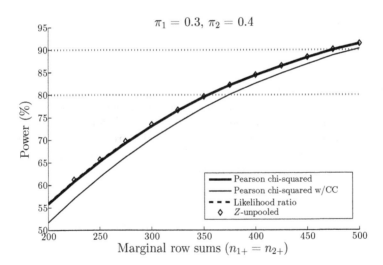

FIGURE 4.14
Power of four asymptotic tests

4.5 Confidence Intervals for the Difference between Probabilities

4.5.1 Introduction and Estimation

The *difference between probabilities* is an important effect measure in many applications, such as randomized controlled trials and cohort studies, where we have sampling under the row margins fixed model or the total sum fixed model. In studies with sampling under the total sum fixed model—such as in cross-sectional studies—we condition on the row sums. We denote the difference between probabilities by Δ:

$$\Delta = \pi_1 - \pi_2. \tag{4.15}$$

For Δ to have a meaningful interpretation, π_1 and π_2 must represent probabilities from two independent samples. This is not the case for the both margins fixed model, for which there is only one parameter (θ). Under the row margins fixed model, π_1 and π_2 are probabilities from two independent samples. The difference between probabilities may also be used for cross-sectional data sampled under the total sum fixed model, if we condition on the row sums. In a case-control study, the effect measure could be $\Delta = \gamma_1 - \gamma_2$, but that is rarely of interest.

The difference between probabilities is also called the *probability difference* or the *risk difference*. The latter name is most often used when the event in question is a harmful one. In epidemiology, Δ is sometimes called the *absolute risk reduction* or the *attributable risk (reduction)*. The difference between probabilities is estimated using the sample proportions, which are the maximum likelihood estimates of π_1 and π_2:

$$\hat{\Delta} = \hat{\pi}_1 - \hat{\pi}_2 = \frac{n_{11}}{n_{1+}} - \frac{n_{21}}{n_{2+}}. \tag{4.16}$$

The effect measures we use for the 2×2 table, except for the number needed to treat, are related to the regression coefficients in generalized linear models (GLMs) for binary data (see Section 1.5). Let Y_2 denote the response variable, with values $Y_2 = 1$ (success) and $Y_2 = 0$ (failure), and let Y_1 denote the explanatory variable: $Y_1 = 1$ (Group 1/Exposed) and $Y_1 = 2$ (Group 2/Unexposed). For a general link function, we formulate the following GLM:

$$\text{link}\big[\Pr(Y_2 = 1 \mid Y_1 = i)\big] = \text{link}(\pi_i) = \alpha + \beta y_i, \tag{4.17}$$

for $i = 1, 2$, where $y_1 = 1$ and $y_2 = 0$. The effect of the grouping (Y_1) on the outcome (Y_2) is described by:

$$\beta = \text{link}(\pi_1) - \text{link}(\pi_2). \tag{4.18}$$

If the link function is the linear link, $\beta = \pi_1 - \pi_2$, which is the difference between probabilities (Δ) in Equation 4.15. When using the log link, $\exp(\beta)$ is the ratio of probabilities (Section 4.7), and when the logit link is used, $\exp(\beta)$ is the odds ratio (Section 4.8).

Sections 4.5.2–4.5.6 present different confidence interval methods for Δ. Section 4.5.7 sets the methods to work on the examples presented in Section 4.2. We evaluate the methods in Section 4.5.8 and present our recommendations in Section 4.9.

4.5.2 The Wald Interval

The traditional Wald confidence interval for Δ is based on the asymptotic normal distribution of $\hat{\Delta}$:

$$\hat{\Delta} \pm z_{\alpha/2} \sqrt{\frac{\hat{\pi}_1(1 - \hat{\pi}_1)}{n_{1+}} + \frac{\hat{\pi}_2(1 - \hat{\pi}_2)}{n_{2+}}}, \qquad (4.19)$$

where $z_{\alpha/2}$ is the upper $\alpha/2$ percentile of the standard normal distribution. The Wald interval has zero-width when (i) $n_{11} = n_{21} = 0$ or $n_{12} = n_{22} = 0$, which gives the interval $(0,0)$; (ii) $n_{11} = n_{22} = 0$, which gives the interval $(-1,-1)$; and (iii) $n_{12} = n_{21} = 0$, which gives the interval $(1,1)$.

A continuity corrected version of the Wald interval, due to Yates (1934), can be expressed as shown in Fleiss et al. (2003, p. 60):

$$\hat{\Delta} \pm \left[z_{\alpha/2} \sqrt{\frac{\hat{\pi}_1(1 - \hat{\pi}_1)}{n_{1+}} + \frac{\hat{\pi}_2(1 - \hat{\pi}_2)}{n_{2+}}} + \frac{1}{2}\left(\frac{1}{n_{1+}} + \frac{1}{n_{2+}} \right) \right]. \qquad (4.20)$$

The Wald interval with continuity correction avoids zero-width but has a higher overshoot rate. Overshoot (intervals outside $[-1,1]$) is also possible with the Agresti-Caffo interval (Section 4.5.3). The problem of overshoot can easily be eliminated by truncation for both the Wald, Wald with continuity correction, and Agresti-Caffo intervals; however, this solution is not entirely satisfactory, as the width of the interval will be narrower than the uncertainty of the data indicates.

4.5.3 The Agresti-Caffo Interval

Agresti and Caffo (2000) proposed a simple, yet effective procedure for computing a confidence interval: add one success and one failure (pseudo-frequencies, see Section 2.4.4) in each sample and calculate the Wald confidence interval on the resulting data:

$$\tilde{\pi}_1 - \tilde{\pi}_2 \pm z_{\alpha/2} \sqrt{\frac{\tilde{\pi}_1(1 - \tilde{\pi}_1)}{\tilde{n}_{1+}} + \frac{\tilde{\pi}_2(1 - \tilde{\pi}_2)}{\tilde{n}_{2+}}}, \qquad (4.21)$$

where

$$\tilde{n}_{1+} = n_{1+} + 2, \quad \tilde{n}_{2+} = n_{2+} + 2, \quad \tilde{\pi}_1 = (n_{11} + 1)/\tilde{n}_{1+}, \quad \tilde{\pi}_2 = (n_{21} + 1)/\tilde{n}_{2+}.$$

Note that our estimate of the difference between probabilities is still given by the difference in sample proportions (Equation 4.16), the calculations of $\tilde{\pi}_1$ and $\tilde{\pi}_2$ go only into the calculations of the confidence interval.

The Agresti-Caffo interval is usually consistent with the results of the Pearson chi-squared test (Section 4.4.2), but it can overshoot the $[-1,1]$ boundaries for Δ. Note that adjustments by adding values to the observed counts—as is done in the Agresti-Caffo interval and several other intervals we consider—are discouraged on general principles by some authors, for instance, Hirji (2006, p. 78).

4.5.4   The Newcombe Hybrid Score Interval

Newcombe (1998b) proposed a confidence interval for Δ formed as a combination of the Wilson score (Section 2.4.3) confidence limits for π_1 and the Wilson score confidence limits for π_2. Denote the interval for π_1 by (l_1, u_1) and the interval for π_2 by (l_2, u_2). The Newcombe hybrid score confidence interval (L, U) for Δ is given by

$$L = \hat{\Delta} - \sqrt{\left(\hat{\pi}_1 - l_1\right)^2 + \left(u_2 - \hat{\pi}_2\right)^2} \tag{4.22}$$

and

$$U = \hat{\Delta} + \sqrt{\left(\hat{\pi}_2 - l_2\right)^2 + \left(u_1 - \hat{\pi}_1\right)^2}. \tag{4.23}$$

The Newcombe hybrid score interval is an example of a square-and-add method, named after the technique of squaring and adding standard errors. This general approach to constructing confidence intervals is also known as the method of variance estimates recovery (MOVER). MOVER intervals exist for both difference-based and ratio-based effect measures in both unpaired and paired 2×2 tables. In principle, any confidence interval for the binomial parameter (see Section 2.4) can be used, although the Wilson score interval is most often the preferred choice. MOVER intervals for the ratio of probabilities and the odds ratio will be considered later in this chapter.

4.5.5   Asymptotic Score Intervals

One method of obtaining an asymptotic score interval is to invert two one-sided $\alpha/2$ level score tests (the tail method). For a specified value of the difference between probabilities, $\Delta_0 \in [-1, 1]$, the score statistic is

$$T_{\text{score}}(\mathbf{n} \mid \Delta_0) = \frac{\hat{\pi}_1 - \hat{\pi}_2 - \Delta_0}{\sqrt{\dfrac{\hat{p}_1(1 - \hat{p}_1)}{n_{1+}} + \dfrac{\hat{p}_2(1 - \hat{p}_2)}{n_{2+}}}}, \tag{4.24}$$

where $\mathbf{n} = \{n_{11}, n_{12}, n_{21}, n_{22}\}$, as usual, denotes the observed table and $\hat{p}_1$ and $\hat{p}_2$ are the maximum likelihood estimates of π_1 and π_2 subject to $\pi_1 - \pi_2 = \Delta_0$. An asymptotic confidence interval based on inverting two one-sided score tests (Equation 4.24) was first proposed by Mee (1984). The *Mee asymptotic score interval* (L, U) for Δ is obtained by solving

$$T_{\text{score}}(\mathbf{n} \mid L) = z_{\alpha/2} \tag{4.25}$$

and

$$T_{\text{score}}(\mathbf{n} \mid U) = -z_{\alpha/2}. \tag{4.26}$$

Mee (1984) marked a disadvantage of the method: iterative computations are required to calculate both $\hat{p}_1$, $\hat{p}_2$, L, and U. Miettinen and Nurminen (1985), however, showed that the restricted maximum likelihood estimates ($\hat{p}_1$ and

$\hat{p}_2$) can be obtained by solving a cubic equation and gave unique closed-form expressions for them. Using the notation for the marginal and total sums (Table 4.1), we obtain the expressions:

$$
\begin{aligned}
L_3 &= N, \\
L_2 &= (n_{1+} + 2n_{2+})\Delta_0 - N - n_{+1}, \\
L_1 &= [n_{2+}\Delta_0 - N - 2n_{21}]\Delta_0 + n_{+1}, \\
L_0 &= n_{21}\Delta_0(1 - \Delta_0), \\
q &= L_2^3/(3L_3)^3 - L_1L_2/6L_3^2 + L_0/2L_3, \\
p &= \operatorname{sign}(q)\sqrt{L_2^2/(3L_3)^2 - L_1/3L_3}, \\
a &= \frac{1}{3}\left[\pi + \cos^{-1}(q/p^3)\right], \\
\hat{p}_2 &= 2p\cos(a) - L_2/3L_3, \\
\hat{p}_1 &= \hat{\pi}_2 + \Delta_0.
\end{aligned}
$$

Almost equal expressions for $\hat{p}_1$ and $\hat{p}_2$ can be found in Farrington and Manning (1990). An iterative algorithm, such as the secant method or the bisection method, is still needed to find the confidence limits L and U.

Miettinen and Nurminen (1985) suggested an asymptotic score interval based on the adjusted score statistic

$$
T_{\mathrm{MN}}(\mathbf{n}\,|\,\Delta_0) = \frac{\hat{\pi}_1 - \hat{\pi}_2 - \Delta_0}{\sqrt{\dfrac{\hat{p}_1(1 - \hat{p}_1)}{n_{1+}} + \dfrac{\hat{p}_2(1 - \hat{p}_2)}{n_{2+}}}} \cdot \sqrt{1 - \frac{1}{n_{1+} + n_{2+}}}, \qquad (4.27)
$$

which is equal to Equation 4.24 multiplied by an additional square-root term. The restricted maximum likelihood estimates $\hat{p}_1$ and $\hat{p}_2$ are the same as for the Mee asymptotic score interval, and the correction term in Equation 4.27 makes a difference only for small sample sizes. The *Miettinen-Nurminen asymptotic score interval* (L, U) for Δ is obtained by solving

$$
T_{\mathrm{MN}}(\mathbf{n}\,|\,L) = z_{\alpha/2} \qquad (4.28)
$$

and

$$
T_{\mathrm{MN}}(\mathbf{n}\,|\,U) = -z_{\alpha/2}. \qquad (4.29)
$$

4.5.6 Exact Unconditional Intervals

In the previous two sections, we inverted asymptotic tests to obtain asymptotic score intervals. Equation 4.24 can also be used to construct exact tests, which can be inverted to obtain exact unconditional score intervals. Under the restriction $\pi_1 - \pi_2 = \Delta_0$, the domain of π_1 given Δ_0 is

$$
D(\Delta_0) = \left\{\pi_1 : \max(0, \Delta_0) \leq \pi_1 \leq \min(1, 1 + \Delta_0)\right\}. \qquad (4.30)
$$

Let $\mathbf{x} = \{x_{11}, n_{1+} - x_{11}, x_{21}, n_{2+} - x_{21}\}$ denote any 2×2 table that might be observed given the fixed row sums n_{1+} and n_{2+} (the row margins fixed model). The probability of observing $\mathbf{x}$ is given by the product of two independent binomial probabilities, as in Equation 4.3. With the restriction $\pi_1 - \pi_2 = \Delta_0$, we re-parameterize Equation 4.3 such that we use π_1 and Δ_0 instead of π_1 and π_2:

$$f(\mathbf{x} \,|\, \pi_1, \Delta_0, \mathbf{n}_+) = \qquad\qquad\qquad\qquad\qquad\qquad\qquad\qquad (4.31)$$

$$\binom{n_{1+}}{x_{11}} \pi_1^{x_{11}} (1 - \pi_1)^{n_{1+} - x_{11}} \binom{n_{2+}}{x_{21}} (\pi_1 - \Delta_0)^{x_{21}} (1 - \pi_1 + \Delta_0)^{n_{2+} - x_{21}},$$

where $\mathbf{n}_+ = \{n_{1+}, n_{2+}\}$.

The interval by Chan and Zhang (1999) is based on inverting two one-sided exact score tests of size at most $\alpha/2$ (the tail method). The lower (L) and upper (U) confidence limits of the *Chan-Zhang exact unconditional interval* for Δ are the two values that satisfy the equations:

$$\max_{\pi_1 \in D(\Delta_0)} \left\{ \sum_{\Omega(\mathbf{x}|\mathbf{n}_+)} I\big[T(\mathbf{x} \,|\, L) \geq T(\mathbf{n} \,|\, L)\big] \cdot f(\mathbf{x} \,|\, \pi_1, L, \mathbf{n}_+) \right\} = \alpha/2 \quad (4.32)$$

and

$$\max_{\pi_1 \in D(\Delta_0)} \left\{ \sum_{\Omega(\mathbf{x}|\mathbf{n}_+)} I\big[T(\mathbf{x} \,|\, U) \leq T(\mathbf{n} \,|\, U)\big] \cdot f(\mathbf{x} \,|\, \pi_1, U, \mathbf{n}_+) \right\} = \alpha/2, \quad (4.33)$$

where $\Omega(\mathbf{x}|\mathbf{n}_+)$ denotes the set of all tables with row sums equal to $\mathbf{n}_+$, $I()$ is the indicator function, and $T()$ is the score statistic in Equation 4.24.

The parameter π_1 in Equations 4.32 and 4.33 is a nuisance parameter (see Section 4.4.7), which we have eliminated by taking the maximum value over the range $D(\Delta_0)$ given in Equation 4.30.

Iterative computations are required to find L and U. The Berger and Boos procedure (see Section 4.4.7) may be used to reduce the domain of π_1 such that the maximizations in Equations 4.32 and 4.33 are performed over a $100(1-\gamma)\%$ confidence interval for π_1 instead of the entire range. No optimal choice of γ has been suggested for exact unconditional intervals. We use $\gamma = 0.0001$, which is the recommended value for the Suissa-Shuster exact unconditional test (Lydersen et al., 2012b).

Instead of inverting two one-sided tests of size at most $\alpha/2$, Agresti and Min (2001) proposed to invert one two-sided test of size at most α. The lower (L) and upper (U) limits of the *Agresti-Min exact unconditional interval* for Δ are the two values that satisfy

$$\max_{\pi_1 \in D(\Delta_0)} \left\{ \sum_{\Omega(\mathbf{x}|\mathbf{n}_+)} I\big[|T(\mathbf{x} \,|\, L)| \geq |T(\mathbf{n} \,|\, L)|\big] \cdot f(\mathbf{x} \,|\, \pi_1, L, \mathbf{n}_+) \right\} = \alpha \quad (4.34)$$

and

$$\max_{\pi_1 \in D(\Delta_0)} \left\{ \sum_{\Omega(\mathbf{x}|\mathbf{n}_+)} I\Big[|T(\mathbf{x}|U)| \geq |T(\mathbf{n}|U)| \Big] \cdot f(\mathbf{x}|\pi_1, U, \mathbf{n}_+) \right\} = \alpha. \quad (4.35)$$

The Berger and Boos procedure may be used to reduce the range over which the maximization in Equations 4.34 and 4.35 is performed.

4.5.7 Examples

A Lady Tasting a Cup of Tea (Table 4.2)

The data in Table 4.2 are sampled under the both margins fixed model and do not represent two independent probabilities. The difference between the observed proportions n_{11}/n_{1+} and n_{21}/n_{2+} does not have a useful interpretation in this case, and neither does the ratio of probabilities. The odds ratio, on the other hand, does, and we shall revisit this example in Section 4.8.9.

A Randomized Clinical Trial of High Versus Standard Dose of Epinephrine (Table 4.3)

For this example, we have two independent proportions: the proportions that survived after 24 hours in the standard and high dose groups. The sample proportions are $\hat{\pi}_1 = 7/34 = 0.206$ in the standard dose group and $\hat{\pi}_2 = 1/34 = 0.0294$ in the high dose group. An estimate of the difference between probabilities is

$$\hat{\Delta} = 0.206 - 0.0294 = 0.177.$$

The 24 hour survival probability is estimated to be 18 percentage points higher in the standard dose group than the high dose group.

The Wald interval is on the form $\hat{\Delta} \pm z_{\alpha/2} \cdot \text{SE}$, where SE is the standard error of the estimate (Equation 4.19). We calculate the standard error as

$$\sqrt{0.206(1 - 0.206)/34 + 0.0294(1 - 0.0294)/34} = 0.0752.$$

For a 95% confidence interval, we have $z_{\alpha/2} = 1.96$, and the Wald interval is computed as

$$0.177 \mp 1.96 \cdot 0.0752 = (0.029 \text{ to } 0.324).$$

If we use the Yates's continuity correction to the Wald interval (Equation 4.20), we add the quantity

$$\frac{1}{2}\left(\frac{1}{34} + \frac{1}{34} \right) = 0.0294$$

to the upper limit and subtract the same quantity from the lower limit. The resulting 95% confidence interval is

$$(0.0292 - 0.0294, 0.324 + 0.0294) = (-0.0002 \text{ to } 0.353),$$

which is considerably wider than the Wald interval without the continuity correction.

To calculate the Agresti-Caffo interval, we add one success ("Yes") and one failure ("No") in each treatment group: $\tilde{\pi}_1 = 8/36 = 0.250$ and $\tilde{\pi}_2 = 2/36 = 0.0556$. We insert these estimates in Equation 4.21 and obtain

$$0.167 \mp 1.96 \cdot 0.0791 = (0.012 \text{ to } 0.322).$$

The Newcombe hybrid score interval is also of closed-form expression and can easily be computed with a simple calculator. There are more terms to compute and add up, and we do not show all the details here. Let $(l_1, u_1) = (0.104, 0.368)$ and $(l_2, u_2) = (0.0052, 0.149)$ be the 95% Wilson score intervals for π_1 and π_2, respectively. By Equations 4.22 and 4.23, the 95% lower (L) and upper (U) limits of the Newcombe hybrid score interval are

$$L = 0.177 - \sqrt{(0.206 - 0.104)^2 + (0.149 - 0.0294)^2} = 0.019$$
$$U = 0.177 + \sqrt{(0.0294 - 0.0052)^2 + (0.368 - 0.206)^2} = 0.340.$$

The Mee and Miettinen-Nurminen asymptotic score intervals require software resources to compute. Figure 4.15 illustrates how the 95% confidence limits of the Mee interval are found. The lower limit $(L = 0.028)$ is the value of Δ_0 for which the score statistic (Equation 4.24) is equal to 1.96 $(= z_{\alpha/2})$, and the upper limit $(U = 0.344)$ is the value of Δ_0 for which the score statistic is equal to -1.96. The corresponding plot for the Miettinen-Nurminen interval is almost identical to Figure 4.15, and the confidence limits are similar: (0.027 to 0.345).

The Chan-Zhang and Agresti-Min exact unconditional intervals are considerably more complex and computer-intensive to calculate than the other intervals for the difference between probabilities. Highly specialized software algorithms are required, and the support of exact unconditional methods in commercial software packages is limited. An exception is the StatXact software (Cytel Inc., Cambridge, MA), which includes many exact unconditional tests and confidence intervals. The exact unconditional intervals and the asymptotic score intervals have one principle in common: the confidence limits are obtained by (iteratively) finding values of Δ_0 that make the P-value of a test for the null hypothesis H_0: $\Delta = \Delta_0$ based on the score (or adjusted score) statistic equal to α (Agresti-Min) or $\alpha/2$ (the other intervals). For the asymptotic score intervals, the tests are asymptotic score tests, for which the observed value of the score statistic can be directly compared with the standard normal distribution, as in Equations 4.25–4.26 and 4.28–4.29. The exact unconditional intervals need to compute exact unconditional tests as intermediate steps toward finding the confidence limits. Figure 4.16 shows how the 95% confidence limits for the Chan-Zhang and Agresti-Min exact unconditional intervals are obtained.

Table 4.16 summarizes the results of the eight different confidence intervals for the difference between probabilities calculated on the epinephrine trial

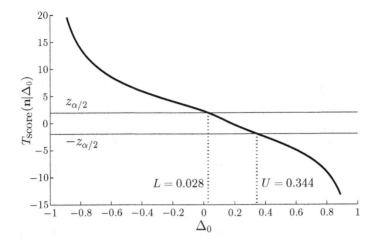

FIGURE 4.15
The score statistic (Equation 4.24) as function of the difference between probabilities (Δ_0) for the epinephrine trial data in Table 4.3

data. The width of each interval is included in the last column. The Wald interval has the shortest width. The Agresti-Caffo, Newcombe hybrid score, and the two asymptotic score intervals have similar widths but somewhat different location. The two exact unconditional intervals are slightly wider than the other intervals, except for the Wald interval with continuity correction, which is the only interval to include the null effect value ($\Delta = 0$).

TABLE 4.16
95% confidence intervals for the difference between probabilities
($\hat{\Delta} = 0.177$) based on the epinephrine trial data in Table 4.3

| Interval | Confidence limits | | |
	Lower	Upper	Width
Wald	0.029	0.324	0.295
Wald with continuity correction	-0.0002	0.353	0.353
Agresti-Caffo	0.012	0.322	0.310
Newcombe hybrid score	0.019	0.340	0.321
Mee asymptotic score	0.028	0.344	0.316
Miettinen-Nurminen asymptotic score	0.027	0.345	0.318
Chan-Zhang exact unconditional*	0.019	0.359	0.339
Agresti-Min exact unconditional*	0.023	0.355	0.331

*Calculated with Berger and Boos procedure ($\gamma = 0.0001$)

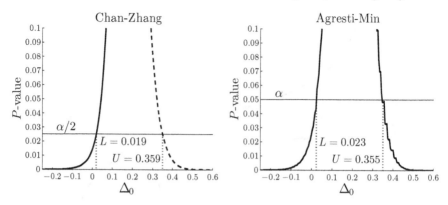

FIGURE 4.16
The exact unconditional intervals obtain their confidence limits by finding the two values of Δ_0 for which the corresponding exact unconditional score tests of H_0: $\Delta = \Delta_0$ have P-values equal to $\alpha/2$ (Chan-Zhang) or α (Agresti-Min)

A Case-Control Study of the Effect of GADA Exposure on IPEX Syndrome (Table 4.4)

The number of cases and controls in a case-control study is not sampled to match the ratio of cases to controls in the source population. Often their numbers are chosen in a 1:1, 1:2, or any such even ratio out of convenience or statistical considerations and not so that the ratio of the sample is equal to that of the source population. The data from a case-control study, therefore, cannot estimate the probability of being a case or a control, and the difference between probabilities and the ratio of probabilities are unavailable as effect measures, see also Section 4.3.1. The data from Table 4.4 is not considered in this section, and we postpone confidence interval estimation for these data until we deal with the odds ratio in Section 4.8.9.

A Cross-Sectional Study of the Association between CHRNA4 Genotypes and XFS (Table 4.5)

The data in Table 4.5 originate from a cross-sectional study, where only the total sum N is fixed beforehand (sampling under the total sum fixed model). A common approach to analyze cross-sectional studies is to condition on the row sums and proceed as if the data were sampled under the row margins fixed model. That is, we assume that the probability model of the data is given by the product of two independent binomial probabilities (Equation 4.3). This is usually a reasonable simplification that leads to parameters and effect measures that has intuitive interpretations. For the data in Table 4.5, π_1 and π_2 represent the probabilities of exfoliation syndrome in the eyes for participants with genotype CHRNA4-CC and CHRNA4-CC/TT, respectively.

We estimate these probabilities by $\hat{\pi}_1 = 0/16 = 0$ and $\hat{\pi}_2 = 15/72 = 0.208$, and the estimate of the difference between probabilities is given as

$$\hat{\Delta} = -0.208.$$

We do not show the details of how to calculate the confidence intervals for these data but give the results in Table 4.17. Perhaps owing to the zero in the upper left of Table 4.5, the intervals differ more for these data than we observed in Table 4.16, even though the total sample size is greater in Table 4.5 than in Table 4.3. Furthermore, we note that the Wald interval with continuity correction now is the second shortest interval and that the Agresti-Min interval is considerably wider than the Chan-Zhang interval. Both incidents—in our experience—happen rarely.

TABLE 4.17
95% confidence intervals for the difference between probabilities ($\hat{\Delta} = -0.208$) based on the cross-sectional data in Table 4.5

Interval	Lower	Upper	Width
Wald	-0.302	-0.115	0.188
Wald with continuity correction	-0.340	-0.076	0.264
Agresti-Caffo	-0.302	-0.019	0.283
Newcombe hybrid score	-0.316	0.0003	0.316
Mee asymptotic score	-0.316	-0.007	0.308
Miettinen-Nurminen asymptotic score	-0.316	-0.006	0.311
Chan-Zhang exact unconditional*	-0.325	-0.006	0.319
Agresti-Min exact unconditional*	-0.317	0.055	0.372

*Calculated with Berger and Boos procedure ($\gamma = 0.0001$)

4.5.8 Evaluation of Intervals

Evaluation Criteria

We evaluate confidence intervals for the difference between probabilities using three indices of performance: coverage probability, width, and location (see Section 1.4). As we did when we evaluated the tests in Section 4.4.9, we assume sampling under the row margins fixed model, such that the probability model is defined by the product of two independent binomial probabilities (Equation 4.3). The coverage probability depends on the two success probabilities, π_1 and π_2, the marginal row sums, n_{1+} and n_{2+}, and the nominal level $1 - \alpha$. We define the exact coverage probability as

$$CP(\pi_1, \pi_2, n_{1+}, n_{2+}, \alpha) =$$

$$\sum_{x_{11}=0}^{n_{1+}} \sum_{x_{21}=0}^{n_{2+}} I(L \leq \Delta \leq U) \cdot f(\mathbf{x} \mid \pi_1, \pi_2, n_{1+}, n_{2+}), \qquad (4.36)$$

where $I()$ is the indicator function, $L = L(\mathbf{x}, \alpha)$ and $U = U(\mathbf{x}, \alpha)$ are the lower and upper $100(1 - \alpha)\%$ confidence limits of an interval for the table $\mathbf{x} = \{x_{11}, n_{1+} - x_{11}, x_{21}, n_{2+} - x_{21}\}$, $\Delta = \pi_1 - \pi_2$, and $f()$ is the probability of observing $\mathbf{x}$, given in Equation 4.3.

The expected width of an interval depends on the same parameters as the coverage probability. It, too, can be calculated exactly with complete enumeration:

$$\text{Width}(\pi_1, \pi_2, n_{1+}, n_{2+}, \alpha) =$$

$$\sum_{x_{11}=0}^{n_{1+}} \sum_{x_{21}=0}^{n_{2+}} (U - L) \cdot f(\mathbf{x} \mid \pi_1, \pi_2, n_{1+}, n_{2+}).$$

To measure location, we use the index MNCP/NCP. As was the case with coverage probability and width, we use complete enumeration to calculate MNCP/NCP exactly. NCP is obtained from the coverage probability as NCP $= 1 -$ CP, where CP is computed using Equation 4.36. MNCP is defined as

$$\text{MNCP}(\pi_1, \pi_2, n_{1+}, n_{2+}, \alpha) =$$

$$\sum_{x_{11}=0}^{n_{1+}} \sum_{x_{21}=0}^{n_{2+}} I\big(L > \Delta \geq 0 \text{ or } U < \Delta \leq 0\big) \cdot f(\mathbf{x} \mid \pi_1, \pi_2, n_{1+}, n_{2+}).$$

Evaluation of Coverage Probability

In Section 4.4.9, we fixed the row margins and plotted the actual significance level as a function of the common success probability. We can plot the coverage probability in a similar fashion; however, we now have $\Delta = \pi_1 - \pi_2$ instead of a common success probability. We thus fix Δ in addition to the row margins and plot the coverage probability as a function of π_2, so that the range of the x-axis is $(0, 1 - \Delta)$. We might as well put π_1 on the x-axis, making the range $(\Delta, 1)$; however, we prefer the former option.

We start by considering the intervals with closed-form expressions: the Wald, Wald with continuity correction, Agresti-Caffo, and Newcombe hybrid score intervals. These intervals are simple to compute; neither interval requires special software resources to be calculated. Figure 4.17 shows the coverage probabilities of the four intervals for $n_{1+} = n_{2+} = 25$, $\Delta = 0.25$, and a nominal level of 95%. The two Wald intervals perform poorly, each in its own fashion. Without continuity correction, the coverage is too low, at about 94%, whereas with continuity correction, the coverage is too high, at about 97%. The Agresti-Caffo and Newcombe hybrid score intervals perform considerably better. Both intervals have coverage close to—and slightly above—the nominal 95%.

The same pattern of results can be observed for larger row margins, as in Figure 4.18, where $n_{1+} = n_{2+} = 100$ and $\Delta = 0.2$. The Wald interval without

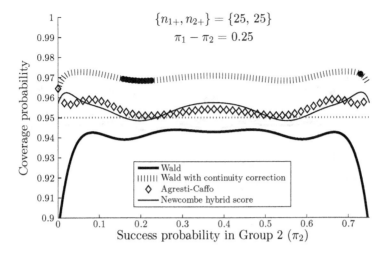

FIGURE 4.17
Coverage probabilities of four confidence intervals of closed-form expression
for the difference between probabilities

continuity correction now has coverage closer to the nominal level, but the
coverage does not reach 95%. The Wald interval with continuity correction
has not improved much and its coverage is still markedly higher than the
nominal level. The Agresti-Caffo and Newcombe hybrid score intervals perform
excellently.

We now turn to the Mee and Miettinen-Nurminen asymptotic score intervals. An iterative algorithm is needed to compute the confidence limits of
these two intervals, and, unfortunately, their support in standard software
packages is limited. As seen in Figure 4.19, the coverage of both intervals is
usually quite close to the nominal level; however, the coverage dips below 95%
more frequently than does the coverage of the Agresti-Caffo and Newcombe
hybrid score intervals. The Miettinen-Nurminen interval is usually slightly
better than the Mee interval.

An example of the coverage probabilities of the two exact unconditional
intervals is shown in Figure 4.20. Because exact intervals guarantee that the
coverage is at least at the nominal level, they can be quite conservative, particularly when π_1 or π_2 is close to zero or one. This is more of a problem
for intervals that invert two one-sided exact unconditional tests, such as the
Chan-Zhang interval, than intervals that invert one two-sided exact unconditional test, such as the Agresti-Min interval. The downside for the Agresti-Min
interval is that only the overall non-coverage is guaranteed to be no more than
α, whereas the Chan-Zhang interval guarantees that the non-coverage in each
tail is limited by $\alpha/2$. If the latter assurance is of no concern, the Agresti-

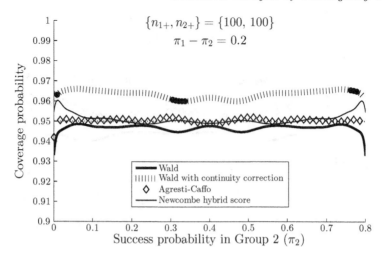

FIGURE 4.18
Coverage probabilities of four confidence intervals of closed-form expression for the difference between probabilities

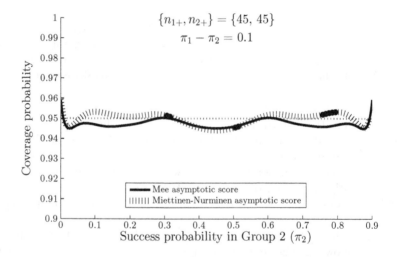

FIGURE 4.19
Coverage probabilities of the Mee and Miettinen-Nurminen asymptotic score intervals for the difference between probabilities

Min interval is superior to the Chan-Zhang interval. For an exact method, the conservativeness of the Agresti-Min interval is small.

We end the evaluation of coverage probabilities by pitting the inter-

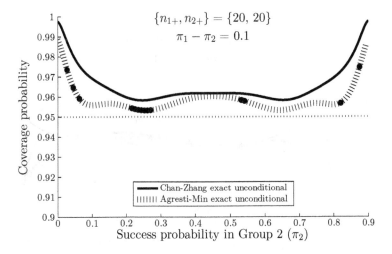

FIGURE 4.20
Coverage probabilities of the Chan-Zhang and Agresti-Min exact unconditional intervals for the difference between probabilities

vals with best coverage properties against each other. Figure 4.21 illustrates the coverage probabilities of the Agresti-Caffo, Newcombe hybrid score, Miettinen-Nurminen asymptotic score, and Agresti-Min exact unconditional intervals for a combination of unequal row margins. This example showcases the excellent coverage of the Newcombe interval. It does not always provide at least 95% coverage, but the infractions of the nominal level are small. The other intervals also perform quite well, although the Miettinen-Nurminen interval has low coverage for parts of the parameter space.

Evaluation of Width

The expected width of an interval can be calculated and plotted in a similar manner as we did for the coverage probability. An example is shown in Figure 4.22. It includes all the intervals except for the Wald interval with continuity correction and the Mee asymptotic score interval. The Wald interval with continuity correction is considerably wider than the other intervals. For the parameters in Figure 4.22, it has a maximum width equal to 0.54. The Mee interval is wider than the Miettinen-Nurminen asymptotic score interval, but narrower than the Agresti-Min exact unconditional interval. Most of the intervals in Figure 4.22 have similar widths. The Newcombe hybrid score interval, and partly the Agresti-Caffo interval, stand out as having shorter widths than the other intervals, whereas the Chan-Zhang exact unconditional interval is generally wider than the other intervals.

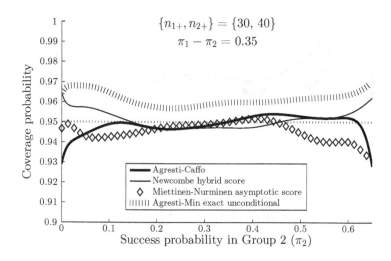

FIGURE 4.21
Coverage probabilities of the best performing intervals for the difference between probabilities

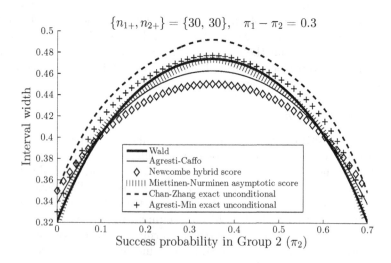

FIGURE 4.22
Expected interval width of six confidence intervals for the difference between probabilities

Evaluation of Location

For most combinations of parameters, all eight intervals for the difference between probabilities are either satisfactorily located ($0.4 \leq$ MNCP/NCP ≤ 0.6) or slightly too mesially located (MNCP/NCP < 0.4). An exception is the two Wald intervals, which sometimes have too distal locations (MNCP/NCP > 0.6) for medium to large effect sizes ($\Delta > 0.25$). Figure 4.23 shows a typical situation. Three intervals are not shown: the Wald interval with continuity correction, which has location similar to the Wald interval without continuity correction; the Mee asymptotic score interval, which has location similar to the Miettinen-Nurminen asymptotic score interval; and the Chan-Zhang exact unconditional interval, which in this case, has identical location to the Agresti-Min exact unconditional interval. The latter equality of locations does not hold in general. If we compare the four intervals with best coverage properties (for a wide range of parameter space points not shown here), we observe that the locations of the Miettinen-Nurminen asymptotic score and Agresti-Min exact unconditional intervals are almost always in the satisfactory range, whereas the Agresti-Caffo and Newcombe hybrid score intervals sometimes also have mesial locations.

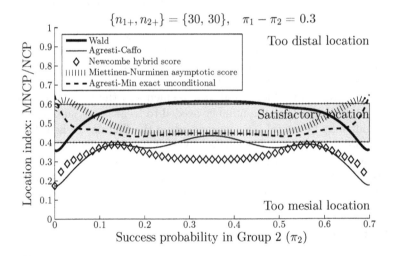

FIGURE 4.23
Location, as measured by the MNCP/NCP index, of five confidence intervals for the difference between probabilities

4.6   Confidence Intervals for the Number Needed to Treat

4.6.1   Introduction

The *number needed to treat* can be a useful effect measure to summarize the results from studies of two treatments or exposures. It was originally proposed by Laupacis et al. (1988) as a measure to inform clinicians and patients how much treatment effort they must expend to prevent one adverse event. The number needed to treat has later gained more widespread use and can, with a few changes of wording, also be used for cohort, case-control, and cross-sectional studies. In epidemiologic studies, the number needed to treat is often called the *number needed to be exposed*. In the following, we present the number needed to treat in the context of a clinical trial of two competing treatments: one new treatment (Group 1) versus one standard treatment (Group 2).

The use of the number needed to treat is associated with some controversy. One view states that the number needed to treat is particularly suitable for clinical decision making, as it incorporates both statistical and clinical significance by dealing with numbers of patients rather than probabilities (Cook and Sackett, 1995). This view is, however, not universally acknowledged, and some authors argue that the number needed to treat can be confusing, particularly for clinicians (Newcombe, 1999). We do not go into the details of this argument but refer the reader to the article by Stang et al. (2010) for a summary of the most common problems and disagreements in the field. An important point worth noting is that the number needed to treat is not suitable for data analysis but may be useful for communicating the findings of health studies. We also note that the number needed to treat, as calculated in this section, can be inaccurate and misleading in studies with varying follow-up times. In such cases, Altman and Andersen (1999) suggest that the number needed to treat can be computed as a function of follow-up time.

4.6.2   Estimation

The number needed to treat is the estimated number of patients that would have to be treated with a new treatment instead of a standard treatment for one additional patient to benefit. In contrast to the other effect measures we consider in this chapter, the number needed to treat does not originate as a natural parameter in a generalized linear model. Instead, the number needed to treat is derived from the estimate of the difference between probabilities. We therefore dispense with the usual notation of putting a "hat" above the estimate and simply denote it by NNT:

$$\text{NNT} = \frac{1}{\hat{\pi}_1 - \hat{\pi}_2}. \tag{4.37}$$

We have now assumed that a positive value of $\hat{\pi}_1 - \hat{\pi}_2$, and thereby a positive value of NNT, indicates that the new treatment is superior to the standard treatment, i.e., that π_1 and π_2 are the probabilities of a beneficial event. If this is not the case, one may simply reverse the order of $\hat{\pi}_1$ and $\hat{\pi}_2$ and define the difference as $\hat{\pi}_2 - \hat{\pi}_1$. A similar issue is the need to distinguish between positive and negative values of NNT. As suggested by Altman (1998), it may be informative to denote positive values of NNT by NNTB: the number of patients needed to be treated for one additional patient to benefit. In like manner, negative values of NNT can be made positive and denoted by NNTH: the number of patients needed to be treated for one additional patient to be harmed.

4.6.3 Confidence Intervals

To calculate a confidence interval for the number needed to treat, we first compute a confidence interval for the difference between probabilities using one of the methods in Section 4.5. Denote the lower and upper limits of that interval by L and U. If the confidence interval for the difference between probabilities does not include zero, the confidence interval for NNTB and NNTH can be obtained by taking the reciprocals of the absolute values of L and U and reversing their order:

$$1/|U| \text{ to } 1/|L|. \tag{4.38}$$

If, on the other hand, the interval (L, U) contains zero, the confidence interval for the number needed to treat should be denoted by (Altman, 1998):

$$\text{NNTH } 1/|L| \text{ to } \infty \text{ to NNTB } 1/U. \tag{4.39}$$

We show through our standard examples how to calculate and interpret both types of intervals in the next section.

4.6.4 Examples

Only two of the examples from Section 4.2 are relevant as examples of how to calculate and interpret confidence intervals for the number needed to treat. These are the same two examples that were used to illustrate the confidence intervals for the difference between probabilities in Section 4.5.7. The reasons are the same: the tea tasting example in Table 4.2 does not represent two independent binomial probabilities, and the case-control study in Table 4.4 does not allow for the estimation of π_1 and π_2, which is needed to estimate NNT (Equation 4.37).

A Randomized Clinical Trial of High Versus Standard Dose of Epinephrine (Table 4.3)

The usual definition of the number needed to treat puts the new treatment as Group 1 and the standard treatment as Group 2. In the example in Table 4.3,

132   *Statistical Analysis of Contingency Tables*

it is the other way around. Moreover, the estimate of the difference between probabilities is positive, indicating that the standard treatment (standard dose epinephrine) may be better than the new treatment (high dose epinephrine). We must thus take care to rephrase the NNT in terms of the number needed to benefit (NNTB) with the standard versus the new treatment, instead of the usual new versus standard treatment comparison. Alternatively, we could swap the rows in Table 4.3, obtain a negative estimate of the difference between probabilities, and rephrase the NNT as the number needed to harm (NNTH). In the following, we choose the first approach.

Because $\hat{\pi}_1 = 0.206$ and $\hat{\pi}_2 = 0.0294$, we have the estimate:

$$\text{NNTB} = \frac{1}{0.177} = 5.7.$$

For every 5.7 patients treated with standard dose epinephrine, one additional patient survives, as compared with high dose epinephrine.

The Newcombe hybrid score interval is a good all-round confidence interval for the difference between probabilities. For the data in Table 4.3, the 95% Newcombe interval is (0.019 to 0.340). Since this interval does not contain zero, we use Equation 4.38 to obtain a 95% confidence interval for NNTB as $(1/0.34, 1/0.019) = (2.9 \text{ to } 53)$. If we instead wanted to base the interval for NNTB on one of the other intervals for the difference between probabilities in Table 4.16 (except for the Wald interval with continuity correction), the lower limit would range from 2.8 to 3.1, and the upper limit would range from 34 to 83.

The Wald interval with continuity correction is the only interval in Table 4.16 that includes zero. If we use this interval to calculate a confidence interval for NNTB, we need to use Equation 4.39 instead of Equation 4.38. We illustrate construction and interpretation of such intervals in the next example.

A Cross-Sectional Study of the Association between CHRNA4 Genotypes and XFS (Table 4.5)

For this example, the estimate of the difference between probabilities is negative ($\hat{\Delta} = -0.208$), as is both the lower and upper confidence limits for most of—but not all—the intervals in Table 4.17. Since $\hat{\Delta}$ is negative, the reciprocal of the absolute value is the number needed to harm: $\text{NNTH} = 1/0.208 = 4.8$. For every 4.8 patient with genotype CHRNA4-TC/TT, one additional patient will have presence of exfoliation syndrome in the eyes, as compared with patients with genotype CHRNA4-CC.

To calculate a confidence interval for NNTH, we use first the 95% Agresti-Caffo interval (-0.302 to -0.019) for the corresponding difference between probabilities as basis. By Equation 4.38, the lower limit is $1/0.019 = 53$, and the upper limit is $1/0.302 = 3.3$. When we calculate with the number needed to harm, the lower limit has greater numeric value than the upper limit. We state with 95% confidence that the effect of genotype CHRNA4-TC/TT may be as low as NNTH 53 and as high as NNTH 3.3. Note that the effect here is an association and not a causal effect.

We end this section with an example where the confidence interval encloses the null value. The Agresti-Min exact unconditional interval for the difference between probabilities is $(-0.317$ to $0.055)$. Since this interval includes zero, we use Equation 4.39, which involves the quantities $1/|L| = 1/0.317 = 3.2$ and $1/U = 1/0.055 = 18$. The confidence interval for NNTH then is NNTH 3.2 to ∞ to NNTB 18. A proper interpretation of this interval requires an understanding of the scale of the NNT and how it relates to the scale of the difference between probabilities. Figure 4.24 illustrates the idea. The scale of the difference between probabilities is on the left-hand side, and the scale of the NNT is on the right-hand side. The unusual thing about the NNT scale is that we put the value ∞, which represents the null effect, in the middle and not at either end. In this way, there is a one-to-one relationship between the difference between probabilities and the NNT.

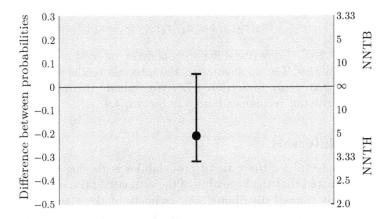

FIGURE 4.24
The correspondence between the estimate of the difference between probabilities and the NNT (with 95% confidence intervals) for the cross-sectional data in Table 4.5

4.7   Confidence Intervals for the Ratio of Probabilities

4.7.1   Introduction and Estimation

The ratio of probabilities is an effect measure of interest in both randomized controlled trials and cohort studies, where we have sampling under the row margins fixed model or the total sum fix model. In cohort studies, the outcome of interest is usually a harmful event, and the ratio of probabilities is often called the *risk ratio* or the *relative risk*. The ratio of probabilities is defined by

$$\phi = \frac{\pi_1}{\pi_2}$$

and estimated by

$$\hat{\phi} = \frac{\hat{\pi}_1}{\hat{\pi}_2} = \frac{n_{11}/n_{1+}}{n_{21}/n_{2+}}. \tag{4.40}$$

As shown in Section 4.5.1, a GLM formulation of the effect of Y_1 (the grouping variable) on Y_2 (the outcome variable) can be described by the regression coefficient

$$\beta = \text{link}(\pi_1) - \text{link}(\pi_2).$$

When the link function is the log link, we get the following relationship between ϕ and the regression coefficient:

$$\beta = \log(\pi_1) - \log(\pi_2) = \log(\pi_1/\pi_2) \quad \Rightarrow \quad \exp(\beta) = \phi.$$

Sections 4.7.2–4.7.7 describe different confidence interval methods for the ratio of probabilities. The application of the intervals to the data examples from Section 4.2 is shown in Section 4.7.8. We evaluate the intervals in Section 4.7.9 and give our recommendations in Section 4.9.

4.7.2   Log Intervals

The simplest interval for the ratio of probabilities is the log interval. Katz et al. (1978) showed that the logarithm of the estimated ratio of probabilities is approximately normal distributed. An estimate of the standard error of $\log(\hat{\phi})$ is thus obtained by the delta method (Agresti, 2013, pp. 72–73), which uses a first-order Taylor expansion of the estimate in Equation 4.40. The *Katz log interval* for ϕ is obtained by exponentiating the endpoints of

$$\log(\hat{\phi}) \pm z_{\alpha/2} \sqrt{\frac{1}{n_{11}} + \frac{1}{n_{21}} - \frac{1}{n_{1+}} - \frac{1}{n_{2+}}}. \tag{4.41}$$

This interval cannot be computed when $n_{11} = 0$ or $n_{21} = 0$. When $n_{11} = n_{1+}$ and $n_{21} = n_{2+}$, the standard error estimate is zero and the interval is $(1, 1)$.

One disadvantage with the Katz log interval is that it cannot be computed when $n_{11} = 0$ or $n_{21} = 0$. A simple solution is to add 1/2 success (a pseudo-frequency) to each group:

$$\hat{\phi}_{1/2} = \frac{(n_{11} + 0.5)/(n_{1+} + 0.5)}{(n_{21} + 0.5)/(n_{2+} + 0.5)}.$$

The adjusted estimate $\hat{\phi}_{1/2}$ is an example of a mesially shrunk estimate: an estimate that is always closer to the null value $\phi = 1$ than the maximum likelihood estimate $\hat{\phi}$. Mesially shrunk estimates are commonly used to avoid problems with zero cell counts for ratio-based effect measures, such as the ratio of probabilities and the odds ratio. One of the first to use a pseudo-frequency method to achieve mesial shift was Haldane (1956), who suggested adding $1/2$ to each cell to estimate odds ratios.

Pettigrew et al. (1986) showed that an estimate of the standard error of $\log(\hat{\phi}_{1/2})$ is obtained by adding $1/2$ to each denominator in the standard error estimate in Equation 4.41. Hence, the *adjusted log interval* for ϕ is given by exponentiating the endpoints of

$$\log(\hat{\phi}_{1/2}) \pm z_{\alpha/2} \sqrt{\frac{1}{n_{11} + 0.5} + \frac{1}{n_{21} + 0.5} - \frac{1}{n_{1+} + 0.5} - \frac{1}{n_{2+} + 0.5}}.$$

Although the adjusted log interval allows for zero events in either group, it produces the zero-width interval $(1, 1)$ when $n_{11} = n_{1+}$ and $n_{21} = n_{2+}$. Furthermore, the adjusted log interval excludes the estimate of the ratio of probabilities ($\hat{\phi}$) in two situations: (i) when $n_{11} = 0$ and $n_{21} \neq 0$, where $\hat{\phi} = 0$ and the lower endpoint (L) is $L > 0$; and (ii) when $n_{11} \neq 0$ and $n_{21} = 0$, where $\hat{\phi} = \infty$ and the upper endpoint is finite.

4.7.3 The Price and Bonett Approximate Bayes Interval

Another method to achieve mesial shrinkage is to use a Bayesian interval (see Section 1.11). Price and Bonett (2008) investigate Bayesian intervals for the ratio of probabilities and model π_1 and π_2 with independent priors having Beta(a, b) distributions. For given values of a and b, a closed-form confidence interval for ϕ is obtained by exponentiating the endpoints of

$$\log(\tilde{\phi}_{a,b}) \pm z_{\alpha/2} \sqrt{\text{var}\left[\log(\tilde{\phi}_{a,b})\right]},$$

where

$$\log(\tilde{\phi}_{a,b}) = \log\left(\frac{n_{11} + a - 1}{n_{1+} + a + b - 2}\right) - \log\left(\frac{n_{21} + a - 1}{n_{2+} + a + b - 2}\right)$$

and

$$\text{var}\left[\log(\tilde{\phi}_{a,b})\right] =$$

$$\frac{1}{n_{11} + a - 1 + \dfrac{(n_{11} + a - 1)^2}{n_{1+} - n_{11} + b - 1}} + \frac{1}{n_{21} + a - 1 + \dfrac{(n_{21} + a - 1)^2}{n_{2+} - n_{21} + b - 1}}$$

Price and Bonett (2008) evaluate the coverage probabilities of intervals resulting from different choices of a and b. They recommend $a = 1.25$ and $b = 2.5$, which we will adopt for the evaluations in Section 4.7.9. As was the case with the adjusted log interval, the Price and Bonett interval excludes the estimate of the ratio of probabilities ($\hat{\phi}$) when $n_{11} = 0$ and $n_{21} \neq 0$ and when $n_{11} \neq 0$ and $n_{21} = 0$ (see the preceding section).

4.7.4 Inverse Hyperbolic Sine Intervals

The inverse hyperbolic sine transformation is defined as

$$\sinh^{-1}(z) = \log\left(z + \sqrt{1 + z^2}\right).$$

The rationale for using it to construct confidence intervals comes from an observation about the logit Wald and Wilson score (Section 2.4.3) confidence intervals for the binomial probability. The logit Wald interval for π is the Wald interval (Section 2.4.1) on $\mathrm{logit}(\pi)$, followed by a transformation back to the original scale. Agresti and Coull (2000) observed that the logit Wald interval always—at least if we ignore cases with zero successes or zero failures—contains the Wilson score interval. Newcombe (2001) then showed that the widths of the two intervals are linked by a simple hyperbolic sine (sinh) relationship. Newcombe further suggested that the inverse sinh transformation could be used to shorten two intervals similar to the logit Wald: the Katz log interval for the ratio of probabilities and the Woolf logit interval for the odds ratio.

The *inverse sinh interval for the ratio of probabilities* is obtained by exponentiating the endpoints of (Newcombe, 2001):

$$\log(\hat{\phi}) \pm 2\sinh^{-1}\left(\frac{z_{\alpha/2}}{2}\sqrt{\frac{1}{n_{11}} + \frac{1}{n_{21}} - \frac{1}{n_{1+}} - \frac{1}{n_{2+}}}\right). \tag{4.42}$$

If $n_{11} = 0$, let the upper limit be $(z_{\alpha/2}^2/n_{1+})/(n_{21}/n_{2+})$. If $n_{21} = 0$, let the lower limit be $(n_{11}/n_{1+})/(z_{\alpha/2}^2/n_{2+})$. If both $n_{11} = n_{1+}$ and $n_{21} = n_{2+}$, the standard error estimate is zero and the interval is $(1, 1)$.

One concern with the inverse sinh interval is that it sometimes provides too little coverage. This concern is more acute for the inverse sinh interval for the odds ratio than for the ratio of probabilities. Fagerland and Newcombe (2013) sought to improve the situation by proposing pseudo-frequency modifications of the inverse sinh intervals. They focused on the odds ratio but considered the ratio of probabilities as well. For the ratio of probabilities, we may add four pseudo-frequencies: ψ_1 to the successes and ψ_2 to the failures in the calculation of the point estimate

$$\tilde{\phi} = \frac{(n_{11} + \psi_1)/(n_{1+} + \psi_1 + \psi_2)}{(n_{21} + \psi_1)/(n_{2+} + \psi_1 + \psi_2)}$$

and ψ_3 to the successes and ψ_4 to the failures in the calculation of the standard error estimate:

$\log(\tilde{\phi})\pm$

$$2\sinh^{-1}\left(\frac{z_{\alpha/2}}{2}\sqrt{\frac{1}{n_{11}+\psi_3}+\frac{1}{n_{21}+\psi_3}-\frac{1}{n_{1+}+\psi_3+\psi_4}-\frac{1}{n_{2+}+\psi_3+\psi_4}}\right).$$

Evoke the substitutions succeeding Equation 4.42 when $\psi_3 = 0$ and either $n_{11} = 0$ or $n_{21} = 0$.

Based on a brief evaluation of coverage probability, width, and location, Fagerland and Newcombe (2013) recommended that $\psi_1 = \psi_3 = 0$, $\psi_4 \approx 1.0$, and ψ_2 can take on any value in the range $[0, 1]$. In the following, we use $\psi_1 = \psi_2 = \psi_3 = 0$ and $\psi_4 = 1$. The resulting interval will be denoted the *adjusted inverse sinh interval*.

4.7.5 The MOVER-R Wilson Interval

Section 4.5.4 described the Newcombe hybrid score interval as the realization of the method of variance estimates recovery (MOVER) for the difference between probabilities. For the ratio of probabilities, we need a slightly more sophisticated method, which we call MOVER-R (MOVER for a ratio). As in Section 4.5.4, we use the Wilson score interval for the binomial parameter to estimate separate confidence intervals for π_1 and π_2 (see Equation 2.5). Let (l_1, u_1) denote the confidence limits of π_1, and let (l_2, u_2) denote the confidence limits of π_2. The MOVER-R Wilson interval (L, U) for ϕ is (Donner and Zou, 2012)

$$L = \frac{\hat{\pi}_1\hat{\pi}_2 - \sqrt{(\hat{\pi}_1\hat{\pi}_2)^2 - l_1 u_2(2\hat{\pi}_1 - l_1)(2\hat{\pi}_2 - u_2)}}{u_2(2\hat{\pi}_2 - u_2)} \quad (4.43)$$

and

$$U = \frac{\hat{\pi}_1\hat{\pi}_2 + \sqrt{(\hat{\pi}_1\hat{\pi}_2)^2 - u_1 l_2(2\hat{\pi}_1 - u_1)(2\hat{\pi}_2 - l_2)}}{l_2(2\hat{\pi}_2 - l_2)}. \quad (4.44)$$

The limits in Equations 4.43 and 4.44 are correct under the assumption that π_1/π_2 is non-negative. If we use MOVER-R in a more general setting, where the aim is to estimate a confidence interval for the ratio of two independent quantities θ_1 and θ_2, θ_1/θ_2 or its limits could be negative. Newcombe (2016) gives the appropriate formulas for this case, with examples of relevant applications.

4.7.6 Asymptotic Score Intervals

Two asymptotic score intervals for the difference between probabilities were presented in Section 4.5.5: one interval due to Mee (1984) and one interval due to Miettinen and Nurminen (1985). The latter article also proposed asymptotic score intervals for the ratio of probabilities and the odds ratio. Under the constraint $\phi_0 = \pi_1/\pi_2$, Miettinen and Nurminen's version of the score statistic for the ratio of probabilities is

$$T_{\text{MN}}(\mathbf{n}\,|\,\phi_0) = \frac{\hat{\pi}_1 - \phi_0\hat{\pi}_2}{\sqrt{\dfrac{\hat{p}_1(1-\hat{p}_1)}{n_{1+}} + \dfrac{\phi_0^2\hat{p}_2(1-\hat{p}_2)}{n_{2+}}}} \cdot \sqrt{1 - \frac{1}{n_{1+}+n_{2+}}}, \quad (4.45)$$

where $\mathbf{n} = \{n_{11}, n_{12}, n_{21}, n_{22}\}$ denotes the observed table and $\hat{p}_1$ and $\hat{p}_2$ are the constrained maximum likelihood estimates of π_1 and π_2, subject to $\phi_0 = \pi_1/\pi_2$. Miettinen and Nurminen (1985) gave closed-form expressions for $\hat{p}_1$ and $\hat{p}_2$:

$$\begin{aligned}
A &= (n_{1+} + n_{2+})\phi_0, \\
B &= -(n_{1+}\phi_0 + n_{11} + n_{2+} + n_{21}\phi_0), \\
C &= n_{+1}, \\
\hat{p}_2 &= (-B - \sqrt{B^2 - 4AC})/2A, \\
\hat{p}_1 &= \hat{p}_2\phi_0.
\end{aligned}$$

We obtain the *Miettinen-Nurminen asymptotic score interval* (L, U) for ϕ by solving the two equations

$$T_{\text{MN}}(\mathbf{n}\,|\,L) = z_{\alpha/2}$$

and

$$T_{\text{MN}}(\mathbf{n}\,|\,U) = -z_{\alpha/2}.$$

As usual with score confidence intervals, we need to use an iterative algorithm to find the confidence limits.

At about the same time as Miettinen and Nurminen published their article on asymptotic score intervals, Koopman (1984) suggested a similar interval for the ratio of probabilities. The Koopman interval is arguably more well known than the Miettinen-Nurminen interval and is available in some standard software packages. The test statistic for the Koopman interval is usually presented as the chi-squared statistic

$$T_{\text{Koopman}}(\mathbf{n}\,|\,\phi_0) = \frac{(n_{11} - n_{1+}\tilde{p}_1)^2}{n_{1+}\tilde{p}_1(1-\tilde{p}_1)} + \frac{(n_{21} - n_{2+}\tilde{p}_2)^2}{n_{2+}\tilde{p}_2(1-\tilde{p}_2)}, \quad (4.46)$$

where

$$\tilde{p}_1 = \frac{\phi_0 \cdot (n_{1+} + n_{21}) + n_{11} + n_{2+}}{2N} - \frac{\sqrt{[\phi_0 \cdot (n_{1+} + n_{21}) + n_{11} + n_{2+}]^2 - 4\phi_0 \cdot N(n_{11} + n_{21})}}{2N}$$

and $\tilde{p}_2 = \tilde{p}_1/\phi_0$ are the maximum likelihood estimates of π_1 and π_2 subject to $\phi_0 = \pi_1/\pi_2$. The confidence limits L and U of the *Koopman asymptotic score interval* for ϕ are the two values that satisfy

$$T_{\text{Koopman}}(\mathbf{n} \,|\, L) = \chi_1^2(\alpha)$$

and

$$T_{\text{Koopman}}(\mathbf{n} \,|\, U) = \chi_1^2(\alpha),$$

such that $L < U$, and $\chi_1^2(\alpha)$ is the upper α percentile of the chi-squared distribution with one degree of freedom. If $n_{11} = 0$, let $L = 0$. If $n_{21} = 0$, let $U = \infty$. The chi-squared statistic in Equation 4.46 equals the traditional Pearson chi-squared statistic (see Section 4.4.2) when $\phi_0 = 1$. The Koopman interval is therefore always consistent with the results of the Pearson chi-squared test.

Although the expressions for the Miettinen-Nurminen and Koopman intervals look quite different, the Koopman interval is identical to the Miettinen-Nurminen interval without the variance correction term in Equation 4.45 (the last square root). A proof of this equality is given in the appendix to Gart and Nam (1988). The correction term is practically negligible for large sample sizes. In small and moderate sample sizes, the Miettinen-Nurminen interval will be slightly wider than the Koopman interval.

By rewriting the expressions for the Koopman interval, Nam (1995) arrived at a cubic equation in $\tilde{p}_1$, which can be solved to obtain closed-form expression for the confidence limits. The expressions are quite elaborate and we do not give them here.

4.7.7 Exact Unconditional Intervals

In this section, we present two exact unconditional intervals: the Chan-Zhang and Agresti-Min intervals. The Chan-Zhang interval uses the tail method and inverts two one-sided exact score tests of size at most $\alpha/2$, whereas the Agresti-Min interval inverts one two-sided exact score test of size at most α. These two intervals are based on the same principles as their namesakes in Section 4.5.6—exact unconditional intervals for the difference between probabilities—only the details differ.

Both the Chan-Zhang and Agresti-Min intervals are based on the score statistic for the ratio of probabilities. This statistic is equal to the one in Equation 4.45 without the variance correction term, and it is the one the Koopman asymptotic score interval would use if we were to formulate that interval in the fashion of Miettinen-Nurminen. With ϕ_0, $\hat{p}_1$, and $\hat{p}_2$ as in the preceding section, the score statistic for the ratio of probabilities is

$$T_{\text{score}}(\mathbf{n} \,|\, \phi_0) = \frac{\hat{\pi}_1 - \phi_0 \hat{\pi}_2}{\sqrt{\dfrac{\hat{p}_1(1 - \hat{p}_1)}{n_{1+}} + \dfrac{\phi_0^2 \hat{p}_2(1 - \hat{p}_2)}{n_{2+}}}}. \tag{4.47}$$

The domain of π_1 given ϕ_0 is

$$D(\phi_0) = \{\pi_1 : 0 \le \pi_1 \le \min(\phi_0, 1)\}. \tag{4.48}$$

Let $\mathbf{x} = \{x_{11}, n_{1+} - x_{11}, x_{21}, n_{2+} - x_{21}\}$ denote any 2×2 table that might be observed given the fixed row sums n_{1+} and n_{2+} (the one margin fixed model). The probability of observing $\mathbf{x}$ is given by the product of two independent binomial probabilities, as in Equation 4.3. Using the constraint $\phi_0 = \pi_1/\pi_2$, we re-parameterize Equation 4.3 such that we use π_1 and ϕ_0 instead of π_1 and π_2:

$$f(\mathbf{x} \,|\, \pi_1, \phi_0, \mathbf{n}_+) =$$

$$\binom{n_{1+}}{x_{11}} \pi_1^{x_{11}} (1 - \pi_1)^{n_{1+} - x_{11}} \binom{n_{2+}}{x_{21}} (\pi_1/\phi_0)^{x_{21}} (1 - \pi_1/\phi_0)^{n_{2+} - x_{21}},$$

where $\mathbf{n}_+ = \{n_{1+}, n_{2+}\}$.

To calculate the interval (L, U) by Chan and Zhang (1999), we solve the two equations

$$\max_{\pi_1 \in D(\phi_0)} \left\{ \sum_{\Omega(\mathbf{x}|\mathbf{n}_+)} I[T(\mathbf{x}\,|\,L) \ge T(\mathbf{n}\,|\,L)] \cdot f(\mathbf{x}\,|\,\pi_1, L, \mathbf{n}_+) \right\} = \alpha/2 \tag{4.49}$$

and

$$\max_{\pi_1 \in D(\phi_0)} \left\{ \sum_{\Omega(\mathbf{x}|\mathbf{n}_+)} I[T(\mathbf{x}\,|\,U) \le T(\mathbf{n}\,|\,U)] \cdot f(\mathbf{x}\,|\,\pi_1, U, \mathbf{n}_+) \right\} = \alpha/2, \tag{4.50}$$

where $\Omega(\mathbf{x}|\mathbf{n}_+)$ denotes the set of all tables with row sums equal to $\mathbf{n}_+$, $I()$ is the indicator function, and $T()$ is the score statistic in Equation 4.47.

The parameter π_1 in Equations 4.49 and 4.50 is a nuisance parameter (see Section 4.4.7), which we have eliminated by taking the maximum value over the range $D(\phi_0)$ given in Equation 4.48.

Iterative computations are required to find L and U. The Berger and Boos procedure (see Section 4.4.7) may be used to reduce the domain of π_1 such that the maximizations in Equations 4.49 and 4.50 are performed over a $100(1-\gamma)\%$ confidence interval for π_1 instead of the entire range. No optimal choice of γ has been suggested for exact unconditional intervals. We use $\gamma = 0.0001$, which is the recommended value for the Suissa-Shuster exact unconditional test (Lydersen et al., 2012b).

For the calculation of the interval by Agresti and Min (2001), we find the confidence limits L and U that satisfy

$$\max_{\pi_1 \in D(\phi_0)} \left\{ \sum_{\Omega(\mathbf{x}|\mathbf{n}_+)} I\big[|T(\mathbf{x}\,|\,L)| \ge |T(\mathbf{n}\,|\,L)|\big] \cdot f(\mathbf{x}\,|\,\pi_1, L, \mathbf{n}_+) \right\} = \alpha \tag{4.51}$$

and

$$\max_{\pi_1 \in D(\phi_0)} \left\{ \sum_{\Omega(\mathbf{x}|\mathbf{n}_+)} I\left[\left|T(\mathbf{x}\,|\,U)\right| \geq \left|T(\mathbf{n}\,|\,U)\right|\right] \cdot f(\mathbf{x}\,|\,\pi_1, U, \mathbf{n}_+) \right\} = \alpha \quad (4.52)$$

The Berger and Boos procedure may be used to reduce the range over which the maximization in Equations 4.51 and 4.52 is performed.

4.7.8   Examples

For reasons set out in Section 4.5.7, it is not appropriate to estimate the ratio of probabilities for the lady tasting a cup of tea (Table 4.2) and the case-control study of the effect of GADA exposure on IPEX syndrome (Table 4.4). We will return to these examples when we illustrate confidence interval estimation for the odds ratio in Section 4.8.9.

A Randomized Clinical Trial of High Versus Standard Dose of Epinephrine (Table 4.3)

The proportion that survived after 24 hours in the standard dose group is $\hat{\pi}_1 = 7/34 = 0.206$, and the proportion that survived after 24 hours in the high dose group is $\hat{\pi}_2 = 1/34 = 0.0294$. Hence, the maximum likelihood estimate of the ratio of probabilities is

$$\hat{\phi} = 7.00.$$

The probability of survival with standard dose treatment is estimated to be 7 times the probability of survival with high dose treatment.

For the Katz log interval, we estimate the standard error of $\log(\hat{\phi})$ as follows:

$$\sqrt{1/7 + 1/1 - 1/34 - 1/34} = 1.041.$$

The log limits are $\log(7.00) \pm z_{\alpha/2} \cdot 1.041$, and a 95% Katz log interval is

$$\exp(-0.0944, 3.99) = (0.91 \text{ to } 53.9).$$

This interval is wide and supports a wide range of possible effects from a small $(\phi = 0.91)$ advantage for the high dose treatment to a large $(\phi = 53.9)$ benefit with the standard dose treatment.

The adjusted log interval uses the mesially shrunk estimate

$$\hat{\phi}_{1/2} = \frac{7.5/34.5}{1.5/34.5} = 5.00.$$

This interval is considerably closer to the null value $(\phi = 1)$ than the maximum likelihood estimate $(\hat{\phi} = 7.00)$. An estimate of the standard error is

$$\sqrt{1/7.5 + 1/1.5 - 1/34.5 - 1/34.5} = 0.861,$$

and the resulting 95% adjusted log interval is

$$\exp(-0.0781, 3.30) = (0.92 \text{ to } 27.1).$$

The lower limits of the Katz log and the adjusted log intervals are similar; however, the upper limit of the adjusted log interval is considerably lower than the upper limit of the Katz log interval.

The calculations of the Price and Bonett approximate Bayes and adjusted inverse sinh intervals follow the same pattern as the calculations of the Katz log and adjusted log intervals: (i) compute point and standard error estimates on the log scale; (ii) compute log limits; and (iii) exponentiate the results. We do not show how to calculate the two intervals here but simply give their results. The 95% Price and Bonett interval is (0.92 to 36.5), which is similar to the two log intervals. The 95% adjusted inverse sinh interval is (1.17 to 42.0), which differs from the other three intervals by having a lower limit above the null value. In the following, we shall see that the adjusted inverse sinh interval is more akin to the MOVER-R Wilson and asymptotic score intervals than the two log intervals and the Price and Bonett interval.

Next, we consider the MOVER-R Wilson interval. A 95% Wilson confidence interval for π_1 is $(l_1, u_1) = (0.104 \text{ to } 0.368)$, and a 95% Wilson confidence interval for π_2 is $(l_2, u_2) = (0.0052 \text{ to } 0.149)$. We insert these limits into Equations 4.43 and 4.44 and obtain (1.15 to 42.0), which is almost the same as the adjusted inverse sinh interval.

The confidence limits of the Miettinen-Nurminen asymptotic score interval are the values of ϕ_0 for which the score statistic in Equation 4.45 is equal to $\pm z_{\alpha/2}$. The idea is illustrated in Figure 4.25, where a logarithmic scale is used for the x-axis. To find L and U, an iterative algorithm, such as the bisection or secant method, is required. The 95% Miettinen-Nurminen interval is (1.21 to 43.0).

The confidence limits of the Koopman asymptotic score interval are found in a similar fashion. Figure 4.26 shows a plot of the chi-squared statistic in Equation 4.46 as function of ϕ_0. This time, the limits are the values of ϕ_0 for which $T_{\text{Koopman}}(\mathbf{n} \,|\, \phi_0)$ equals $\chi_1^2(\alpha)$. The 95% Koopman interval is (1.22 to 42.6). If we instead calculated the Koopman interval using the score statistic in Equation 4.46 without the variance correction term—which is equal to the score statistic used by the exact unconditional intervals (Equation 4.47)—we would obtain a plot almost identical to that in Figure 4.26, but with $L = 1.22$ and $U = 42.6$.

The Chan-Zhang and Agresti-Min exact unconditional intervals add a layer (or two) of complexity to the calculations of the confidence limits. Section 4.5.7 discussed and illustrated the analogous Chan-Zhang and Agresti-Min exact unconditional intervals for the difference between probabilities. The concepts and computational challenges considered therein apply here as well, and we do not go into the specifics.

Table 4.18 summarizes the results of applying nine intervals for the ratio of probabilities to the epinephrine trial data. The log widths of the intervals, that

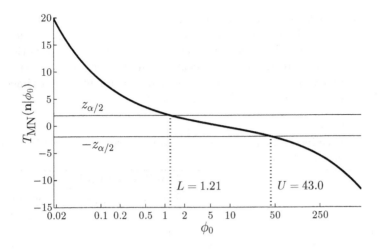

FIGURE 4.25
The Miettinen-Nurminen score statistic (Equation 4.45) as function of the
ratio of probabilities (ϕ_0) for the epinephrine trial data in Table 4.3

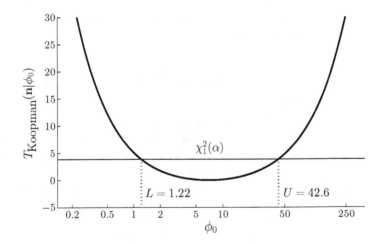

FIGURE 4.26
The Koopman chi-squared statistic (Equation 4.46) as function of the ratio of
probabilities (ϕ_0) for the epinephrine trial data in Table 4.3

is log(Upper) − log(Lower), are shown in the rightmost column. The first three
intervals include the null value ($\phi = 1$), whereas the other intervals do not.
In that regard, the first three intervals indicate greater uncertainty for these
data than do the intervals for the difference between probabilities (Table 4.16).

The adjusted inverse sinh, MOVER-R Wilson, and the two asymptotic score intervals have quite similar confidence limits (and widths). The exact intervals are considerably wider than the other intervals. It is particularly noticeable for the Chan-Zhang interval, which indicates that the data supports that the probability of survival with standard dose treatment might be 181 times the probability of survival with high dose treatment.

TABLE 4.18

95% Confidence intervals for the ratio of probabilities ($\hat{\phi} = 7.00$) based on the epinephrine trial data in Table 4.3

	Confidence limits		
Interval	Lower	Upper	Log width
Katz log	0.910	53.9	4.08
Adjusted log	0.924	27.1	3.38
Price and Bonett approximate Bayes	0.921	36.5	3.68
Adjusted inverse sinh	1.17	42.0	3.59
MOVER-R Wilson	1.15	42.0	3.59
Miettinen-Nurminen asymptotic score	1.21	43.0	3.57
Koopman asymptotic score	1.22	42.6	3.55
Chan-Zhang exact unconditional*	1.22	181	5.00
Agresti-Min exact unconditional*	1.17	90.3	4.34

*Calculated with Berger and Boos procedure ($\gamma = 0.0001$)

A Cross-Sectional Study of the Association between CHRNA4 Genotypes and XFS (Table 4.5)

As discussed in Section 4.5.7, it can be reasonable to analyze this example as if the data were sampled under the row margins fixed model, even though only the total sum N is fixed by the study design. Hence, we compare the probability of exfoliation syndrome in the eyes (XFS) for participants with genotype CHRNA4-CC (π_1) with the probability of XFS for participants with genotype CHRNA4-CC/TT (π_2). The estimates of π_1 and π_2 are $\hat{\pi}_1 = 0/16 = 0$ and $\hat{\pi}_2 = 15/72 = 0.208$, and the maximum likelihood estimate of the ratio of probabilities is $\hat{\phi} = 0$. The fact that none of the participants with genotype CHRNA4-CC has XFS ($n_{11} = 0$) is interesting, because it allows us to examine how the intervals cope with boundary cases.

The calculation of $1/n_{11}$ is required to estimate the standard error for the Katz log interval. When $n_{11} = 0$, we cannot estimate the standard error, and no confidence interval can be computed. The traditional remedy for such cases is to add 1/2 success to each group and calculate the adjusted log interval. The result is the interval (0.009 to 2.25), which does not include $\hat{\phi} = 0$. Similar limits are obtained with the Price and Bonett interval: (0.001 to 3.54). If we use any of these two intervals, we implicitly make the assumption that a value of $\phi = 0$ is not possible, that is, the probability of XFS for persons with genotype CHRNA4-CC cannot be zero. This seems reasonable; it is likely that there is at least one person in the source population (healthy adults 50–75 years in central Norway) with both genotype CHRNA4-CC and XFS. Otherwise, genotype CHRNA4-CC would completely protect against XFS in this population, a proposition that is difficult to advocate.

All the remaining intervals for the ratio of probabilities have a lower limit of zero and a computable (finite) upper limit (Table 4.19). The asymptotic score and exact unconditional intervals have upper limits below the null value ($\phi = 1$), whereas the other intervals include the null value. There is considerable variation in results between the intervals, perhaps so that different conclusions could be reached for different choices of interval method. Both the adjusted log and the Koopman asymptotic score intervals are methods in common use, but they indicate quite different levels of uncertainty.

When we calculated confidence intervals for the difference between probabilities for these data (Table 4.17, Section 4.5.7), we noted the unusual event that the Chan-Zhang interval was shorter than the Agresti-Min interval. This is also the case here, although to a lesser degree.

TABLE 4.19

95% Confidence intervals for the ratio of probabilities ($\hat{\phi} = 0$) based on the cross-sectional data in Table 4.5

Interval	Confidence limits		Log width
	Lower	Upper	
Katz log	0	∞	–
Adjusted log	0.009	2.25	5.53
Price and Bonett approximate Bayes	0.001	3.54	7.84
Adjusted inverse sinh	0	1.15	–
MOVER-R Wilson	0	1.00	–
Miettinen-Nurminen asymptotic score	0	0.971	–
Koopman asymptotic score	0	0.962	–
Chan-Zhang exact unconditional*	0	0.935	–
Agresti-Min exact unconditional*	0	0.974	–

*Calculated with Berger and Boos procedure ($\gamma = 0.0001$)

4.7.9 Evaluation of Intervals

Evaluation Criteria

We use three indices of performance to evaluate confidence intervals for the ratio of probabilities: coverage probability, width, and location. Section 1.4 gave general descriptions of these performance characteristics, and we do not repeat them here. In the following, we show how coverage, width, and location for the ratio of probabilities can be calculated with complete enumeration.

The succeeding expressions are simple modifications of the formulas in Section 4.5.8.

We assume sampling under the row margins fixed model, such that the probability model is two independent binomial samples (Equation 4.3). The exact coverage probability for the ratio of probabilities is

$$\mathrm{CP}(\pi_1, \pi_2, n_{1+}, n_{2+}, \alpha) =$$

$$\sum_{x_{11}=0}^{n_{1+}} \sum_{x_{21}=0}^{n_{2+}} I(L \leq \phi \leq U) \cdot f(\mathbf{x} \mid \pi_1, \pi_2, n_{1+}, n_{2+}), \qquad (4.53)$$

where $I()$ is the indicator function, $L = L(\mathbf{x}, \alpha)$ and $U = U(\mathbf{x}, \alpha)$ are the lower and upper $100(1 - \alpha)\%$ confidence limits of an interval for the table $\mathbf{x} = \{x_{11}, n_{1+} - x_{11}, x_{21}, n_{2+} - x_{21}\}$, $\phi = \pi_1/\pi_2$, and $f()$ is the probability of observing $\mathbf{x}$, given in Equation 4.3. The exact expected width (on the logarithmic scale) is

$$\mathrm{Width}(\pi_1, \pi_2, n_{1+}, n_{2+}, \alpha) =$$

$$\sum_{x_{11}=0}^{n_{1+}} \sum_{x_{21}=0}^{n_{2+}} \left[\log(U) - \log(L) \right] \cdot f(\mathbf{x} \mid \pi_1, \pi_2, n_{1+}, n_{2+}).$$

To calculate the location index MNCP/NCP, we compute $\mathrm{NCP} = 1 - \mathrm{CP}$ with Equation 4.53 and

$$\mathrm{MNCP}(\pi_1, \pi_2, n_{1+}, n_{2+}, \alpha) =$$

$$\sum_{x_{11}=0}^{n_{1+}} \sum_{x_{21}=0}^{n_{2+}} I\left[\log(L) > \log(\phi) \geq 0 \ \text{or} \ \log(U) < \log(\phi) \leq 0 \right] \cdot f(\mathbf{x}),$$

where $f(\mathbf{x})$ is a shorthand for $f(\mathbf{x} \mid \pi_1, \pi_2, n_{1+}, n_{2+})$.

Evaluation of Coverage Probability

Consider the intervals with closed-form expressions: the Katz log, adjusted log, Price and Bonett approximate Bayes, adjusted inverse sinh, and MOVER-R Wilson intervals. Figure 4.27 shows the coverage probabilities of the five intervals as functions of π_1 for the fixed values $n_{1+} = n_{2+} = 40$ and $\phi = 2.5$. The Katz log interval is the most conservative interval for small values of π_1 (for which π_2 is even smaller), but it has coverage slightly below the nominal level for large values of π_1. The adjusted log interval is usually less conservative than the Katz log interval, although the difference between the two intervals is often small. The Price and Bonett approximate Bayes interval performs much like the two log intervals. A more obvious advantage can be gained with the adjusted inverse sinh and MOVER-R Wilson intervals. Both intervals have coverage close to and slightly above the nominal level for most of the parameter space seen in Figure 4.27. For other choices of n_{1+}, n_{2+}, and ϕ, the coverage of the adjusted inverse sinh and MOVER-R Wilson intervals may violate the nominal level to a greater extent than seen here; however, that happens only for small subsets of the parameter space (Fagerland and Newcombe, 2013).

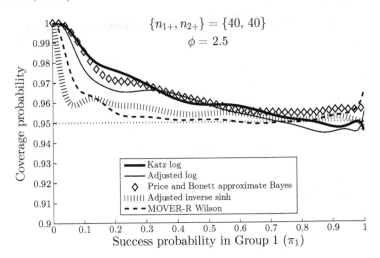

FIGURE 4.27
Coverage probabilities of five confidence intervals of closed-form expression for the ratio of probabilities

Figure 4.28 shows the coverage of the intervals of closed-form expression for row margins equal to $n_{1+} = n_{2+} = 100$. When $\pi_1 > 0.5$, the five intervals perform similarly; however, for small values of π_1, there is still a clear advantage for the adjusted inverse sinh and MOVER-R Wilson intervals. Note in particular the excellent coverage of the adjusted inverse sinh interval for $\pi_1 < 0.1$.

The coverage probabilities of the Miettinen-Nurminen and Koopman asymptotic score intervals are illustrated in Figure 4.29. Recall that the two intervals are nearly equal: the Miettinen-Nurminen statistic equals the Koopman statistic multiplied by a variance correction term that is asymptotically zero. The correction term makes the Miettinen-Nurminen interval slightly wider than the Koopman interval. Hence, the Koopman interval has lower coverage probabilities than the Miettinen-Nurminen interval, which can be clearly observed in Figure 4.29. As was the case with the adjusted inverse sinh and MOVER-R Wilson intervals, the asymptotic score intervals sometimes have coverage considerably lower than the nominal level. This will usually happen for combinations of $\phi \geq 4$ and small values of π_1 and π_2. Notwithstanding, the overall performance of the asymptotic score intervals is very good.

Figure 4.30 illustrates the differences in coverage probabilities between the

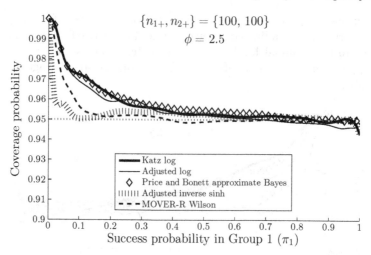

FIGURE 4.28
Coverage probabilities of five confidence intervals of closed-form expression
for the ratio of probabilities

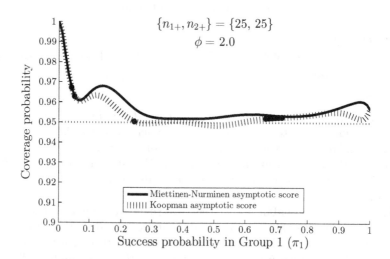

FIGURE 4.29
Coverage probabilities of the Miettinen-Nurminen and Koopman asymptotic
score intervals for the ratio of probabilities

Chan-Zhang and Agresti-Min exact unconditional intervals. These differences
echo those seen for the two corresponding exact unconditional intervals for
the difference between probabilities (see Figure 4.20 and the discussion in

the associated paragraph): the Agresti-Min interval has coverage closer to the nominal level than the Chan-Zhang interval. The Chan-Zhang interval guarantees that the non-coverage in each tail is limited by $\alpha/2$, whereas the Agresti-Min interval only guarantees that the total non-coverage is no more than α. The Agresti-Min interval is thus, in general, shorter than the Chan-Zhang interval, although the opposite may occur for probabilities close to zero or one. This can be seen in Figure 4.30 for $\pi_1 > 0.9$ and for the data from the cross-sectional study of the association between CHRNA4 genotypes and XFS (Table 4.19).

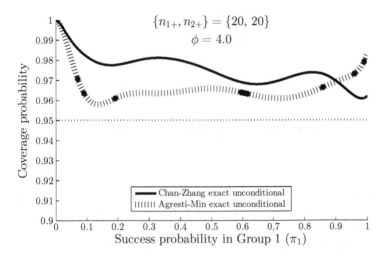

FIGURE 4.30
Coverage probabilities of the Chan-Zhang and Agresti-Min exact unconditional intervals for the ratio of probabilities

The best performing intervals for the ratio of probabilities are compared in Figure 4.31. The overall impression is that the four intervals perform quite similarly. The coverage probability of the Koopman asymptotic score interval is below the nominal level for parts of the parameter space, although not by much. The Agresti-Min exact unconditional interval is slightly more conservative than the other intervals. Both the adjusted inverse sinh interval and the MOVER-R Wilson interval have excellent coverage probabilities, particularly in light of the fact that they have closed-form expressions, whereas the other two intervals require specialized software resources to calculate.

Evaluation of Width

An example of the expected widths—plotted on a logarithmic scale—of seven intervals for the ratio of probabilities is shown in Figure 4.32. Two intervals are not shown: the MOVER-R Wilson and the Miettinen-Nurminen asymptotic

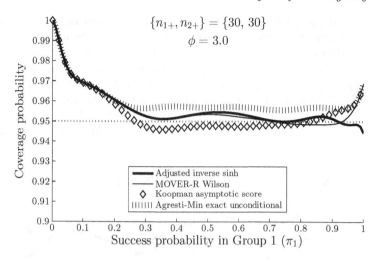

FIGURE 4.31
Coverage probabilities of the best performing intervals for the ratio of probabilities

score intervals, both of which have widths similar to the adjusted inverse sinh and Koopman asymptotic score intervals in this case. For values of $\pi_1 < 0.4$, there can be considerable differences in the widths of the intervals. The two asymptotic score, the adjusted inverse sinh, and the MOVER-R Wilson intervals are noticeably shorter than the other intervals. The Chan-Zhang exact unconditional interval is generally wider than the other intervals, although the adjusted log and the Price and Bonett approximate Bayes intervals are also quite wide when the probabilities are small.

Evaluation of Location

Figures 4.33 and 4.34 show examples of the location index MNCP/NCP for six confidence intervals for the ratio of probabilities. The sample size is the same in the two plots, only the value of ϕ differs. When ϕ is small (Figure 4.33), most of the intervals have satisfactory location ($0.4 \leq$ MNCP/NCP ≤ 0.6); however, the Katz log interval is too mesially located (MNCP/NCP < 0.4). For larger values of ϕ (Figure 4.34), all the intervals except for the Chan-Zhang exact unconditional interval are too mesially located. If we increase the sample size, the location of the intervals improves. The adjusted log and Price-Bonett approximate Bayes intervals—they are not shown in the figures—have slightly more mesial locations than the Katz log interval. The Miettinen-Nurminen asymptotic score interval—also not shown in the figures—has location almost identical to that of the Koopman asymptotic score interval.

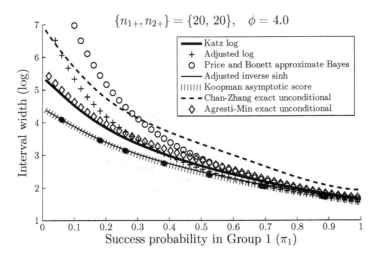

FIGURE 4.32
Expected width of seven confidence intervals for the ratio of probabilities

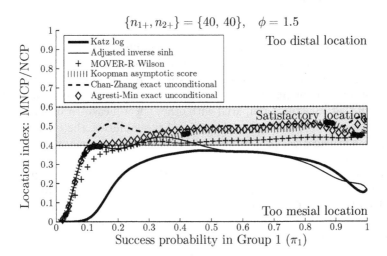

FIGURE 4.33
Location, as measured by the MNCP/NCP index, of six confidence intervals
for the ratio of probabilities

4.8 Confidence Intervals for the Odds Ratio

4.8.1 Introduction and Estimation

The odds ratio is a versatile effect measure that can be used for all the probability models described in Section 4.3. It is the natural effect measure in

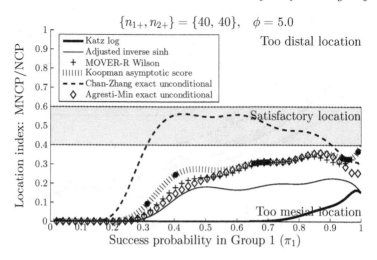

FIGURE 4.34
Location, as measured by the MNCP/NCP index, of six confidence intervals
for the ratio of probabilities

case-control studies. In case-control studies, the difference between probabilities and the ratio of probabilities are unavailable as effect measures. The odds ratio also plays an important role in logistic regression, where the regression coefficient for an explanatory variable equals the log odds ratio for a one-unit increase in that variable. Compared with other effect measures such as the difference between probabilities or the ratio of probabilities, the odds ratio has agreeable mathematical properties and is much used as a summary measure in meta-analysis.

The odds of an event is the probability that the event occurs divided by the probability that it does not occur. Under the row margins fixed model, the odds of success for Group 1 is $\pi_1/(1 - \pi_1)$, and the odds of success for Group 2 is $\pi_2/(1 - \pi_2)$. The ratio of the two odds is the *success odds ratio*, and it is denoted by

$$\theta = \frac{\pi_1/(1 - \pi_1)}{\pi_2/(1 - \pi_2)}$$

and estimated by

$$\hat{\theta} = \frac{n_{11}/n_{12}}{n_{21}/n_{22}} = \frac{n_{11}n_{22}}{n_{12}n_{21}}, \tag{4.54}$$

which is also called the *cross-product*.

In a case-control study, the odds ratio θ is defined as the odds of being exposed for cases divided by the odds of being exposed for the controls. This odds ratio is called the *exposure odds ratio*. The exposure odds ratio equals the success odds ratio, and it is estimated by the cross-product $\hat{\theta}$. This is what makes the odds ratio so useful in case-control studies.

If we use the logit link in the GLM formulation in Equations 4.17 and 4.18, we get the following relationship between θ and the regression coefficient:

$$\begin{aligned}
\beta &= \mathrm{logit}(\pi_1) - \mathrm{logit}(\pi_2) \\
&= \log\left[\pi_1/(1-\pi_1)\right] - \log\left[\pi_2/(1-\pi_2)\right] \\
&= \log\left[\frac{\pi_1/(1-\pi_1)}{\pi_2/(1-\pi_2)}\right] \\
\Rightarrow \quad \exp(\beta) &= \theta.
\end{aligned}$$

4.8.2 Logit Intervals

A confidence interval based on the approximate normal distribution of the logarithm of the odds ratio was first proposed by Woolf (1955) and is often referred to as the *logit interval* or the *Woolf logit interval*. As for the Katz log interval for the ratio of probabilities, we estimate the standard error of $\log(\theta)$ by the delta method (Agresti, 2013, pp. 72–73), which uses a first order Taylor expansion of the estimate in Equation 4.54. The Woolf logit interval for θ is obtained by exponentiating the endpoints of

$$\log(\hat{\theta}) \pm z_{\alpha/2} \sqrt{\frac{1}{n_{11}} + \frac{1}{n_{12}} + \frac{1}{n_{21}} + \frac{1}{n_{22}}}.$$

If one of the cell counts is zero, the standard error will be infinite, and the interval cannot be computed.

Gart (1966) suggested an adjustment to the logit interval: add 0.5 pseudo-frequencies to all cell counts and proceed as above. The resulting confidence interval for θ is obtained by exponentiating the endpoints of

$$\log(\tilde{\theta}) \pm z_{\alpha/2} \sqrt{\frac{1}{\tilde{n}_{11}} + \frac{1}{\tilde{n}_{12}} + \frac{1}{\tilde{n}_{21}} + \frac{1}{\tilde{n}_{22}}},$$

where

$$\tilde{\theta} = \frac{\tilde{n}_{11}\tilde{n}_{22}}{\tilde{n}_{12}\tilde{n}_{21}}$$

and

$$\tilde{n}_{11} = n_{11} + 0.5, \quad \tilde{n}_{12} = n_{12} + 0.5, \quad \tilde{n}_{21} = n_{21} + 0.5, \quad \tilde{n}_{22} = n_{22} + 0.5.$$

The *Gart adjusted logit interval* solves the problem of estimating a standard error for zero cell counts; however, it does not include the estimate of the odds ratio ($\hat{\theta}$) in the following cases: if $n_{11} = 0$ and $n_{21} \neq 0$, the estimate is $\hat{\theta} = 0$, but the lower endpoint is $L > 0$; if $n_{11} = n_{1+}$ and $n_{21} \neq n_{2+}$, or if $n_{11} \neq 0$ and $n_{21} = 0$, the estimate is $\hat{\theta} = \infty$, but the upper endpoint is finite.

The Gart adjustment is a special case of a general class of adjustments.

Suppose that we add a non-negative pseudo-frequency c to each cell count and proceed as above. For $c = 0$ and $c = 0.5$, we obtain the Woolf and Gart adjusted logit intervals, respectively. But, we may also add a different quantity to each cell. Agresti (1999) proposed the following:

$$c_{ij} = 2n_{i+}n_{+j}/N^2, \quad i, j = 1, 2. \tag{4.55}$$

Under the assumption that $\pi_1 = \pi_2$, the expected cell counts are denoted by $m_{ij} = n_{i+}n_{+j}/N$, and we have that $c_{ij} = 2m_{ij}/N$. The adjustment in Equation 4.55 thus results in a mesial shift toward the null value ($\theta = 1$). Agresti calls this smoothing toward independence, and we denote the resulting confidence interval the *independence-smoothed logit interval*. The independence-smoothed logit interval excludes the estimate of the odds ratio for the same situations as does the Gart adjusted logit interval.

4.8.3 Inverse Hyperbolic Sine Intervals

Section 4.7.4 explained the role of the sinh transformation in shortening the widths of delta log and logit intervals. In that section, the inverse sinh interval for the ratio of probabilities was introduced, and it was noted that the interval tends to provide coverage below the nominal level. A solution that involved adding pseudo-frequencies to the observed cell counts was suggested. Here, we present the corresponding inverse sinh and adjusted inverse sinh intervals for the odds ratio.

We obtain the *inverse sinh interval for the odds ratio* (Newcombe, 2001) by exponentiating the endpoints of

$$\log(\hat{\theta}) \pm 2 \sinh^{-1}\left(\frac{z_{\alpha/2}}{2}\sqrt{\frac{1}{n_{11}} + \frac{1}{n_{12}} + \frac{1}{n_{21}} + \frac{1}{n_{22}}}\right). \tag{4.56}$$

If $n_{11} = 0$ or $n_{22} = 0$, let the upper limit be $(z_{\alpha/2}^2 n_{22})/(n_{12}n_{21})$ and $(n_{11}z_{\alpha/2}^2)/(n_{12}n_{21})$, respectively. Correspondingly, if $n_{12} = 0$ or $n_{21} = 0$, let the lower limit be $(n_{11}n_{22})/(z_{\alpha/2}^2 n_{21})$ and $(n_{11}n_{22})/(n_{12}z_{\alpha/2}^2)$, respectively.

The proposed modifications of the inverse sinh interval for the ratio of probabilities (Section 4.7.4) consisted of adding four pseudo-frequencies: one to the successes and one to the failures in the calculation of the point estimate and two others to the successes and failures in the calculation of the standard error estimate. Four pseudo-frequencies were initially needed because the successes and failures affect both the point and standard error estimates of the ratio of probabilities differently. For the odds ratio, however, the successes and failures are interchangeable, and it thus makes sense to add the same pseudo-frequency (ψ_1) to all cells in the calculation of the point estimate and another pseudo-frequency (ψ_2) to all cells in the calculation of the standard error estimate. The adjusted point estimate then is

$$\tilde{\theta} = \frac{(n_{11} + \psi_1)(n_{22} + \psi_1)}{(n_{12} + \psi_1)(n_{21} + \psi_1)},$$

and the adjusted standard error estimate is obtained by adding ψ_2 to each n_{ij} in the standard error estimate given by the square root in Equation 4.56. The *adjusted inverse sinh interval for the odds ratio* (Fagerland and Newcombe, 2013) is given by exponentiating

$$\log(\tilde{\theta}) \pm 2\sinh^{-1}\left(\frac{z_{\alpha/2}}{2}\sqrt{\frac{1}{n_{11}+\psi_2}+\frac{1}{n_{12}+\psi_2}+\frac{1}{n_{21}+\psi_2}+\frac{1}{n_{22}+\psi_2}}\right).$$

If we constrain the pseudo-frequencies to be positive ($\psi_1 > 0$ and $\psi_2 > 0$), the adjusted inverse sinh interval copes with zero cell entries without additional specifications.

To find the optimal values of ψ_1 and ψ_2, Fagerland and Newcombe (2013) carried out a wide-ranging evaluation of coverage probability, width, and location, in which a total of 1600 different combinations of ψ_1 and ψ_2 were compared across a broad range of different scenarios. Because increasing ψ_2 leads to narrower intervals, it is desirable to select as large a ψ_2 as possible; however, the coverage needs to remain at a sufficient level. A good balance was obtained with the two choices $(\psi_1, \psi_2) = (0.45, 0.25)$ and $(\psi_1, \psi_2) = (0.6, 0.4)$.

4.8.4 The MOVER-R Wilson Interval

The expressions for the MOVER-R interval for the ratio of probabilities in Section 4.7.5 can be used to construct a MOVER-R interval for the odds ratio as well. Let (l_1, u_1) and (l_2, u_2) denote the Wilson score intervals for π_1 and π_2, respectively (see Equation 2.5). Further, let $q_i = \pi_i/(1 - \pi_i)$, $i = 1, 2$, be the logit of π_i with estimate given by $\hat{q}_i = \hat{\pi}_i/(1 - \hat{\pi}_i)$. The confidence limits of q_i are $L_i = l_i/(1 - l_i)$ and $U_i = u_i/(1 - u_i)$. Then, the lower (L) and upper (U) limits of the MOVER-R Wilson interval for the odds ratio is

$$L = \frac{\hat{q}_1\hat{q}_2 - \sqrt{(\hat{q}_1\hat{q}_2)^2 - L_1U_2(2\hat{q}_1 - L_1)(2\hat{q}_2 - U_2)}}{U_2(2\hat{q}_2 - U_2)}$$

and

$$U = \frac{\hat{q}_1\hat{q}_2 + \sqrt{(\hat{q}_1\hat{q}_2)^2 - U_1L_2(2\hat{q}_1 - U_1)(2\hat{q}_2 - L_2)}}{L_2(2\hat{q}_2 - L_2)}.$$

4.8.5 Asymptotic Score Intervals

We have previously seen asymptotic score intervals for the difference between probabilities and the ratio of probabilities. Similar asymptotic score intervals can also be derived for the odds ratio. The asymptotic score intervals by Miettinen and Nurminen (1985) use a score statistic that includes a variance correction term. For the odds ratio, the Miettinen-Nurminen score statistic is

$$T_{\mathrm{MN}}(\mathbf{n}\,|\,\theta_0) = \frac{n_{1+}(\hat{\pi}_1 - \hat{p}_1)}{\left[\dfrac{1}{n_{1+}\hat{p}_1(1-\hat{p}_1)} + \dfrac{1}{n_{2+}\hat{p}_2(1-\hat{p}_2)}\right]^{-1/2}} \cdot \sqrt{1 - \frac{1}{n_{1+}+n_{2+}}},$$

where $\mathbf{n} = \{n_{11}, n_{12}, n_{21}, n_{22}\}$, as usual, denotes the observed table and $\hat{p}_1$ and $\hat{p}_2$ are the maximum likelihood estimates of π_1 and π_2 subject to $\theta_0 = [\pi_1/(1-\pi_1)]/[\pi_2/(1-\pi_2)]$. The maximum likelihood estimates can be expressed in closed form as

$$
\begin{aligned}
A &= n_{2+}(\theta_0 - 1),\\
B &= n_{1+}\theta_0 + n_{2+} - n_{+1}(\theta_0 - 1),\\
C &= -n_{+1},\\
\hat{p}_2 &= (-B - \sqrt{B^2 - 4AC})/2A,\\
\hat{p}_1 &= \frac{\hat{p}_2\theta_0}{1 + \hat{p}_2(\theta_0 - 1)}.
\end{aligned}
$$

We obtain the *Miettinen-Nurminen asymptotic score interval* (L, U) for θ by solving the two equations

$$T_{\mathrm{MN}}(\mathbf{n}\,|\,L) = z_{\alpha/2}$$

and

$$T_{\mathrm{MN}}(\mathbf{n}\,|\,U) = -z_{\alpha/2}.$$

An iterative algorithm is needed to find the confidence limits.

We may also derive asymptotic score intervals based on score statistics without the Miettinen-Nurminen variance correction term. For the difference between probabilities, this was the Mee interval, and for the ratio of probabilities, it was the Koopman interval. The corresponding interval for the odds ratio does not have a name, but we shall call it the *uncorrected asymptotic score interval* (for the odds ratio). The score statistic then is

$$T_{\mathrm{score}}(\mathbf{n}\,|\,\theta_0) = \frac{n_{1+}(\hat{\pi}_1 - \hat{p}_1)}{\left[\dfrac{1}{n_{1+}\hat{p}_1(1-\hat{p}_1)} + \dfrac{1}{n_{2+}\hat{p}_2(1-\hat{p}_2)}\right]^{-1/2}}, \qquad (4.57)$$

where $\hat{p}_1$ and $\hat{p}_2$ are calculated as above. The lower (L) and upper (U) limits of the uncorrected asymptotic score interval are the two values that satisfy

$$T_{\mathrm{score}}(\mathbf{n}\,|\,L) = z_{\alpha/2}$$

and

$$T_{\mathrm{score}}(\mathbf{n}\,|\,U) = -z_{\alpha/2}.$$

4.8.6   Exact Conditional Intervals

In Sections 4.5.6 and 4.7.7, we presented exact unconditional intervals for the difference between probabilities and the ratio of probabilities. Exact unconditional methods eliminate the nuisance parameter by maximizing the exact P-value over the range of the nuisance parameter space. Another approach to cope with the nuisance parameter is to condition on a sufficient statistic for the parameter. We call these methods exact conditional methods. The Fisher exact test (see Section 4.4.5) is an example of an exact conditional test. Sufficient statistics exist only for GLMs that use the canonical link, which, for binary data, are log odds models, see Section 1.6 and Agresti (2013, p. 609). Exact conditional intervals can thus be constructed for the odds ratio but not for the difference between probabilities nor the ratio of probabilities. Exact conditional methods are much simpler and computationally easier than exact unconditional methods; however, they are hampered by discreteness and conservatism: exact conditional tests have low power and exact conditional intervals are wide.

Let θ_0 be a fixed value of the odds ratio. If we condition on the total number of successes (n_{+1}) and the total number of failures (n_{+2}), such that all marginal totals in Table 4.1 are fixed, any one table is completely characterized by the count of one cell. The probability of observing a table with x_{11} successes in Group 1 follows the Fisher non-central hypergeometric distribution in Equation 4.1, which we repeat here:

$$f(x_{11} \mid \theta_0, \mathbf{n}_{++}) = \frac{\dbinom{n_{1+}}{x_{11}} \dbinom{n_{2+}}{n_{+1} - x_{11}} \theta_0^{x_{11}}}{\sum\limits_{t=n_0}^{n_1} \dbinom{n_{1+}}{t} \dbinom{n_{2+}}{n_{+1} - t} \theta_0^{t}},$$

where $\mathbf{n}_{++} = \{n_{1+}, n_{2+}, n_{+1}, n_{+2}\}$ denotes the fixed row and column sums, $n_0 = \max(0, n_{+1} - n_{2+})$, and $n_1 = \min(n_{1+}, n_{+1})$. We obtain the *Cornfield exact conditional interval* (L, U) by iteratively solving

$$\sum_{x_{11}=n_{11}}^{n_1} f(x_{11} \mid L, \mathbf{n}_{++}) = \alpha/2 \qquad (4.58)$$

and

$$\sum_{x_{11}=n_0}^{n_{11}} f(x_{11} \mid U, \mathbf{n}_{++}) = \alpha/2, \qquad (4.59)$$

see Cornfield (1956). This procedure corresponds to inverting two one-sided exact conditional tests of the null hypothesis H_0: $\theta = \theta_0$, which, for $\theta_0 = 1$, are Fisher-Irwin exact conditional tests (see Section 4.4.5). The Cornfield exact conditional interval is guaranteed to have coverage probabilities at least to the nominal level.

Instead of inverting two one-sided tests, the interval by Baptista and Pike

(1977) inverts one two-sided test and uses an acceptance region formed by ordered null probabilities, after a method by Sterne (1954). The lower (L) and upper (U) confidence limits of the *Baptista-Pike exact conditional interval* are the solutions to

$$\sum_{x_{11}=n_0}^{n_1} f(x_{11} \mid L, \mathbf{n}_{++}) \cdot I\big[f(x_{11} \mid L, \mathbf{n}_{++}) \leq f(n_{11} \mid L, \mathbf{n}_{++})\big] = \alpha \qquad (4.60)$$

and

$$\sum_{x_{11}=n_0}^{n_1} f(x_{11} \mid U, \mathbf{n}_{++}) \cdot I\big[f(x_{11} \mid U, \mathbf{n}_{++}) \leq f(n_{11} \mid U, \mathbf{n}_{++})\big] = \alpha, \qquad (4.61)$$

where $I()$ is the indicator function and $L < U$.

4.8.7 Mid-*P* Intervals

Mid-*P* versions of both the Cornfield and Baptista-Pike intervals can be constructed by simple alterations of Equations 4.58–4.61. The *Cornfield mid-P interval* is obtained by substituting Equations 4.58 and 4.59 with

$$\sum_{x_{11}=n_{11}}^{n_1} f(x_{11} \mid L, \mathbf{n}_{++}) - \frac{1}{2} f(n_{11} \mid L, \mathbf{n}_{++}) = \alpha/2$$

and

$$\sum_{x_{11}=n_0}^{n_{11}} f(x_{11} \mid U, \mathbf{n}_{++}) - \frac{1}{2} f(n_{11} \mid U, \mathbf{n}_{++}) = \alpha/2.$$

The *Baptista-Pike mid-P interval* is obtained by substituting Equations 4.60 and 4.61 with

$$\sum_{x_{11}=n_0}^{n_1} f(x_{11} \mid L, \mathbf{n}_{++}) \cdot I\big[f(x_{11} \mid L, \mathbf{n}_{++}) \leq f(n_{11} \mid L, \mathbf{n}_{++})\big]$$
$$- \frac{1}{2} f(n_{11} \mid L, \mathbf{n}_{++}) \quad = \quad \alpha$$

and

$$\sum_{x_{11}=n_0}^{n_1} f(x_{11} \mid U, \mathbf{n}_{++}) \cdot I\big[f(x_{11} \mid U, \mathbf{n}_{++}) \leq f(n_{11} \mid U, \mathbf{n}_{++})\big]$$
$$- \frac{1}{2} f(n_{11} \mid U, \mathbf{n}_{++}) = \alpha.$$

4.8.8 Exact Unconditional Intervals

Exact unconditional intervals for the odds ratio have received less attention in the literature than have exact unconditional intervals for the difference between probabilities and the ratio of probabilities. This might be because exact conditional intervals for the odds ratio, such as the Cornfield exact interval, have been available for a long time. Both conditional and unconditional intervals are based on inverting exact tests. Conditional intervals invert conditional tests, whereas unconditional intervals invert unconditional tests. Unconditional intervals are thus based on probability distributions that are less discrete than those of conditional intervals, and they are shorter.

In this section, we present two exact unconditional intervals. Both use the score statistic in Equation 4.57 and invert exact unconditional tests. One interval inverts two one-sided tests of size at most $\alpha/2$ (the tail method) and the other interval inverts one two-sided test of size at most α. We refer to the first interval as the *Chan-Zhang interval* because of its conceptual similarities with the Chan-Zhang interval for the difference between probabilities. The second interval was suggested by Agresti and Min (2002), and we refer to it as the *Agresti-Min interval*.

Let $\mathbf{x} = \{x_{11}, n_{1+} - x_{11}, x_{21}, n_{2+} - x_{21}\}$ denote any 2×2 table that might be observed given the fixed row sums n_{1+} and n_{2+}. Under the restriction $\theta_0 = [\pi_1/(1 - \pi_1)]/[\pi_2/(1 - \pi_2)]$, we have that $\pi_2 = \pi_1/(\pi_1 + \theta_0 - \pi_1\theta_0)$, and the probability of observing $\mathbf{x}$ is given by

$$f(\mathbf{x} \mid \pi_1, \theta_0, \mathbf{n}_+) =$$

$$\binom{n_{1+}}{x_{11}} \pi_1^{x_{11}} (1 - \pi_1)^{n_{1+} - x_{11}} \binom{n_{2+}}{x_{21}} \pi_2^{x_{21}} (1 - \pi_2)^{n_{2+} - x_{21}}.$$

where $\mathbf{n}_+ = \{n_{1+}, n_{2+}\}$. The above equation is a re-parameterization of Equation 4.3.

To calculate the Chan-Zhang interval (L, U), we (iteratively) solve the two equations

$$\max_{0 \leq \pi_1 \leq 1} \left\{ \sum_{\Omega(\mathbf{x}|\mathbf{n}_+)} I\big[T(\mathbf{x} \mid L) \geq T(\mathbf{n} \mid L)\big] \cdot f(\mathbf{x} \mid \pi_1, L, \mathbf{n}_+) \right\} = \alpha/2 \qquad (4.62)$$

and

$$\max_{0 \leq \pi_1 \leq 1} \left\{ \sum_{\Omega(\mathbf{x}|\mathbf{n}_+)} I\big[T(\mathbf{x} \mid U) \leq T(\mathbf{n} \mid U)\big] \cdot f(\mathbf{x} \mid \pi_1, U, \mathbf{n}_+) \right\} = \alpha/2, \qquad (4.63)$$

where $\Omega(\mathbf{x}|\mathbf{n}_+)$ denotes the set of all tables with row sums equal to $\mathbf{n}_+$, $I()$ is the indicator function, and $T()$ is the score statistic in Equation 4.57. As usual, π_1 is a nuisance parameter, which we eliminate by maximization.

The Berger and Boos procedure (see Section 4.4.7) may be used to reduce the domain of π_1 such that the maximizations in Equations 4.62 and 4.63 are performed over a $100(1 - \gamma)\%$ confidence interval for π_1 instead of the entire range. No optimal choice of γ has been suggested for exact unconditional intervals. We use $\gamma = 0.0001$, which is the recommended value for the Suissa-Shuster exact unconditional test (Lydersen et al., 2012b).

To calculate the Agresti-Min interval, we find the two values that satisfy

$$\max_{0 \leq \pi_1 \leq 1} \left\{ \sum_{\Omega(\mathbf{x}|\mathbf{n}_+)} I\Big[\big|T(\mathbf{x}\,|\,L)\big| \geq \big|T(\mathbf{n}\,|\,L)\big|\Big] \cdot f(\mathbf{x}\,|\,\pi_1, L, \mathbf{n}_+) \right\} = \alpha$$

and

$$\max_{0 \leq \pi_1 \leq 1} \left\{ \sum_{\Omega(\mathbf{x}|\mathbf{n}_+)} I\Big[\big|T(\mathbf{x}\,|\,U)\big| \geq \big|T(\mathbf{n}\,|\,U)\big|\Big] \cdot f(\mathbf{x}\,|\,\pi_1, U, \mathbf{n}_+) \right\} = \alpha$$

The range over which the maximization is done may be reduced with the Berger and Boos procedure.

4.8.9   Examples

A Lady Tasting a Cup of Tea (Table 4.2)

We revisit the hypothetical results of Fisher's tea tasting experiment. In Section 4.4.8, we calculated the Fisher exact test and the Fisher mid-P test for these data, and neither test indicated a significant association between Muriel Bristol's guess of whether milk or tea was added first and the actual order they were poured. That result did not come as a surprise, as the sample size is very small. We have not previously in this chapter estimated an effect measure for the association between guessed and poured because the difference between probabilities and the ratio of probabilities are inappropriate for data that do not represent two independent binomials. The odds ratio, however, may be used, and we estimate it as

$$\hat{\theta} = \frac{3 \cdot 3}{1 \cdot 1} = 9.00.$$

The association between guessed and poured is estimated to be quite strong; however, with such a small sample size, the uncertainty is considerable, and we expect wide confidence intervals.

The sampling model for these data is the both margins fixed model, which treats all marginal sums in the 2×2 table as fixed. Exact conditional intervals are then the natural choices, because they are derived under just that assumption. In this case, we may argue that it is incorrect to label these methods as "conditional" methods. The usual meaning of a conditional method is one that is derived by conditioning on quantities (such as row or column margins) that is not fixed by design. But for these data, the conditional methods (the Fisher exact test, Cornfield/Baptista-Pike intervals) are naturally derived from the manner in which the data were collected. Nevertheless, the important issues of discreteness and conservatism remain, and the mid-P approach is—as usual—a good alternative.

The 95% Cornfield exact (conditional) interval for the data in Table 4.2 is (0.212 to 626). This interval is very wide and includes large associations in both directions. The Baptista-Pike interval (0.314 to 306) is about half as wide as the Cornfield interval, but it, too, indicates a wide range of possible effects. Figure 4.35 illustrates how the confidence limits are found by iteratively calculating exact (conditional) tests of the null hypothesis H_0: $\theta = \theta_0$, until the P-value equals $\alpha/2$ (Cornfield) or α (Baptista-Pike).

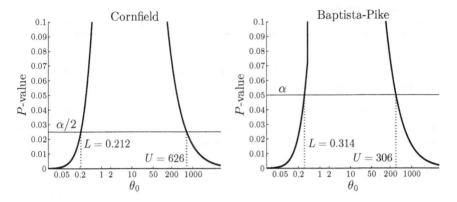

FIGURE 4.35
The exact conditional intervals obtain their confidence limits by finding the two values of θ_0 for which the corresponding exact conditional tests of H_0: $\theta = \theta_0$ have P-values equal to $\alpha/2$ (Cornfield) or α (Baptista-Pike)

Discreteness is a considerable problem with these exact intervals. The sample size is small, and the number of possible tables—whose probabilities make up the P-values—is also small. This leads to coverage probabilities that can be much larger than the nominal level and unnecessary wide intervals. Two alternatives that reduce discreteness are mid-P intervals and exact unconditional intervals. Exact unconditional intervals are not appropriate in this case because they assume sampling under the row margins fixed model (two independent binomial samples). The Cornfield mid-P interval is (0.310 to 309). This improves the Cornfield exact interval but is quite similar to the Baptista-Pike exact interval, which we might as well use. We derive a more obvious benefit from the Baptista-Pike mid-P interval, which is (0.397 to 149). We still have a wide interval, however, it is considerably better than the Cornfield exact interval. Table 4.20 summarizes the results.

The small sample size in this example does not allow for asymptotic intervals to have much credibility. We do not recommend that any asymptotic method is used for such a small sample size, and as such, we do not provide the results. We show how to calculate the asymptotic intervals for the case-control example later in this section.

TABLE 4.20

95% Confidence intervals for the odds ratio ($\hat{\theta} = 9.00$) based on the tea tasting data in Table 4.2

Interval	Confidence limits		Log width
	Lower	Upper	
Cornfield exact conditional	0.212	626	7.99
Baptista-Pike exact conditional	0.314	306	6.88
Cornfield mid-P	0.310	309	6.90
Baptista-Pike mid-P	0.397	149	5.93

A Randomized Clinical Trial of High Versus Standard Dose of Epinephrine (Table 4.3)

An estimate of the odds ratio for the data in Table 4.3 is $\hat{\theta} = 7{\cdot}33/1{\cdot}27 = 8.56$. The odds of survival with standard dose epinephrine is thus estimated to be 8.56 times the odds of survival with high dose epinephrine. (The estimated ratio of probabilities was shown in Section 4.7.8 to be $\hat{\phi} = 7.00$.)

The sample size in this example is greater than that in the tea tasting example above; however, with expected counts equal to four in two cells, the accuracy of the asymptotic intervals is in doubt (see Cochran's criterion on page 100). Therefore, we calculate only exact and mid-P intervals for these data and return to the asymptotic intervals in the next example.

The results of two exact conditional intervals, two mid-P intervals, and two exact unconditional intervals are shown in Table 4.21. There is considerable variation between the widest interval, which is the Cornfield exact interval of (0.973 to 397), and the shortest interval, which is the Baptista-Pike mid-P interval of (1.33 to 98.8). The Baptista-Pike mid-P and Agresti-Min exact unconditional intervals are similar, as are the Cornfield mid-P and the Chan-Zhang exact unconditional intervals. This is typical for the mid-P approach. A mid-P interval based on an exact conditional interval that inverts two one-sided tests will be similar to an exact unconditional interval that inverts two one-sided tests. Analogously, a mid-P interval based on an exact conditional interval that inverts one two-sided test will be similar to an exact unconditional interval that inverts one two-sided test. Also note that the intervals that invert two one-sided tests (Cornfield and Chan-Zhang intervals) have upper limits about twice the upper limits of the intervals that invert one two-sided tests (Baptista-Pike and Agresti-Min intervals). A similar result can be observed for the Chan-Zhang and Agresti-Min exact unconditional intervals for the ratio of probabilities on the same data (Table 4.18). For the difference between probabilities (Table 4.16), the discrepancies between the Chan-Zhang and Agresti-Min exact unconditional intervals are much smaller.

TABLE 4.21

95% Confidence intervals for the odds ratio ($\hat{\theta} = 8.56$) based on the epinephrine trial data in Table 4.3

Interval	Confidence limits		Log width
	Lower	**Upper**	
Cornfield exact conditional	0.973	397	6.01
Baptista-Pike exact conditional	1.00	195	5.28
Cornfield mid-P	1.19	200	5.12
Baptista-Pike mid-P	1.33	98.8	4.31
Chan-Zhang exact unconditional*	1.19	219	5.22
Agresti-Min exact unconditional*	1.18	108	4.52

*Calculated with Berger and Boos procedure ($\gamma = 0.0001$)

A Case-Control Study of the Effect of GADA Exposure on IPEX Syndrome (Table 4.4)

We have previously considered tests for association for this example but have not estimated an effect size nor a confidence interval. The estimated odds ratio is now $\hat{\theta} = 9 \cdot 10/(4 \cdot 4) = 5.63$. In Section 4.4.8, we calculated the expected cell counts under the null hypothesis of no association between cases/controls and GADA exposure. Although the sample size is small, the distribution of cell counts is only slightly unbalanced and all expected counts are larger than five. We thereby expect asymptotic intervals to perform adequately.

The unadjusted standard error estimate of $\log(\theta)$ is

$$\sqrt{1/9 + 1/4 + 1/4 + 1/10} = 0.843.$$

The log limits of the Woolf logit interval are $\log(5.63) \pm z_{\alpha/2} \cdot 0.843$, and a 95% confidence interval for θ is

$$\exp(0.0758, 3.38) = (1.08 \text{ to } 29.4).$$

This interval is much narrower than those for the epinephrine trial data (Table 4.21), even though the total sample size of the case-control study ($N = 27$) is less than half the sample size of the epinephrine trial ($N = 68$). The high degree of uncertainty in the epinephrine trial data is due to the small number of surviving patients ($n_{21} = 1$) in the high dose group. Increase this cell count by one and observe that the width of the confidence intervals decrease considerably.

To compute the Gart adjusted logit interval, we add 0.5 to all cell counts. We then obtain a mesially shifted estimate—an estimate closer to the null value—of the odds ratio: $\tilde{\theta} = 9.5 \cdot 10.5/4.5 \cdot 4.5 = 4.93$. The standard error estimate is now

$$\sqrt{1/9.5 + 1/4.5 + 1/4.5 + 1/10.5} = 0.803,$$

which is noticeably smaller than the standard error estimate of the Woolf logit interval. The resulting 95% Gart adjusted logit interval is

$$\exp(0.0215, 3.17) = (1.02 \text{ to } 23.8).$$

This interval is shorter than the Woolf logit interval but also more mesially located; its lower limit is almost equal to the null value ($\theta = 1.0$).

Another mesially shifted interval for the odds ratio is the independence-smoothed logit interval. We now add c_{ij} to cell n_{ij}, where $i, j = 1, 2$ and c_{ij} is given in Equation 4.55. We have that $c_{11} = 0.464$, $c_{12} = 0.499$, $c_{21} = 0.499$, and $c_{22} = 0.538$. For these data, all c_{ij} are close to 0.5, thus we expect the independence-smoothed logit interval to be similar to the Gart adjusted logit interval. In fact, both the two lower limits and the two upper limits are equal to four significant digits.

The adjusted inverse sinh intervals with $(\psi_1, \psi_2) = (0.45, 0.25)$ and $(\psi_1, \psi_2) = (0.6, 0.4)$ give 95% confidence intervals of (1.14 to 21.8) and (1.12 to 20.6), respectively. We do not show all the computational details but mention that these intervals, too, are mesially shifted with adjusted estimates of $\tilde{\theta} = 4.99$ and $\tilde{\theta} = 4.81$, which are similar to the adjusted estimates of the Gart and the independence-smoothed logit intervals, which both are $\tilde{\theta} = 4.93$.

The example sections for the difference between probabilities (Section 4.5.7) and the ratio of probabilities (Section 4.7.8) have shown how to calculate MOVER and asymptotic score intervals. The approaches are the same for the corresponding intervals for the odds ratio, and we do not go into details here. Table 4.22 shows a summary of the results of eight asymptotic intervals and the Baptista-Pike mid-P interval. Note the similar limits of the MOVER-R Wilson, asymptotic score, and Baptista-Pike mid-P intervals.

TABLE 4.22
95% Confidence intervals for the odds ratio ($\hat{\theta} = 5.63$) based on the case-control data in Table 4.4

Interval	Confidence limits		
	Lower	Upper	Log width
Woolf logit	1.08	29.4	3.31
Gart adjusted logit	1.02	23.8	3.15
Independence-smoothed logit	1.02	23.8	3.15
Adjusted inverse sinh (0.45, 0.25)	1.14	21.8	2.95
Adjusted inverse sinh (0.6, 0.4)	1.12	20.6	2.91
MOVER-R Wilson	1.17	27.1	3.15
Miettinen-Nurminen asymptotic score	1.09	28.9	3.28
Uncorrected asymptotic score	1.13	28.1	3.22
Baptista-Pike mid-P	1.15	29.1	3.23

A Cross-Sectional Study of the Association between CHRNA4 Genotypes and XFS (Table 4.5)

One of the interesting features of this example is the zero ($n_{11} = 0$) in the upper left cell of Table 4.5. This zero cell count results in an estimated odds ratio of zero and provides a good illustration of how the different intervals perform for data on the border of the sample space. The results are shown in Table 4.23.

TABLE 4.23

95% Confidence intervals for the odds ratio ($\hat{\theta} = 0$) based on the cross-sectional data in Table 4.5

| | Confidence limits | | |
Interval	Lower	Upper	Log width
Woolf logit	0	∞	–
Gart adjusted logit	0.0064	1.98	5.74
Independence-smoothed logit	0.000005	39.5	15.8
Adjusted inverse sinh (0.45, 0.25)	0.0057	1.82	5.77
Adjusted inverse sinh (0.6, 0.4)	0.011	1.61	4.98
MOVER-R Wilson	0	1.01	–
Miettinen-Nurminen asymptotic score	0	0.965	–
Uncorrected asymptotic score	0	0.953	–
Cornfield exact conditional	0	1.14	–
Baptista-Pike exact conditional	0	1.09	–
Cornfield mid-P	0	0.886	–
Baptista-Pike mid-P	0	0.868	–
Chan-Zhang exact unconditional*	0	0.948	–
Agresti-Min exact unconditional*	0	1.07	–

*Calculated with Berger and Boos procedure ($\gamma = 0.0001$)

The estimate of the standard error of the Woolf logit interval is infinite, and the interval cannot be computed. The mesially shifted intervals, which add pseudo-frequencies to the observed cell counts, avoid the problem of an infinite standard error estimate; however, the resulting intervals do not contain $\hat{\theta} = 0$. As discussed on page 145, it is reasonable to assume that the source population contains at least one person with genotype CHRNA4-CC and XFS; hence, $\theta > 0$. For these data, we thereby argue that a lower limit > 0 is acceptable.

The upper limit of the independence-smoothed logit interval is very large compared with the upper limits of the other intervals. This is because the quantity c_{11}, which is added to $n_{11} = 0$, is only 0.062, which results in a high standard error estimate and a wide interval.

All the other intervals give similar results, with a lower limit of zero and an upper limit slightly above or slightly below the null value ($\theta = 1$). Note that the Baptista-Pike mid-P interval is the shortest interval and that the Chan-Zhang exact unconditional interval is shorter than the Agresti-Min exact unconditional interval. The latter observation is an exception to the maxim that an exact interval that inverts one two-sided test is shorter than the exact interval that inverts two corresponding one-sided tests.

4.8.10   Evaluation of Intervals

Evaluation Criteria

As in previous sections, we use three indices of performance to evaluate confidence intervals: coverage probability, width, and location. We refer the reader to Section 1.4 for general descriptions of these performance characteristics. In the following, we show how coverage, width, and location for the odds ratio can be calculated with complete enumeration. The succeeding expressions are simple modifications of the formulas in Section 4.5.8.

We assume sampling under the row margins fixed model, such that the probability model is two independent binomial samples (Equation 4.3). The exact coverage probability for the odds ratio is

$$\text{CP}(\pi_1, \pi_2, n_{1+}, n_{2+}, \alpha) =$$

$$\sum_{x_{11}=0}^{n_{1+}} \sum_{x_{21}=0}^{n_{2+}} I(L \leq \theta \leq U) \cdot f(\mathbf{x} \mid \pi_1, \pi_2, n_{1+}, n_{2+}), \qquad (4.64)$$

where $I()$ is the indicator function, $L = L(\mathbf{x}, \alpha)$ and $U = U(\mathbf{x}, \alpha)$ are the lower and upper $100(1 - \alpha)\%$ confidence limits of an interval for the table $\mathbf{x} = \{x_{11}, n_{1+} - x_{11}, x_{21}, n_{2+} - x_{21}\}$, $\theta = [\pi_1/(1 - \pi_1)]/[\pi_2/(1 - \pi_2)]$, and $f()$ is the probability of observing $\mathbf{x}$, given in Equation 4.3. The exact expected width (on the logarithmic scale) is

$$\text{Width}(\pi_1, \pi_2, n_{1+}, n_{2+}, \alpha) =$$

$$\sum_{x_{11}=0}^{n_{1+}} \sum_{x_{21}=0}^{n_{2+}} \left[\log(U) - \log(L) \right] \cdot f(\mathbf{x} \mid \pi_1, \pi_2, n_{1+}, n_{2+}).$$

To calculate the location index MNCP/NCP, we compute $\text{NCP} = 1 - \text{CP}$ using Equation 4.64 and

$$\text{MNCP}(\pi_1, \pi_2, n_{1+}, n_{2+}, \alpha) =$$

$$\sum_{x_{11}=0}^{n_{1+}} \sum_{x_{21}=0}^{n_{2+}} I\left[\log(L) > \log(\theta) \geq 0 \text{ or } \log(U) < \log(\theta) \leq 0 \right] \cdot f(\mathbf{x}),$$

where $f(\mathbf{x})$ is a shorthand for $f(\mathbf{x} \mid \pi_1, \pi_2, n_{1+}, n_{2+})$.

Evaluation of Coverage Probability

Figure 4.36 shows the coverage probabilities of five intervals of closed-form expression for the sample sizes $n_{1+} = n_{2+} = 30$ and a fixed odds ratio of $\theta = 5.0$.

The Woolf logit interval—the prevailing interval for odds ratios in standard software packages—has coverage well above the nominal level for most values of π_1, and the independence-smoothed logit interval performs similarly. A greater improvement on the Woolf interval is obtained with the Gart adjusted logit interval, which has very good coverage for probabilities close to zero or one. Yet, the best performing intervals of closed-form expression are the adjusted inverse sinh and MOVER-R Wilson intervals. The adjusted inverse sinh interval performs particularly well in this example, with coverage close to the nominal level for almost all values of π_1. Although not shown here, the adjusted inverse sinh and MOVER-R Wilson interval sometimes have coverage probabilities considerably below the nominal level; however, this happens only for small subsets of the parameter space. Figure 4.36 does not show the coverage probabilities of the adjusted inverse sinh interval with $(\psi_1, \psi_2) = (0.45, 0.25)$. This interval has coverage probabilities slightly above the shown adjusted inverse sinh interval with $(\psi_1, \psi_2) = (0.6, 0.4)$ and slightly below the Gart adjusted logit interval.

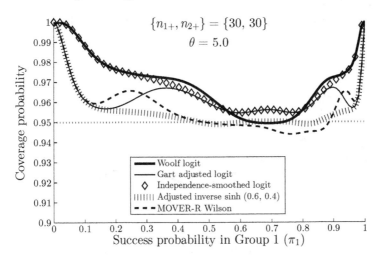

FIGURE 4.36
Coverage probabilities of five confidence intervals of closed-form expression for the odds ratio

Figure 4.37 illustrates how the intervals of closed-form expression perform when we increase the sample size to $n_{1+} = n_{2+} = 100$. There are still interesting differences between the intervals, although they have similar coverage probabilities when π_1 is in the range 0.4–0.8.

An example of the coverage probabilities of the two asymptotic score intervals is shown in Figure 4.38. As expected, the uncorrected asymptotic score interval has coverage probabilities slightly lower than the Miettinen-Nurminen asymptotic score interval; however, the overall impression is that the two inter-

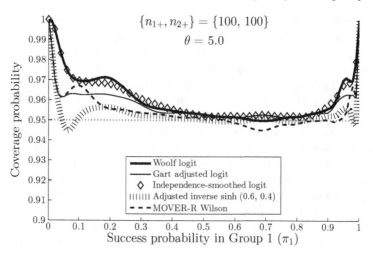

FIGURE 4.37
Coverage probabilities of five confidence intervals of closed-form expression for the odds ratio

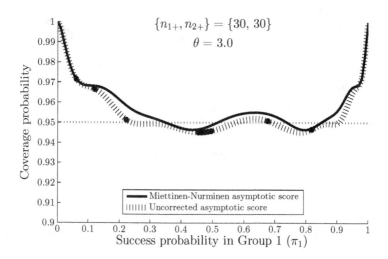

FIGURE 4.38
Coverage probabilities of the Miettinen-Nurminen and uncorrected asymptotic score intervals for the odds ratio

vals perform quite similarly. This is consistent with previous observations for the difference between probabilities (Figure 4.19) and the ratio of probabilities (Figure 4.29).

For most combinations of sample sizes and odds ratios, the two asymptotic score intervals have coverage probabilities much like those in Figure 4.38, where the coverage probabilities fluctuate around the nominal level with small deflections. Sometimes, however, the coverage probabilities of the asymptotic score intervals dip substantially below the nominal level. An example is seen in Figure 4.39, where the only change in parameter settings from the previous figure is that $\theta = 4.0$ instead of $\theta = 3.0$. Fortunately, the occurrence of such low coverage probabilities is limited to cases with small probabilities of success. For $\theta = 4.0$ and $\pi_1 = 0.05$, $\pi_2 = 0.013$, and with a sample size of 30 in each group, the expected numbers of successes are 1.6 in Group 1 and 0.39 in Group 2. It is unlikely that many studies with that low number of expected events are carried out. This goes to show that the problem of low coverage of the asymptotic score intervals might not be of any important practical concern.

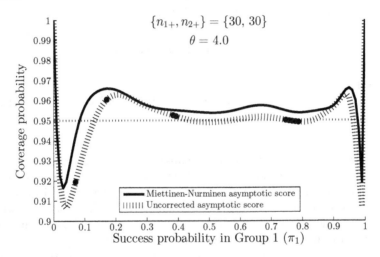

FIGURE 4.39
Coverage probabilities of the Miettinen-Nurminen and uncorrected asymptotic score intervals for the odds ratio

We now turn to the two exact conditional intervals and their associated mid-P versions. In the example section (Section 4.8.9), we observed that the Cornfield exact conditional interval was very wide, that the Cornfield mid-P interval was similar to the Baptista-Pike exact conditional interval, and that the Baptista-Pike mid-P interval was considerably shorter than the other three intervals. The same pattern of results is seen in Figure 4.40, which contains coverage probabilities of the four intervals. The Baptista-Pike mid-P interval often performs similarly to the asymptotic score intervals; however, the Baptista-Pike interval does not exhibit low dips in coverage, as is the occasional case with the asymptotic score intervals. The Baptista-Pike mid-P in-

terval sometimes has coverage probabilities below the nominal level, although
the infringements are generally small.

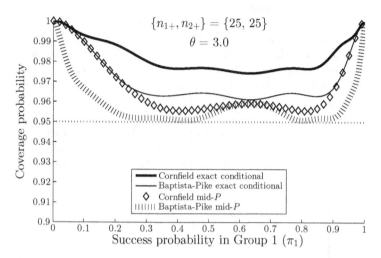

FIGURE 4.40
Coverage probabilities of the Cornfield and Baptista-Pike exact conditional
and mid-P intervals for the odds ratio

The coverage probabilities of the Chan-Zhang and Agresti-Min exact unconditional intervals are illustrated in Figure 4.41. The differences between the
two intervals resemble those for the corresponding exact unconditional intervals for the ratio of probabilities: the coverage probabilities of the Agresti-Min
interval are considerably closer to the nominal level than those of the Chan-
Zhang interval. Infrequently, however, the opposite occurs.

Figure 4.42 shows how the best performing intervals for the odds ratio
perform head-to-head. The two intervals of closed-form expression—the adjusted inverse sinh and MOVER-R Wilson intervals—do not perform as well
as the other three intervals; however, considering their simplicity and ease of
computation, they perform more than satisfactorily. The three other intervals
perform excellently, and in particular the Agresti-min exact unconditional interval, which is the only interval in Figure 4.42 that always has coverage at
least to the nominal level. Unfortunately, as far as we know, the Agresti-Min
interval for the odds ratio is not available in any standard software package.

Evaluation of Width

An example of the expected width of a selection of six confidence intervals
for the odds ratio is shown in Figure 4.43. For small probabilities ($\pi_1 <$
0.2), the uncorrected asymptotic score and the MOVER-R Wilson interval
is shorter than the other intervals; however, for larger values of π_1, it is the

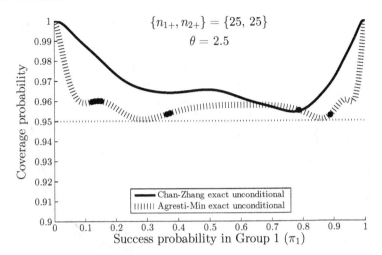

FIGURE 4.41
Coverage probabilities of the Chan-Zhang and Agresti-Min exact unconditional intervals for the odds ratio

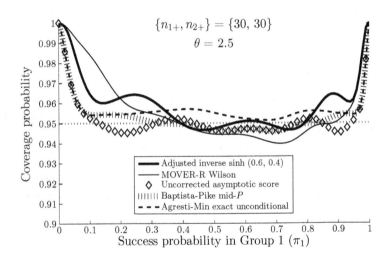

FIGURE 4.42
Coverage probabilities of the best performing intervals for the odds ratio

adjusted inverse sinh interval that is the narrowest one. The Woolf logit and
Agresti-Min exact unconditional intervals are the widest intervals for most
situations, although Figure 4.43 does not show the width of the Chan-Zhang

exact unconditional interval, which is considerably wider than all the other intervals.

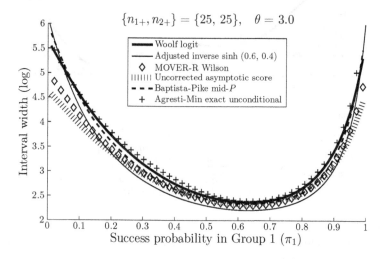

FIGURE 4.43
Expected interval width of six confidence intervals for the odds ratio

Evaluation of Location

Figure 4.44 illustrates the location properties of six confidence intervals for the odds ratio. For small and moderate values of the odds ratio, most of the intervals have MNCP/NCP values within the satisfactory range. The adjusted inverse sinh interval, which uses the most mesially shifted odds ratio estimate of all the intervals, is, unsurprisingly, too mesially located. As was the case with the intervals for the ratio of probabilities, the effect of increasing the effect size is that the intervals become more mesially located. Too distal location is seldom a problem, although the Chan-Zhang exact unconditional interval (not shown in Figure 4.44) sometimes has MNCP/NCP values above 0.6.

4.9   Recommendations

4.9.1   Summary

Literally hundreds of statistical methods exist for the analysis of 2×2 tables. This chapter has considered a subset of methods, selected for their common usage, consistency with general theory, computational simplicity, availability in software packages, and good performance in published evaluations. The

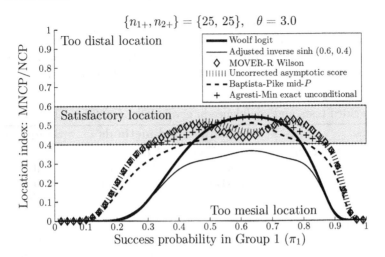

FIGURE 4.44
Location, as measured by the MNCP/NCP index, of six confidence intervals for the odds ratio

ideal method has strong theoretical grounds; it is simple to describe, simple to calculate, available in every standard software package, and has excellent properties. Moreover, the ideal method is based on general principles that can be used to construct tests and confidence intervals across different outcomes and analyses. With these criteria in mind, we make the following general observations:

- Exact unconditional methods are superior to exact conditional methods

- Mid-P versions of exact conditional methods perform similarly to exact unconditional methods

- Asymptotic score intervals perform well

- MOVER intervals (including the Newcombe hybrid score interval) perform well

- Traditional asymptotic methods need a large sample size to perform well

- Adding pseudo-frequencies to the observed data (and other techniques to achieve mesial shift) can make traditional asymptotic methods perform well for small and moderate sample sizes

Table 4.24 provides a summary of the recommended tests and confidence intervals, and gives the sample sizes for which the recommended methods are appropriate. The labels small, medium, and large cannot be given precise definitions, they will vary from one analysis to the other, and some subjectivity needs to be applied. As a rule of thumb, small may be taken as less than 50 in each group, medium as between 50 and 200 in each group, and large as 200 or more in each group. Sections 4.9.2–4.9.6 discuss the recommendations in more detail and summarize the merits of the different methods.

TABLE 4.24
Recommended tests and confidence intervals (CIs) for 2 × 2 tables

Analysis	Recommended methods	Sample sizes
Tests for association	Fisher mid-P*	all
	Suissa-Shuster exact unconditional[†]	small/medium
	Fisher-Boschloo exact uncond.[†]	small/medium
	Pearson chi-squared*	large
CIs for difference between probabilities	Agresti-Min exact unconditional[†]	small/medium
	Agresti-Caffo*	medium/large
	Newcombe hybrid score*	medium/large
	Miettinen-Nurminen asympt. score	medium/large
	Wald*	large
CIs for number needed to treat	The reciprocals of the limits of the recommended intervals for the difference between probabilities	
CIs for ratio of probabilities	Adjusted inverse sinh*	all
	MOVER-R Wilson*	all
	Koopman asymptotic score	all
	Agresti-Min exact unconditional[†]	small/medium
	Katz log*	large
CIs for odds ratio	Adjusted inverse sinh*	all
	MOVER-R Wilson*	all
	Baptista-Pike mid-P	all
	Agresti-Min exact unconditional[†]	small/medium
	Woolf logit*	large

*These methods have closed-form expression
[†]Preferably with the Berger and Boos procedure ($\gamma = 0.0001$)

4.9.2 Tests for Association

For small sample sizes, none of the asymptotic tests perform well. The Fisher exact test is the commonly used alternative; however, its actual significance levels are usually very low, often at about half the nominal level, and hence, unnecessarily low power. A better option is to use an exact unconditional test. These tests have actual significance levels much closer to the nominal level and has the nominal level as an upper bound. When calculating exact unconditional tests, we recommend use of the Berger and Boos procedure, which has the effect of reducing the computational effort and increasing the power. A value of $\gamma = 0.0001$ in the Berger and Boos procedure is recommended for significance levels $\alpha = 0.01$ or higher (Lydersen et al., 2012b). The best test statistics to use in an exact unconditional test are the Pearson chi-squared statistic and the Fisher exact test, which leads to the Suissa-Shuster and Fisher-Boschloo exact unconditional tests, respectively. This recommendation is in line with the recommendations in Mehrotra et al. (2003) and Lydersen et al. (2009).

The exact unconditional tests require complex calculations and their support in software packages is limited. An excellent alternative is the Fisher mid-P test. It is quite easy to calculate and performs much like an exact unconditional test, except that it sometimes violates the nominal significance level. Since it does not require much computational effort, we recommend it for all sample sizes.

For large sample sizes, asymptotic tests can be used. The Cochran's criterion (see page 100) is a good rule of thumb to decide when the sample size is sufficiently large. We have presented and evaluated four different asymptotic tests but only one can be recommended: the Pearson chi-squared test. Both the likelihood ratio and Z-unpooled tests may violate the nominal significance level by a large amount, even for moderately large sample sizes (Figure 4.8). The Pearson chi-squared test with continuity correction is even more conservative than the Fisher exact test, and we find no reason ever to use it.

One situation we have not considered in this chapter is that of extremely unbalanced designs, where one of the row margins is much greater than the other. Kang and Ahn (2008) report that in such cases, exact conditional tests can be more powerful than exact unconditional tests; however, exact unconditional tests that use the Berger and Boos procedure are slightly more powerful than exact conditional tests.

4.9.3 Confidence Intervals for the Difference between Probabilities

The Agresti-Min exact unconditional interval is an excellent option for small sample sizes. It can be a bit conservative; however, its conservativeness is less than we usually expect from an exact method. If the interest is to bound the overall non-coverage and not the non-coverage in both tails, the Agresti-Min interval is superior to the Chan-Zhang exact unconditional interval, which is considerably more conservative. As with exact unconditional tests, we recommend the Berger and Boos procedure, for instance, with $\gamma = 0.0001$. One disadvantage with exact unconditional intervals is that they are computationally complex and require specialized software to be calculated.

Two simple intervals that also perform well are the Agresti-Caffo and Newcombe hybrid score intervals. The Agresti-Caffo interval is particularly easy to calculate, we only need to add one success and one failure to both groups before calculating the Wald interval on the modified data. The Newcombe interval is slightly more elaborate but does not require complex computations and is quite well supported in standard software packages, such as R, SAS, and Stata (see Table 8 in Fagerland et al. (2015)). We recommend both intervals for moderate and large sample sizes. They can also be used for small sample sizes should an exact unconditional interval be unavailable.

The Miettinen-Nurminen asymptotic score interval is our last recommended interval for the difference between probabilities. We prefer it ahead of the Mee asymptotic score interval because the coverage of the Mee interval is more frequently below the nominal level than the coverage of the Miettinen-Nurminen interval. An iterative algorithm is required to find the confidence limits of the asymptotic score intervals. Unfortunately, these intervals are poorly supported in standard software packages, which hampers their widespread use.

We end this section with some notes on the two Wald intervals. The Wald interval with continuity correction is always very conservative, even for large sample sizes, such as 500 in each group. We do not recommend it for any situation. Without continuity correction, the Wald interval has coverage below the nominal level. For sample sizes with more than 100 in each group, this difference is small; however, it is still clearly visible in a plot of coverage probability as in Figure 4.18 but with $n_{1+} = n_{2+} = 500$. Nevertheless, we can recommend the Wald interval (without continuity correction) for large sample sizes, say, 200 or more in each group.

4.9.4   Confidence Intervals for the Number Needed to Treat

The calculation of a confidence interval for the number needed to treat is based on the confidence limits for the associated difference between probabilities. Hence, the recommended intervals for the difference between probabilities (Table 4.24) apply for the number needed to treat as well. The Newcombe hybrid score interval deserves particular attention: it is of closed-form expression, has good support in software packages, and was the recommended interval in Bender (2001) for the explicit purpose of constructing a confidence interval for the number needed to treat.

4.9.5   Confidence Intervals for the Ratio of Probabilities

Two intervals of closed-form expression perform well in most situations: the adjusted inverse sinh and the MOVER-R Wilson intervals. One particular advantage with the adjusted inverse sinh interval is that the coverage of the interval is close to the nominal level even for small probabilities, when most other intervals are quite conservative. Both the adjusted inverse sinh and MOVER-R Wilson intervals can have coverage considerably below the nominal level when ϕ is large, but this is also a problem with other intervals, such as the more complex Miettinen-Nurminen and Koopman asymptotic score intervals. The two asymptotic score intervals perform well and can be recommended for general use. We rank the Koopman interval ahead of the Miettinen-Nurminen interval because the Koopman interval is slightly shorter. Moreover, the Koopman interval is more frequently implemented in standard software packages.

If an exact interval is needed, we recommend the Agresti-Min exact unconditional interval, which is considerably shorter than the Chan-Zhang exact unconditional interval. The Chan-Zhang interval should only be used if it is important to control the non-coverage in each tail. For both the Agresti-Min and Chan-Zhang intervals, we recommend use of the Berger and Boos procedure, for instance, with $\gamma = 0.0001$.

The Katz log interval is usually quite conservative; it has coverage probabilities further from the nominal level than the other intervals, except for the Chan-Zhang exact unconditional interval. This difference is reduced with increasing sample size, and at $n_{1+} = n_{2+} = 200$, the difference is small and of limited consequence.

4.9.6   Confidence Intervals for the Odds Ratio

There are several good all-round confidence intervals for the odds ratio: the adjusted inverse sinh interval with $\psi_1 = 0.6$ and $\psi_2 = 0.4$, the MOVER-R Wilson interval, the two asymptotic score intervals, and the Baptista-Pike mid-P interval. Among these, the Baptista-Pike mid-P interval performs best. Even though it does not guarantee coverage to the nominal level, the infringements of the nominal level are smaller for the Baptista-Pike mid-P interval than for the other four intervals. The only drawback of the Baptista-Pike mid-P interval is that it requires iterative computations. It is not widely supported in software packages; however, a Stata version exists (Fagerland, 2012). When the Baptista-Pike mid-P interval is not available, the simple adjusted inverse sinh and MOVER-R Wilson intervals are good alternatives. Both intervals are of closed-form expression and can be calculated without the aid of specialized software resources. The two asymptotic score intervals perform much like the adjusted inverse sinh and MOVER-R Wilson intervals but require iterative computations and are not widely available in software packages.

Exact intervals for the odds ratio are available in both conditional and unconditional varieties. The most commonly used exact interval is the Cornfield exact conditional interval—a standard method in many software packages, such as R, SAS, and Stata. It is, however, very conservative and can give wide and almost uninformative intervals for small sample sizes, which is exactly the kind of situations in which exact intervals are most often used. The similar Baptista-Pike exact conditional interval is an improvement, but the Agresti-Min exact unconditional interval is, undoubtedly, the superior exact interval. Unfortunately, the Agresti-Min interval is not available in any standard software package. The Baptista-Pike exact conditional interval is available in Stata (Fagerland, 2012).

The Woolf logit interval, which is a standard—and often only available—confidence interval for the odds ratio in most software packages, is conservative for most parameter space points, with occasional dips in coverage below the nominal level. It needs a large sample size to perform as well as the best performing intervals. When $n_{1+} = n_{2+} = 200$, the Woolf logit interval has coverage probabilities close to the nominal level, except for some situations involving small success probabilities, for which it is still somewhat conservative.

5

The Ordered r × 2 Table

5.1 Introduction

In Chapter 4, we considered the 2×2 table for unpaired data, where the columns represent the outcomes, denoted as success and failure, and the rows comprise a dichotomous grouping of the data. In this chapter, we extend this setup to include r groups, and our interest lies in the association between the outcome and the classification into the r groups. Thus, an $r \times 2$ table gives the observed counts for the cross-classification of r groups and two outcomes, as shown in Table 5.1. Often, $r \times 2$ tables are analyzed without assuming that the groups are ordered; this situation will be covered in Chapter 7, which deals with the general case of $r \times c$ tables.

Ordered $r \times 2$ tables may occur when we simply have constraints that imply an ordering of the probabilities of the outcome of interest—usually called success—over the r groups. In that case, a trend means that the grouping is consistent with an ordering from the lowest to the highest probability of success, or the other way around, from the highest to the lowest probability of success. Sometimes, it is useful to assign scores to the groups. These scores are either strictly increasing or strictly decreasing with the row numbers.

The content of this chapter is as follows. We introduce two examples of ordered $r \times 2$ tables in Section 5.2. Section 5.3 deals with notation and statistical models, and Section 5.4 gives relevant study designs and sampling distributions. A brief discussion on assigning scores to the row categories is provided in Section 5.5. Section 5.6 considers the general case of tests for unspecific ordering, whereas Section 5.7 considers the more commonly used tests for trend in the linear and logit models. Confidence interval estimation for the trend is the topic of Section 5.8. Finally, we summarize the results and provide recommendations in Section 5.9.

TABLE 5.1

The observed counts of an $r \times 2$ table

	Score	Success	Failure	Total
Group 1	a_1	n_{11}	n_{12}	n_{1+}
Group 2	a_2	n_{21}	n_{22}	n_{2+}
$\vdots$	$\vdots$	$\vdots$	$\vdots$	$\vdots$
Group r	a_r	n_{r1}	n_{r2}	n_{r+}
Total		n_{+1}	n_{+2}	N

Additional notation:

$\mathbf{n} = \{n_{11}, n_{12}, n_{21}, n_{22}, \ldots, n_{r1}, n_{r2}\}$: the observed table
$\mathbf{x} = \{x_{11}, x_{12}, x_{21}, x_{22}, \ldots, x_{r1}, x_{r2}\}$: any possible table

5.2   Examples

5.2.1   Alcohol Consumption and Malformations

The potential risks of drinking during pregnancy are well documented. In a study reported in Graubard and Korn (1987) and Mills and Graubard (1987), the authors asked the following research question: at what level of consumption does the risk of congenital anomalies begin to increase. The study population consisted of more than 32 000 women.

A long list of malformations are reported in Table 4 in Mills and Graubard (1987). Here, we consider malformations of the sex organs (yes/no), which we have cross-classified with five categories of alcohol consumption in Table 5.2. The rows are ordered with increasing number of drinks per day. This table has also been analyzed by Agresti (2013, p. 89) and in the manual of the StatXact software (StatXact 11, 2015, p. 726).

TABLE 5.2

Malformations of the sex organs by alcohol consumption during pregnancy (Mills and Graubard, 1987)

Number of drinks per day	Malformation Yes	No	Total
0	48 (0.3%)	17066 (99.7%)	17114
< 1	38 (0.3%)	14464 (99.7%)	14502
1–2	5 (0.6%)	788 (99.4%)	793
3–5	1 (0.8%)	126 (99.2%)	127
> 5	1 (2.6%)	37 (97.4%)	38
Total	93 (0.3%)	32481 (99.7%)	32574

5.2.2 Elevated Troponin T Levels in Stroke Patients

In a study from the stroke unit at the University Hospital of Trondheim, Norway, 489 patients were included and followed up with assessment of 16 prespecified complications (Indredavik et al., 2008). During the first week, 312 of the 489 patients experienced one or more of the 16 complications listed in Table 1 in Indredavik et al. (2008). The severity of stroke at admission is considered to be the most important risk factor for complications. At admission, the severity of the stroke was assessed by the Scandinavian Stroke Scale (SSS). SSS was graded into five categories, from very mild to very severe.

One of the 16 complications assessed during the first week of follow-up was elevated troponin T levels (> 0.06 mmol/L), a marker for myocardial damage, without other criteria necessary for the diagnosis of acute myocardial infarction (Indredavik et al., 2008, Table 1). Table 5.3 shows the cross-classification of stroke severity with elevated troponin T levels, based on the information in Table 5 in Indredavik et al. (2008).

TABLE 5.3
Elevated troponin T levels by stroke severity (Indredavik et al., 2008)

Stroke severity	SSS	Elevated troponin T Yes	No	Total
Very severe	0–14	8 (13%)	53 (87%)	61
Severe	15–29	10 (17%)	48 (83%)	58
Moderate	30–44	11 (10%)	100 (90%)	111
Mild	45–51	22 (18%)	102 (82%)	124
Very mild	52–59	6 (4%)	129 (96%)	135
Total		57 (12%)	432 (88%)	489

5.3 Notation and Statistical Models

The notation for ordered $r \times 2$ tables follows the notation for the general two-way $r \times c$ table (Table 1.2), except for the naming of the categories. Here, as in Section 1.3, Variable 1 (Y_1) equals the grouping, and Variable 2 (Y_2) equals the outcome. Let the two possible outcomes be success ($Y_2 = 1$) and failure ($Y_2 = 0$). For group number i and observation number k, we have that $Y_{2k} = 1$ if the outcome is a success, with $n_{i1} = \sum_{k=1}^{n_{i+}} Y_{2k}$. The grouping variable Y_1 is categorical with r categories, and $Y_1 = i$ indicates that the observed outcome belongs to Group i.

Let the probability of success for Group i $(i = 1, 2, \ldots, r)$ be

$$\Pr(Y_2 = 1 \mid Y_1 = i) = \pi_i,$$

and let the probability of failure for Group i be

$$\Pr(Y_2 = 0 \mid Y_1 = i) = 1 - \pi_i.$$

Note that these are conditional probabilities, which we from time to time in this book also denote by $\pi_{1|i} = \pi_i$. In this chapter, we will only use the notation π_i.

The null hypothesis of interest is

$$H_0 : \pi_1 = \pi_2 = \ldots = \pi_r. \tag{5.1}$$

The alternative hypothesis reflects a trend (or an ordering) in the probabilities, either

$$H_A : \pi_1 \leq \pi_2 \leq \ldots \leq \pi_r,$$

or

$$H_A : \pi_1 \geq \pi_2 \geq \ldots \geq \pi_r,$$

with at least one strict inequality.

5.3.1 Models When We Have No Information on the Grouping

With no ordering of the grouping variable Y_1, we will typically use the methods for testing hypotheses of no association and the various measures of association from Chapter 4 (2×2 tables) and Chapter 7 (general $r \times c$ tables). In this chapter, we will use the linear and logit links to model the probability of the binary outcome. Without using the information on the scores, we can generalize the notation in Equation 4.17 to the case with r categories, rather than just two. The general link function is given by

$$\text{link}\big[\Pr(Y_2 = 1 \mid Y_1 = i)\big] = \text{link}(\pi_i) = \alpha + \beta_i, \tag{5.2}$$

for $i = 1, 2, \ldots, r$, where we for simplicity assume that $\beta_1 = 0$. Here, β_i is the effect of the grouping.

Throughout this chapter, we consider two link functions: the linear and the logit. The linear link is simply given as

$$\pi_i = \alpha_{\text{linear}} + \beta_{i,\text{linear}},$$

which gives

$$\beta_{i,\text{linear}} = \pi_i - \pi_1.$$

Thus, for the linear link, the $\beta_{i,\text{linear}}$ are simply the differences in probabilities, which we discussed in Section 4.5 for the 2×2 table.

The logit link is given by

$$\log\left(\frac{\pi_i}{1 - \pi_i}\right) = \alpha_{\text{logit}} + \beta_{i,\text{logit}},$$

which gives

$$\beta_{i,\text{logit}} = \log\left[\frac{\pi_i/(1 - \pi_i)}{\pi_1/(1 - \pi_1)}\right].$$

Thus, $\beta_{i,\text{logit}}$ is the log odds for success in Group i relative to Group 1, and $\exp(\beta_{i,\text{logit}})$ is the odds ratio that we studied in Section 4.8 for the 2×2 table.

5.3.2 Models with Scores Assigned to the Groups

Assume now that we have assigned scores $a_1, a_2, \ldots, a_r$ with an ordering $a_1 < a_2 < \cdots < a_r$ to the grouping variable Y_1. The grouping may, for instance, represent exposure categories, grouped according to severity, or it may represent increasing doses of a drug treatment. Equally spaced scores, i.e., $a_i = i$ for $i = 1, 2, \ldots, r$ are common, see Breslow and Day (1980) or Senn (2007). Sections 5.5 and 5.7.14 discuss the issue of assigning scores in more detail.

When scores are assigned to the groups, we are interested in the trend in the probability of success, as a function of the scores. We are interested in tests of no trend, and in the estimation of the trend itself. The most commonly used trend test is the *Cochran-Armitage test*, which has a related trend estimate and confidence interval. We will describe the Cochran-Armitage test in Section 5.7.2 and the related trend estimate and confidence interval in Section 5.8.1.

With scores assigned to the groups, a relevant link function is

$$\text{link}\big[\Pr(Y_2 = 1 \mid Y_1 = i)\big] = \text{link}(\pi_i) = \alpha + \beta a_i, \tag{5.3}$$

for $i = 1, 2, \ldots, r$. The regression coefficient β is defined as the trend, and the link function defines the scale on to which the trend is interpreted. The linear link is given by

$$\pi_i = \alpha_{\text{linear}} + \beta_{\text{linear}} a_i, \tag{5.4}$$

and the logit link is given by

$$\text{logit}(\pi_i) = \alpha_{\text{logit}} + \beta_{\text{logit}} a_i. \tag{5.5}$$

5.3.3 Trend Analysis in Case-Control Studies

Trend analysis for case-control studies is frequently used in epidemiology. Note that in Table 5.1, the cases take the place of the successes, and the controls take the place of the failures. In a case-control design, we have to redefine the problem in terms of probabilities:

$$\Pr(Y_1 = i \mid Y_2 = j) = \gamma_{i|j}$$

for $i = 1, 2, \ldots, r$ and $j = 0, 1$, where $Y_2 = 1$ denotes the cases, and $Y_2 = 0$ denotes the controls. As noted in Buonaccorsi et al. (2014), the $\gamma_{i|j}$ can be expressed by the π_i in the following way:

$$\gamma_{i|1} = \frac{\pi_i \Pr(Y_1 = i)}{\sum_k \pi_k \Pr(Y_1 = i)} \tag{5.6}$$

and

$$\gamma_{i|0} = \frac{(1 - \pi_i)\Pr(Y_1 = i)}{\sum_k (1 - \pi_k)\Pr(Y_1 = i)}, \tag{5.7}$$

which are the conditional probabilities for being in (exposure) Group i conditional on being a case (Equation 5.6) or a control (Equation 5.7). The null hypothesis $H_0 : \pi_1 = \pi_2 = \ldots = \pi_r$ is equivalent to $H_0 : \gamma_{i|0} = \gamma_{i|1}$ for all i. Thus, with case-control data, the null hypothesis of equal π_i can be tested.

Note that for a case-control design with the logit link function in Equation 5.5, the logit can—by straightforward insertion of Equations 5.6 and 5.7 into Equation 5.5—be expressed as

$$\log\left(\frac{\gamma_{i|1}}{\gamma_{i|0}}\right) = \alpha^*_{\text{logit}} + \beta_{\text{logit}} a_i,$$

where the constant α^* equals α + an additional constant (Hosmer et al., 2013, p. 232). Thus, for the logit link, the trend is invariant with respect to the sampling design (random, stratified, or case-control, see Section 5.4).

5.3.4 Choice of Link Function

There is no general consensus about which link function to use. Note that once the link function is chosen, there is an effect measure connected to that link function. The link functions studied in this chapter will all give models that are within the family of generalized linear models, see McCullagh and Nelder (1989). For the binomial distribution, the logit link is the canonical link function, which is an advantage, as we shall see when it comes to the analysis of $r \times 2$ tables, see also Section 1.5. Social scientists, see, for instance Wooldridge (2013, pp. 283–243) and Hellevik (2009), or epidemiologists, see, for instance, Rothman et al. (2008), often prefer linear links or log links, rather than logit links. The interpretation of the effect measures is easier for the linear than for the logit link. It is also argued that the linear link is more in line with causal interpretations, see Rothman et al. (2008) or Hellevik (2009).

As mentioned in Section 5.3.3, the trend with logit link is invariant with respect to the sampling design. This is one major reason why the logit link is preferred to the linear link, see also the seminal articles by Cornfield (1956) and Breslow (1996).

Here, we will discuss both the linear and logit link functions, but in our final recommendations (Section 5.9), we will conclude that the logit link is the most appropriate link function for testing and estimating trend.

5.4 Designs and Sampling Distributions

As discussed in Section 4.2 for the 2×2 table, there are three sampling models of special interest: the row margins fixed model (with fixed group sizes, see Section 4.2.2), the column margins fixed model (with fixed numbers of successes and failures, see Section 4.2.3), and the total sum fixed model (Section 4.2.4). A summary is given in Table 5.4.

TABLE 5.4
Designs in $r \times 2$ tables

Design	Specification
Random sample	Only N fixed
Fixed group sizes or stratification on Y_1	$n_{1+}, n_{2+}, \ldots, n_{r+}$ fixed
Case-control or stratification on Y_2	n_{+1} and n_{+2} fixed

A total sum fixed design occurs, for instance, in cross-sectional studies (for an example, see Section 4.2.4). Fixed group sizes may occur either if the group sizes are set a priori, like in clinical studies of say r doses of a specific medicine, or by post hoc stratification on the variable Y_1. In the latter case, we may have a total sum fixed design, but the statistical analysis is done for the strata, formed by post hoc stratification on Y_1. The case with n_{+1} and n_{+2} fixed occurs in case-control studies. The cases ($Y_2 = 1$) are selected and compared with the controls ($Y_2 = 0$) with respect to the grouping variable, which may be exposure status.

The sampling distributions for $r \times 2$ tables are given as straightforward extensions of those given in Section 4.3 for 2×2 tables. For the case of fixed group sizes $n_{1+}, n_{2+}, \ldots, n_{r+}$, the sampling distribution is given as

$$f(\mathbf{x} \mid \pi_1, \ldots, \pi_r; \mathbf{n}_+) = \prod_{i=1}^{r} \binom{n_{i+}}{x_{i1}} \pi_i^{x_{i1}} (1 - \pi_i)^{n_{i+} - x_{i1}}, \qquad (5.8)$$

where $\mathbf{x}$ denotes any possible table with fixed group sizes $\mathbf{n}_+ = \{n_{1+}, n_{2+}, \ldots, n_{r+}\}$.

Exact conditional tests for the null hypothesis $H_0 : \pi_1 = \pi_2 = \ldots \pi_r$ are relevant for $r \times 2$ tables. Thus, we need the conditional distribution under the null hypothesis, given the marginal row and column sums $\mathbf{n}_{++} = \{n_{1+}, n_{2+}, \ldots, n_{r+}; n_{+1}, n_{+2}\}$, which is

$$f(\mathbf{x} \mid \mathbf{n}_{++}) = \frac{\prod_{i=1}^{r} \binom{n_{i+}}{x_{i1}}}{\binom{N}{n_{+1}}}. \tag{5.9}$$

This is the multiple hypergeometric distribution, a generalization of the hypergeometric distribution given in Equation 4.2. The distribution in Equation 5.9 is free of unknown parameters and can be used to calculate exact conditional P-values. Note that the conditional distribution is always valid, regardless of the design.

Our main interest in ordered $r \times 2$ tables lies in the estimation of the trend and testing of trend over the rows. Now, assume the logit model in Equation 5.5. Here, α is a nuisance parameter, and the total number of successes, $T_\alpha = n_{+1} = \sum_{i=1}^{r} n_{i1}$ is a sufficient statistic for α. Then it is appropriate to use the principle of conditional maximum likelihood estimation, which was pioneered by Cox and Reid (1987) for the case of nuisance parameters. The conditional distribution of $x_{11}, x_{21}, \ldots, x_{r1}$ given $T_\alpha = n_{+1}$, i.e., given the total number of successes, equals

$$f(\mathbf{x}, \beta \mid n_{+1}) = \frac{\exp\left(\beta \sum_{i=1}^{r} x_{i1} a_i\right)}{\sum_{\Omega(\mathbf{x}\mid n_{+1})} \exp\left(\beta \sum_{i=1}^{r} x_{i1} a_i\right)}, \tag{5.10}$$

where $\Omega(\mathbf{x}\mid n_{+1}) = \{\mathbf{x} : \sum_{i=1}^{r} x_{i1} = n_{+1}\}$, i.e., the sum in the denominator in Equation 5.10 is over all samples $\{x_{11}, x_{21}, \ldots, x_{r1}\}$ with the total number of successes equal to n_{+1}. Then, $T_\beta = \sum_{i=1}^{r} n_{i1} a_i$ is a sufficient statistic for β. The conditional distribution of T_β given $T_\alpha = n_{+1}$ is

$$f(t, \beta \mid n_{+1}) = \frac{c(t, n_{+1}) \exp(\beta t)}{\sum_u c(u, n_{+1}) \exp(\beta u)}, \tag{5.11}$$

In Equation 5.11, $c(t, n_{+1})$ in the numerator is the sum of $\prod_{i=1}^{r} \binom{n_{i+}}{x_{i1}}$ for all tables in $\Omega(\mathbf{x}\mid n_{+1})$ with $T_\beta = t$. The denominator is a normalizing constant, where the summation is over all tables in $\Omega(\mathbf{x}\mid n_{+1})$. See also Agresti (2013, p. 267) for derivations of the related conditional distribution in logistic regression.

5.5 Assigning Scores

The trend tests are generally defined with increasing scores assigned to the r groups. The scores should be assigned a priori; however, it is not always clear how the scores should be assigned, and it has been pointed out by many that different sets of scores may lead to different conclusions, see Breslow and Day (1980), Graubard and Korn (1987), Ivanova and Berger (2001), Senn (2007), and Zheng (2008). In Section 5.7.14, we confirm this in the analysis of our examples.

When there is no indication what the scores might be, there are arguments for equally spaced integer scores. The approach is simple, and the use of more complex non-integer scores may be inferior to the straightforward approach, see Senn (2007). It was, however, pointed out by Ivanova and Berger (2001), that the use of equally spaced integer scores may entail loss of power, and uniformly more powerful tests can be obtained with only slightly different scores. Zheng (2008) suggests procedures to remedy the problem of choosing scores. He suggests two robust test statistics, called MAX and BASE. MAX is the maximum of the test statistic over a predefined set of scores. BASE measures how often the test statistic is significant at a given significance level for the predefined set of scores. Any such non-standard procedure for assigning scores will never fully remedy the problem of scores, and our recommendation is therefore to follow the advice of Graubard and Korn (1987) and Senn (2007) and use equally spaced scores, unless there is information about the categories that clearly advises us not to.

5.6 Tests for Unspecific Ordering

Testing for *unspecific ordering*—also called testing for ordered alternatives—was proposed by Bartholomew (1959a,b), and later described in the seminal textbook by Barlow et al. (1972). Agresti and Coull (2002) also give an excellent review of the analysis of contingency tables with inequality constraints. Here, we will use a similar notation as in Agresti and Coull (2002).

5.6.1 The Null and Alternative Hypotheses

Assume that our model is given in Equation 5.2. The null hypothesis of no association between the outcome and the grouping is then

$$H_0 : \beta_1 = \beta_2 = \ldots = \beta_r = 0. \tag{5.12}$$

If there were no ordering of the categories in the r groups, the alternative hypothesis would be H_A: at least one $\beta_i \neq 0$ for at least one i. When testing for ordering, however, the alternative hypothesis of interest is

$$H_A : \beta_1 \leq \beta_2 \leq \ldots \leq \beta_r$$

or

$$H_A : \beta_1 \geq \beta_2 \geq \ldots \geq \beta_r,$$

with at least one strict inequality. Note that this hypothesis setup is more general—and less commonly used—than the one for testing trend in linear and logit models (Section 5.7).

5.6.2   The Pearson Chi-Squared Test for Unspecific Ordering

Under the null hypothesis in Equation 5.12—or the equivalent null hypothesis in Equation 5.1—the maximum likelihood estimate of π_i is n_{+1}/N for all $i = 1, 2, \ldots, r$. Let the sample proportions be $p_i = n_{i1}/n_{i+}$. If the sample proportions are ordered in accordance with the alternative hypothesis $p_1 \leq p_2 \leq \ldots \leq p_r$ or $p_1 \geq p_2 \geq \ldots \geq p_r$, the maximum likelihood estimates of the π_i (under the alternative hypothesis) are equal to the sample proportions $\hat{\pi}_i = p_i = n_{i1}/n_{i+}$. If any pair of p_i and p_{i+1} is out of order according to the alternative hypothesis, we pool the adjacent sample proportions: $p_i = p_{i+1} = (n_{i1} + n_{i+1,1})/(n_{i+} + n_{i+1,+})$. Continue pooling out-of-order proportions until all the proportions are monotonically increasing (or decreasing). Then, the maximum likelihood estimates of the π_i under the alternative are the pooled estimates for adjacent categories with proportions out of order, denoted by $\hat{\pi}_i$.

Bartholomew (1959a) proposed to use the standard Pearson test statistic with pooled estimates for proportions out of order. Let $n_{i1}^* = n_{i+}\hat{\pi}_i$ and $n_{i2}^* = n_{i+}(1 - \hat{\pi}_i)$. Then, the Pearson test statistic is

$$T_{\text{Pearson}}(\mathbf{n}) = \sum_{i,j} \frac{(n_{ij}^* - m_{ij})^2}{m_{ij}}, \tag{5.13}$$

where $m_{ij} = n_{i+}n_{+j}/N$ are the expected cell counts under H_0. T_{Pearson} has an asymptotic distribution that is a mixture of independent chi-squared variables. This distribution is called the *chi-bar-squared distribution*. The P-value for the Pearson chi-squared test is then given by

$$P\text{-value} = \sum_{i=1}^{r} \rho(r, i) \cdot \Pr\left[\chi_{i-1}^2 \geq T_{\text{Pearson}}(\mathbf{n})\right]. \tag{5.14}$$

Here, $\chi_0^2 = 0$, and the $\rho(r, i)$ are probabilities. Barlow et al. (1972, Corollary B, p. 145) give a recursive algorithm to calculate the probabilities $\rho(r, i)$:

$$\rho(r, 1) = \frac{1}{r},$$

$$\rho(r, r) = \frac{1}{r!},$$

$$\rho(r, i) = \frac{1}{r}\rho(r - 1, i - 1) + \frac{r - 1}{r}\rho(r - 1, i), \quad i = 2, \ldots, r - 1.$$

The probabilities for $r \le 8$ are shown in Table 5.5. A larger table with probabilities for $r \le 12$ is given in Barlow et al. (1972, Table A5, p. 363).

TABLE 5.5
The probabilities $\rho(r, i)$

				i				
r	1	2	3	4	5	6	7	8
2	0.5000	0.5000						
3	0.3333	0.5000	0.1667					
4	0.2500	0.4583	0.2500	0.0417				
5	0.2000	0.4167	0.2917	0.0833	0.0083			
6	0.1667	0.3806	0.3125	0.1181	0.0208	0.0014		
7	0.1429	0.3500	0.3222	0.1458	0.0347	0.0042	0.0002	
8	0.1250	0.3241	0.3257	0.1679	0.0486	0.0080	0.0007	0.00002

5.6.3 The Likelihood Ratio Test for Unspecific Ordering

Robertson et al. (1988) derived a likelihood ratio test based on the maximum likelihood estimates under the null hypothesis and under the alternative hypothesis. The likelihood ratio test statistic has the form

$$T_{\text{LR}}(\mathbf{n}) = 2 \sum_{i,j} n_{ij}^* \log\left(\frac{n_{ij}^*}{m_{ij}}\right), \tag{5.15}$$

where a term is zero if $n_{ij}^* = 0$. The asymptotic distribution of T_{LR} is—like the Pearson test statistic in the proceeding section—chi-bar-squared, and we obtain P-values for the likelihood ratio test from Equation 5.14, with T_{LR} instead of T_{Pearson}.

For more details concerning these tests, see Barlow et al. (1972), Robertson et al. (1988), or Agresti and Coull (2002).

5.6.4 Exact and Mid-P Tests for Unspecific Ordering

The tests derived above are asymptotic tests that rely on the chi-bar-squared distribution. These tests might be inappropriate for the analysis of tables with sparse data, for which the asymptotics do not hold. One alternative approach is to derive exact P-values by conditioning to eliminate the nuisance parameters (Agresti, 1992; Agresti and Coull, 1998a). In Section 5.4, the conditional distribution, given all the marginals, was presented in Equation 5.9. Here, we will use the two test statistics in Equations 5.13 and 5.15 to produce two exact conditional and two mid-P tests for unspecific ordering. Exact and mid-P tests for the ordered $r \times 2$ table will be presented in more detail in Sections 5.7.4 and 5.7.5. Here, we merely give the necessary formulas for calculating the P-values. The P-value for the exact conditional test with the Pearson chi-squared statistic is given by

$$\text{exact cond. } P\text{-value} = \sum_{\Omega(\mathbf{x}|\mathbf{n}_{++})} I\big[T_{\text{Pearson}}(\mathbf{x}) \geq T_{\text{Pearson}}(\mathbf{n})\big] \cdot f(\mathbf{x}\,|\,\mathbf{n}_{++}),$$

where $\Omega(\mathbf{x}|\mathbf{n}_{++})$ is the set of all tables $\mathbf{x}$ with marginals equal to $\mathbf{n}_{++}$, $I()$ is the indicator function, $T_{\text{Pearson}}()$ is the Pearson chi-squared test statistic, and $f(\mathbf{x}\,|\,\mathbf{n}_{++})$ is the conditional probability distribution in Equation 5.9. The corresponding mid-P value is given by

$$\text{mid-}P \text{ value } = \sum_{\Omega(\mathbf{x}|\mathbf{n}_{++})} I\big[T_{\text{Pearson}}(\mathbf{x}) > T_{\text{Pearson}}(\mathbf{n})\big] \cdot f(\mathbf{x}\,|\,\mathbf{n}_{++})$$

$$+ \; 0.5 \cdot \sum_{\Omega(\mathbf{x}|\mathbf{n}_{++})} I\big[T_{\text{Pearson}}(\mathbf{x}) = T_{\text{Pearson}}(\mathbf{n})\big] \cdot f(\mathbf{x}\,|\,\mathbf{n}_{++}).$$

The exact conditional and mid-P tests with the likelihood ratio statistic are obtained by substituting $T_{\text{LR}}()$ for $T_{\text{Pearson}}()$ in the two equations above.

5.6.5   Examples

Alcohol Consumption and Malformations (Table 5.2)

In this section, we consider the example of the association between alcohol consumption during pregnancy and malformations of the sex organs, first introduced in Section 5.2.1. We will calculate the Pearson chi-squared and likelihood ratio tests for unspecific ordering, as well as the two exact conditional tests and the two mid-P tests, that were presented in the preceding sections. We also calculate the Pearson chi-squared and likelihood ratio tests for unordered grouping (unordered alternatives). These methods will be presented in more detail in the chapter on $r \times c$ tables (Chapter 7); we include them here merely to illustrate the difference between tests for ordered and unordered alternative hypotheses, and we do not advocate the use of unordered tests for ordered data.

Under the null hypothesis of no association between alcohol consumption and malformations, the maximum likelihood estimate of the probability of malformations of the sex organs is $93/32574 = 0.00286$. The sample proportions, however, are nearly monotone increasing: $(p_1, p_2, p_3, p_4, p_5) = (0.00280,$

0.00262, 0.00631, 0.00787, 0.02632). Because $p_2 < p_1$, we pool p_1 and p_2: $p_1 = p_2 = (48+38)/(17114+14502) = 0.00272$. Then, the maximum likelihood estimates under the alternative hypothesis are $(\hat{\pi}_1, \hat{\pi}_2, \hat{\pi}_3, \hat{\pi}_4, \hat{\pi}_5) = (0.00272, 0.00272, 0.00631, 0.00787, 0.02632)$. We can now calculate the n_{ij}^*s as described in Section 5.6.2. The observed values of the Pearson and likelihood ratio test statistics are $T_{\text{Pearson}} = 11.99$ and $T_{\text{LR}} = 6.10$. The probabilities $\rho(5, i)$ can be found from Table 5.5. The P-value for the Pearson chi-squared test can be calculated by

$$
\begin{aligned}
P\text{-value} \;=\; & 0.2000 \cdot \Pr(\chi_0^2 \geq 11.99) + 0.4167 \cdot \Pr(\chi_1^2 \geq 11.99) \\
+\; & 0.2917 \cdot \Pr(\chi_2^2 \geq 11.99) + 0.0833 \cdot \Pr(\chi_3^2 \geq 11.99) \\
+\; & 0.0083 \cdot \Pr(\chi_4^2 \geq 11.99),
\end{aligned}
$$

which gives

$$
\begin{aligned}
P\text{-value} \;=\; & 0.2000 \cdot 0 + 0.4167 \cdot 0.0005 + 0.2917 \cdot 0.0025 + 0.0833 \cdot 0.0074 \\
+\; & 0.0083 \cdot 0.0174 \\
=\; & 0.0017.
\end{aligned}
$$

Similarly, the P-value for the likelihood ratio test can be calculated by

$$
\begin{aligned}
P\text{-value} \;=\; & 0.2000 \cdot \Pr(\chi_0^2 \geq 6.10) + 0.4167 \cdot \Pr(\chi_1^2 \geq 6.10) \\
+\; & 0.2917 \cdot \Pr(\chi_2^2 \geq 6.10) + 0.0833 \cdot \Pr(\chi_3^2 \geq 6.10) \\
+\; & 0.0083 \cdot \Pr(\chi_4^2 \geq 6.10),
\end{aligned}
$$

which gives

$$
\begin{aligned}
P\text{-value} \;=\; & 0.2000 \cdot 0 + 0.4167 \cdot 0.0135 + 0.2917 \cdot 0.0473 + 0.0833 \cdot 0.1067 \\
+\; & 0.0083 \cdot 0.1916 \\
=\; & 0.0299.
\end{aligned}
$$

The results are summarized in Table 5.6. First, note the difference in P-values for the Pearson chi-squared test and the likelihood ratio test. Since there are three cells with expected counts less than five, $m_{31} = 2.26$, $m_{41} = 0.36$, and $m_{51} = 0.11$, we do not expect the Pearson chi-squared test to perform well, and we are inclined to prefer the likelihood ratio test in this case. Second, the P-values are smaller when we assume an ordered grouping of the data, rather than an unordered grouping. This illustrates that for ordered data, tests that account for the ordering through constraints on the alternative hypothesis have higher power than tests that do not put any constraint on the alternative hypothesis.

Table 5.6 also includes the results of the two exact conditional tests and the two mid-P tests. The P-values for the tests based on the Pearson chi-squared statistic are smaller than those of the likelihood ratio tests. In both cases, the mid-P value is only slightly smaller than its corresponding exact P-value.

TABLE 5.6

Results of tests for unordered and ordered alternatives (unspecific ordering) on the data in Table 5.2

Test	Statistic	df*	*P*-value
Unordered alternatives			
Pearson chi-squared[†]	12.08	4	0.0168
Likelihood ratio[†]	6.20	4	0.185
Ordered alternatives			
Pearson chi-squared	11.99	(chi-bar-squared)	0.0017
Likelihood ratio	6.10	(chi-bar-squared)	0.0299
Exact conditional (with T_{Pearson} as test statistic)			0.0232
Mid-*P* (with T_{Pearson} as test statistic)			0.0229
Exact conditional (with T_{LR} as test statistic)			0.0719
Mid-*P* (with T_{LR} as test statistic)			0.0716

* Degrees of freedom for the chi-squared distribution
[†] These methods are presented in Chapter 7

Based on the results above, we conclude that there seems to be an increase in the probability of malformations of the sex organ for increasing alcohol consumption during pregnancy; however, not all methods support this conclusion unequivocally. We return to this example in Section 5.7.14, where we test the data for linear trend.

5.6.6 Evaluation of Tests

Evaluation Criteria

We evaluate the tests by calculating their actual significance levels and power. We assume sampling under the row margins fixed model (with fixed group sizes), such that the probability for an arbitrary $r \times 2$ table is the product of r independent binomial probabilities (Equation 5.8). The actual significance level depends on the common success probability under the null hypothesis $\pi = \pi_1 = \pi_2 = \ldots = \pi_r$, the fixed group sizes $\mathbf{n}_+ = \{n_{1+}, n_{2+}, \ldots, n_{r+}\}$, and the nominal significance level α:

$$\text{ASL}(\pi, \mathbf{n}_+, \alpha \mid H_0) = \sum_{x_{11}=0}^{n_{1+}} \sum_{x_{21}=0}^{n_{2+}} \cdots \sum_{x_{r1}=0}^{n_{r+}} I\big[P(\mathbf{x}) \le \alpha\big] \cdot f(\mathbf{x} \mid \pi, \mathbf{n}_+),$$

where $I()$ is the indicator function, $P(\mathbf{x})$ is the *P*-value for a test on $\mathbf{x} = \{x_{11}, n_{1+} - x_{11}, x_{21}, n_{2+} - x_{21}, \ldots, x_{r1}, n_{r+} - x_{r1}\}$, and $f(\mathbf{x} \mid \pi, \mathbf{n}_+)$ is given in Equation 5.8 with $\pi_1 = \pi_2 = \ldots = \pi_r = \pi$.

Power is calculated in a similar manner; however, we now assume that

$\pi_1 \leq \pi_2 \leq \ldots \leq \pi_r$ or $\pi_1 \geq \pi_2 \geq \ldots \geq \pi_r$, with at least one strict inequality:

$$\text{Power}(\pi_1, \ldots, \pi_r, \mathbf{n}_+, \alpha \mid H_A) =$$

$$\sum_{x_{11}=0}^{n_{1+}} \sum_{x_{21}=0}^{n_{2+}} \cdots \sum_{x_{r1}=0}^{n_{r+}} I[P(\mathbf{x}) \leq \alpha] \cdot f(\mathbf{x} \mid \pi_1, \ldots, \pi_r; \mathbf{n}_+).$$

Evaluation of Actual Significance Level

We will let $r = 4$, such that we have a 4×2 table, and we assume that the row sums are fixed. Figure 5.1 shows the actual significance levels of the Pearson chi-squared, likelihood ratio, exact conditional, and mid-P tests as functions of the common success probability (π) under the null hypothesis: $\pi = \pi_1 = \pi_2 = \pi_3 = \pi_4$, when each of the four row sums equal 25. Here, we use the exact and mid-P tests with the Pearson chi-squared statistic, and their actual significance levels are calculated with simulations and shown in Figure 5.1 with moving average smoothing.

The Pearson chi-squared and likelihood ratio tests are liberal with actual significance levels mostly at about 6–7% for a nominal level of 5%. The Pearson chi-squared test is a little bit better than the likelihood ratio test. These results hold for sample sizes in the range of 10 to 100 in each row. The exact test has actual significance levels very close to the nominal level for all values of π in the range 0.1–0.9. The mid-P test has somewhat higher actual significance levels than the exact test, and the values sometimes exceed the nominal level slightly. If we use the likelihood ratio statistic instead of the Pearson chi-squared statistic to calculate the exact and mid-P tests, the results are almost indistinguishable from those shown in Figure 5.1. Because the exact test performs so well here, there does not seem to be much to gain by using the mid-P test. This result stands in sharp contrast to those of Chapter 4, where the Fisher mid-P test is a considerable improvement over the Fisher exact test.

In cases with unequal row sums, the actual significance levels of the Pearson chi-squared test may be further inflated, as shown in Figure 5.2, for which the row sums are 10, 10, 20, and 30. This is, however, mostly a problem for probabilities close to zero or one. If such small or large probabilities are expected, the sample size in each row should be much larger than in the current example, which would probably attenuate—or even eliminate—the problem. Note that the exact and mid-P tests perform well also in this situation, although the mid-P test is slightly more often above the nominal level than in Figure 5.1.

Evaluation of Power

The power of the tests for unspecific ordering is illustrated in Figure 5.3. The success probabilities for these calculations, which we have labeled Power Scenario #1, are given by $\{\pi_1, \pi_2, \pi_3, \pi_4\} = \{0.7, 0.5, 0.45, 0.4\}$. This is a strict

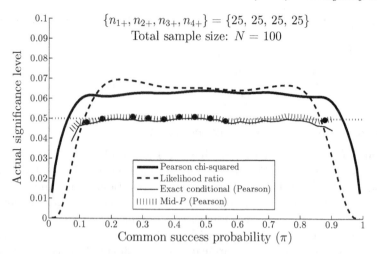

FIGURE 5.1
Actual significance levels of four tests for unspecific ordering as functions of the common success probability under H_0

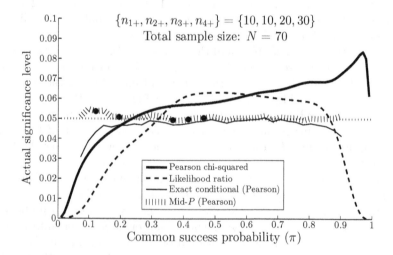

FIGURE 5.2
Actual significance levels of four tests for unspecific ordering as functions of the common success probability under H_0

decrease in the probabilities, but it is not a linear trend. Exact calculations are used for the Pearson chi-squared and likelihood ratio tests for $n_{i+} \leq 50$, and simulations are used to calculate the remaining values. There is practically nothing to separate the power of the Pearson chi-squared and likelihood ratio tests. Both tests have higher power than the exact and mid-P tests. This is not surprising, because the Pearson chi-squared and likelihood ratio tests have actual significance levels above the nominal level (Figures 5.1 and 5.2). The power of the exact and mid-P tests seem to be quite similar.

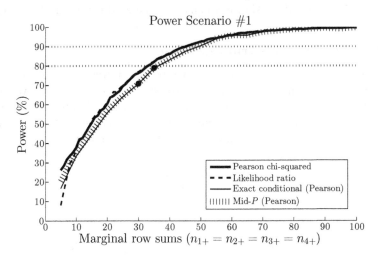

FIGURE 5.3
Power of four tests for unspecific ordering

5.7 Tests for Trend in the Linear and Logit Models

5.7.1 The Null and Alternative Hypotheses

Assume now that we have assigned scores to the grouping categories, and that our model is given in Equation 5.3. Regardless of the link function, the null hypothesis of no trend is equivalent to the null hypothesis

$$H_0 : \beta = 0,$$

which we will test against the alternative hypothesis

$$H_A : \beta > 0 \quad \text{or} \quad \beta < 0.$$

Testing for linear trend has a long history, as it goes back to Cochran (1954b) and Armitage (1955). They proposed a test for trend in the linear model (Equation 5.4), and estimated the trend by the standard linear regression coefficient. This test for trend is called the *Cochran-Armitage test*, and it is without doubt the most common test for linear trend in $r \times 2$ tables. We will see below that the Cochran-Armitage test is quite versatile; it is both a score test and an exact conditional test for trend in a logit model.

In addition to the Cochran-Armitage-based tests, we also consider some alternative tests for testing both linear trend (Equation 5.4) and logit trend (Equation 5.5), including Mantel-Haenszel, likelihood ratio, and Wald tests.

5.7.2   The Cochran-Armitage Test

The motivation behind the Cochran-Armitage test was originally based on the estimated slope in a simple linear regression (Cochran, 1954b; Armitage, 1955). We calculate the test statistic by comparing the estimated slope with its standard error under the null hypothesis. When we assume the linear model in Equation 5.4, the linear slope can be estimated by the ordinary least squares estimate

$$\hat{\beta}_{CA} = \frac{\sum_{i=1}^{r} n_{i+}(a_i - \bar{a})\hat{\pi}_i}{\sum_{i=1}^{r} n_{i+}(a_i - \bar{a})^2} = \frac{U}{S_{aa}}, \tag{5.16}$$

where

$$\bar{a} = \sum_{i=1}^{r} n_{i+}a_i/N, \quad \hat{\pi}_i = n_{i1}/n_{i+}, \quad U = \sum_{i=1}^{r} n_{i+}(a_i - \bar{a})\hat{\pi}_i,$$

and

$$S_{aa} = \sum_{i=1}^{r} n_{i+}(a_i - \bar{a})^2. \tag{5.17}$$

If we have a random sample design or a fixed group sizes design, or if we stratify on Y_1, we know that the n_{i1} are binomially distributed with expectation $n_{i+}\pi_i$ and variance equal to $n_{i+}\pi_i(1 - \pi_i)$. The expectation and variance of U then follow as

$$E(U) = \sum_{i=1}^{r} n_{i+}(a_i - \bar{a})\pi_i,$$

and

$$\text{var}(U) = \sum_{i=1}^{r} n_{i+}(a_i - \bar{a})^2 \pi_i(1 - \pi_i). \tag{5.18}$$

Under the null hypothesis, $H_0 : \beta = 0$, with the joint probability denoted by π, the expectation and the variance are given by

$$E(U) = 0 \quad \text{and} \quad \text{var}(U) = S_{aa}\pi(1 - \pi).$$

Now, let $\hat{\pi} = n_{+1}/N$ be the estimate of the joint probability π. Then

$$s_0^2 = \hat{\pi}(1 - \hat{\pi})S_{aa} \tag{5.19}$$

is an estimate of the variance of U under the null hypothesis, and we can formulate the Cochran-Armitage test statistic as

$$Z_{\text{CA}}(\mathbf{n}) = \frac{U}{s_0} = \frac{\hat{\beta}_{\text{CA}}}{[\hat{\pi}(1 - \hat{\pi})/S_{aa}]^{1/2}}. \tag{5.20}$$

For the details in these derivations, see Buonaccorsi et al. (2014). Z_{CA} is approximately standard normal distributed, and we obtain P-values for the Cochran-Armitage test as

$$P\text{-value} = \Pr\left[Z \geq |Z_{\text{CA}}(\mathbf{n})|\right],$$

where Z is a standard normal variable. The Cochran-Armitage test statistic is sometimes formulated as a chi-square statistic, $T_{\text{CA}} = Z_{\text{CA}}^2$, which is approximately chi-squared distributed with one degree of freedom.

Note that the Cochran-Armitage test statistics Z_{CA} can also be expressed by means of the sample Pearson correlation coefficient, $\hat{r}_{\text{P}}$ (see Section 7.9.1), between Y_1 and Y_2:

$$Z_{\text{CA}} = \sqrt{N}\hat{r}_{\text{P}}.$$

The Cochran-Armitage test was originally derived for the situation with a random sample design, a fixed group sizes design, or a stratified analysis on the group variable (Y_1). It is, however, also in common use for the analysis of case-control studies. In case-control studies, the sample sizes n_{+1} and n_{+2} are fixed by design, and

$$E(U) = N \cdot n_{+1}(1 - n_{+1}) \sum_{i=1}^{r} a_i(\gamma_{i|1} - \gamma_{i|0}).$$

where the $\gamma_{i|j}$ were defined in Section 5.3.3. As noted below Equation 5.7, the null hypothesis $H_0 : \pi_1 = \pi_2 = \ldots = \pi_r$ is equivalent to $H_0 : \gamma_{i|0} = \gamma_{i|1}$ for all i. Under the null hypothesis $E(U) = 0$, it can be shown that the estimated variance of U under the null hypothesis is given by s_0^2 (Equation 5.19) also for the case-control design (Zheng and Gastwirth, 2006; Buonaccorsi et al., 2014).

It is interesting to note that the Cochran-Armitage test is the score test for a logit model, assuming that

$$\text{logit}(\pi_i) = \alpha + \beta a_i,$$

for $i = 1, 2, \ldots, r$. Based on this model, we can formulate an expression for the score test for $\beta = 0$, see for instance Hosmer et al. (2013, p. 15) or Zheng et al. (2012, p. 68). The score test statistic has the form of Z_{CA} in Equation 5.20.

5.7.3 The Modified Cochran-Armitage Test

In the previous section, we discussed the standard Cochran-Armitage test, which is derived with the estimated variance of U under the null hypothesis, s_0^2, in the denominator. The use of a variance estimate under the null hypothesis is a standard method to avoid large-scale violations of the nominal significance level. As an alternative, we may use the general estimate of the variance (s^2) in the denominator, which gives the following test statistic

$$Z_{\text{CAMOD}}(\mathbf{n}) = \frac{U}{s}. \tag{5.21}$$

We expect that Z_{CAMOD} might be more powerful than Z_{CA} in some situations, so it is of interest in itself to study the modified test statistic.

Note that the formula for the general variance will take different forms, depending on the design. Since the Cochran-Armitage test was developed for the row margins fixed model (with fixed group sizes, or stratification on Y_1), we concentrate on that. It follows from Equation 5.18, that the variance of U in this case is estimated by

$$s^2 = \sum_{i=1}^r n_{i+}(a_i - \bar{a})^2 \hat{\pi}_i(1 - \hat{\pi}_i). \tag{5.22}$$

For the case-control design (the column margins fixed model), the variance of U can be estimated by s^2 given in Equation 13 of Buonaccorsi et al. (2014). Then Z_{CAMOD} in Equation 5.21 coincides with Z_{CC} in Zheng and Gastwirth (2006).

For the random sample design (the total sum fixed model), the estimated variance of U is given by the sum of the variance for the case-control sample, plus a term given by the variance of the conditional expectation of U given the proportion of cases $p = n_{+1}/N$. For the derivations, see Section 2.2 of Buonaccorsi et al. (2014).

5.7.4 The Cochran-Armitage Exact Conditional Test

The background and rationale for exact conditional tests are provided in Sections 1.9 and 4.4.5. Recall that if T is a test statistic, $\mathbf{n}$ denotes the observed table, and $\mathbf{x}$ denotes any possible table, the exact P-value is given as

$$\text{exact } P\text{-value} = \Pr[T(\mathbf{x}) \geq T(\mathbf{n}) \mid H_0],$$

which is the sum of the probabilities of all possible tables that agree less than or equally with the null hypothesis—as measured by T—than does the observed table. When we condition on both the row and column marginals $(\mathbf{n}_{++})$, the exact conditional P-value is given by

$$\text{exact cond. } P\text{-value} = \Pr[T(\mathbf{x}) \geq T(\mathbf{n}) \mid H_0, \mathbf{n}_{++}].$$

Section 5.4 gave the sampling distributions of $r \times 2$ tables under different sampling models. When we condition on all the marginals $(\mathbf{n}_{++})$, the relevant sampling distribution is the multiple hypergeometric distribution $f(\mathbf{x} \,|\, \mathbf{n}_{++})$ in Equation 5.9. If we now let T be the linear rank statistic

$$T_{\text{linrank}}(\mathbf{n}) = \sum_{i=1}^{r} a_i n_{i1},$$

which—under conditioning on $\mathbf{n}_{++}$—gives an equivalent ordering of tables as the Cochran-Armitage statistic, we obtain the one-sided Cochran-Armitage exact conditional P-value as

$$\text{one-sided } P\text{-value} = \sum_{\Omega(\mathbf{x}|\mathbf{n}_{++})} I\big[T_{\text{linrank}}(\mathbf{x}) \geq T_{\text{linrank}}(\mathbf{n})\big] \cdot f(\mathbf{x} \,|\, \mathbf{n}_{++}), \quad (5.23)$$

where $\Omega(\mathbf{x}|\mathbf{n}_{++})$ is the set of all tables $\mathbf{x}$ with marginals equal to $\mathbf{n}_{++}$ and $I()$ is the indicator function. As discussed in Section 1.3.4, we use the principle of twice the smallest tail for computing two-sided P-values, see page 8.

5.7.5 The Cochran-Armitage Mid-P Test

As described in Section 1.10, the mid-P value is the sum of the probabilities of all possible tables that agree less with the null hypothesis than does the observed table, plus half the point probability of the tables that agree equally with the null hypothesis as the observed table. The one-sided Cochran-Armitage mid-P value is given by

$$\text{one-sided mid-}P\text{ value} = \sum_{\Omega(\mathbf{x}|\mathbf{n}_{++})} I\big[T_{\text{linrank}}(\mathbf{x}) > T_{\text{linrank}}(\mathbf{n})\big] \cdot f(\mathbf{x} \,|\, \mathbf{n}_{++})$$

$$+\ 0.5 \cdot \sum_{\Omega(\mathbf{x}|\mathbf{n}_{++})} I\big[T_{\text{linrank}}(\mathbf{x}) = T_{\text{linrank}}(\mathbf{n})\big] \cdot f(\mathbf{x} \,|\, \mathbf{n}_{++}).$$

In a similar manner as in the previous section, we define the two-sided mid-P value as twice the smallest of the two tail probabilities, that is, twice the one-sided mid-P value above, assuming that this is the smallest of the two tails.

5.7.6 The Mantel-Haenszel Test for Trend

The Mantel-Haenszel test for stratified 2×2 tables was proposed by Mantel and Haenszel (1959), see Section 10.13.3. The test was extended to $r \times c$ tables by Mantel (1963), for testing linear-by-linear association. Consequently, the Mantel-Haenszel test can be used for testing trend in $r \times 2$ tables. In Section 5.1, we defined row scores $\{a_1, a_2, \ldots, a_r\}$ for the r groups, and we now introduce outcome scores $\{b_1, b_2\}$ to the columns; see also Section 6.8 for ordering of the outcomes in $2 \times c$ tables, and Section 7.8 for ordering of the outcomes in $r \times c$ tables.

In an ordered $r \times 2$ table, we will usually let $b_1 = 1$ and $b_2 = 0$. A linear function of the scores is given by

$$\bar{d} = \sum_{i=1}^{r} \sum_{j=1}^{2} a_i b_j n_{ij}/N.$$

Under the null hypothesis,

$$\mathrm{E}(\bar{d}) = \sum_{i=1}^{r} a_i \frac{n_{i+}}{N} \cdot \sum_{j=1}^{2} b_j \frac{n_{+j}}{N} = \bar{a} \cdot \bar{b},$$

and

$$\mathrm{Var}(\bar{d}) = \sum_{i=1}^{r}(a_i - \bar{a})^2 \frac{n_{ij}}{N} \cdot \sum_{j=1}^{2}(b_j - \bar{b})^2 \frac{n_{ij}/N}{N-1}.$$

Then, the Mantel-Haenszel test is given as

$$Z_{\mathrm{MH}}^2(\mathbf{n}) = \frac{[\bar{d} - \mathrm{E}(\bar{d})]^2}{\mathrm{Var}(\bar{d})} = \frac{(N-1)\left[\sum_{i=1}^{r}\sum_{j=1}^{2}(a_i - \bar{a})(b_j - \bar{b})n_{ij}\right]^2}{\sum_{i=1}^{r}(a_i - \bar{a})^2 n_{i+} \cdot \sum_{j=1}^{2}(b_j - \bar{b})^2 n_{+j}}, \quad (5.24)$$

which simply is $N - 1$ times the sample Pearson correlation coefficient (see Section 7.9.1) between Y_1 and Y_2. For the derivations, see also Stokes et al. (2012, p. 89). Thus, the Mantel-Haenszel test is closely related to the Cochran-Armitage test, since the Mantel-Haenszel test statistic Z_{MH} can be written as

$$Z_{\mathrm{MH}} = \sqrt{\frac{N}{N-1}} Z_{\mathrm{CA}}.$$

For testing the null hypothesis of no trend, we use that Z_{MH} is approximately standard normal distributed, or equivalently, that $T_{\mathrm{MH}} = Z_{\mathrm{MH}}^2$ is approximately chi-squared distributed with one degree of freedom. Thus, P-values for the Mantel-Haenszel test for trend are obtained as

$$P\text{-value} = \mathrm{Pr}\left[\chi_1^2 \geq T_{\mathrm{MH}}(\mathbf{n})\right].$$

5.7.7   The Wald Test for the Linear Model

Let $\hat{\beta}_{\mathrm{ML, linear}}$ be the maximum likelihood estimate for β in the linear model, and let $\widehat{\mathrm{SE}}(\hat{\beta}_{\mathrm{ML, linear}})$ be its estimated standard error. The Wald test statistic for the linear model (see Section 1.7 for a general description of the Wald test) is

$$Z_{\text{Wald, linear}}(\mathbf{n}) = \frac{\hat{\beta}_{\text{ML, linear}}}{\widehat{\text{SE}}(\hat{\beta}_{\text{ML, linear}})}.$$

$Z_{\text{Wald,linear}}$ is asymptotically standard normal distributed. Alternatively, $T_{\text{Wald,linear}} = Z^2_{\text{Wald,linear}}$ is asymptotically chi-squared distributed with one degree of freedom.

5.7.8 The Likelihood Ratio Test for the Linear Model

Let l_0 denote the maximized likelihood of the linear model under the null hypothesis, and let l_1 denote the maximized likelihood of the linear model under the alternative hypothesis. The likelihood ratio test statistic for the linear model is

$$T_{\text{LR, linear}}(\mathbf{n}) = -2\log(\Lambda) = -2\log(l_0/l_1) = -2(L_0 - L_1),$$

where L_0 and L_1 denote the maximized log-likelihood functions under the linear model. The test statistic $T_{\text{LR, linear}}$ is asymptotically chi-squared distributed with one degree of freedom.

5.7.9 The Wald Test for the Logit Model

Let $\hat{\beta}_{\text{ML, logit}}$ be the maximum likelihood estimate for β in the logit model (Equation 5.5), and let $\widehat{\text{SE}}(\hat{\beta}_{\text{ML, logit}})$ be its estimated standard error. The Wald test statistic for the logit model is

$$Z_{\text{Wald, logit}}(\mathbf{n}) = \frac{\hat{\beta}_{\text{ML, logit}}}{\widehat{\text{SE}}(\hat{\beta}_{\text{ML, logit}})}.$$

$Z_{\text{Wald,logit}}$ is asymptotically standard normal distributed. Alternatively, $T_{\text{Wald,logit}} = Z^2_{\text{Wald,logit}}$ is asymptotically chi-squared distributed with one degree of freedom.

5.7.10 The Likelihood Ratio Test for the Logit Model

Let l_0 denote the maximized likelihood of the logit model under the null hypothesis, and let l_1 denote the maximized likelihood of the logit model under the alternative hypothesis. The likelihood ratio test statistic for the logit model is

$$T_{\text{LR, logit}}(\mathbf{n}) = -2\log(l_0/l_1) = -2(L_0 - L_1),$$

where L_0 and L_1 denote the maximized log-likelihood functions under the logit model. The test statistic $T_{\text{LR, logit}}$ is asymptotically chi-squared distributed with one degree of freedom.

5.7.11   The Exact Conditional and Mid-*P* Tests for the Logit Model

The conditional distribution of T_β given n_{+1} is given in Equation 5.11. To test the null hypothesis, we order the tables by the values of T_β with the marginal equal to n_{+1}. This is equivalent to ordering the tables by the Cochran-Armitage statistic. Thus, the exact conditional test is equivalent to the Cochran-Armitage exact conditional test in Section 5.7.4.

Following the same argument, we also note that the mid-*P* conditional test is equivalent to the Cochran-Armitage mid-*P* test. In like manner, we get that the exact conditional likelihood ratio test also equals the Cochran-Armitage test.

5.7.12   An Exact Unconditional Test

In Section 5.7.4, we outlined the Cochran-Armitage exact conditional test, which is based on the conditional distribution in Equation 5.9. We obtain the exact conditional *P*-value by summing the conditional probabilities over all tables for which the linear rank statistic, $T_{\mathrm{linrank}}(\mathbf{x})$, is greater than or equal to the observed statistic, $T_{\mathrm{linrank}}(\mathbf{n})$. We used the linear rank statistic instead of the Cochran-Armitage statistic because the linear rank statistic has a simpler formulation and the two statistics give identical ordering of tables when the marginals n_{+1} and n_{+2} (the total number of successes and failures, see Table 5.1) are fixed.

Exact unconditional tests were introduced for 2×2 tables in Section 4.4.7, where we pointed out that exact conditional tests can be overly conservative due to discreteness of the possible values of the test statistic. An exact unconditional test is a test that does not assume that n_{+1} and n_{+2} are fixed by design, and does not condition on them. The probability distribution of interest is thus the one in Equation 5.8. An exact unconditional test will not suffer from the conservatism of an exact conditional test.

Under the null hypothesis

$$H_0 : \pi_1 = \pi_2 = \ldots = \pi_r = \pi, \tag{5.25}$$

π is a nuisance parameter. An exact unconditional *P*-value is obtained as the maximum value over the range of the nuisance parameter π over the set of tables where the Cochran-Armitage test statistic is greater than or equal to the observed test statistic, but now multiplied by the unconditional probabilities. The exact unconditional *P*-value then is

$$P\text{-value} = \max_{0 \le \pi \le 1} \left\{ \sum_{\Omega(\mathbf{x}|\mathbf{n}_+)} I\big[Z_{\mathrm{CA}}(\mathbf{x}) \ge Z_{\mathrm{CA}}(\mathbf{n})\big] \cdot f(\mathbf{x} \,|\, \pi, \mathbf{n}_+) \right\},$$

where $\Omega(\mathbf{x}|\mathbf{n}_+)$ denotes the set of all tables with row sums equal to $\mathbf{n}_+$, $I()$ is the indicator function and $f()$ is the unconditional distribution given in

Equation 5.8 with $\pi_1 = \pi_2 = \ldots = \pi_r = \pi$. The exact unconditional test is very computer-intensive, and it is only feasible to use it for quite small data sets.

5.7.13 Testing the Fit of a Model with Linear or Logit Trend

We consider two models: the linear model

$$\pi_i = \alpha_{\text{linear}} + \beta_{\text{linear}} a_i,$$

and the logit model

$$\text{logit}(\pi_i) = \alpha_{\text{logit}} + \beta_{\text{logit}} a_i,$$

where $i = 1, 2, \ldots, r$. Let $\hat{\pi}_i$ be the estimated success probability for Group i, based on either the linear or the logit model. Then, $m_{i1} = n_{i+} \hat{\pi}_i$ and $m_{i2} = n_{i+}(1 - \hat{\pi}_i)$ are the estimated expected counts in Group i.

The Pearson Goodness-of-Fit Test

For both the linear model and the logit model, the Pearson goodness-of-fit statistic is of the form

$$\chi^2(\mathbf{n}) = \sum_{i=1}^{r} \sum_{j=1}^{2} \frac{(n_{ij} - m_{ij})^2}{m_{ij}}.$$

The Pearson goodness-of-fit statistic follows a chi-square distribution with $r - 2$ degrees of freedom:

$$P\text{-value} = \Pr\left[\chi_{r-2}^2 \geq \chi^2(\mathbf{n})\right].$$

The Pearson Residuals

If the P-value of the Pearson goodness-of-fit test is small, which is an indication of lack of fit, the next step would be to investigate which cells in the table contribute most to the lack of fit. For this, we can use Pearson residuals, defined as

$$r_{ij} = \frac{n_{ij} - m_{ij}}{\sqrt{m_{ij}}}.$$

Alternatively, we can use the estimated standard error $\widehat{\text{SE}}(n_{ij} - m_{ij})$ in the denominator, which gives standardized Pearson residuals, see Section 7.5.7.

The Likelihood Ratio (Deviance) Test

The likelihood ratio (also called the *deviance*) test statistic is given by

$$D(\mathbf{n}) = 2 \sum_{i=1}^{r} \sum_{j=1}^{2} n_{ij} \log\left(\frac{n_{ij}}{m_{ij}}\right).$$

The likelihood ratio statistic, like the Pearson goodness-of-fit statistic, follows a chi-square distribution with $r - 2$ degrees of freedom.

The Score Test for Fit

The score statistic would be an alternative for testing the fit of the linear probability model; however, Smyth (2003) has shown that for any generalized linear model, the Pearson goodness-of-fit test is a score test for testing a model against the saturated model.

5.7.14   Examples

In this section, we consider the two examples introduced in Section 5.2.

Alcohol Consumption and Malformations (Table 5.2)

For this example, we assign equally spaced scores to the five exposure categories (number of drinks per day), so that row number i is given score i, for $i = 1, 2, \ldots, 5$. Table 5.7 summarizes the results of applying the tests from Section 5.7 to the data in Table 5.2.

TABLE 5.7

Results of tests for the linear and logit models on the data in Table 5.2, with equally spaced scores

Test	Statistic	df*	P-value
Tests for trend in the linear model			
Cochran-Armitage	1.35		0.176
Modified Cochran-Armitage	1.06		0.289
Mantel-Haenszel	1.35		0.176
Wald	0.96		0.339
Likelihood ratio	1.27	1	0.260
Testing the fit of a linear model			
Pearson goodness-of-fit	7.31	3	0.063
Likelihood ratio (deviance)	4.93	3	0.177
Tests for trend in the logit model			
Wald	1.35		0.176
Likelihood ratio	1.76	1	0.185
Cochran-Armitage exact cond.			0.209
Cochran-Armitage mid-P			0.181
Testing the fit of a logit model			
Pearson goodness-of-fit	5.68	3	0.128
Likelihood ratio (deviance)	4.45	3	0.217

*Degrees of freedom for the chi-squared distribution

We know that the Cochran-Armitage test is equivalent to the score test in a logit model. The score test is asymptotically equivalent to the Wald test and the log likelihood test. Moreover, we know that the Cochran-Armitage exact conditional test is equivalent to the exact conditional likelihood test, which also links the linear and the logit approaches. We have therefore listed the Cochran-Armitage exact test and the Cochran-Armitage mid-P test under the logit model in Table 5.7.

It is notable that the P-values for the Cochran-Armitage test, the Mantel-Haenszel test, and the Wald test for the logit model are equal to at least three significant digits. The Wald P-value for the linear model is much larger than the P-values for the other tests. We also note that the Pearson goodness-of-fit test has a much lower P-value than the deviance test. Since there are three cells with expected counts less than five ($m_{31} = 2.26$, $m_{41} = 0.36$, and $m_{51} = 0.11$), under the null hypothesis of no trend, we do not expect the chi-squared approximation to the Pearson statistic to hold.

Overall, the different tests for linear and logit trend give quite similar P-values, and none of the tests indicate a significant trend in the probability of malformations of the sex organ with increasing alcohol consumption during pregnancy. In Section 5.6.5, however, the tests for unspecific ordering all gave $P < 0.031$. It might be that the increase in probabilities is not linear in the exposure, but follows another shape. The group sizes and number of malformations in the two highest exposure categories are, however, too small to explore this further.

Elevated Troponin T Levels in Stroke Patients (Table 5.3)

We will consider two different strategies of assigning scores to the categories of stroke severity. First, we assign equally spaced scores $1, 2, \ldots, 5$ to the rows. This results in the P-values shown in Table 5.8. Under the null hypothesis of no trend, no cell has expected counts less than five.

As in the previous example, we note that the tests for trend in the linear model give similar results as the tests for trend in the logit model. In particular, we see that the Wald test for the logit model agrees very well with the Cochran-Armitage and Mantel-Haenszel tests. The Cochran-Armitage exact test is somewhat conservative. The tests for fit indicate that neither the linear nor the logit model fits the data well, which is not surprising when we look at the oscillating proportions of "Yes" in Table 5.3. If there is a trend in the probabilities across stroke severity, it is not an obvious one.

Next, we assign mid-interval scores to the categories of stroke severity, based on the Scandinavian Stroke Scale (SSS). The scores we use are 7, 22, 37, 48, and 55. The resulting P-values are shown in Table 5.9.

With the mid-interval scores, we see that the P-values are consistently higher than for equally spaced scores. The reason is that for mid-interval scores, a severe stroke has a score that is relatively lower than for a mild stroke. From a very severe to a moderate stroke, there is a more than five times increase in score, but from moderate to very mild, there is only an 1.5 times increase. We notice that the difference in the observed proportions of complications is highest between moderate and mild complications. If we assign relatively higher scores to these categories, we can easily obtain P-

TABLE 5.8
Results of tests for the linear and logit models on the data in
Table 5.3, with equally spaced scores

Test	Statistic	df*	P-value
Tests for trend in the linear model			
Cochran-Armitage	-1.79		0.074
Modified Cochran-Armitage	-1.89		0.058
Mantel-Haenszel	-1.79		0.074
Wald	-2.17		0.030
Likelihood ratio	3.86	1	0.050
Testing the fit of a linear model			
Pearson goodness-of-fit	11.47	3	0.009
Likelihood ratio (deviance)	10.77	3	0.013
Tests for trend in the logit model			
Wald	-1.78		0.075
Likelihood ratio	3.12	1	0.077
Cochran-Armitage exact cond.			0.086
Cochran-Armitage mid-P			0.077
Testing the fit of a logit model			
Pearson goodness-of-fit	11.89	3	0.008
Likelihood ratio (deviance)	11.50	3	0.009

*Degrees of freedom for the chi-squared distribution

values significantly less than 0.05. Perhaps needless to say, such a P-value-driven assignment of scores is not recommended.

5.7.15 Evaluation of Tests for Trend

Evaluation Criteria

The evaluation criteria for ordered $r \times 2$ tables were laid out in Section 5.6.6. Throughout this section, we will use a nominal significance level of 5%.

Evaluation of Actual Significance Level

First, we consider the tests for trend in the linear model. We use a four-group setup, such that we have an ordered 4×2 table, and we assign equally spaced scores $\mathbf{a} = \{1, 2, 3, 4\}$ to the groups. Figure 5.4 shows the results for a total sample size of $N = 100$, distributed equally to the four groups. The Cochran-Armitage and Mantel-Haenszel tests perform similarly, with actual significance levels close to the nominal level. Both tests violate the nominal level; however, the infringements on the nominal level are quite small. The Mantel-Haenszel test violates the nominal level slightly more often than the Cochran-Armitage test. The likelihood ratio test performs similarly to the

TABLE 5.9

Results of tests for the linear and logit models on the data in Table 5.3, when mid-interval scores have been assigned to the categories of stroke severity

Test	Statistic	df*	P-value
Tests for trend in the linear model			
Cochran-Armitage	-1.48		0.139
Modified Cochran-Armitage	-1.50		0.135
Mantel-Haenszel	-1.48		0.138
Wald	-1.63		0.103
Likelihood ratio	2.45	1	0.117
Testing the fit of a linear model			
Pearson goodness-of-fit	12.56	3	0.006
Likelihood ratio (deviance)	12.17	3	0.007
Tests for trend in the logit model			
Wald	-1.47		0.141
Likelihood ratio	2.11	1	0.147
Cochran-Armitage exact cond.			0.145
Cochran-Armitage mid-P			0.144
Testing the fit of a logit model			
Pearson goodness-of-fit	12.64	3	0.005
Likelihood ratio (deviance)	12.52	3	0.006

*Degrees of freedom for the chi-squared distribution

Cochran-Armitage and Mantel-Haenszel tests when the common success probability is in the range 0.3–0.7; however, its actual significance levels for small and large π are too high to be acceptable. The modified Cochran-Armitage test has actual significance levels close to 6% for most values of π. We find this result poor, particularly in light of the excellent performance of the Cochran-Armitage and Mantel-Haenszel tests. Yet, the worst performance comes from the Wald test, which sometimes has actual significance levels more than double the nominal level.

Figure 5.5 shows an example in which the total sample size of $N = 70$ is distributed unequally to the four groups. The actual significance levels of the tests follow the same patterns as in Figure 5.4; however, the violations of the nominal significance level are now a bit greater, particularly for the modified Cochran-Armitage test. The performance of the Cochran-Armitage and Mantel-Haenszel tests are still very good. The results in Figure 5.5 are similar to what we get if we have a smaller total sample size of about $N = 50$–60 but with an equal sample size in each group.

Second, we consider the tests for trend in the logit model. Figures 5.6 and 5.7 show the results for the same two situations we considered above for the tests of trend in the linear model. For the logit model, we find that

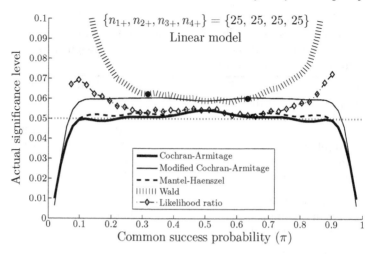

FIGURE 5.4
Actual significance levels of five tests for trend in the linear model as functions of the common success probability under H_0

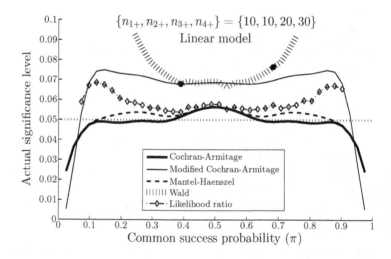

FIGURE 5.5
Actual significance levels of five tests for trend in the linear model as functions of the common success probability under H_0

the Wald test is—somewhat surprisingly considering the results of the linear model—quite acceptable, and not that much off the Cochran-Armitage mid-P test, which is the best test in both the balanced (Figure 5.6) and unbalanced

(Figure 5.7) examples. The Cochran-Armitage exact conditional test is, as expected, a bit conservative, with actual significance levels about 3–4%. The likelihood ratio test is in general slightly too liberal.

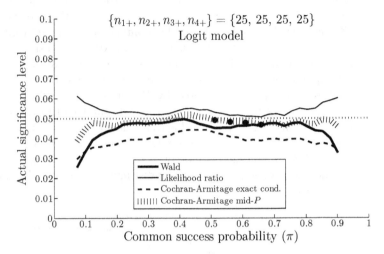

FIGURE 5.6
Actual significance levels of four tests for trend in the logit model as functions of the common success probability under H_0

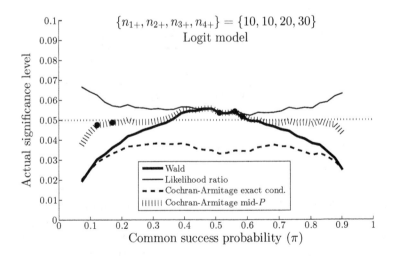

FIGURE 5.7
Actual significance levels of four tests for trend in the logit model as functions of the common success probability under H_0

Evaluation of Power

Figure 5.8 illustrates the power of the Cochran-Armitage, modified Cochran-Armitage, and Mantel-Haenszel tests for trend in the linear model. We do not consider the Wald and likelihood ratio tests because these tests have actual significance levels too far above the nominal level. That argument can also be made to exclude the modified Cochran-Armitage test; however, we have included the modified test to illustrate the potential power gain of using it compared with the standard Cochran-Armitage test. The success probabilities in Figure 5.8, denoted by Power Scenario #2, are given by $\{\pi_1, \pi_2, \pi_3, \pi_4\} = \{0.2, 0.3, 0.4, 0.5\}$, which is a linear trend in the probabilities across four groups. As expected, due to its increased actual significance levels, the modified Cochran-Armitage test has higher power than the other two tests, although the difference is mostly indicated for small sample sizes. The power of the Cochran-Armitage test is indistinguishable from that of the Mantel-Haenszel test.

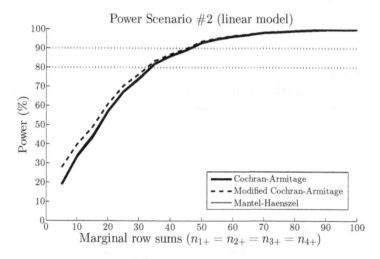

FIGURE 5.8

Power of three tests for trend in the linear model

An example of the power of the four tests for trend in the logit model is shown in Figure 5.9, where the success probabilities—labeled Power Scenario #3—are given by $\{\pi_1, \pi_2, \pi_3, \pi_4\} = \{0.17, 0.23, 0.31, 0.40\}$, which is an approximate linear trend in the logit. The Cochran-Armitage exact test has slightly lower power than the other tests. Apart from that, there is not much to separate the power of the tests.

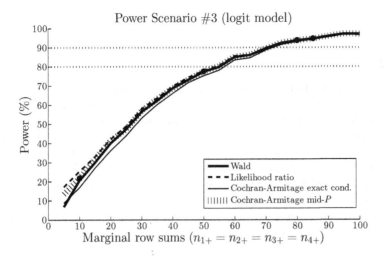

FIGURE 5.9
Power of four tests for trend in the logit model

5.8 Confidence Intervals for the Trend

In the literature on $r \times 2$ tables, there is emphasis on testing for trend, rather than estimation of trends (with confidence intervals). Testing for trend and confidence interval estimation are evidently closely related. In this section, we consider confidence intervals for two trend estimates: the Cochran-Armitage estimate and the maximum likelihood estimate.

5.8.1 The Cochran-Armitage Interval

In Section 5.7.2, the slope in the linear model (Equation 5.4) was estimated by the ordinary least squares estimate (Equation 5.16). In the general case, i.e., not under the null hypothesis, the expectations of $\hat{\beta}_{CA}$ is β, and the estimated variance of $\hat{\beta}_{CA}$ is s^2/S_{aa}^2, where s^2 is given in Equation 5.22 and S_{aa} is given in Equation 5.17.

Note that the form of the estimated variance s^2 depends on the design, and that the one given in Equation 5.22 is for a fixed group sizes design. For more details, we refer the reader to Section 5.7.2 and to Buonaccorsi et al. (2014).

For the fixed group sizes design, or stratification on the rows (the variable Y_1), the Cochran-Armitage confidence interval for β is

$$\hat{\beta}_{CA} \pm z_{\alpha/2}\frac{s}{S_{aa}},$$

where $z_{\alpha/2}$ is the upper $\alpha/2$ percentile of the standard normal distribution.

5.8.2 Wald Intervals

We have two alternative maximum likelihood confidence intervals. Both are Wald-type intervals (see Section 1.7 for a general description), and are given by the parameter estimate $\pm$ the standard normal percentile $\times$ the estimated standard error of the parameter estimate. One interval is obtained under the assumption of the linear model (Equation 5.4), and the other interval is obtained under the assumption of the logit model (Equation 5.5).

First, we consider the linear model. The maximum likelihood estimate of the trend is denoted by $\hat{\beta}_{\mathrm{ML,\,linear}}$ and obtained by maximizing Equation 5.4. Let the estimated standard error be $\widehat{\mathrm{SE}}(\hat{\beta}_{\mathrm{ML,\,linear}})$. The Wald confidence interval with linear link is given by

$$\hat{\beta}_{\mathrm{ML,\,linear}} \pm z_{\alpha/2}\widehat{\mathrm{SE}}(\hat{\beta}_{\mathrm{ML,\,linear}}).$$

Second, we consider the logit model. The maximum likelihood estimate now is $\hat{\beta}_{\mathrm{ML,\,logit}}$, and it is obtained by maximizing Equation 5.5. Denote the estimated standard error by $\widehat{\mathrm{SE}}(\hat{\beta}_{\mathrm{ML,\,logit}})$. The Wald confidence interval with logit link is given by

$$\hat{\beta}_{\mathrm{ML,\,logit}} \pm z_{\alpha/2}\widehat{\mathrm{SE}}(\hat{\beta}_{\mathrm{ML,\,logit}}).$$

The estimates of the standard error, $\widehat{\mathrm{SE}}(\hat{\beta}_{\mathrm{ML,\,linear}})$ and $\widehat{\mathrm{SE}}(\hat{\beta}_{\mathrm{ML,\,logit}})$, are obtained by inverting the second-order partial derivatives of the log likelihood (the information, see Section 1.7).

5.8.3 The Profile Likelihood Interval

The Wald-type confidence interval is one of the standard procedures for estimating confidence intervals. But the coverage of the Wald-type intervals relies on the assumption of a maximum likelihood estimate that is normally distributed, and a correct estimate of its standard error. For the linear and the logit links, the standard errors are given by the inverse of the information matrix, and may be valid only for large samples. For small to moderate sample sizes, the Wald-type intervals may have coverage lower than the nominal coverage probability. It is well known that the likelihood ratio test may perform better than the Wald test, for reasons just stated. In like manner, we may construct likelihood-based confidence intervals as an alternative to Wald-type intervals. Cox (1970, p. 88) studied likelihood methods, which generated interest in profile likelihood methods. Venzon and Moolgavkar (1988) described an algorithm for constructing profile likelihood confidence intervals, which again inspired the article by Royston (2007), who presented the `pllf` command that computes profile likelihood intervals in Stata (StataCorp LP, College Station, TX).

Suppose now that we are interested in a parameter β, which may, in general, be a vector. Let α denote the nominal significance level. Then, a $100(1 - \alpha)\%$ profile likelihood confidence interval for β consists of all parameter values β_0 that would not reject the null hypothesis $H_0 : \beta = \beta_0$ at a significance level of α. The lower (L) and upper (U) end points of the profile likelihood interval for β are given by

$$2\left[l(\hat{\beta}) - l_p(U)\right] = 2\left[l(\hat{\beta}) - l_p(L)\right] = \chi_1^2(\alpha),$$

where $l(\hat{\beta})$ is the log-likelihood of the full model, $l_p()$ is the profile log-likelihood function, see Royston (2007) or Hosmer et al. (2013, p. 19), and $\chi_1^2(\alpha)$ is the upper α percentile of the chi-squared distribution with one degree of freedom.

5.8.4 The Exact Conditional Interval

In this section, we use the sufficient statistic for β

$$T_\beta(\mathbf{n}) = \sum_{i=1}^{r} n_{i1} a_i,$$

and its exact conditional distribution (Equation 5.11) to construct a confidence interval for β. The method is an analogue to the Cornfield exact conditional interval for the 2×2 table considered in Section 4.8.6. The exact conditional interval (L, U) is obtained by solving the two equations

$$\sum_{\Omega(\mathbf{x}|n_{+1})} I\left[T_\beta(\mathbf{x}) \geq T_\beta(\mathbf{n})\right] \cdot f\left[T_\beta(\mathbf{x}), L \,|\, n_{+1}\right] = \alpha/2 \qquad (5.26)$$

and

$$\sum_{\Omega(\mathbf{x}|n_{+1})} I\left[T_\beta(\mathbf{x}) \leq T_\beta(\mathbf{n})\right] \cdot f\left[T_\beta(\mathbf{x}), U \,|\, n_{+1}\right] = \alpha/2, \qquad (5.27)$$

where $\Omega(\mathbf{x}|n_{+1})$ is the set of all tables $\mathbf{x}$ with the total number of successes equal to n_{+1}, $I()$ is the indicator function, and $f()$ is the conditional probability function in Equation 5.11. If the observed value of the statistic, $T_\beta(\mathbf{n})$, equals its minimum (maximum) value, then $L = -\infty$ $(U = \infty)$. Equations 5.26 and 5.27 must be solved iteratively.

5.8.5 The Mid-P Interval

A mid-P version of the exact conditional interval is obtained by a simple modification of Equations 5.26 and 5.27:

$$\sum_{\Omega(\mathbf{x}|n_{+1})} I\left[T_\beta(\mathbf{x}) > T_\beta(\mathbf{n})\right] \cdot f\left[T_\beta(\mathbf{x}), L \,|\, n_{+1}\right]$$

$$+ \; 0.5 \cdot \sum_{\Omega(\mathbf{x}|n_{+1})} I\left[T_\beta(\mathbf{x}) = T_\beta(\mathbf{n})\right] \cdot f\left[T_\beta(\mathbf{x}), L \,|\, n_{+1}\right] \; = \; \alpha/2$$

and

$$\sum_{\Omega(\mathbf{x}|n_{+1})} I\big[T_\beta(\mathbf{x}) < T_\beta(\mathbf{n})\big] \cdot f\big[T_\beta(\mathbf{x}), U \,|\, n_{+1}\big]$$

$$+ \; 0.5 \cdot \sum_{\Omega(\mathbf{x}|n_{+1})} I\big[T_\beta(\mathbf{x}) = T_\beta(\mathbf{n})\big] \cdot f\big[T_\beta(\mathbf{x}), U \,|\, n_{+1}\big] \;=\; \alpha/2.$$

The analogue interval for the 2×2 table is the Cornfield mid-P interval in Section 4.8.7.

5.8.6   Examples

Alcohol Consumption and Malformations (Table 5.2)

Here, we analyze Table 5.2, after assigning equally spaced scores, $1, 2, \ldots, 5$ to the exposure categories (number of drinks per day). Table 5.10 shows the estimation results.

TABLE 5.10

Estimation of the trend parameter with 95% confidence intervals (CIs) on the data in Table 5.2, with equally spaced scores

Interval	Estimate	95% CI	Width
Linear model			
Cochran-Armitage	0.0007	-0.0006 to 0.0020	0.0025
Wald	0.0005	-0.0005 to 0.0015	0.0021
Profile likelihood	0.0005	-0.0003 to 0.0015	0.0018
Logit model			
Wald	0.2278	-0.1022 to 0.5578	0.6599
Profile likelihood	0.2278	-0.1126 to 0.5465	0.6591
Exact conditional	0.2278	-0.1249 to 0.5616	0.6865
Mid-P	0.2278	-0.1107 to 0.5499	0.6606

The exact conditional interval is wider than the others, and the location of the intervals for the logit model is somewhat different; however, the overall impression is that the intervals agree well. The sample size in this example is fairly large and we do not expect big differences. For small sample sizes, on the other hand, the difference between the intervals, such as between the exact conditional and mid-P intervals, can be more substantial.

Elevated Troponin T Levels in Stroke Patients (Table 5.3)

When we analyzed the data on elevated troponin T levels in stroke patients with tests for trend in Section 5.7.14, we considered two strategies for assigning scores: equal spaced scores and mid-interval scores. Here, we follow the first strategy, that of assigning equally spaced scores. Table 5.11 shows the results.

TABLE 5.11

Estimation of the trend parameter with 95% confidence
intervals (CIs) on the data in Table 5.3, with equally spaced
scores

Interval	Estimate	95% CI	Width
Linear model			
Cochran-Armitage	-0.019	-0.0395 to 0.0007	0.0402
Wald	-0.024	-0.0458 to -0.0023	0.0435
Profile likelihood	-0.024	-0.0487 to -0.0001	0.0486
Logit model			
Wald	-0.183	-0.384 to 0.019	0.403
Profile likelihood	-0.183	-0.384 to 0.020	0.404
Exact conditional	-0.182	-0.389 to 0.025	0.414
Mid-P	-0.182	-0.384 to 0.020	0.404

The estimates for both the linear and the logit model suggest a small decrease in the probability of elevated troponin T for increasing stroke severity;
however, only the Wald and profile likelihood intervals for the linear model do
not include $\beta = 0$. The intervals give similar results, particularly for the logit
model, with the exception of the exact conditional interval, which is slightly
wider than the other intervals.

5.8.7 Evaluation of Intervals

Evaluation Criteria

We use coverage probability, interval width, and location to evaluate the performance of confidence intervals (see Section 1.4). The exact coverage probability of intervals for the ordered $r \times 2$ table is defined by

$$
\mathrm{CP}(\boldsymbol{\pi}, \mathbf{n}_+, \alpha) = \sum_{x_{11}=0}^{n_{1+}} \sum_{x_{21}=0}^{n_{2+}} \cdots \sum_{x_{r1}=0}^{n_{r+}} I(L \leq \beta \leq U) \cdot f(\mathbf{x} \,|\, \boldsymbol{\pi}, \mathbf{n}_+),
$$

where $\boldsymbol{\pi} = \{\pi_1, \pi_2, \ldots, \pi_r\}$, $I()$ is the indicator function, $L = L(\mathbf{x}, \alpha)$ and
$U = U(\mathbf{x}, \alpha)$ are the lower and upper $100(1 - \alpha)\%$ confidence limits of an
interval for the table $\mathbf{x} = \{x_{11}, n_{1+} - x_{11}, x_{21}, n_{2+} - x_{21}, \ldots, x_{r1}, n_{r+} - x_{r1}\}$,
and $f()$ is the probability distribution in Equation 5.8.

The exact expected interval width is given by

$$
\mathrm{Width}(\boldsymbol{\pi}, \mathbf{n}_+, \alpha) = \sum_{x_{11}=0}^{n_{1+}} \sum_{x_{21}=0}^{n_{2+}} \cdots \sum_{x_{r1}=0}^{n_{r+}} (U - L) \cdot f(\mathbf{x} \,|\, \boldsymbol{\pi}, \mathbf{n}_+).
$$

We measure the location of an interval with the MNCP/NCP index. The non-coverage probability (NCP) is computed as $1 - \text{CP}$, and the mesial non-coverage probability (MNCP) is defined as

$$\text{MNCP}(\boldsymbol{\pi}, \mathbf{n}_+, \alpha) =$$

$$\sum_{x_{11}=0}^{n_{1+}} \sum_{x_{21}=0}^{n_{2+}} \cdots \sum_{x_{r1}=0}^{n_{r+}} I(L > \beta \geq 0 \text{ or } U < \beta \leq 0) \cdot f(\mathbf{x} \,|\, \boldsymbol{\pi}, \mathbf{n}_+).$$

Evaluation of Coverage Probability

Figure 5.10 illustrates the coverage probabilities of the Cochran-Armitage and Wald intervals for the trend parameter (β) in the linear model. In this scenario, we have four groups with equally spaced scores assigned to them, and we have set $\beta = 0.15$, which indicates a fairly large increasing trend in the probabilities of success for increasing row numbers. In the figure, the coverage probabilities are plotted as functions of the sample size, which is equally distributed across the four groups. We only show the results of one scenario here; other scenarios, with for instance, a smaller β, tend to give very similar results. Both intervals have coverage probabilities below the nominal level. Except for sample sizes in the range 70–90, the drop in coverage below the nominal level is moderate to substantial. The Cochran-Armitage interval is slightly better than the Wald interval.

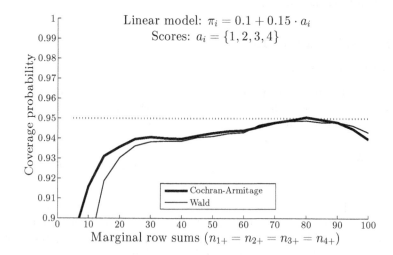

FIGURE 5.10
Simulated coverage probabilities of the Cochran-Armitage and Wald intervals for the trend parameter in the linear model

The profile likelihood interval is not included in these calculations because of its computational burden. The calculation of individual intervals is of no concern; however, to calculate or simulate coverage probabilities, thousands of intervals need to be computed, and it is outside the scope of this book to perform those calculations. We note that the profile likelihood interval for the linear model tends to produce confidence limits that are similar to the Wald interval or the Cochran-Armitage interval or both. It is therefore likely that the coverage probability of the profile likelihood interval for the linear model is also too low; however, we state this with considerable caution.

We now turn our attention to the logit model. The previous comments on the feasibility of calculating coverage probabilities of the profile likelihood interval for the linear model also applies to the profile likelihood, exact conditional, and mid-P intervals for the logit model. That leaves the Wald interval for the logit model, for which we illustrate the coverage probabilities in Figure 5.11. The performance of the Wald interval is excellent. The coverage probabilities are slightly above the nominal level, even for quite small sample sizes, and this performance is repeated in other scenarios that we do not show here.

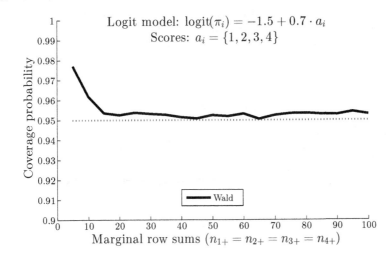

FIGURE 5.11
Coverage probabilities of the Wald interval for the trend parameter in the logit model

Based on the results of calculating the profile likelihood, exact conditional, and mid-P intervals for the examples (see Section 5.8.6), we expect the profile likelihood and mid-P intervals to have similar coverage probabilities as the Wald interval, and that the coverage probabilities of the exact conditional interval is somewhat higher.

Evaluation of Width

For the linear model, there is not much to report about the width of the intervals. In our evaluations of the widths of the Cochran-Armitage and Wald intervals, we find no meaningful difference between them. The width of the profile likelihood interval also seems to be quite similar to the other two intervals.

For the logit model, the Wald, profile likelihood, and mid-P intervals tend to produce similar results; however, we have not investigated this in any detail. Because of its conservative nature, the exact conditional interval is somewhat wider than the other three intervals.

Evaluation of Location

Figures 5.12 and 5.13 show the location index MNCP/NCP for the Cochran-Armitage and Wald intervals for the linear model (Figure 5.12), and the Wald interval for the logit model (Figure 5.13). These two figures provide more evidence for the preference of using the logit model and not the linear model for estimating the trend with a confidence interval; the two intervals for the linear model are too distally located, whereas the Wald interval for the logit model is satisfactorily located for all but the smallest sample sizes.

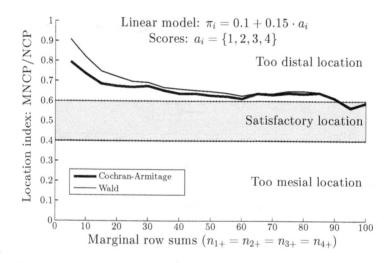

FIGURE 5.12

Location, as measured by the MNCP/NCP index, of the Cochran-Armitage and Wald intervals for the trend parameter in the linear model

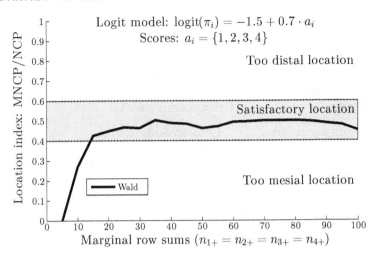

FIGURE 5.13
Location, as measured by the MNCP/NCP index, of the Wald interval for the trend parameter in the logit model

5.9 Recommendations

5.9.1 Summary

Table 5.12 contains our main recommendations for testing and estimation in ordered $r \times 2$ tables. In the following subsections, we go into more detail concerning the methods for each type of analysis, but first, we give some overall thoughts on when each type of analysis should be used.

We have a soft spot for the tests for unspecific ordering. In our experience, model-based trends can be too restrictive, and the more general alternative hypothesis of an unspecific ordering is often a good match to the subject matter research question. Because we do not assume a specific type of trend, there is no trend parameter to estimate, and some may consider this—on general terms—to be a disadvantage. We do not think so, and unless there is ground to believe that the true underlying probabilities can be well fitted by a simple model, we prefer to test for unspecific ordering, and, if indicated, describe the differences between the groups that make up the rows in the table with descriptive statistics or combine appropriate groups and analyze the data in more detail with methods for the 2×2 table. A further advantage with testing for unspecific ordering is the excellent performance of the exact conditional tests. The obvious disadvantage with these tests is their lack of support in statistical software packages.

If a model-based approach is to be used, we have a choice between the linear

TABLE 5.12

Recommended tests and confidence intervals (CIs) for ordered $r \times 2$ tables

Analysis	Recommended methods	Sample sizes
Tests for unspecific ordering	Exact conditional (Pearson)	all
	Exact conditional (LR)	all
Tests for linear trend	Cochran-Armitage	all
	Mantel-Haenszel	all
CIs for linear trend	We do not recommend this analysis	
Testing the fit of a linear model	Pearson goodness-of-fit	all
	Likelihood ratio (deviance)	all
Tests for logit trend	Cochran-Armitage mid-P	small/medium
	Wald	medium/large
CIs for logit trend	Wald	all
Testing the fit of a logit model	Pearson goodness-of-fit	all
	Likelihood ratio (deviance)	all

model and the logit model. For testing trend in the probabilities, both the linear and logit models have tests that perform well, including the ubiquitous Cochran-Armitage test. For estimating the trend with a confidence interval, however, we do not recommend the linear model, simply because we are not aware of any confidence interval method for the linear model that performs well. The good news is that the simple Wald interval for the logit model is a method that performs excellently in most situations. The logit model is therefore our model of choice for ordered $r \times 2$ tables. This recommendation is strengthened by the fact that the Cochran-Armitage test, which originally was derived for the linear model, also is a score test for trend in the logit model.

5.9.2   Tests for Unspecific Ordering

The asymptotic Pearson chi-squared and likelihood ratio tests for unspecific ordering do not perform well. For a nominal significance level of 5%, they tend to have actual significance levels in the range 6–7%. This problem does not seem to become less with increasing sample size. At $N = 4000$, equally distributed to four groups, the actual significance levels of both tests are still in the 6–7% range.

The exact conditional tests, on the other hand, regardless of whether the Pearson chi-squared or likelihood ratio statistic is used, perform excellently, with actual significance levels slightly below the nominal level. Because the tests are exact and obtain P-values by summing probabilities over all possible tables, the computational burden increases with increasing sample size. Thus, the exact tests may not be feasible for very large sample sizes; however, an ordinary laptop computer has no problem with the two examples in Table 5.2 ($N = 32574$) and Table 5.3 ($N = 489$).

If the exact conditional tests are unavailable or if the sample size is too big for exact calculations, the Pearson chi-squared and likelihood ratio tests can be used with the provision that the results are interpreted with an increased probability of type I error in mind.

5.9.3 Tests for Trend

The Cochran-Armitage test is the workhorse for testing trend in ordered binomial tables. It goes back to Cochran (1954b) and Armitage (1955) who proposed a test for trend in the linear model, based on the estimate of the standard linear regression coefficient. Both the Cochran-Armitage and Mantel-Haenszel tests—but not the modified Cochran-Armitage test and definitely not the Wald test—can be recommended for the linear model.

In general, however, we recommend use of the logit model. These conclusions are in line with those of Wellek and Ziegler (2012) for genetic association studies. It turns out that the Cochran-Armitage test also is the score test for testing trend in the logit model, see Section 5.7.2, which explains why it is adequate also for the logit model.

Another benefit of the logit model is that the simple Wald test performs surprisingly well for all but the smallest sample sizes. The corresponding Wald confidence interval also performs well and together they provide a unified approach to the analysis of ordered $r \times 2$ tables. An alternative test for the logit model that performs even better than the Wald test, at least for small and moderately large tables, is the Cochran-Armitage mid-P test. The mid-P test would be our first choice based on its performance; however, it is not widely available in statistical software packages. The Cochran-Armitage exact conditional test, upon which the mid-P test is based, is somewhat conservative, and we recommend it only in situations where no violations of the nominal significance level is acceptable. Note that the exact conditional likelihood tests are equivalent to the Cochran-Armitage tests.

5.9.4 Confidence Intervals for the Trend

For confidence interval estimation, we do not recommend use of the linear model, simply because the confidence interval methods available for the linear model do not perform well. For the logit model, however, the simple Wald interval has excellent performance even for quite small sample sizes. It is likely that the profile likelihood, exact conditional, and mid-P intervals for the logit model also perform well; however, the Wald interval is considerably simpler to calculate, so we do not think it is worth the extra effort to calculate any of the other three intervals.

6

The Ordered $2 \times c$ Table

6.1 Introduction

In Chapter 3, we studied $1 \times c$ tables, which occur when we have observations of a nominal categorical variable with c possible outcomes. The probability of outcome i is π_i. If the N observations are independent, the cell counts are multinomially distributed with parameters N and $\pi_1, \pi_2, \ldots, \pi_c$.

In this chapter, we extend the situation in Chapter 3 in two directions. First, we extend from $1 \times c$ tables to $2 \times c$ tables, that is from one group to two groups. Such tables occur if we, for instance, study the effect of one medicine in comparison with placebo in randomized controlled trials, or in cohort studies, where we compare an exposed with an unexposed group. Second, we extend from nominal to ordinal outcomes. Ordinal outcomes are quite common, and examples include degrees of severity for diseases, which may range from none to very severe, or ranking of travelers' rating of hotels and restaurants on TripAdvisor, where the range is from terrible to excellent. We assume that the two groups are independent, with two multinomial distributions for the outcomes.

An ordered $2 \times c$ table gives the observed counts for a cross-classification of two groups and c ordered outcomes, as shown in Table 6.1.

TABLE 6.1
The observed counts of a $2 \times c$ table

	Category 1	Category 2	...	Category c	Total
Group 1	n_{11}	n_{12}	...	n_{1c}	n_{1+}
Group 2	n_{21}	n_{22}	...	n_{2c}	n_{2+}
Total	n_{+1}	n_{+2}	...	n_{+c}	N

Additional notation:
$\mathbf{n} = \{n_{11}, n_{12}, \ldots, n_{1c}, n_{21}, n_{22}, \ldots, n_{2c}\}$: the observed table
$\mathbf{x} = \{x_{11}, x_{12}, \ldots, x_{1c}, x_{21}, x_{22}, \ldots, x_{2c}\}$: any possible table

There is a vast literature on analysis of ordinal data. Seminal texts in this area include the review paper by McCullagh (1980) and the textbook by McCullagh and Nelder (1989). The standard text book for many, however, is Agresti (2010), and we will refer to this book throughout this chapter. Other relevant texts include Hosmer et al. (2013, Section 8.2) and Kleinbaum et al. (2014, Section 23.7).

This chapter is organized as follows. Section 6.2 introduces two examples of ordered $2 \times c$ tables, which will be analyzed later in the chapter. Section 6.3 defines the notation we will use, and presents two sampling distributions for $2 \times c$ tables. In Section 6.4, we discuss the local, cumulative, and continuation odds ratios as association measures used to account for the ordering. Instead of using the odds ratios to account for the ordering, we may assign scores to the columns in the $2 \times c$ table, which is the topic of Sections 6.8 and 6.9. Section 6.10 presents three models for ordinal data that use the logit link. Our main interest for the remainder of this chapter will be on the cumulative logit model (Sections 6.11–6.14) and the cumulative probit model (Section 6.15), wherein we present methods to test for effects and model fit. In Section 6.16, we return to the examples and use the methods from the previous sections to test for effects and model fit. Section 6.17 evaluates the tests. Methods for confidence interval estimation are presented in Section 6.18. Finally, we summarize and give recommendations in Section 6.19.

6.2   Examples

6.2.1   The Adolescent Placement Study

In an article by Fontanella et al. (2008), results from a study of determinants of aftercare placement for psychiatric hospitalized adolescent patients are presented. These data are also analyzed with multinomial logistic regression by Hosmer et al. (2013). For a code sheet of the variables, see Hosmer et al. (2013, p. 26). Here, we will analyze the association between danger to others as an ordered variable with four outcomes and gender as the group variable. We then have an ordered 2×4, and the observed counts are shown in Table 6.2.

TABLE 6.2

Danger to others for males and females (Fontanella et al., 2008)

	Danger to others				
	Unlikely	**Possible**	**Probable**	**Likely**	**Total**
Male	8 (3.4%)	28 (12%)	72 (31%)	126 (54%)	234
Female	46 (17%)	73 (27%)	69 (25%)	86 (31%)	274
Total	54 (11%)	101 (20%)	141 (28%)	212 (42%)	508

Note that there is no cell in Table 6.2 with expected count less than five, under the null hypothesis of no association between the outcome and the grouping by gender.

6.2.2   Postoperative Nausea

We use an example from Lydersen et al. (2012a, p. 76). In this randomized
controlled trial, the outcome postoperative nausea was recorded with a four-
category variable, as shown in Table 6.3. The sample size in this example
($N = 59$) is much smaller than that in the previous example ($N = 508$), and
two cells have expected counts less than five, under the null hypothesis of no
association between the outcome and the treatment.

TABLE 6.3
Postoperative nausea by treatment (Lydersen et al., 2012a, p. 76)

	Postoperative nausea				
	Not at all	A little	Quite a bit	Very much	Total
VT*	14 (48%)	10 (34%)	3 (10%)	2 (6.9%)	29
Control	11 (37%)	7 (23%)	8 (27%)	4 (13%)	30
Total	25 (42%)	17 (29%)	11 (19%)	6 (10%)	59

*Ventricular tube

6.3   Notation and Sampling Distributions

Here, as in Sections 1.3 and 5.3, we let Variable 1 (Y_1) equal the grouping,
and we let Variable 2 (Y_2) equal the outcome. The outcome variable Y_2 has
c categories, and $Y_2 = j$ if the observation is in category j. The grouping
variable Y_1 has two categories, and we let $Y_1 = 1$ for observations in Group 1,
and $Y_1 = 2$ for observations in Group 2.

The conditional probability of the outcome variable $Y_2 = j$ given $Y_1 = i$ is

$$\Pr(Y_2 = j \mid Y_1 = i) = \pi_{j|i},$$

for $i = 1, 2$ and $j = 1, 2, \ldots, c$. For each $i = 1, 2$, we have that $\sum_{j=1}^{c} \pi_{j|i} = 1$.
A simple generalization of the logit model from the 2×2 table to the $2 \times c$
table would now be

$$\log\left(\frac{\pi_{j|i}}{\pi_{1|i}}\right) = \alpha_j + \beta_j x_i,$$

where j is now indexed as $2, 3, \ldots, c$, and $x_1 = 1$ denotes Group 1 and $x_2 = 0$
denotes Group 2. Then,

$$\beta_j = \log\left(\frac{\pi_{j|1}/\pi_{1|1}}{\pi_{j|2}/\pi_{1|2}}\right), \tag{6.1}$$

or—equivalently—that

$$\exp(\beta_j) = \frac{\pi_{j|1} \cdot \pi_{1|2}}{\pi_{j|2} \cdot \pi_{1|1}},$$

which gives the ratio of odds for outcome in category j compared with category 1 for Group 1 relative to Group 2.

Note that since $\pi_{ij} = \pi_{j|i} \cdot \pi_{i+}$, Equation 6.1 is equivalent to

$$\beta_j = \log \left(\frac{\pi_{1j}/\pi_{11}}{\pi_{2j}/\pi_{21}} \right),$$

and an estimate of β_j is

$$\hat{\beta}_j = \log \left(\frac{n_{1j}/n_{11}}{n_{2j}/n_{21}} \right). \tag{6.2}$$

This is simply the log of the odds ratio that we are familiar with from the 2×2 table in Chapter 4.

In this chapter, we assume that the two groups that make up the rows of an ordered $2 \times c$ table are independent, and that the group sizes (n_{1+} and n_{2+}) are fixed by design. Each row is thus multinomially distributed, and the sampling distribution for the whole table is the product of two multinomial distributions:

$$f(\mathbf{x} \mid \boldsymbol{\pi}, \mathbf{n}_+) = \prod_{i=1}^{2} \frac{n_{i+}!}{x_{i1}!x_{i2}! \cdots x_{ic}!} \pi_{1|i}^{x_{i1}} \cdot \pi_{2|i}^{x_{i2}} \cdots \pi_{c|i}^{x_{ic}}, \tag{6.3}$$

where $\boldsymbol{\pi} = \{\pi_{1|1}, \pi_{2|1}, \ldots, \pi_{c|1}, \pi_{1|2}, \pi_{2|2}, \ldots, \pi_{c|2}\}$ and $\mathbf{n}_+ = \{n_{1+}, n_{2+}\}$ denotes the fixed group sizes. We denote the distribution in Equation 6.3 for the unconditional (sampling) distribution, since we have not conditioned on any marginals—expect those fixed by design—to obtain it. The unconditional distribution gives rise to exact unconditional tests (Section 6.14.6), whose calculations are complicated due to the presence of several nuisance parameters (the elements of $\boldsymbol{\pi}$). To simplify the calculations of exact tests, we may condition on the marginals $n_{+1}, n_{+2}, \ldots, n_{+c}$. Let the vector $\mathbf{n}_{++}$ denote the set of both row and column marginals. Under the null hypothesis

$$H_0 : \pi_{j|1} = \pi_{j|2},$$

for $j = 1, 2, \ldots, c$, the conditional distribution of the ordered $2 \times c$ table, given fixed row and column marginals, is

$$f(\mathbf{x} \mid \mathbf{n}_{++}) = \frac{\prod_{j=1}^{c} \binom{n_{+j}}{x_{1j}}}{\binom{N}{n_{1+}}}. \tag{6.4}$$

This is the multiple hypergeometric distribution, which we also obtained in a similar manner for the $r \times 2$ table (Equation 5.9, Section 5.4). The conditional distribution in Equation 6.4 is free of unknown parameters, and as such, makes calculations of exact conditional P-values straightforward (see Section 6.5 and 6.14.4).

6.4   Accounting for the Ordering with Odds Ratios

The parameterization in Section 6.3 does not take the ordering of the categories of Y_2 into account. We will now describe three odds ratios that take the ordering directly into account: local odds ratios, cumulative odds ratios, and continuation odds ratios (Agresti and Coull, 2002). The local odds ratios are defined as

$$\theta_j^{\mathrm{LO}} = \frac{\pi_{j+1|1}/\pi_{j|1}}{\pi_{j+1|2}/\pi_{j|2}}; \tag{6.5}$$

the cumulative odds ratios are defined as

$$\theta_j^{\mathrm{CU}} = \frac{(\pi_{1|1} + \pi_{2|1} + \cdots + \pi_{j|1})/(\pi_{j+1|1} + \pi_{j+2|1} + \cdots + \pi_{c|1})}{(\pi_{1|2} + \pi_{2|2} + \cdots + \pi_{j|2})/(\pi_{j+1|2} + \pi_{j+2|2} + \cdots + \pi_{c|2})}; \tag{6.6}$$

and the continuation odds ratios are defined as

$$\theta_j^{\mathrm{CO}} = \frac{\pi_{j|1}/(\pi_{j+1|1} + \pi_{j+2|1} + \cdots + \pi_{c|1})}{\pi_{j|2}/(\pi_{j+1|2} + \pi_{j+2|2} + \cdots + \pi_{c|2})}, \tag{6.7}$$

for $j = 1, 2, \ldots, c-1$. Figure 6.1 illustrates the three odds ratios.

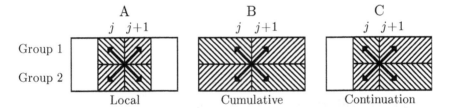

FIGURE 6.1
Three different types of odds ratios in $2 \times c$ tables: local odds ratios (A), cumulative odds ratios (B), and continuation odds ratios (C)

Given any of the above odds ratios, a positive association between the outcome and the grouping is defined as an odds ratio greater than 1. We have that

$$\theta_j^{\mathrm{LO}} \geq 1 \ \Rightarrow \ \theta_j^{\mathrm{CU}} \geq 1, \tag{6.8}$$

and

$$\theta_j^{\mathrm{LO}} \geq 1 \ \Rightarrow \ \theta_j^{\mathrm{CO}} \geq 1, \tag{6.9}$$

see, for instance, Agresti and Coull (2002). Negative associations are defined with odds ratios less than 1. For any of the three odds ratios above, the null hypothesis to be tested is

$$H_0 : \theta_j = 1 \tag{6.10}$$

versus

$$H_A : \theta_j \geq 1,$$

with strict inequality for at least one j. Because of the ordering of the odds ratios in Equations 6.8 and 6.9, the alternatives are nested. When we use the likelihood ratio test, there will be an ordering of the P-values, such that the smallest P-value will be for testing local odds ratios.

In the following three sections, we will briefly summarise some results and tests for the three odds ratios above, in the case of a positive association between the outcome and the grouping.

6.5   Local Odds Ratios

Equation 6.5 defined the local odds ratios via conditional probabilities of Y_2 given Y_1. The "opposite" conditional probabilities, those of Y_1 given Y_2, are denoted by

$$\Pr(Y_1 = i \mid Y_2 = j) = \gamma_{i|j}.$$

In a similar manner as in Section 5.3.3, we can express the $\gamma_{i|j}$ by the $\pi_{j|i}$ as follows:

$$\gamma_{1|j} = \frac{\pi_{j|1}\Pr(Y_1 = 1)}{\Pr(Y_2 = j)} \tag{6.11}$$

and

$$\gamma_{2|j} = \frac{\pi_{j|2}\Pr(Y_1 = 2)}{\Pr(Y_2 = j)}. \tag{6.12}$$

We combine Equations 6.5, 6.11, and 6.12 and get

$$\theta_j^{\mathrm{LO}} = \frac{\gamma_{1|j+1}/\gamma_{1|j}}{\gamma_{2|j+1}/\gamma_{2|j}}.$$

We have that $\gamma_{1|j} + \gamma_{2|j} = 1$ and $\gamma_{1|j+1} + \gamma_{2|j+1} = 1$, which gives that if $\theta_j^{\mathrm{LO}} \geq 1$ for all $j = 1, 2, \ldots, c-1$, then $\gamma_{1|j+1} \geq \gamma_{1|j}$ for all $j = 1, 2, \ldots, c-1$.

This means that when we have a positive association defined by the local odds ratios, we have a similar ordering of odds ratios based on the conditional probabilities of Y_1 given Y_2. Thus, we have the same ordering of probabilities as in an $r \times 2$ table, and we can analyze the $2 \times c$ table as if it were an $r \times 2$ table, with the methods described in Section 5.6. For the sake of completeness, we briefly repeat the methods here, and illustrate the calculations on the example in Table 6.3.

6.5.1 The Pearson Chi-Squared and Likelihood Ratio Tests for Local Odds Ratios

For simplicity, we now let $\gamma_j = \gamma_{1|j}$. Under the null hypothesis in Equation 6.10, the maximum likelihood estimate of γ_j is n_{1+}/N for all $j = 1, 2, \ldots, c$. Let the sample proportions be $p_j = n_{1j}/n_{+j}$. If the sample proportions are ordered in accordance with the alternative hypothesis $p_1 \leq p_2 \leq \ldots \leq p_c$ or $p_1 \geq p_2 \geq \ldots \geq p_c$, the maximum likelihood estimates of the γ_j (under the alternative hypothesis) are equal to the sample proportions $\hat{\gamma}_j = p_j = n_{1j}/n_{+j}$. If any pair of p_j and p_{j+1} is out of order according to the alternative hypothesis, we pool the adjacent sample proportions: $p_j = p_{j+1} = (n_{1j} + n_{1,j+1})/(n_{+j} + n_{+,j+1})$. Continue pooling out-of-order proportions until all the proportions are monotonically increasing (or decreasing). Then, the maximum likelihood estimates of the γ_j are equal to the sample proportions, with pooling of the out-of-order proportions: $\hat{\gamma}_j = p_j$.

Let $n_{1j}^* = n_{+j}\hat{\gamma}_j$ and $n_{2j}^* = n_{+j}(1 - \hat{\gamma}_j)$. The Pearson test statistic has the form

$$T_{\text{Pearson}}(\mathbf{n}) = \sum_{i,j} \frac{(n_{ij}^* - m_{ij})^2}{m_{ij}}, \tag{6.13}$$

and the likelihood ratio test statistic is

$$T_{\text{LR}}(\mathbf{n}) = 2 \sum_{i,j} n_{ij}^* \log\left(\frac{n_{ij}^*}{m_{ij}}\right), \tag{6.14}$$

where $m_{ij} = n_{i+}n_{+j}/N$ are the expected cell counts under H_0, and a term in Equation 6.14 is zero if $n_{ij}^* = 0$.

Both T_{Pearson} and T_{LR} are asymptotically chi-bar-squared distributed (see Section 5.6). The P-value for the Pearson chi-squared test is then given by

$$P\text{-value} = \sum_{j=1}^{c} \rho(c, j) \cdot \Pr\left[\chi_{j-1}^2 \geq T_{\text{Pearson}}(\mathbf{n})\right]. \tag{6.15}$$

Here, $\chi_0^2 = 0$, and the $\rho(c, j)$ are probabilities, see Section 5.6. To obtain P-values for the likelihood ratio test, substitute T_{LR} for T_{Pearson} in Equation 6.15.

6.5.2 Exact and Mid-P Tests for Local Odds Ratios

As in Section 5.6, we may use the Pearson chi-squared and likelihood ratio test statistics to derive exact conditional and mid-P tests. The P-value for the exact conditional test with the Pearson chi-squared statistic is given by

$$\text{exact cond. } P\text{-value} = \sum_{\Omega(\mathbf{x}|\mathbf{n}_{++})} I\left[T_{\text{Pearson}}(\mathbf{x}) \geq T_{\text{Pearson}}(\mathbf{n})\right] \cdot f(\mathbf{x} \,|\, \mathbf{n}_{++}),$$

where $\Omega(\mathbf{x}|\mathbf{n}_{++})$ is the set of all tables $\mathbf{x}$ with marginals equal to $\mathbf{n}_{++}$, $I()$ is the indicator function, $T_{\text{Pearson}}()$ is the Pearson chi-squared statistic in

Equation 6.13, and $f(\mathbf{x} \,|\, \mathbf{n}_{++})$ is the conditional probability distribution in Equation 6.4. The corresponding mid-P value is given by

$$\text{mid-}P \text{ value} = \sum_{\Omega(\mathbf{x}|\mathbf{n}_{++})} I\big[T_{\text{Pearson}}(\mathbf{x}) > T_{\text{Pearson}}(\mathbf{n})\big] \cdot f(\mathbf{x} \,|\, \mathbf{n}_{++})$$

$$+ \; 0.5 \cdot \sum_{\Omega(\mathbf{x}|\mathbf{n}_{++})} I\big[T_{\text{Pearson}}(\mathbf{x}) = T_{\text{Pearson}}(\mathbf{n})\big] \cdot f(\mathbf{x} \,|\, \mathbf{n}_{++}).$$

The exact conditional and mid-P tests with the likelihood ratio statistic are obtained by substituting $T_{\text{LR}}()$ for $T_{\text{Pearson}}()$ in the two equations above.

We return to exact and mid-P tests and the conditional distribution in Sections 6.14.4 and 6.14.5, where we derive an exact conditional linear rank test and a mid-P linear rank test.

6.5.3   Example

Postoperative Nausea (Table 6.3)

Here, we consider the randomized controlled trial of ventricular tube versus control treatment for postoperative nausea, first introduced in Section 6.2. We will calculate six tests for ordered local odds ratios: the Pearson chi-squared test, the likelihood ratio test, two exact conditional tests, and two mid-P tests. We also calculate the Pearson chi-squared and likelihood ratio tests for unordered alternatives. These methods will be presented in more detail in the chapter on $r \times c$ tables (Chapter 7); we include them here merely to illustrate the difference between tests for ordered and unordered alternative hypotheses, and we do not advocate use of unordered tests for ordered data.

Under the null hypothesis of no association between treatment and postoperative nausea, the maximum likelihood estimate of γ_j is $29/59 = 0.4915$ for all $j = 1, 2, \ldots, c$. The sample proportions, however, seem to be decreasing: $(p_1, p_2, p_3, p_4) = (0.5600, 0.5882, 0.2727, 0.3333)$. Because $p_2 > p_1$, we pool p_1 and p_2: $p_1 = p_2 = (14 + 10)/(25 + 17) = 0.5714$. Moreover, $p_4 > p_3$, so we pool p_3 and p_4: $p_3 = p_4 = (3 + 2)/(11 + 6) = 0.2941$. Then, the maximum likelihood estimates under the alternative hypothesis are $(\hat{\gamma}_1, \hat{\gamma}_2, \hat{\gamma}_3, \hat{\gamma}_4)$ $= (0.5714, 0.5714, 0.2941, 0.2941)$. We can now calculate the n_{ij}^*s as described in Section 6.5. The observed values of the Pearson chi-squared and likelihood ratio statistics are $T_{\text{Pearson}} = 3.724$ and $T_{\text{LR}} = 3.813$. The probabilities $\rho(4, j)$ can be found from Table 5.5. The P-value for the Pearson chi-squared test can be calculated by

$$\begin{aligned} P\text{-value} \; = \; & 0.2500 \cdot \Pr(\chi_0^2 \geq 3.724) + 0.4583 \cdot \Pr(\chi_1^2 \geq 3.724) \\ & + \; 0.2500 \cdot \Pr(\chi_2^2 \geq 3.724) + 0.0417 \cdot \Pr(\chi_3^2 \geq 3.724), \end{aligned}$$

which gives

$$\begin{aligned} P\text{-value} \; = \; & 0.2500 \cdot 0 + 0.4583 \cdot 0.0536 + 0.2500 \cdot 0.1554 + 0.0417 \cdot 0.2929 \\ = \; & 0.0756. \end{aligned}$$

The likelihood ratio test can be calculated in a similar manner, by substituting the observed value of the likelihood ratio statistic, $T_{LR} = 3.813$, for the observed value of the Pearson chi-squared statistic, $T_{Pearson} = 3.724$, in the equations above.

The results are summarized in Table 6.4. We note two things. First, that the Pearson chi-squared and likelihood ratio tests agree quite well on these data. That was not the case with the data on alcohol consumption and malformations in Section 5.6.5. Second, and this result agrees well with that in Section 5.6.5, the P-values for the ordered tests are markedly smaller than the P-values for the unordered tests. As noted in Section 5.6.5, this illustrates the general result that tests that account for ordered data through constraints on the alternative hypothesis have higher power than tests that do not put any constraint on the alternative hypothesis.

TABLE 6.4

Results of tests for unordered alternatives and tests for ordered local odds ratios on the data in Table 6.3

Test	Statistic	df*	*P*-value
H_A: *Unordered*			
Pearson chi-squared[†]	3.81	3	0.282
Likelihood ratio[†]	3.91	3	0.271
H_A: *Ordered local odds ratios*			
Pearson chi-squared	3.72	(chi-bar-squared)	0.0756
Likelihood ratio	3.81	(chi-bar-squared)	0.0722
Exact conditional (with $T_{Pearson}$ as test statistic)			0.134
Mid-P (with $T_{Pearson}$ as test statistic)			0.128
Exact conditional (with T_{LR} as test statistic)			0.134
Mid-P (with T_{LR} as test statistic)			0.128

*Degrees of freedom for the chi-squared distribution
[†]These methods are presented in Chapter 7

Also included in Table 6.4 are the results for the two exact conditional tests and the two mid-P tests. Both the exact P-values and the mid-P values are considerably larger than the Pearson chi-squared and likelihood ratio tests, with the mid-P values only slightly smaller than the exact P-values. In Chapter 5 (Section 5.6.6), we evaluated tests for unspecific ordering in $r \times 2$ tables that are equivalent to the tests for ordered local odds ratios here. In that evaluation, we observed that the asymptotic Pearson chi-squared and likelihood ratio tests had actual significance levels above the nominal level, mostly between 6% and 7% for a nominal level of 5%. Hence, we prefer to treat the results of the exact conditional and mid-P tests with considerable more confidence than the other four tests in Table 6.4.

We return to this example in Section 6.16, where we apply the data to the proportional odds and cumulative probit models.

6.5.4   Evaluation of Tests

The tests for ordered local odds ratios in the $2 \times c$ table are equal to the tests for unspecific ordering in the $r \times 2$ table, which we presented in Section 5.6. In that section, we also evaluated the tests and found that the two asymptotic tests (Pearson chi-squared and likelihood ratio) have actual significance levels at about 6–7% for a nominal level of 5%, but that the exact and mid-P tests have actual significance levels very close to the nominal level. These results also apply for the tests in the $2 \times c$ table, and we do not repeat the evaluations here. For more details, see Section 5.6.6.

6.6   Cumulative Odds Ratios

The association between the outcome and the grouping can also be measured by the cumulative odds ratios (θ_j^{CU}) in Equation 6.6. When there is a positive association, the θ_j^{CU} are all greater than or equal to 1, which means that

$$\pi_{1|1} + \pi_{2|1} + \ldots + \pi_{j|1} \geq \pi_{1|2} + \pi_{2|2} + \ldots + \pi_{j|2}, \quad \text{for all } j,$$

and the cumulative conditional probabilities in Group 1 are all larger than the cumulative conditional probabilities in Group 2 for all outcome categories.

In the following, we derive Pearson chi-squared, likelihood ratio, exact, and mid-P tests of the null hypothesis in Equation 6.10. As always, we need the maximum likelihood estimates under the null and alternative hypotheses.

6.6.1   The Pearson Chi-Squared and Likelihood Ratio Tests for Cumulative Odds Ratios

The maximum likelihood estimates under the alternative hypotheses was derived by Brunk et al. (1966) and reviewed by Agresti and Coull (2002). Here, we will briefly summarize the derivations. Let $r_j = (n_{11}+n_{12}+\ldots+n_{1j})/(n_{21}+ n_{22} + \ldots + n_{2j})$ for $j = 1, 2, \ldots, c$. In the following, the columns are divided into subsets so that there is an ordering. The first subset consists of columns number $1, 2, \ldots, v_1$, for which r_{v_1} is the minimum of $\{r_1, r_2, \ldots, r_c\}$. We denote this subgroup by J_1. If $v_1 = c$, there is just one subgroup. If $v_1 < c$, the next subset consists of columns number $v_1 + 1, v_1 + 2, \ldots, v_2$, for which r_{v_2} is the minimum of $\{r_{v_1+1}, r_{v_1+2}, \ldots, r_c\}$. We continue along these lines until we have reached the last column, and we have obtained the subgroups $J_1, J_2, \ldots, J_m$. For details, see Grove (1980) and Agresti and Coull (2002). As in Agresti and Coull (2002), we denote the subset of columns in, say, subset J_h by $\{a, a + 1, \ldots, b\}$. The expected count for cell (i, j) in the columns that make up J_h is then

$$m_{ijh} = \frac{n_{+j}(n_{ia} + n_{ia+1} + \ldots + n_{ib})}{n_{+a} + n_{+a+1} + \ldots + n_{+b}},$$

see also Grove (1980). Note that the n_{+j} in the preceding equation are the column sums in J_h, and not the column sums in the original $2 \times c$ table.

The Pearson chi-squared test statistic can now be written as

$$T_{\text{Pearson}}(\mathbf{n}) = \sum_{h=1}^{m} \sum_{i=1}^{2} \sum_{j \in J_h} \frac{(n_{ij} - m_{ijh})^2}{m_{ijh}}, \tag{6.16}$$

and the likelihood test statistic can be written as

$$T_{\text{LR}}(\mathbf{n}) = 2 \sum_{h=1}^{m} \sum_{i=1}^{2} \sum_{j \in J_h} n_{ij} \log\left(\frac{n_{ij}}{m_{ijh}}\right). \tag{6.17}$$

The asymptotic distribution of T_{Pearson} and T_{LR} under the null hypothesis is given in Robertson et al. (1988, p. 253), and we have that

$$P\text{-value} = \sum_{j=1}^{c} \rho(c, j, \boldsymbol{\pi}) \cdot \Pr\left[\chi^2_{c-j} \geq T_{\text{Pearson}}(\mathbf{n})\right], \tag{6.18}$$

where $\boldsymbol{\pi} = \{\pi_1, \pi_2, \ldots, \pi_c\}$. Here, $\boldsymbol{\pi}$ is the vector of joint probabilities in the two groups under the null hypothesis, and $\chi^2_0 = 0$. The P-value for the likelihood ratio test is obtained by substituting T_{LR} for T_{Pearson} in Equation 6.18. The calculation of this P-value is quite complicated, since it is based on a chi-bar-squared distribution, with weights that depend on common, unknown multinomial probabilities. When the column totals $\{n_{+1}, n_{+2}, \ldots, n_{+c}\}$ are equal, we can use the probabilities of Table 5.5 as the weights in Equation 6.18. We follow the proposal of Grove (1980) and calculate P-values under the assumption of equal marginal probabilities, hoping that "in practice a wide range of situations will be covered in this way".

6.6.2   Exact and Mid-P Tests for Cumulative Odds Ratios

As for the local odds ratios, we may also derive exact conditional and mid-P tests based on the Pearson chi-squared and likelihood ratio test statistics for the cumulative odds ratios. The exact conditional test with the Pearson chi-squared statistic is

$$\text{exact cond. } P\text{-value} = \sum_{\Omega(\mathbf{x}|\mathbf{n}_{++})} I\left[T_{\text{Pearson}}(\mathbf{x}) \geq T_{\text{Pearson}}(\mathbf{n})\right] \cdot f(\mathbf{x} \mid \mathbf{n}_{++}),$$

where $\Omega(\mathbf{x}|\mathbf{n}_{++})$ is the set of all tables $\mathbf{x}$ with marginals equal to $\mathbf{n}_{++}$, $I()$ is the indicator function, $T_{\text{Pearson}}()$ is given in Equation 6.16, and $f(\mathbf{x} \mid \mathbf{n}_{++})$

is the conditional probability distribution in Equation 6.4. The corresponding mid-P value is given by

$$\text{mid-}P\text{ value} \;=\; \sum_{\Omega(\mathbf{x}|\mathbf{n}_{++})} I\big[T_{\text{Pearson}}(\mathbf{x}) > T_{\text{Pearson}}(\mathbf{n})\big] \cdot f(\mathbf{x}\,|\,\mathbf{n}_{++})$$

$$+ \; 0.5 \cdot \sum_{\Omega(\mathbf{x}|\mathbf{n}_{++})} I\big[T_{\text{Pearson}}(\mathbf{x}) = T_{\text{Pearson}}(\mathbf{n})\big] \cdot f(\mathbf{x}\,|\,\mathbf{n}_{++}).$$

The exact conditional and mid-P tests with the likelihood ratio statistic in Equation 6.17 are obtained by substituting $T_{\text{LR}}()$ for $T_{\text{Pearson}}()$ in the two equations above.

6.6.3   Example

Postoperative Nausea (Table 6.3)

We return to the example with data on postoperative nausea. We start by calculating r_j for $j = 1, 2, 3, 4$. Because the minimum of the r_j is obtained for $j = 4$, there is only one subgroup, and T_{Pearson} and T_{LR} from Equations 6.16 and 6.17 reduce to the (unordered) test statistics for testing independence in a $2 \times c$ table:

$$T_{\text{Pearson}}(\mathbf{n}) = 3.813 \quad \text{and} \quad T_{\text{LR}}(\mathbf{n}) = 3.914,$$

with $P = 0.282$ (unordered Pearson chi-squared test) and $P = 0.271$ (unordered likelihood ratio test), see also Table 6.4.

An approximate P-value for the Pearson chi-squared test for ordered alternatives, with weights given in Table 5.5, can be calculated by

$$\begin{aligned}
P\text{-value} \;=\;& 0.2500 \cdot \Pr(\chi_0^2 \geq 3.813) + 0.4583 \cdot \Pr(\chi_1^2 \geq 3.813) \\
+\;& 0.2500 \cdot \Pr(\chi_2^2 \geq 3.813) + 0.0417 \cdot \Pr(\chi_3^2 \geq 3.813),
\end{aligned}$$

which gives

$$\begin{aligned}
P\text{-value} \;=\;& 0.2500 \cdot 0 + 0.4583 \cdot 0.0509 + 0.2500 \cdot 0.1486 + 0.0417 \cdot 0.2824 \\
=\;& 0.0722.
\end{aligned}$$

Similar calculations can be done for the likelihood ratio test with T_{LR} in place of T_{Pearson}.

We sum up the results in Table 6.5, including the results for the exact conditional and mid-P tests. First, note that the only thing that separates the tests for ordered cumulative odds ratios from the tests for ordered odds ratios is the way the two test statistics T_{Pearson} and T_{LR} are calculated. Hence, if we obtain similar observed test statistics, the results will agree quite well. For the cumulative odds ratios with the data in Table 6.3, the observed test statistics are slightly higher than those for the local odds ratios in Section 6.5. The P-values in Table 6.5 are thus slightly lower than the P-values in Table 6.4; however, the differences between the tests within each table are quite similar, and we refer to the discussion following Table 6.4.

TABLE 6.5

Results of tests for unordered alternatives and tests for ordered
cumulative odds ratios on the data in Table 6.3

Test	Statistic	df*	*P*-value
H$_A$: *Unordered*			
Pearson chi-squared[†]	3.81	3	0.282
Likelihood ratio[†]	3.91	3	0.271
H$_A$: *Ordered cumulative odds ratios*			
Pearson chi-squared	3.81	(chi-bar-squared)	0.0722
Likelihood ratio	3.91	(chi-bar-squared)	0.0686
Exact conditional (with T_{Pearson} as test statistic)			0.120
Mid-*P* (with T_{Pearson} as test statistic)			0.119
Exact conditional (with T_{LR} as test statistic)			0.114
Mid-*P* (with T_{LR} as test statistic)			0.112

*Degrees of freedom for the chi-squared distribution
[†]These methods are presented in Chapter 7

6.6.4   Evaluation of Tests

The tests for ordered cumulative odds ratios are very similar to the tests for
ordered local odds ratios (Section 6.5), which in turn are equal to the tests
for unspecific ordering in the $r \times 2$ table (Section 5.6). We therefore do not
present any evaluations for the tests for ordered cumulative odds ratios here;
however, we will briefly discuss differences and similarities with the evaluations
in Section 5.6.6.

The exact and mid-*P* tests for ordered cumulative odds ratios perform very
well, as do those in Section 5.6.6, with actual significance levels close to the
nominal level. The asymptotic Pearson chi-squared and likelihood ratio tests,
however, perform much worse for ordered cumulative odds ratios than for unspecific ordering in the $r \times 2$ table, with actual significance levels fluctuating
around 10–12% (for a 5% nominal level). The reason for this deterioration
of results might have to do with the additional assumption of equal column
totals that we put on the data to obtain a simple method—as suggested by
Grove (1980)—of determining the weights for the chi-bar-squared distribution. Because the exact and mid-*P* tests, which perform very well, use the
same test statistics as the asymptotic tests (Equations 6.16 and 6.17), there
must be something wrong with the chi-bar-squared reference distribution. In
all likelihood, the weights associated with the chi-bar-squared distribution
cannot, in general, be approximated with the probabilities in Table 5.5 with
any acceptable precision.

6.7   Continuation Odds Ratios

Finally, we consider the association between the outcome and the grouping as measured by the continuation odds ratios (θ_j^{CO}) in Equation 6.7. First note that

$$\theta_j^{CO} \geq 1$$

is equivalent to

$$\frac{\pi_{j+1|1} + \pi_{j+2|1} + \ldots + \pi_{c|1}}{\pi_{1|1} + \pi_{2|1} + \ldots + \pi_{j|1}} \geq \frac{\pi_{j+1|2} + \pi_{j+2|2} + \ldots + \pi_{c|2}}{\pi_{1|2} + \pi_{2|2} + \ldots + \pi_{j|2}},$$

which again gives that

$$\pi_{1|1} + \pi_{2|1} + \ldots + \pi_{j|1} \geq \pi_{1|2} + \pi_{2|2} + \ldots + \pi_{j|2},$$

for all j, see Grove (1980) and Oluyede (1993). This is the ordering we considered in Section 6.6 on the cumulative odds ratios, and the maximum likelihood estimation and the tests will follow from what was derived there, see also Robertson et al. (1988, p. 262).

6.8   Ranking of Observations

In ordered $2 \times c$ tables, the categories of the column variable are ordered. In Table 6.2, for instance, the categories are ordered from Unlikely to Likely. In Sections 6.4–6.7, we introduced three odds ratios that take the ordering into account. Alternatively, we may assign scores that relate the ordering to numbers. We let the columns scores be $\{b_1, b_2, \ldots, b_c\}$, where $b_1 < b_2 < \ldots < b_c$. In a $2 \times c$ table, it is common to use the midranks as category scores. The midranks are simply given by averaging the ranks of the observation within each cell. For outcome number $j = 1, 2, \ldots, c$, the midranks are given by

$$m_j = \frac{1}{2}\left(\sum_{k=1}^{j-1} n_{+k} + 1 + \sum_{k=1}^{j} n_{+k}\right). \tag{6.19}$$

Alternatively, we can define the ridits by averaging the cumulative proportions in the cells:

$$r_j = \frac{1}{2N}\left(\sum_{k=1}^{j-1} n_{+k} + \sum_{k=1}^{j} n_{+k}\right).$$

The midranks and the ridits are related by

$$m_j = Nr_j + 1/2.$$

Assume now that the outcome variable Y_2 is the result of an underlying continuous variable, with cumulative distribution F. Let the scores v_j be defined such that

$$m_j = F(v_j),$$

or, equivalently, that

$$v_j = F^{-1}(m_j),$$

see also Agresti (2010, p. 11). Let F be the cumulative distribution function of the standard logistic distribution. Then,

$$v_j = \frac{n_{+1} + n_{+2} + \ldots + n_{+j-1} + 1/2 n_{+j}}{1/2 n_{+j} + n_{+j+1} + \ldots + n_{+c}}.$$

In Section 6.14, we will show that

$$\hat{\alpha}_j = \frac{n_{+1} + n_{+2} + \ldots + n_{+j}}{n_{+j+1} + n_{+j+2} + \ldots + n_{+c}}$$

are the maximum likelihood estimates of the thresholds in the proportional odds model, when $\beta = 0$. We see that there is an appealing relationship between the ridits, the midranks, and the thresholds in the proportional odds model when there is no group difference.

Alternatively, we can let F^{-1} be the inverse of the cumulative standard normal distribution. Then,

$$v_j = \Phi^{-1}(m_j),$$

and the scores are estimated by the percentiles of the cumulative normal distribution. Here, the relationship is with the thresholds in the probit model, which we will discuss in Section 6.15.

6.9   Mantel-Haenszel Test of Association with Column Scores

In Section 5.7.6, we used the Mantel-Haenszel test for linear-by-linear association to derive a test for trend in ordered $r \times 2$ tables. In Section 7.8.1, we will describe the Mantel-Haenszel test for $r \times c$ tables. Here, we use the Mantel-Haenszel test for linear-by-linear association to derive a test for the hypothesis of no association in ordered $2 \times c$ tables. Let

$$\bar{b}_1 = \sum_{j=1}^{c} b_j n_{1j} / n_{1+}$$

denote the mean value in Group 1, where $\{b_1, b_2, \ldots, b_c\}$ are the column scores. Under the null hypothesis, the mean value is

$$\bar{b} = \sum_{j=1}^{c} b_j \frac{n_{+j}}{N},$$

and the variance is

$$\frac{N - n_{1+}}{n_{1+}(N - 1)} \sum_{j=1}^{c} (b_j - \bar{b})^2 n_{+j}/N.$$

The Mantel-Haenszel test statistic is

$$T_{\mathrm{MH}}(\mathbf{n}) = \frac{(\bar{b}_1 - \bar{b})^2}{\left\{ (N - n_{1+})/[n_{1+}(N - 1)] \right\} \left[\sum_{j=1}^{c} (b_j - \bar{b})^2 n_{+j}/N \right]},$$

which for $\bar{b}_2 = \sum_{j=1}^{c} b_j n_{2j}/n_{2+}$ can be expressed as

$$T_{\mathrm{MH}}(\mathbf{n}) = \frac{(n_{1+}n_{2+}/N)(\bar{b}_1 - \bar{b}_2)^2}{[N/(N - 1)] \left[\sum_{j=1}^{c} (b_j - \bar{b})^2 n_{+j}/N \right]},$$

where T_{MH} is approximately chi-squared distributed with one degree of freedom:

$$P\text{-value} = \Pr\left[\chi_1^2 \geq T_{\mathrm{MH}}(\mathbf{n}) \right].$$

As a general guideline, the chi-squared approximation to the Mantel-Haenszel test statistic is satisfactory when all expected cell counts are greater than or equal to five, see also Stokes et al. (2012, p. 75).

When midranks are used as scores—see Equation 6.19—the Mantel-Haenszel statistic, T_{MH}, is identical to the square of the Wilcoxon-Mann-Whitney statistic, see Davis and Chung (1995). In Section 6.14.3, we note that the Wilcoxon-Mann-Whitney test is also the score test for $\beta = 0$ in a proportional odds model.

6.9.1   Example

Postoperative Nausea (Table 6.3)

We return to the randomized controlled trial of postoperative nausea. With midranks, as defined in Equation 6.19, we calculate the Mantel-Haenszel test statistic and get $T_{\mathrm{MH}} = 2.13$. With one degree of freedom for the chi-squared distribution, we get $P = 0.144$. If we compare this result with the results of the tests for unordered alternatives in Tables 6.4 and 6.5, we note that the

P-value for the Mantel-Haenszel test is much smaller than those of the Pearson chi-squared test ($P = 0.282$) and the likelihood ratio test ($P = 0.271$). This difference illustrates the advantage of using scores to account for the ordering. The Mantel-Haenszel *P*-value is, however, higher than the *P*-values in Tables 6.4 and 6.5 that take alternatives of ordered odds ratios into consideration, including the *P*-values of the exact tests. Note also that the Mantel-Haenszel test is sensitive to the choice of scores. If we use equally spaced scores, for instance $\{0, 1, 2, 3\}$, on the same data, we obtain $P = 0.123$.

6.10 Logit Models for Ordinal Data

There are many alternative models for ordinal data that takes the ordering of the categories of Y_2 into consideration (McCullagh, 1980; McCullagh and Nelder, 1989). Of these, linear models with the logistic link have been found to work well (McCullagh and Nelder, 1989, p. 151). The logistic link gives models for the logits. In this section, we present three logit models that can be used for ordered $2 \times c$ tables. These models are connected to the three odds ratios that were presented in Section 6.4.

Later in this chapter, in Sections 6.11–6.14, we will elaborate on the cumulative logit model, and in Section 6.15, we will consider the cumulative probit model, which uses the inverse of the standard normal distribution as link function.

6.10.1 Cumulative Logits

The cumulative logits are defined as

$$\text{logit}\left[\Pr(Y_2 \le j \mid Y_1 = 1)\right] = \log\left[\frac{\Pr(Y_2 \le j \mid Y_1 = 1)}{\Pr(Y_2 > j \mid Y_1 = 1)}\right]$$
$$= \log\left(\frac{\pi_{1|1} + \pi_{2|1} + \ldots + \pi_{j|1}}{\pi_{j+1|1} + \pi_{j+2|1} + \ldots + \pi_{c|1}}\right)$$

and

$$\text{logit}\left[\Pr(Y_2 \le j \mid Y_1 = 2)\right] = \log\left[\frac{\Pr(Y_2 \le j \mid Y_1 = 2)}{\Pr(Y_2 > j \mid Y_1 = 2)}\right]$$
$$= \log\left(\frac{\pi_{1|2} + \pi_{2|2} + \ldots + \pi_{j|2}}{\pi_{j+1|2} + \pi_{j+2|2} + \ldots + \pi_{c|2}}\right),$$

for $j = 1, 2, \ldots, c - 1$. Then, the cumulative odds ratios for $Y_1 = 1$ relative to $Y_1 = 2$ are given by

$$\frac{(\pi_{1|1} + \pi_{2|1} + \ldots + \pi_{j|1})/(\pi_{j+1|1} + \pi_{j+2|1} + \ldots + \pi_{c|1})}{(\pi_{1|2} + \pi_{2|2} + \ldots + \pi_{j|2})/(\pi_{j+1|2} + \pi_{j+2|2} + \ldots + \pi_{c|2})}.$$

We have previously seen that the log odds ratios in Equation 6.1 can be estimated by the log odds ratios in Equation 6.2, and, in like manner, we get that the estimated cumulative odds ratios are

$$\frac{(n_{11} + n_{12} + \ldots + n_{1j})/(n_{1j+1} + n_{1j+2} + \ldots + n_{1c})}{(n_{21} + n_{22} + \ldots + n_{2j})/(n_{2j+1} + n_{2j+2} + \ldots + \pi_{2c})}.$$

Note that we define the cumulative logits by $\text{logit}[\Pr(Y_2 \leq j \mid Y_1 = i)]$, for $i = 1, 2$. Some authors define the cumulative logits by $\text{logit}[\Pr(Y_2 \geq j \mid Y_1 = i)]$. In the latter case, the odds ratios are the inverse of those of the former, and the log odds ratios are simply the negative versions of each other.

The cumulative odds model can alternatively be written as

$$\text{logit}[\Pr(Y_2 \leq j \mid x_i)] = \alpha_j - \beta_j x_i,$$

where $x_1 = 1$ denotes Group 1 and $x_2 = 0$ denotes Group 2, and

$$-\beta_j = \log \left[\frac{(\pi_{1|1} + \pi_{2|1} + \ldots + \pi_{j|1})/(\pi_{j+1|1} + \pi_{j+2|1} + \ldots + \pi_{c|1})}{(\pi_{1|2} + \pi_{2|2} + \ldots + \pi_{j|2})/(\pi_{j+1|2} + \pi_{j+2|2} + \ldots + \pi_{c|2})} \right].$$

Then,

$$\exp(-\beta_j) = \theta_j^{\text{CU}},$$

which directly relates the cumulative odds ratios to the one we considered in Section 6.4. Note that $\beta < 0$ means that $\exp(-\beta) > 1$, which again means that θ_j^{CU} as defined in Equation 6.6 is greater than 1.

When the grouping has the same effect for each logit, i.e., $\beta_1 = \beta_2 = \ldots = \beta_{c-1}$, we get the proportional odds model:

$$\text{logit}[\Pr(Y_2 \leq j \mid x_i)] = \alpha_j - \beta x_i. \tag{6.20}$$

The model above simplifies further, and we get that the proportional odds model implies that

$$-\beta = \log \left[\frac{(\pi_{1|1} + \pi_{2|1} + \ldots + \pi_{j|1})/(\pi_{j+1|1} + \pi_{j+2|1} + \ldots + \pi_{c|1})}{(\pi_{1|2} + \pi_{2|2} + \ldots + \pi_{j|2})/(\pi_{j+1|2} + \pi_{j+2|2} + \ldots + \pi_{c|2})} \right].$$

When the model is given as in Equation 6.20, we see that the cumulative odds ratio for $Y_1 = 1$ relative to $Y_1 = 2$ is equal to $\exp(-\beta)$. Thus, if $\beta > 0$, we are more likely to get observations in the upper end of the Y_2 scale for Group 1 than for Group 2. On the other hand, if $\beta < 0$, we are more likely to get observations in the lower end of the Y_2 scale for Group 1 than for Group 2.

6.10.2 Adjacent Category

The logits for the adjacent category model are given by

$$\log\left[\frac{\Pr(Y_2 = j \mid x_i)}{\Pr(Y_2 = j+1 \mid x_i)}\right] = \alpha_j - \beta_j x_i,$$

for $j = 1, 2, \ldots, c-1$, where $x_1 = 1$ denotes Group 1 and $x_2 = 0$ denotes Group 2. Here,

$$-\beta_j = \log\left[\frac{\pi_{j|1}/\pi_{j+1|1}}{\pi_{j|2}/\pi_{j+1|2}}\right].$$

Note that

$$\exp(\beta_j) = \theta_j^{\text{LO}},$$

where θ_j^{LO} is defined in Equation 6.5. Note also that the local odds ratios now are called *adjacent category odds ratios*, which is the common notation in $2 \times c$ tables.

If the effects (β_j) are the same for each logit, we get a model that is similar to the proportional odds model:

$$\log\left[\frac{\Pr(Y_2 = j \mid x_i)}{\Pr(Y_2 = j+1 \mid x_i)}\right] = \alpha_j - \beta x_i.$$

Note that β has an interpretation similar to that of the proportional odds model. If $\beta > 0$, we have a positive effect of the grouping, and we expect higher outcome values for Group 1 than for Group 2.

The adjacent category logits are defined via the probability of each category—except the highest category—relative to the probability of the next category. In specific cases, this might be of particular interest; however, in most cases, it is preferable to use a model that has an interpretation via a continuum, such as the proportional odds model.

6.10.3 Continuation Ratio

We can also model ordinal data via continuation ratio logits. Let

$$\omega_{ij} = \Pr(Y_2 = j \mid Y_2 \geq j, Y_1 = i) = \frac{\pi_{j|i}}{\pi_{j|i} + \pi_{j+1|i} + \cdots + \pi_{c|i}},$$

for $i = 1, 2$ and $j = 1, 2, \ldots, c-1$. The continuation ratio model is given by

$$\text{logit}(\omega_j) = \alpha_j - \beta_j x_i,$$

where, as usual, $x_1 = 1$ denotes Group 1 and $x_2 = 0$ denotes Group 2. Here,

$$-\beta_j = \log\left[\frac{\pi_{j|1}/(\pi_{j|1} + \pi_{j+1|1} + \ldots + \pi_{c|1})}{\pi_{j|2}/(\pi_{j|2} + \pi_{j+1|2} + \ldots + \pi_{c|2})}\right],$$

and we have that

$$\exp(-\beta_j) = \theta_j^{CO},$$

where θ_j^{CO} is defined in Equation 6.7.

If we assume the same effect over the $c - 1$ categories, we get

$$\text{logit}(\omega_j) = \alpha_j - \beta x_i.$$

Again, we interpret β in a similar manner as before: if $\beta > 0$, we expect higher values of Y_2 for Group 1 than for Group 2.

As pointed out by Agresti (2010, pp. 96–97), the continuation ratio model is of special interest when the ordinal data are realizations of a development or duration scale, where we would interpret the group effect via the conditional probabilities $\Pr(Y_2 = j \mid Y_2 \geq j, Y_1 = i)$.

The adjacent category and continuation ratio models have been implemented in Stata (StataCorp LP, College Station, TX) by Fagerland (2014).

6.10.4   Example

The Adolescent Placement Study (Table 6.2)

Consider now the data on danger to others (outcome) and gender (group) in Table 6.2. The three cumulative odds ratios for males relative to females are

$$\text{OR}(1, 2\text{–}4) = \frac{8/(28 + 72 + 126)}{46/(73 + 69 + 86)} = 0.18, \tag{6.21}$$

$$\text{OR}(1\text{–}2, 3\text{–}4) = \frac{(8 + 28)/(72 + 126)}{(46 + 73)/(69 + 86)} = 0.24, \tag{6.22}$$

and

$$\text{OR}(1\text{–}3, 4) = \frac{(8 + 28 + 72)/126}{(46 + 73 + 69)/86} = 0.39, \tag{6.23}$$

where $\text{OR}(a, b)$ denotes the odds ratio comparing category a with category b. These three odds ratios can be calculated by fitting three logistic regressions: one for category Unlikely versus Possible or more likely; one for Unlikely and Possible versus Probable or more likely; and finally, one for Probable or less likely versus Likely.

Note that the cumulative odds ratios have to be interpreted with some care. For instance, we find that the odds of being possible or less likely a danger to others is 0.24 for males relative to females, and the odds of being probable or less likely a danger to others is 0.39 for males relative to females. At first glance, these cumulative odds ratios may seem to be quite different, but when we take the confidence intervals into consideration, it may not be inappropriate to assume that these odds ratios actually are estimates of a common odds ratio. In the following section, we will study models for cumulative logits with proportional odds. In Section 6.13, we present tests of the proportional odds assumption, and in Section 6.16, we apply those tests to this example.

The three adjacent category odds ratios for males relative to females are

$$OR(1,2) = \frac{8/28}{46/73} = 0.45,$$

$$OR(2,3) = \frac{28/72}{73/69} = 0.37,$$

and

$$OR(3,4) = \frac{72/126}{69/86} = 0.71,$$

As for the cumulative odds ratios, there are notable differences between the adjacent category odds ratios.

The three continuation odds ratios are

$$OR(1,1\text{--}4) = \frac{8/(8 + 28 + 72 + 126)}{46/(46 + 73 + 69 + 86)} = 0.17,$$

$$OR(2,2\text{--}4) = \frac{28/(28 + 72 + 126)}{73/(73 + 69 + 86)} = 0.33,$$

and

$$OR(3,3\text{--}4) = \frac{72/(72 + 126)}{69/(69 + 86)} = 0.68,$$

Again, we note some interesting differences between the odds ratios.

Table 6.6 shows the results of fitting proportional odds, adjacent category, and continuation ratio models to the data on danger to others for males versus females. We see that the estimated difference between males and females (the "effect" of being male) is substantially the same across the three models. The estimated coefficients have the same sign, and similar P-values, which gives the same qualitative conclusion.

TABLE 6.6

Results of fitting three ordinal logistic models to the data in Table 6.2

Model	Coefficient	SE	z	P-value	95% CI
Proportional odds model					
Males	1.145	0.170	6.74	< 0.001	0.812 to 1.478
Adjacent category model					
Males	0.673	0.010	6.76	< 0.001	0.478 to 0.868
Continuation ratio model					
Males	1.018	0.142	7.17	< 0.001	0.740 to 1.296

As we know, the estimated common effects for males relative to females are $\exp(-\hat{\beta})$, which gives that the effects (with 95% confidence intervals) are 0.318 (0.228 to 0.444) for the proportional odds model, 0.512 (0.420 to 0.620) for the adjacent category model, and 0.361 (0.273 to 0.477) for the continuation ratio model.

6.11   The Proportional Odds Model

The proportional odds model is based on cumulative odds, with the assumption that the grouping has the same effect for each logit. Cumulative odds are the odds of being in category j or less, relative to being in category $j + 1$ or above. Thus, cumulative odds take the ordering of the data into full consideration. Adjacent category odds ratios and continuation odds ratios may be appropriate for specific problems; however, for ordinal data, the use of cumulative odds can be considered a standard procedure.

When we now impose the assumption of proportionality, the cumulative odds are given by

$$\log \left[\frac{(\pi_{1|1} + \pi_{2|1} + \ldots + \pi_{j|1})/(\pi_{j+1|1} + \pi_{j+2|1} + \ldots + \pi_{c|1})}{(\pi_{1|2} + \pi_{2|2} + \ldots + \pi_{j|2})/(\pi_{j+1|2} + \pi_{j+2|2} + \ldots + \pi_{c|2})} \right] = -\beta, \quad (6.24)$$

for $j = 1, 2, \ldots, c - 1$.

The cumulative logit model with proportional odds is usually denoted as the proportional odds model. When "cumulative logit model" is deleted, the name "proportional odds model" is ambiguous. Nevertheless, because this term is so commonly used, we will use it here as well.

In Equations 6.21, 6.22, and 6.23, we calculated the three cumulative odds ratios for males relative to females for the outcome danger to others for the data from the Adolescent Placement Study. If we now assume that the cumulative logits have proportional odds, we know that these can all be estimated by one odds ratio, namely $\exp(-\beta)$, see Section 6.10.1. We find that $\hat{\beta} = 1.145$, and the cumulative odds ratio is 0.318 for males relative to females.

The parameterization in Equation 6.24 coincides with the logistic model. In this case, $c = 2$, and we have

$$\text{logit}\big[\Pr(Y_2 = 1 \,|\, x_i)\big] = \text{logit}\big[\Pr(Y_2 \le 1 \,|\, x_i)\big] = \alpha_1 - \beta x_i. \quad (6.25)$$

We now let the successes (cases) and the failures (controls) be given by the observed values $Y_2 = 2$ and $Y_2 = 1$, respectively. We can formulate the logistic model by inverting Equation 6.25:

$$\text{logit}\big[\Pr(Y_2 = 2 \,|\, x_i)\big] = -\alpha_1 + \beta x_i, \quad (6.26)$$

which is the standard formulation of a simple logistic model. Note the opposite signs in Equations 6.25 and 6.26. When $\beta > 0$ in Equation 6.26, we are more likely to get observations of $Y_2 = 2$ for $Y_1 = 1$ than for $Y_1 = 2$, which is in accordance with the interpretation of $\beta > 0$ in the proportional odds model in Equation 6.24. Moreover, we see that

$$\frac{\pi_{2|1} \cdot \pi_{1|2}}{\pi_{2|2} \cdot \pi_{1|1}} = \exp(\beta),$$

which can be estimated by the cross-product in a 2 × 2 table.

The formulation of the proportional odds model in Equations 6.20 and 6.24 goes back to the path breaking article of McCullagh (1980). Further, as described in Agresti (2010, pp. 53–55), the observed variable Y_2 in a proportional odds model can be described by a latent variable Y^* with a cumulative distribution function G. The ordering of the categories of Y_2 follows the latent variable Y^*. The link between Y_2 and Y^* is given by

$$\Pr(Y_2 \leq j \,|\, x_i) = \Pr(Y^* \leq \alpha^* \,|\, x_i) = G(\alpha_j - \beta x_i).$$

When G is the standard logistic distribution, we get the proportional odds model

$$\text{logit}\big[\Pr(Y_2 \leq j \,|\, x_i)\big] = \alpha_j - \beta x_i.$$

If we use the standard normal distribution for G, we will get the probit link instead of the logit link. We return to the probit link and the associated cumulative probit model in Section 6.15.

With the model in Equation 6.20, we have a negative sign in front of β. Alternatively, the proportional odds model can be parameterized by

$$\text{logit}\big[\Pr(Y_2 \leq j \,|\, x_i)\big] = \alpha_j + \beta^* x_i. \tag{6.27}$$

With this model, the interpretations of $\beta^* > 0$ and $\beta^* < 0$ will be the opposite of those for β in Equation 6.20. We prefer the representation in Equation 6.20, which also is the parameterization used in many statistical software packages, such as Stata (StataCorp LP, College Station, TX); however, the formulation in Equation 6.27 is also quite common, see, for instance, Kleinbaum and Klein (2010).

6.12   Estimating Proportional Odds Models

The maximum likelihood principle is the standard procedure for estimating the αs and β. The likelihood function for the proportional odds model takes the form of

$$
\begin{aligned}
L(\alpha_1, \alpha_2, \dots, \alpha_{c-1}, \beta) &= \prod_{i=1}^{2} \prod_{j=1}^{c} \big[\Pr(Y_2 \leq j \,|\, x_i) - \Pr(Y_2 \leq j - 1 \,|\, x_i)\big]^{n_{ij}} \\
&= \prod_{i=1}^{2} \prod_{j=1}^{c} \left[\frac{\exp(\alpha_j - \beta x_i)}{1 + \exp(\alpha_j - \beta x_i)} - \frac{\exp(\alpha_{j-1} - \beta x_i)}{1 + \exp(\alpha_{j-1} - \beta x_i)} \right]^{n_{ij}}, \tag{6.28}
\end{aligned}
$$

where $\alpha_0 = -\infty$ and $\alpha_c = 0$. We now maximize the log of the likelihood function with respect to the parameters. The maximum likelihood estimates are obtained by differentiating the log-likelihood function and setting this expression to zero. Let

$$G(z) = \frac{\exp(z)}{1 + \exp(z)}$$

and

$$g(z) = \frac{\exp(z)}{\left[1 + \exp(z)\right]^2}.$$

Note that $G'(z) = g(z)$. The log-likelihood function is (and here we have used that $x_1 = 1$ denotes Group 1 and $x_2 = 0$ denotes Group 2):

$$l(\alpha_1, \alpha_2, \ldots, \alpha_{c-1}, \beta) = \sum_{j=1}^{c} n_{1j} \log\left[G(\alpha_j - \beta) - G(\alpha_{j-1} - \beta)\right]$$

$$+ \sum_{j=1}^{c} n_{2j} \log\left[G(\alpha_j) - G(\alpha_{j-1})\right].$$

Then we have that

$$\frac{\partial l}{\partial \beta} = -\sum_{j=1}^{c} n_{1j} \frac{g(\alpha_j - \beta) - g(\alpha_{j-1} - \beta)}{G(\alpha_j - \beta) - G(\alpha_{j-1} - \beta)}$$

and

$$\frac{\partial l}{\partial \alpha_k} = \sum_{j=1}^{c} n_{1j} \frac{\delta_{jk}\, g(\alpha_j - \beta) - \delta_{j-1k}\, g(\alpha_{j-1} - \beta)}{G(\alpha_j - \beta) - G(\alpha_{j-1} - \beta)}$$

$$+ \sum_{j=1}^{c} n_{2j} \frac{\delta_{jk}\, g(\alpha_j) - \delta_{j-1k}\, g(\alpha_{j-1})}{G(\alpha_j) - G(\alpha_{j-1})},$$

where $\delta_{jk} = 1$ if $j = k$ and $\delta_{jk} = 0$ otherwise (δ_{jk} is called the *Kronecker delta*). The equations to solve for β and the α_k are then $\partial l / \partial \beta = 0$ and $\partial l / \partial \alpha_k = 0$ for $k = 1, 2, \ldots, c - 1$. For similar equations for the more general proportional odds model, see Agresti (2010, pp. 59, 120).

McCullagh (1980) proposed the *Fisher scoring method* for solving the likelihood equations. For large N, there exists a unique solution to this problem (McCullagh, 1980). Let the maximum likelihood estimates of β and α_k in the proportional odds model be $\hat{\beta}_{\mathrm{ML,\,logit}}$ and $\hat{\alpha}_{k,\,\mathrm{ML,\,logit}}$. The covariance matrix of the maximum likelihood estimates are given by the negatives of the expected second partial derivatives of $L(\alpha_1, \alpha_2, \ldots, \alpha_{c-1}, \beta)$, which is the information matrix. By replacing the parameters of the information matrix with their maximum likelihood estimates, and taking the inverse of the resulting matrix, we get an estimate of the asymptotic covariance matrix. Let the estimated standard error of $\hat{\beta}_{\mathrm{ML,\,logit}}$ be $\widehat{\mathrm{SE}}(\hat{\beta}_{\mathrm{ML,\,logit}})$.

The ordinal proportional odds model is not a member of the exponential family, and the expected second partial derivatives will not be identical to the observed. As noted by Agresti (2010, p. 60), some statistical software packages use the expected second partial derivatives and others use the observed. Because these will differ and produce different results, it is important to be aware of which method that is used.

6.13 Testing the Fit of a Proportional Odds Model

Above, we introduced the cumulative logit model with proportional odds (Equation 6.20). This is a rather restrictive model, and we would like to know how well such a model actually fits to an observed $2 \times c$ table. There are many alternative models to the proportional odds model, see for instance Peterson and Harrell (1990). The most general alternative model is

$$\text{logit}\big[\text{Pr}(Y_2 \leq j \,|\, x_i)\big] = \alpha_j - \beta_j x_i,$$

for $j = 1, 2, \ldots, c - 1$. Then, the β_j are given as

$$\beta_j = \log \left[\frac{\text{Pr}(Y_2 \leq j \,|\, Y_1 = 1)/\text{Pr}(Y_2 > j \,|\, Y_1 = 1)}{\text{Pr}(Y_2 \leq j \,|\, Y_1 = 2)/\text{Pr}(Y_2 > j \,|\, Y_1 = 2)} \right],$$

which are the cumulative odds ratios what we studied in Section 6.10.1.

It should be noted here that the alternative model $\text{logit}\big[\text{Pr}(Y_2 \leq j \,|\, x_i)\big] = \alpha_j - \beta_j x_i$, with separate β_j, is a saturated model in the sense that the β_j can be expressed uniquely by the $\pi_{j|i}$.

The null hypothesis for testing proportional odds is now

$$H_0 : \beta_1 = \beta_2 = \ldots = \beta_{c-1} = \beta. \tag{6.29}$$

We will use the Pearson chi-squared test and the log likelihood ratio test for testing the fit of the model (i.e., the null hypothesis in Equation 6.29). We start by calculating the estimated probabilities. Because

$$\text{Pr}\big(Y_2 \leq j \,|\, x_i\big) = \frac{\exp(\alpha_j - \beta x_i)}{1 + \exp(\alpha_j - \beta x_i)} - \frac{\exp(\alpha_{j-1} + \beta x_i)}{1 + \exp(\alpha_{j-1} + \beta x_i)}, \tag{6.30}$$

we find $\hat{\pi}_{j|i}$ by inserting the maximum likelihood estimates of the α_j and β in Equation 6.30. Let m_{ij} denote the estimated expected counts:

$$m_{ij} = n_{i+} \cdot \hat{\pi}_{j|i}.$$

6.13.1 The Pearson Goodness-of-Fit Test

The Pearson goodness-of-fit statistic is of the form

$$\chi^2(\mathbf{n}) = \sum_{i=1}^{2} \sum_{j=1}^{c} \frac{(n_{ij} - m_{ij})^2}{m_{ij}}.$$

Asymptotically, the Pearson test statistic follows a chi-squared distribution with $c - 2$ degrees of freedom, and P-values are obtained as

$$P\text{-value} = \Pr\left[\chi^2_{c-2} \geq \chi^2(\mathbf{n})\right].$$

Pulkstenis and Robinson (2004) and later Fagerland and Hosmer (2013) extended the Pearson goodness-of-fit test to handle more general proportional odds models beyond the contingency table setting.

6.13.2   The Pearson Residuals

If the P-value of the Pearson goodness-of-fit test is small, we can use the Pearson residuals to study where the model fits poorly:

$$r_{ij} = \frac{n_{ij} - m_{ij}}{\sqrt{m_{ij}}}.$$

The sum of the r_{ij}^2 equals the Pearson goodness-of-fit statistic. We can also calculate the standardized Pearson residuals:

$$r_{ij}^* = \frac{n_{ij} - m_{ij}}{\widehat{\text{SE}}(n_{ij} - m_{ij})}.$$

where $\widehat{\text{SE}}(n_{ij} - m_{ij})$ is the estimated standard error under the null hypothesis, see McCullagh and Nelder (1989, p. 396) and Lang (1996). In those references, the form of the standardized residuals is given for the generalized linear model. Little work seems to have been done specifically for the proportional odds model, and the use of the (unstandardized) Pearson residuals seems appropriate, see also McCullagh (1980). We do not pursue the standardized residuals for the proportional odds model any further.

6.13.3   The Likelihood Ratio (Deviance) Test

The log likelihood ratio test statistic, also called the *deviance test statistic*, is of the form

$$D(\mathbf{n}) = 2\sum_{i=1}^{2}\sum_{j=1}^{c} n_{ij} \log\left(\frac{n_{ij}}{m_{ij}}\right).$$

Like the Pearson goodness-of-fit statistic, the deviance statistic is asymptotically chi-squared distributed with $c - 2$ degrees of freedom.

6.13.4   The Brant Test

Brant (1990) suggested a test for the proportional odds assumption based on standard logistic regression. We run $c - 1$ binary logistic regressions; first, category 1 versus categories $2, 3, \ldots, c$, then categories 1 and 2 versus categories

$3, 4, \ldots, c$, and finally, categories $1, 2, \ldots, c - 1$ versus category c. Following the notation in Brant (1990), we let

$$z_j = \begin{cases} 1 & \text{if } Y_2 > j \\ 0 & \text{otherwise} \end{cases}$$

for $j = 1, 2, \ldots, c - 1$, and

$$\pi_j(x_{ki}) = \Pr(z_j = 1 \mid x_{ki}),$$

where $k = 1, 2, \ldots, N$ indexes the observations, with $x_{k1} = 1$ and $x_{k2} = 0$. Then, the $c - 1$ separate binary logistic regressions are given by

$$\text{logit}\left[\pi_j(x_{ki})\right] = -\alpha_j + \beta_j x_{ki},$$

The idea is to compare the $c - 1$ regression coefficients. Let $\mathbf{X}$ denote an $N \times 2$ matrix that has x_{ki} as the second column, augmented by a column of 1s as the first column. For binary regression number j, we obtain an estimate $\hat{\beta}_j$ with estimated variance $\widehat{\text{Var}}(\hat{\beta}_j)$. The estimated probabilities from regression number j is

$$\hat{\pi}_j(x_{ki}) = \frac{\exp(-\hat{\alpha}_j + \hat{\beta}_j x_{ki})}{1 + \exp(-\hat{\alpha}_j + \hat{\beta}_j x_{ki})}.$$

To find the covariance between $\hat{\beta}_m$ and $\hat{\beta}_l$, we need

$$w_{iml} = \hat{\pi}_l(x_{ki}) - \hat{\pi}_m(x_{ki})\hat{\pi}_l(x_{ki}), \quad \text{for } m \leq l.$$

Let $\mathbf{W}_{ml}$ be the diagonal $N \times N$ matrix with ith element w_{iml}. Then, $\text{Cov}(\hat{\beta}_m, \hat{\beta}_l)$, the covariance between $\hat{\beta}_m$ and $\hat{\beta}_l$, is obtained by deleting the first row and the first column in the 2×2 matrix

$$\left(\mathbf{X}^{\mathrm{T}}\mathbf{W}_{mm}\mathbf{X}\right)^{-1}\left(\mathbf{X}^{\mathrm{T}}\mathbf{W}_{ml}\mathbf{X}\right)\left(\mathbf{X}^{\mathrm{T}}\mathbf{W}_{ll}\mathbf{X}\right)^{-1}.$$

Let now $\hat{\boldsymbol{\beta}} = \{\hat{\beta}_1, \hat{\beta}_2, \ldots, \hat{\beta}_{c-1}\}$ be the vector of estimated coefficients from the binary logistic regressions, and let $\widehat{\mathbf{Var}}(\hat{\boldsymbol{\beta}})$ be the matrix of estimated variances and covariances of these coefficients. The null hypothesis to be tested is

$$H_0 : \beta_1 = \beta_2 = \ldots = \beta_{c-1}.$$

Let $\mathbf{D}$ be the $(c - 2) \times (c - 1)$ matrix

$$\mathbf{D} = \begin{bmatrix} 1 & -1 & 0 & \ldots & 0 \\ 1 & 0 & -1 & \ldots & 0 \\ \vdots & \vdots & \vdots & \ddots & 0 \\ 1 & 0 & 0 & \ldots & -1 \end{bmatrix}.$$

The Brant test uses the Wald-type test statistic

$$T_{\text{Brant}}(\mathbf{n}) = \left(\mathbf{D}\hat{\boldsymbol{\beta}}\right)^{\mathrm{T}}\left[\mathbf{D}\widehat{\mathbf{Var}}(\hat{\boldsymbol{\beta}})\mathbf{D}^{\mathrm{T}}\right]^{-1}\left(\mathbf{D}\hat{\boldsymbol{\beta}}\right),$$

which is asymptotically chi-squared distributed with $c - 2$ degrees of freedom under the null hypothesis. For more details concerning the Brant test, see Brant (1990) and Long (1997, p. 144).

Finally, we note that it is also possible to derive a score test for the null hypothesis in Equation 6.29. We do not go into the details here and refer the interested reader to Peterson and Harrell (1990).

6.14   Testing for Effect in a Proportional Odds Model

The proportional odds model is given by Equation 6.20, where β is the effect measure. Due to the negative sign in the equation, higher values of x_1 give smaller values of the logit for $\beta > 0$. Thus, being in Group 1 with $x_1 = 1$, gives a smaller value of the logit, and the odds ratio is less than one, which means that being in Group 1 is associated with higher values of Y_2 compared with being in Group 2 (for which $x_2 = 0$). If $\beta < 0$, on the other hand, the odds ratio for $Y_2 \leq j$ versus $Y_2 > j$ for Group 1 relative to Group 2 is greater than one, and being in Group 1 is associated with lower values of Y_2.

The null value of no association between the groups and the ordinal outcome is $\beta = 0$. Thus,

$$H_0 : \beta = 0$$

is the hypothesis of no group difference in the cumulative odds. Note, however, that the assumption is proportional odds. Moreover, if $\beta = 0$, there are $c - 1$ parameters α_j to estimate, and due to the one-to-one relationship with the logits, the parameters are estimated by the sample logits in the $c - 1$ marginals in the $2 \times c$ table; see also Section 6.8.

6.14.1   The Wald Test

In Section 6.12, we discussed the maximum likelihood estimate of β in the proportional odds model. The Wald test statistic then is given by

$$Z_{\text{Wald}}(\mathbf{n}) = \frac{\hat{\beta}_{\text{ML, logit}}}{\widehat{\text{SE}}(\hat{\beta}_{\text{ML, logit}})}.$$

For large samples, Z_{Wald} follows a standard normal distribution under H_0, and we obtain P-values for the Wald test as

$$P\text{-value} = \Pr\Big[Z \geq \big|Z_{\text{Wald}}(\mathbf{n})\big|\Big],$$

where Z is a standard normal variable. Note that some statistical software packages calculate $T_{\text{Wald}} = Z_{\text{Wald}}^2$, which under H_0 is chi-squared distributed with one degree of freedom:

$$P\text{-value} = \Pr\big[\chi_1^2 \geq T_{\text{Wald}}(\mathbf{n})\big].$$

6.14.2 The Likelihood Ratio Test

Maximum likelihood estimation for the proportional odds model is described in Section 6.12. Let $L_0 = L(\hat{\alpha}_1^*, \hat{\alpha}_2^*, \ldots, \hat{\alpha}_{c-1}^*, 0)$ be the maximum likelihood value under $H_0 : \beta = 0$, and let $L_1 = L(\hat{\alpha}_1, \hat{\alpha}_2, \ldots, \hat{\alpha}_{c-1}, \hat{\beta}_{\text{ML, logit}})$ be the maximum likelihood value under the more general proportional odds model. Then, a likelihood ratio test statistic for H_0 is

$$T_{\text{LR}}(\mathbf{n}) = -2(L_0 - L_1),$$

which is asymptotically chi-squared distributed with one degree of freedom under the null hypothesis.

6.14.3 The Score Test (the Wilcoxon-Mann-Whitney Test)

The score test of the null hypothesis $H_0 : \beta = 0$ is given by the derivatives of the log likelihood with respect to β calculated under the null hypothesis. McCullagh (1980) and Agresti (2010, pp. 80–81) have shown that the test can be expressed as the difference between the midranks in the two groups. Thus, the score test is equivalent to the midrank version of the Wilcoxon-Mann-Whitney test (also known as the *Wilcoxon test* or the *Wilcoxon rank sum test*).

The *Wilcoxon form* of the Wilcoxon-Mann-Whitney test statistic is

$$W(\mathbf{n}) = \sum_{j=1}^{c} m_j n_{1j}, \tag{6.31}$$

where the m_j are the midranks in Equation 6.19. The connection between the Wilcoxon statistic and the Mann-Whitney U statistic is given by $U = W - n_{1+}(n_{1+}+1)/2$. When n_{1+} and n_{2+} are not too small, we can approximate the distribution of W with the normal distribution. Because we have data in a $2 \times c$ table, observations in the same cell will have the same assigned rank, and we have to make *correction for tied observations* (in the expression for the variance). The expectation and the variance of W are given by

$$\mathrm{E}(W) = \frac{1}{2} n_{1+}(N+1)$$

and

$$\mathrm{Var}(W) = \frac{n_{1+}n_{2+}(N+1)}{12} - \frac{n_{1+}n_{2+}\sum_{j=1}^{c}(n_{+j}^3 - n_{+j})}{12N(N-1)},$$

where the last term is the correction for ties. For details, see Lehmann (1975, p. 20).

We define the Wilcoxon-Mann-Whitney test via the normalized Wilcoxon test statistic:

$$Z_{\text{W}}(\mathbf{n}) = \frac{W(\mathbf{n}) - \mathrm{E}(W)}{\sqrt{\mathrm{Var}(W)}},$$

which has an approximate standard normal distribution.

The Wilcoxon test statistic in Equation 6.31 is an example of a linear rank statistic. A linear rank statistic is defined as the weighted sum of the observations in one of the groups (for instance, Group 1):

$$T_{\text{linear rank}}(\mathbf{n}) = \sum_{j=1}^{c} b_j n_{1j}, \qquad (6.32)$$

where $b_1, b_2, \ldots, b_c$ are scores (or weights) connected to the columns in the $2 \times c$ table. We may use either the asymptotic or the exact distribution of $T_{\text{linear rank}}$ to define a test. Above, we used the midranks as scores and the asymptotic normal distribution. In Section 6.14.4, we will discuss the exact distribution.

As pointed out by McCullagh (1980), there is no most powerful test for $H_0 : \beta = 0$. Thus, there is no canonical choice of scores for a test based on the linear rank statistic. For the logit model, the Wilcoxon-Mann-Whitney test is the locally most powerful rank test.

Non-parametric tests based on ranks—for this and many other problems—are considered in detail in Neuhäuser (2012) and Hollander et al. (2014).

6.14.4 The Exact Conditional Linear Rank Test

The linear rank test statistic is given in Equation 6.32. To get the exact distribution of $T_{\text{linear rank}}$, we condition on the marginal sums n_{1+}, n_{2+} and $n_{+1}, n_{+2}, \ldots, n_{+c}$. We denote the vector of these marginals by $\mathbf{n}_{++}$. The conditional distribution under the null hypothesis is the multiple hypergeometric distribution, $f(\mathbf{x} \mid \mathbf{n}_{++})$, in Equation 6.4.

If T is an arbitrary test statistic, $\mathbf{n}$ is the observed table, and $\mathbf{x}$ is any possible table, the general expression for the exact P-value is given as

$$\text{exact } P\text{-value} = \Pr\big[T(\mathbf{x}) \geq T(\mathbf{n}) \mid H_0\big], \qquad (6.33)$$

which is the sum of the probabilities of all possible tables that agree less than or equally with the null hypothesis—as measured by the test statistic T—than the observed table. When we condition on the marginals ($\mathbf{n}_{++}$), the exact conditional P-value is given by

$$\text{exact cond. } P\text{-value} = \Pr\big[T(\mathbf{x}) \geq T(\mathbf{n}) \mid H_0, \mathbf{n}_{++}\big].$$

Thus, the one-sided exact conditional linear rank P-value, with midranks as scores, is

$$\text{one-sided } P\text{-value} = \sum_{\Omega(\mathbf{x}|\mathbf{n}_{++})} I\big[W(\mathbf{x}) \geq W(\mathbf{n})\big] \cdot f(\mathbf{x} \mid \mathbf{n}_{++}), \qquad (6.34)$$

where $\Omega(\mathbf{x}|\mathbf{n}_{++})$ is the set of all tables with marginals equal to $\mathbf{n}_{++}$, $I()$ is the indicator function and $W()$ is the Wilcoxon test statistic in Equation 6.31. As discussed in Section 1.3.4, we use the principle of twice the smallest tail for computing two-sided P-values. Provided Equation 6.34 gives the smallest of the two tail probabilities, by summation over all tables for which $\{W(\mathbf{x}) \geq W(\mathbf{n})\}$ and $\{W(\mathbf{x}) \leq W(\mathbf{n})\}$, respectively, the two-sided P-value is twice the one in Equation 6.34.

6.14.5   The Mid-P Linear Rank Test

The mid-P approach was introduced in Section 1.10, and the arguments for using it will not be repeated here. The mid-P value is the sum of the probabilities of all possible tables that agree less with the null hypothesis than does the observed table, plus one-half the point probability of the tables that agree equally with the null hypothesis as the observed table. The one-sided linear rank mid-P value (with midranks as scores) is given by

$$\text{one-sided mid-}P\text{ value} = \sum_{\Omega(\mathbf{x}|\mathbf{n}_{++})} I[W(\mathbf{x}) > W(\mathbf{n})] \cdot f(\mathbf{x}|\mathbf{n}_{++})$$

$$+ \ 0.5 \cdot \sum_{\Omega(\mathbf{x}|\mathbf{n}_{++})} I[W(\mathbf{x}) = W(\mathbf{n})] \cdot f(\mathbf{x}|\mathbf{n}_{++}).$$

In a similar manner as in the previous section, we define the two-sided mid-P value as twice the smallest of the two tail probabilities, that is, twice the one-sided mid-P value above, assuming that this is the smallest of the two tails.

6.14.6   An Exact Unconditional Test

Exact unconditional tests were introduced for the 2×2 table in Section 4.4.7, and an exact unconditional test was derived for the $r \times 2$ table in Section 5.7.12. Exact unconditional tests do not assume that all the marginals are fixed by the design, and do not condition on them. Exact unconditional tests will not suffer from the conservatism of exact conditional test.

   In Section 6.14.4, we considered an arbitrary test statistic T, with observed value $T(\mathbf{n})$, and gave the general expression for the exact unconditional P-value in Equation 6.33. The calculation of the exact unconditional P-value requires the unconditional probability distribution of $\mathbf{x} = \{x_{11}, x_{12}, \ldots, x_{1c}, x_{21}, x_{22}, \ldots, x_{2c}\}$ under the null hypothesis of no group differences, which is given by

$$f(\mathbf{x}|\pi_1, \pi_2, \ldots, \pi_c; \mathbf{n}_+) = \prod_{i=1}^{2} \frac{n_{i+}!}{x_{i1}!x_{i2}! \cdots x_{ic}!} \pi_1^{x_{i1}} \pi_2^{x_{i2}} \cdots \pi_c^{x_{ic}},$$

where $\mathbf{n}_+ = \{n_{1+}, n_{2+}\}$ denotes the fixed group sizes. This distribution is obtained from Equation 6.3 after setting $\pi_{j|1} = \pi_{j|2} = \pi_j$ for all $j = 1, 2, \ldots, c$,

which is the consequence of assuming the null hypothesis. The exact unconditional P-value can then be expressed as

$$P\text{-value} = \max_{\{\pi_1, \pi_2, \ldots, \pi_c\}} \left\{ \sum_{\Omega(\mathbf{x}|\mathbf{n}_+)} I\big[T(\mathbf{x}) \geq T(\mathbf{n})\big] \cdot f(\mathbf{x} \,|\, \pi_1, \pi_2, \ldots, \pi_c; \mathbf{n}_+) \right\},$$

where $\Omega(\mathbf{x}|\mathbf{n}_+)$ is the set of all tables with row sums equal to $\mathbf{n}_+$. Computation
of this P-value is quite complicated, due to the need for maximization over
the unknown nuisance parameters $\pi_1, \pi_2, \ldots, \pi_c$. We do not consider the exact
unconditional test further in this chapter; however, we refer the reader to Shan
and Ma (2016) for more information about this and other related tests.

6.15 The Cumulative Probit Model

The proportional odds model for a $2 \times c$ table is given by

$$\text{logit}\big[\Pr(Y_2 \leq j \,|\, x_i)\big] = \alpha_j - \beta x_i,$$

for $j = 1, 2, \ldots, c - 1$. As usual, $x_1 = 1$ denotes Group 1 and $x_2 = 0$ denotes
Group 2. An ordered $2 \times c$ table can be considered as a result of observations
from a latent variable Y^* on a continuous scale, and defined with thresholds
$-\infty = \alpha_0 < \alpha_1 < \ldots < \alpha_c = \infty$, such that

$$Y_2 = j \quad \text{when } \alpha_{j-1} \leq Y^* \leq \alpha_j,$$

for $j = 1, 2, \ldots, c$. In general, let

$$Y^* = \beta x_i + \epsilon,$$

where ϵ has a cumulative distribution G with $\mathrm{E}(\epsilon) = 0$, see Agresti (2010, pp.
53–55) or Skrondal and Rabe-Hesketh (2004, p. 16). Then, we get that

$$\Pr(Y_2 \leq j \,|\, x_i) = \Pr(Y^* \leq \alpha_j \,|\, x_i) = G(\alpha_j - \beta x_i),$$

for $j = 1, 2, \ldots, c - 1$. We now use G^{-1} as a link function to get a linear
predictor function

$$G^{-1}\big[\Pr(Y_2 \leq j \,|\, x_i)\big] = \alpha_j - \beta x_i.$$

If we let G be the cumulative distribution function of the logistic distribution,
which gives that G^{-1} is the logit link function, we obtain the proportional
odds model

$$\text{logit}\big[\Pr(Y_2 \leq j \,|\, x_i)\big] = \alpha_j - \beta x_i.$$

Alternatively, G can be the standard normal distribution, which we denote by Φ. Then,

$$\Pr(Y_2 \leq j \,|\, x_i) = \Phi(\alpha_j - \beta x_i), \tag{6.35}$$

or equivalently,

$$\Phi^{-1}\big[\Pr(Y_2 \leq j \,|\, x_i)\big] = \alpha_j - \beta x_i.$$

As noted by many, the results from a proportional odds model and a cumulative probit model are quite similar. It should be added, however, that the parameter estimates of β are on different scales, because the standard normal distribution has mean 0 and standard deviation 1, whereas the standard logistic distribution has mean 0 and standard deviation $\pi/\sqrt{3}$, see Agresti (2010, p. 123). Typically, the maximum likelihood estimate from the proportional odds model is 1.6 to 1.8 times the size of the maximum likelihood estimate from the cumulative probit model.

The logistic link function for cumulative probabilities is tightly connected to standard logistic regression, and per se much used for ordinal data, see Hosmer et al. (2013, Chapter 8) and Agresti (2010, Chapters 2–4). Moreover, the logistic link function is commonly used in the medical and biological sciences. The probit link, on the other hand, seems to be more common in the social sciences and in economics, see, for instance, Wooldridge (2013, Chapter 17) and Angrist and Pische (2009, Chapter 3).

6.15.1 Estimating Cumulative Probit Models

As for the proportional odds model, the standard procedure for estimating the αs and the β in the cumulative probit model is the maximum likelihood principle. The likelihood function for the cumulative probit model takes the form of

$$L(\alpha_1, \alpha_2, \ldots, \alpha_{c-1}, \beta) \;=\; \prod_{i=1}^{2} \prod_{j=1}^{c} \big[\Phi(\alpha_j - \beta x_i) - \Phi(\alpha_{j-1} - \beta x_i)\big]^{n_{ij}},$$

where $\alpha_0 = -\infty$ and $\alpha_c = \infty$. The maximum likelihood estimates are obtained by differentiating the log-likelihood function and setting the results to zero. Similarly to the calculations in Section 6.12 for the proportional odds model, we have that

$$\frac{\partial l}{\partial \beta} = -\sum_{j=1}^{c} n_{1j} \frac{\phi(\alpha_j - \beta) - \phi(\alpha_{j-1} - \beta)}{\Phi(\alpha_j - \beta) - \Phi(\alpha_{j-1} - \beta)} \tag{6.36}$$

and

$$\frac{\partial l}{\partial \alpha_k} = \sum_{j=1}^{c} n_{1j} \frac{\delta_{jk}\,\phi(\alpha_j - \beta) - \delta_{j-1\,k}\,\phi(\alpha_{j-1} - \beta)}{\Phi(\alpha_j - \beta) - \Phi(\alpha_{j-1} - \beta)}$$

$$+ \sum_{j=1}^{c} n_{2j} \frac{\delta_{jk}\,\phi(\alpha_j) - \delta_{j-1\,k}\,\phi(\alpha_{j-1})}{\Phi(\alpha_j) - \Phi(\alpha_{j-1})}, \tag{6.37}$$

where $\delta_{jk} = 1$ if $j = k$ and $\delta_{jk} = 0$ otherwise, and $\Phi()$ and $\phi()$ are the cumulative normal distribution function and normal density function, respectively. The equations to solve for β and the αs are then $\partial l / \partial \beta = 0$ and $\partial l / \partial \alpha_k = 0$ for $k = 1, 2, \ldots, c - 1$. For details, see McKelvey and Zavoina (1975).

Fisher scoring is used for maximum likelihood estimation. Let the maximum likelihood estimate of β be $\hat{\beta}_{\mathrm{ML,\,probit}}$. Again, the covariance matrix of the maximum likelihood estimates are given by the expected second partial derivatives of $L(\alpha_1, \alpha_2, \ldots, \alpha_{c-1}, \beta)$. Let

$$\phi_j = \phi(\alpha_j - \beta), \quad \phi_{j-1} = \phi(\alpha_{j-1} - \beta)$$

and

$$\Phi_j = \Phi(\alpha_j - \beta), \quad \Phi_{j-1} = \Phi(\alpha_{j-1} - \beta).$$

The second partial derivatives are given as

$$
\frac{\partial^2 l}{\partial \beta^2} = \sum_{j=1}^{c} n_{1j} \left\{ \frac{(\Phi_j - \Phi_{j-1})\left[(\alpha_{j-1} - \beta)\phi_{j-1} - (\alpha_j - \beta)\phi_j\right]}{(\Phi_j - \Phi_{j-1})^2} \right.
$$
$$
\left. - \frac{(\phi_{j-1} - \phi_j)^2}{(\Phi_j - \Phi_{j-1})^2} \right\},
\tag{6.38}
$$

$$
\frac{\partial^2 l}{\partial \beta \partial \alpha_l} = \sum_{j=1}^{c} n_{1j} \left\{ \frac{(\Phi_j - \Phi_{j-1})\left[\delta_{jl}(\alpha_j - \beta)\phi_j - \delta_{j-1,l}(\alpha_{j-1} - \beta)\phi_{j-1}\right]}{(\Phi_j - \Phi_{j-1})^2} \right.
$$
$$
\left. - \frac{(\phi_{j-1} - \phi_j)(\delta_{jl}\phi_j - \delta_{j-1,l}\phi_{j-1})}{(\Phi_j - \Phi_{j-1})^2} \right\},
\tag{6.39}
$$

and

$$
\frac{\partial^2 l}{\partial \alpha_k \partial \alpha_l} =
$$
$$
\sum_{j=1}^{c} n_{1j} \left\{ \frac{(\Phi_j - \Phi_{j-1})\left[\delta_{j-1,k}\delta_{j-1,l}(\alpha_{j-1} - \beta)\phi_{j-1} - \delta_{jk}\delta_{jl}(\alpha_j - \beta)\phi_j\right]}{(\Phi_j - \Phi_{j-1})^2} \right.
$$
$$
\left. - \frac{(\delta_{jk}\phi_j - \delta_{j-1,k}\phi_{j-1})(\delta_{jl}\phi_j - \delta_{j-1,l}\phi_{j-1})}{(\Phi_j - \Phi_{j-1})^2} \right\}
$$
$$
+ \sum_{j=1}^{c} n_{2j} \left\{ \frac{\left[\Phi(\alpha_j) - \Phi(\alpha_{j-1})\right]\left[\delta_{j-1,k}\delta_{j-1,l}\alpha_{j-1}\phi(\alpha_{j-1}) - \delta_{jk}\delta_{jl}\alpha_j\phi(\alpha_j)\right]}{\left[\Phi(\alpha_j) - \Phi(\alpha_{j-1})\right]^2} \right.
$$
$$
\left. - \frac{\left[\delta_{jk}\phi(\alpha_j) - \delta_{j-1,k}\phi(\alpha_{j-1})\right]\left[\delta_{jl}\phi(\alpha_j) - \delta_{j-1,l}\phi(\alpha_{j-1})\right]}{\left[\Phi(\alpha_j) - \Phi(\alpha_{j-1})\right]^2} \right\},
\tag{6.40}
$$

see also McKelvey and Zavoina (1975). By replacing the parameters by their maximum likelihood estimates, and taking the inverse of the resulting matrix, we get an estimate of the asymptotic covariance matrix. The estimated standard error of $\hat{\beta}_{\text{ML, probit}}$ is $\widehat{\text{SE}}(\hat{\beta}_{\text{ML, probit}})$. For details, see again McCullagh (1980) or Agresti (2010, pp. 119–121).

6.15.2 Testing the Fit of a Cumulative Probit Model

For testing goodness of fit, we use the Pearson chi-squared test and the log likelihood ratio test. As we did for the proportional odds model, we find $\hat{\pi}_{j|i}$ by inserting the maximum likelihood estimates of the αs and β in Equation 6.35. The estimated cell counts are

$$m_{ij} = n_{i+} \cdot \hat{\pi}_{j|i}.$$

The Pearson Goodness-of-Fit Test

The Pearson goodness-of-fit statistic is of the form

$$\chi^2(\mathbf{n}) = \sum_{i=1}^{c} \sum_{j=1}^{2} \frac{(n_{ij} - m_{ij})^2}{m_{ij}},$$

and it follows a chi-squared distribution with $c - 2$ degrees of freedom under the null hypothesis. P-values can be obtained as

$$P\text{-value} = \Pr\left[\chi_{c-2}^2 \geq \chi^2(\mathbf{n})\right].$$

The Likelihood Ratio (Deviance) Test

The log likelihood ratio test statistic (also called the *deviance*) is of the form

$$D(\mathbf{n}) = 2 \sum_{i=1}^{c} \sum_{j=1}^{2} n_{ij} \log\left(\frac{n_{ij}}{m_{ij}}\right),$$

and, like the Pearson goodness-of-fit statistic, it is asymptotically chi-squared distributed with $c - 2$ degrees of freedom.

6.15.3 Testing for Effect in a Cumulative Probit Model

We now test the null hypothesis of no association between the groups and the ordinal outcome in the cumulative probit model (Equation 6.35):

$$H_0 : \beta = 0. \tag{6.41}$$

The Wald Test

The Wald test statistic for testing the effect in the cumulative probit model is given by

$$Z_{\text{Wald}}(\mathbf{n}) = \frac{\hat{\beta}_{\text{ML, probit}}}{\widehat{\text{SE}}(\hat{\beta}_{\text{ML, probit}})}.$$

Z_{Wald} is standard normal distributed, such that P-values are obtained by

$$P\text{-value} = \Pr\left[Z \geq \left| Z_{\text{Wald}}(\mathbf{n}) \right| \right],$$

where Z is a standard normal variable.

The Likelihood Ratio Test

Let L_0 be the likelihood function under $H_0 : \beta = 0$, and let L_1 be the likelihood function under the cumulative probit model in Equation 6.35 with $\hat{\beta}_{\text{ML, probit}}$ inserted. A likelihood ratio test statistic for the effect in the cumulative probit model is given by

$$T_{\text{LR}}(\mathbf{n}) = -2(L_0 - L_1),$$

which is chi-squared distributed with one degree of freedom.

The Score Test

In Section 6.14.3, we discussed the score test for a proportional odds model, which is equivalent to the midrank version of the Wilcoxon-Mann-Whitney test. For the cumulative probit model, the score test for the null hypothesis in Equation 6.41 is given by the score divided by the second-order derivatives of the log likelihood, calculated under the null hypothesis. The first-order derivatives are given in Equations 6.36 and 6.37, and the second-order derivatives are given in Equations 6.38–6.40. Now, let $\boldsymbol{\theta}$ denote the vector of first-order derivatives,

$$\boldsymbol{\theta} = \left\{ \frac{\partial l}{\partial \alpha_1}, \frac{\partial l}{\partial \alpha_2}, \ldots, \frac{\partial l}{\partial \alpha_{c-1}}, \frac{\partial l}{\partial \beta} \right\}^{\text{T}},$$

and let I denote the matrix of second-order derivatives (the information matrix):

$$I = \begin{bmatrix} \dfrac{\partial^2 l}{\partial \alpha_1 \partial \alpha_1} & \dfrac{\partial^2 l}{\partial \alpha_1 \partial \alpha_2} & \cdots & \dfrac{\partial^2 l}{\partial \alpha_1 \partial \alpha_{c-1}} & \dfrac{\partial^2 l}{\partial \beta \partial \alpha_1} \\[2ex] \dfrac{\partial^2 l}{\partial \alpha_2 \partial \alpha_1} & \dfrac{\partial^2 l}{\partial \alpha_2 \partial \alpha_2} & \cdots & \dfrac{\partial^2 l}{\partial \alpha_2 \partial \alpha_{c-1}} & \dfrac{\partial^2 l}{\partial \beta \partial \alpha_2} \\[2ex] \vdots & \vdots & \ddots & \vdots & \vdots \\[2ex] \dfrac{\partial^2 l}{\partial \alpha_{c-1} \partial \alpha_1} & \dfrac{\partial^2 l}{\partial \alpha_{c-1} \partial \alpha_2} & \cdots & \dfrac{\partial^2 l}{\partial \alpha_{c-1} \partial \alpha_{c-1}} & \dfrac{\partial^2 l}{\partial \beta \partial \alpha_{c-1}} \\[2ex] \dfrac{\partial^2 l}{\partial \beta \partial \alpha_1} & \dfrac{\partial^2 l}{\partial \beta \partial \alpha_2} & \cdots & \dfrac{\partial^2 l}{\partial \beta \partial \alpha_{c-1}} & \dfrac{\partial^2 l}{\partial \beta^2} \end{bmatrix}.$$

We now evaluate $\boldsymbol{\theta}$ and I under the null hypothesis $H_0 : \beta = 0$. We obtain $\widehat{\boldsymbol{\theta}}_0$ and $\widehat{I}_0$ by substituting $\beta = 0$ and $\hat{\alpha}_j$ into the equations for $\boldsymbol{\theta}$ and I, where $\hat{\alpha}_j$

are the maximum likelihood estimates under H_0. Then, the score test statistic is

$$T_{\text{Score}}(\mathbf{n}) = \widehat{\boldsymbol{\theta}}_0^{\text{T}}(-\widehat{I}_0)^{-1}\widehat{\boldsymbol{\theta}}_0.$$

The score test statistic is chi-squared distributed with one degree of freedom under the null hypothesis.

6.16 Examples of Testing for Fit and Effect in the Proportional Odds and Cumulative Probit Models

The Adolescent Placement Study (Table 6.2)

Here, we study the association between gender and dangers to others from the Adolescent Placement Study (Fontanella et al., 2008). Table 6.7 shows the results of the tests for fit and effect presented in Sections 6.13, 6.14, and 6.15.

TABLE 6.7
Results of tests for fit and effect for the association between gender and danger to others based on the data in Table 6.2

Test	Statistic	df*	P-value
Testing the fit of a proportional odds model			
Pearson goodness-of-fit	7.49	2	0.0237
Likelihood ratio (deviance)	7.67	2	0.0216
Brant	7.19	2	0.0275
Testing the effect in a proportional odds model			
Wald	6.72		< 0.0001
Likelihood ratio	47.16	1	< 0.0001
Score (Wilcoxon-Mann-Whitney)	6.78		< 0.0001
Exact conditional linear rank			< 0.0001
Mid-P linear rank			< 0.0001
Testing the fit of a cumulative probit model			
Pearson goodness-of-fit	5.61	2	0.0604
Likelihood ratio (deviance)	5.62	2	0.0601
Testing the effect in a cumulative probit model			
Wald	6.99		< 0.0001
Likelihood ratio	49.21	1	< 0.0001
Score	48.83	1	< 0.0001

*Degrees of freedom for the chi-squared distribution

This is an analysis of a $2 \times c$ table with a large sample size ($N = 508$), and there is no cell with expected count less than five under the null hypothesis $H_0 : \beta = 0$.

For testing for a group effect in the proportional odds model or in the cumulative probit model, we observe only minor differences in the size of the test statistics, and due to the size of them, no difference in the P-values.

For testing the fit of a proportional odds model, we note that the three statistics are similar, with the Brant statistic somewhat smaller than the Pearson and the likelihood ratio statistics. We consider the Pearson statistic to be the most appropriate here, especially because it is related to Pearson residuals that can be used to study the discrepancies between the observed and predicted cell values, which can add further information about potential lack of fits.

We observe here that the model with proportional odds fits poorly. Such a result, however, must be interpreted with caution. Because of the large sample size, we know that the goodness-of-fit statistics increase with increasing sample size, and at $N = 508$, they are sensitive to small deviations from a perfect fit. We might thereby expect low P-values, regardless of whether the model fit is satisfactory or not. If the sample size is half the one we have in Table 6.2 (obtained by halving every cell count), the P-value for the Pearson goodness-of-fit test is $P = 0.141$. For this sample ($N = 254$), we calculate the Pearson residuals and find the results in Table 6.8.

TABLE 6.8
Pearson residuals for a sample of half the data in Table 6.2

	Danger to others			
	Unlikely	**Possible**	**Probable**	**Likely**
Male	-0.92	-0.39	1.14	-0.29
Female	0.45	0.05	-0.98	0.61

Note first that the square of the Pearson residuals adds to the test statistic in Table 6.7. Second, no residuals are greater than 1.14 in absolute value. Thus, we find that the model fit is satisfactory, regardless of the size of the test statistics.

There are many options for alternative analyses when the proportional odds assumption is in doubt. One could, as proposed by Peterson and Harrell (1990) or by Capuano and Dawson (2013), extend the proportional odds model to one with increasing or decreasing odds across the thresholds, which is a special case of the *partial proportional odds model*, see also Agresti (2010, pp. 77–80). For the $2 \times c$ table, we would, however, rather go for the simple solution that is related to the Wald test for goodness of fit. The Wald test, proposed by Brant (1990), is based on fitting three separate cumulative logistic regressions that in this case give the results in Table 6.9.

We observe substantial differences between the odds ratios, which increase with the categories. These odds ratios are simple to interpret, and since there

TABLE 6.9
Results of three cumulative logistic regressions

	Unlikely	Unlikely or possible	Unlikely, possible or probable
Coefficient	1.74	1.44	0.94
Odds ratio	0.18	0.24	0.39

are only $c - 1$ of them, it is manageable to interpret the differences between them.

Note that there are some differences in the goodness-of-fit statistics for the proportional odds model and the cumulative probit model in Table 6.7. We consider these differences to be small, and we find no reason to choose the cumulative probit model over the proportional odds model.

Postoperative Nausea (Table 6.3)

Now, we analyze the data on postoperative nausea in a randomized clinical trial of two treatments (Lydersen et al., 2012a, p. 76). The sample size in this example ($N = 59$) is much smaller than that in the previous example ($N = 508$). Two cells in the $2 \times c$ table have expected counts less than five, under the null hypothesis of no association between the outcome and the treatment. Table 6.10 shows the results of the tests for fit and treatment effect. None of the tests indicates a significant difference between the treatments, and there are only minor differences between the tests. Moreover, the tests for fit also agree very well, and no lack of fit is indicated.

6.17 Evaluation of Tests for Effect in the Proportional Odds and Cumulative Probit Models

Evaluation Criteria

As usual, we evaluate tests for effect in the proportional odds and cumulative probit models by calculating their actual significance levels and power. We assume that the two rows in the $2 \times c$ table represent two independent groups with fixed group sizes n_{1+} and n_{2+}, with a multinomial distribution for each row. The sampling distribution for the whole table is thus the product of two multinomial distributions (see Section 6.3).

TABLE 6.10
Results of tests for fit and treatment effect based on the data
on postoperative nausea in Table 6.3

Test	Statistic	df*	P-value
Testing the fit of a proportional odds model			
Pearson goodness-of-fit	1.68	2	0.432
Likelihood ratio (deviance)	1.72	2	0.424
Brant	1.67	2	0.434
Testing the effect in a proportional odds model			
Wald	-1.47		0.141
Likelihood ratio	2.20	1	0.138
Score (Wilcoxon-Mann-Whitney)	-1.46		0.144
Exact conditional linear rank			0.150
Mid-P linear rank			0.144
Testing the fit of a cumulative probit model			
Pearson goodness-of-fit	1.72	2	0.423
Likelihood ratio (deviance)	1.77	2	0.414
Testing the effect in a cumulative probit model			
Wald	-1.47		0.142
Likelihood ratio	2.15	1	0.143
Score	2.15	1	0.143

*Degrees of freedom for the chi-squared distribution

The actual significance level can be calculated exactly by

$$\text{ASL}(\boldsymbol{\pi}, \mathbf{n}_+, \alpha \mid H_0) = \sum_{x_{11}=0}^{n_{1+}} \sum_{x_{12}=0}^{n_{1+}-x_{11}} \cdots \sum_{x_{1,c-1}=0}^{n_{1+}-(x_{11}+\ldots+x_{1,c-2})}$$

$$\sum_{x_{21}=0}^{n_{2+}} \sum_{x_{22}=0}^{n_{2+}-x_{21}} \cdots \sum_{x_{2,c-1}=0}^{n_{2+}-(x_{21}+\ldots+x_{2,c-2})} I\big[P(\mathbf{x}) \leq \alpha\big] \cdot f(\mathbf{x} \mid \boldsymbol{\pi}, \mathbf{n}_+), \quad (6.42)$$

where $\mathbf{n}_+ = \{n_{1+}, n_{2+}\}$ denotes the fixed group sizes, α is the nominal significance level, $I()$ is the indicator function, $P(\mathbf{x})$ is the P-value for a test on $\mathbf{x} = \{x_{11}, \ldots, x_{1,c-1}, n_{1+} - (x_{11} + \ldots + x_{1,c-1}), x_{21}, \ldots, x_{2,c-1}, n_{2+} - (x_{21} + \ldots + x_{2,c-1})\}$, and $f()$ is given in Equation 6.3.

Power is calculated in the same manner as the actual significance level, except that we substitute H_A for H_0 in Equation 6.42. This change is reflected in the probabilities. The actual significance level is calculated under the null hypothesis, for which we assume that $\beta = 0$ and thereby $\pi_{j|1} = \pi_{j|2}$ for $j = 1, 2, \ldots, c$. Power, on the other hand, is calculated under the alternative hypothesis, where $\beta \neq 0$, and, in general, $\pi_{j|1} \neq \pi_{j|2}$.

The relationship between the parameters of the models and the probabilities $\boldsymbol{\pi}$ in Equations 6.3 and 6.42 can be shown as follows. The proportional odds model is given by

$$\text{logit}\left[\Pr(Y_2 \leq j \,|\, x_i)\right] = \alpha_j - \beta x_i,$$

and the cumulative probit model is given by

$$\phi^{-1}\left[\Pr(Y_2 \leq j \,|\, x_i)\right] = \alpha_j - \beta x_i,$$

where $x_1 = 1$ denotes Group 1 and $x_2 = 0$ denotes Group 2, and ϕ is the standard normal cumulative distribution function. The cumulative probabilities are

$$\Pr(Y_2 \leq j \,|\, x_i) = \frac{\exp(\alpha_j - \beta x_i)}{1 + \exp(\alpha_j - \beta x_i)} \tag{6.43}$$

for the proportional odds model, and

$$\Pr(Y_2 \leq j \,|\, x_i) = \phi(\alpha_j - \beta x_i) \tag{6.44}$$

for the cumulative probit model. The cell probabilities in the $2 \times c$ table are thus

$$
\begin{aligned}
\pi_{1|i} &= \Pr(Y_2 \leq 1 \,|\, x_i) \\
\pi_{2|i} &= \Pr(Y_2 \leq 2 \,|\, x_i) - \Pr(Y_2 \leq 1 \,|\, x_i) \\
&\ \ \vdots \\
\pi_{c-1|i} &= \Pr(Y_2 \leq c - 1 \,|\, x_i) - \Pr(Y_2 \leq c - 2 \,|\, x_i) \\
\pi_{c|i} &= 1 - \Pr(Y_2 \leq c - 1 \,|\, x_i),
\end{aligned}
$$

where $i = 1, 2$ indexes the rows, and where we use Equation 6.43 for the proportional odds model and Equation 6.44 for the cumulative probit model.

Evaluation of Actual Significance Level

The computational burden of calculating actual significance levels (and power) exactly for the $2 \times c$ table is quite heavy, except for very small sample sizes (n_{1+} and n_{2+}). The number of possible tables to consider can be huge, as we can see in Equation 6.42. This is different from the $r \times 2$ table in Chapter 5, because for the $2 \times c$ table, we only condition on n_{1+} and n_{2+} and have c free (unconditioned) marginals ($n_{+1}, n_{+2}, \ldots, n_{+c}$), whereas for the $r \times 2$ table in Chapter 5, we conditioned on $n_{1+}, n_{2+}, \ldots, n_{r+}$ and had only two free marginals (n_{+1} and n_{+2}). We thereby use simulations instead of exact calculations throughout this evaluation.

To calculate actual significance levels, we set $\beta = 0$ and fix the α_j of the model. First, we consider the proportional odds model. The results for one example, with $c = 4$ and $\alpha_j = \{-1.5, 0.5, 1.0\}$, are shown in Figure 6.2. These α_j-values correspond to the probabilities given in Table 6.11. The likelihood ratio test is too liberal for small sample sizes (marginals less than 50). The exact conditional test is somewhat conservative for small sample sizes; however, not by much. The Wald, score, and mid-P tests all perform excellently, even for very small sample sizes. These results are replicated for other scenarios (other values of α_j).

TABLE 6.11

The cell probabilities (not the cumulative probabilities) of the proportional odds model with $\alpha_j = \{-1.5, 0.5, 1.0\}$ and $\beta = 0$

	Cat. 1	Cat. 2	Cat. 3	Cat. 4
Group 1	0.1824	0.4400	0.1086	0.2689
Group 2	0.1824	0.4400	0.1086	0.2689

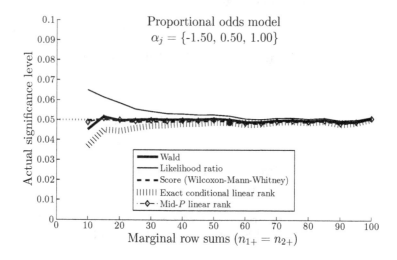

FIGURE 6.2

Actual significance levels of five tests for effect in the proportional odds model

We now turn to the cumulative probit model, and consider an example with $c = 4$ and $\alpha_j = \{0.0, 0.5, 1.1\}$. The probabilities associated with this model are shown in Table 6.12. The results for the three tests for effect in the cumulative probit model are shown in Figure 6.3. All tests are liberal, with actual significance levels between 5% and 6% for a nominal significance level of 5%. Other values of α_js give similar results.

Evaluation of Power

Figure 6.4 shows an example of the power of the five tests for effect in the proportional odds model. The parameters of the model are given by $\alpha_j = \{-0.7, 0.2, 0.7\}$ and $\beta = 1.1$. The cell probabilities of this model are shown in

TABLE 6.12
The cell probabilities (not the cumulative probabilities) of the cumulative probit model with $\alpha_j = \{0.0, 0.5, 1.1\}$ and $\beta = 0$

	Cat. 1	Cat. 2	Cat. 3	Cat. 4
Group 1	0.500	0.1915	0.1729	0.1357
Group 2	0.500	0.1915	0.1729	0.1357

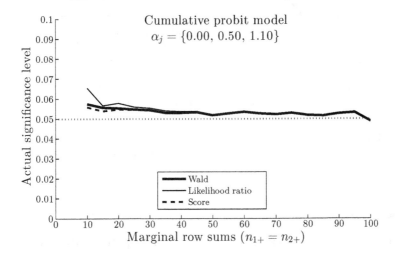

FIGURE 6.3
Actual significance levels of three tests for effect in the cumulative probit model

Table 6.13. For small sample sizes, the likelihood ratio test has a small power gain on the other tests; however, to achieve 80% or 90% power, there is no practical difference between the tests.

TABLE 6.13
The cell probabilities (not the cumulative probabilities) of the proportional odds model with $\alpha_j = \{-0.7, 0.2, 0.7\}$ and $\beta = 1.1$

	Cat. 1	Cat. 2	Cat. 3	Cat. 4
Group 1	0.3318	0.2180	0.1184	0.3318
Group 2	0.1419	0.1472	0.1123	0.5987

For the cumulative probit model, there is no substantial difference in the power of the three tests for effect.

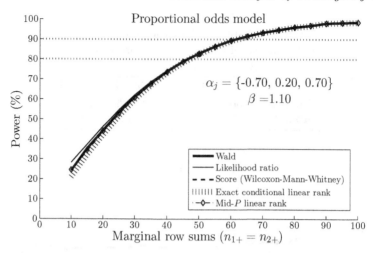

FIGURE 6.4
Power of five tests for effect in the proportional odds model

6.18   Confidence Intervals for the Proportional Odds and Cumulative Probit Models

Section 6.12 described the maximum likelihood estimate of the effect, β, in the proportional odds model, and Section 6.15.1 described the maximum likelihood estimate of the effect in the cumulative probit model (also called β). In this section, we consider two methods to construct confidence intervals for β, and the methods are the same regardless of whether it is the effect in a proportional odds or cumulative probit model.

6.18.1   The Wald Interval

Let $\hat{\beta}_{\mathrm{ML}}$ denote the maximum likelihood estimate of β. For the proportional odds model, $\hat{\beta}_{\mathrm{ML}}$ equals $\hat{\beta}_{\mathrm{ML, logit}}$ from Section 6.12, and for the cumulative probit model, $\hat{\beta}_{\mathrm{ML}}$ equals $\hat{\beta}_{\mathrm{ML, probit}}$ from Section 6.15.1. Let the standard error be given by $\widehat{\mathrm{SE}}(\hat{\beta}_{\mathrm{ML}})$, which may be found as described in Sections 6.12 and 6.15.1. A Wald interval for β may then be constructed as

$$\hat{\beta}_{\mathrm{ML}} \pm z_{\alpha/2}\widehat{\mathrm{SE}}(\hat{\beta}_{\mathrm{ML}}),$$

where $z_{\alpha/2}$ is the upper $\alpha/2$ percentile of the standard normal distribution.

6.18.2 The Profile Likelihood Interval

The method of constructing confidence intervals by the profile likelihood method was introduced in Section 5.8.3. Recall that a $100(1 - \alpha)\%$ profile likelihood interval for β consists of all parameter values β_0 that would not reject the null hypothesis $H_0 : \beta = \beta_0$ at a significance level of α. The lower (L) and upper (U) end points of the profile likelihood interval for β are given by

$$2\left[l(\hat{\beta}) - l_p(U)\right] = 2\left[l(\hat{\beta}) - l_p(L)\right] = \chi_1^2(\alpha),$$

where $l(\hat{\beta})$ is the log-likelihood of the full model, $l_p()$ is the profile log-likelihood function, see Royston (2007) or Hosmer et al. (2013, p. 19), and $\chi_1^2(\alpha)$ is the upper α percentile of the chi-squared distribution with one degree of freedom.

6.18.3 Examples

The Adolescent Placement Study (Table 6.2)

The Wald and profile likelihood intervals for β, the association between gender and danger to others—in both the proportional odds and cumulative probit models—are shown in Table 6.14. Since the sample size is quite large ($N = 508$), we do not expect much difference between the Wald and profile likelihood intervals.

TABLE 6.14
Estimation of the effect parameter β of the association between gender and danger to others with 95% confidence intervals (CIs), based on the data in Table 6.2

Interval	$\hat{\beta}$ (95% CI)	$\widehat{OR}^*$ (95% CI)
Proportional odds model		
Wald	1.145 (0.811 to 1.479)	0.318 (0.228 to 0.444)
Profile likelihood	1.145 (0.814 to 1.481)	0.318 (0.227 to 0.443)
Cumulative probit model		
Wald	0.7044 (0.5069 to 0.9019)	
Profile likelihood	0.7044 (0.5069 to 0.9025)	

*Cumulative odds ratio

Note that the estimate of β in the proportional odds model is positive: $\hat{\beta}_{\text{ML, logit}} = 1.145$. In our definition of the proportional odds model (Equation 6.20), we have a negative sign for β, and if we substitute the estimate for β in Equation 6.24, we get that

$$\frac{\Pr(Y_2 \leq j \mid Y_1 = 1)/\Pr(Y_2 > j \mid Y_1 = 1)}{\Pr(Y_2 \leq j \mid Y_1 = 2)/\Pr(Y_2 > j \mid Y_1 = 2)} = \exp(-1.145) = 0.318.$$

Because of the positive sign of $\hat{\beta}_{\mathrm{ML,\,logit}}$, we are more likely to get observations in higher outcomes for males than for females. Thus, males are more likely to be a danger to others than females, and the cumulative odds ratio is 0.318.

Postoperative Nausea (Table 6.3)

The Wald and profile likelihood intervals of the treatment effect (β) on postoperative nausea are shown in Table 6.15. The estimated treatment effect is negative: $\hat{\beta}_{\mathrm{ML,\,logit}} = -0.716$. If we insert the estimated treatment effect into Equation 6.24, we get

$$\frac{\Pr(Y_2 \le j \mid Y_1 = 1)/\Pr(Y_2 > j \mid Y_1 = 1)}{\Pr(Y_2 \le j \mid Y_1 = 2)/\Pr(Y_2 > j \mid Y_1 = 2)} = \exp(0.716) = 2.05.$$

Since $\hat{\beta}_{\mathrm{ML,\,logit}}$ now is negative, we are more likely to get observations in lower outcomes (less nausea) for those treated with ventricular tubes compared with those in the control group. The cumulative odds ratio for those with ventricular tubes relative to the controls is 2.05.

TABLE 6.15
Estimation of the effect parameter β (the treatment effect on postoperative nausea) with 95% confidence intervals (CIs), based on the data in Table 6.3

Interval	$\hat{\beta}$ (95% CI)	$\widehat{\mathrm{OR}}^*$ (95% CI)
Proportional odds model		
Wald	-0.716 (-1.67 to 0.236)	2.05 (0.790 to 5.30)
Profile likelihood	-0.716 (-1.68 to 0.229)	2.05 (0.795 to 5.39)
Cumulative probit model		
Wald	-0.425 (-0.992 to 0.142)	
Profile likelihood	-0.425 (-0.995 to 0.143)	

*Cumulative odds ratio

6.18.4   Evaluations of Intervals

Evaluation Criteria

We use coverage probability, interval width, and location to evaluate the performance of confidence intervals (see Section 1.4). The sampling distribution for the $2 \times c$ table is the product of two multinomial distributions (see Section 6.3), and is derived from the assumption that the two rows in the $2 \times c$ table represent two independent groups with fixed group sizes n_{1+} and n_{2+}.

The exact coverage probability of intervals for the ordered $2 \times c$ table is defined by

$$\text{CP}(\boldsymbol{\pi}, \mathbf{n}_+, \alpha) = \sum_{x_{11}=0}^{n_{1+}} \sum_{x_{12}=0}^{n_{1+}-x_{11}} \cdots \sum_{x_{1,c-1}=0}^{n_{1+}-(x_{11}+\ldots+x_{1,c-2})}$$

$$\sum_{x_{21}=0}^{n_{2+}} \sum_{x_{22}=0}^{n_{2+}-x_{21}} \cdots \sum_{x_{2,c-1}=0}^{n_{2+}-(x_{21}+\ldots+x_{2,c-2})} I(L \le \beta \le U) \cdot f(\mathbf{x} \,|\, \boldsymbol{\pi}, \mathbf{n}_+),$$

where $\mathbf{n}_+ = \{n_{1+}, n_{2+}\}$ denotes the fixed group sizes, $I()$ is the indicator function, $L = L(\mathbf{x}, \alpha)$ and $U = U(\mathbf{x}, \alpha)$ are the lower and upper $100(1 - \alpha)\%$ confidence limits of an interval for the table $\mathbf{x} = \{x_{11}, \ldots, x_{1,c-1}, n_{1+} - (x_{11} + \ldots + x_{1,c-1}), x_{21}, \ldots, x_{2,c-1}, n_{2+} - (x_{21} + \ldots + x_{2,c-1})\}$, and $f()$ is given in Equation 6.3.

The exact expected interval width is given by

$$\text{Width}(\boldsymbol{\pi}, \mathbf{n}_+, \alpha) = \sum_{x_{11}=0}^{n_{1+}} \sum_{x_{12}=0}^{n_{1+}-x_{11}} \cdots \sum_{x_{1,c-1}=0}^{n_{1+}-(x_{11}+\ldots+x_{1,c-2})}$$

$$\sum_{x_{21}=0}^{n_{2+}} \sum_{x_{22}=0}^{n_{2+}-x_{21}} \cdots \sum_{x_{2,c-1}=0}^{n_{2+}-(x_{21}+\ldots+x_{2,c-2})} (U - L) \cdot f(\mathbf{x} \,|\, \boldsymbol{\pi}, \mathbf{n}_+).$$

Location is measured by the MNCP/NCP index. The non-coverage probability (NCP) is computed as $1 - \text{CP}$, and the mesial non-coverage probability (MNCP) is defined as

$$\text{MNCP}(\boldsymbol{\pi}, \mathbf{n}_+, \alpha) = \sum_{x_{11}=0}^{n_{1+}} \sum_{x_{12}=0}^{n_{1+}-x_{11}} \cdots \sum_{x_{1,c-1}=0}^{n_{1+}-(x_{11}+\ldots+x_{1,c-2})}$$

$$\sum_{x_{21}=0}^{n_{2+}} \sum_{x_{22}=0}^{n_{2+}-x_{21}} \cdots \sum_{x_{2,c-1}=0}^{n_{2+}-(x_{21}+\ldots+x_{2,c-2})} I(L > \beta \ge 0 \text{ or } U < \beta \le 0) \cdot f(\mathbf{x} \,|\, \boldsymbol{\pi}, \mathbf{n}_+).$$

Evaluation of Coverage Probability

Because of the heavy computational burden of using exact calculations for the $2 \times c$ table (see Section 6.42, page 263), we use simulations instead of exact calculations to evaluate coverage probabilities. The only two confidence interval methods to consider are the Wald interval and the profile likelihood interval. For reasons equal to those stated on page 216 (for the profile likelihood interval for the $r \times 2$ table), we do not include the profile likelihood interval in these simulations. That leaves only the Wald interval; however, we have two models to consider: the proportional odds model and the cumulative probit model. An example of how the Wald interval performs in these two models is shown in Figure 6.5. Note that the two Wald intervals are not directly comparable, because they are computed on data based on different models. Still, we have put the two plots in the same figure because they represent typical performance across different parameter values. Hence, the coverage probability of the Wald interval for the proportional odds model is usually very close to the nominal level, whereas the coverage probability of the Wald interval for the cumulative probit model is usually a bit lower than the nominal level, often at about 94.5% for a nominal level of 95%. This result agrees well with the actual significance levels of the tests for effect in the proportional odds and cumulative probit models (Section 6.17).

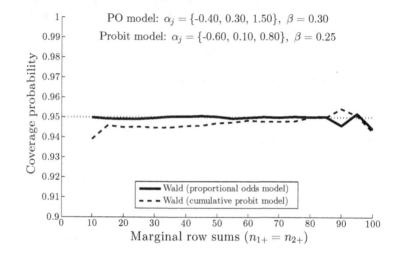

FIGURE 6.5
Simulated coverage probabilities of the Wald interval for the proportional odds and cumulative probit models

Evaluation of Width

Based on our experience, the Wald interval and the profile likelihood interval tend to produce confidence intervals of similar widths, although the location of the intervals may differ slightly (see next paragraph).

Evaluation of Location

The location of the Wald interval—for both the proportional odds and cumulative probit models—is excellent, with MNCP/NCP index in the middle of the satisfactory region at about 0.5, as exemplified in Figure 6.6. Although we have not calculated the MNCP/NCP index for the profile likelihood interval, we have observed that the profile likelihood interval tends to be slightly more distally located than the Wald interval.

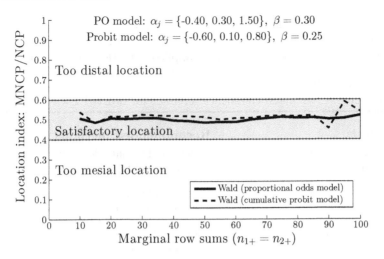

FIGURE 6.6
Location, as measured by the MNCP/NCP index, of the Wald interval for the proportional odds and cumulative probit models

6.19   Recommendations

6.19.1   Summary

We summarize our main recommendations for analysis of ordered $2 \times c$ tables in Table 6.16. In this subsection, we discuss the different types of analysis that are possible in the $2 \times c$ table, and which ones we prefer. In the following subsections, we go into more detail concerning some of the analyses.

In Section 5.9, we wrote warmly of the tests for unspecific ordering in the $r \times 2$ table, which are equal to the tests for ordered local odds ratios in the $2 \times c$ table. Moreover, the tests for ordered cumulative odds ratios—which also test for ordering in the continuation odds ratios—are very similar to the tests for unspecific ordering in Chapter 5. Hence, we like this approach also for the $2 \times c$ table. The pros and cons of analyzing unspecific ordering versus model-based analysis were discussed in Section 5.9.1, and that discussion also applies to the $2 \times c$ table. On the choice between testing for ordered local odds ratios and testing for ordered cumulative—and continuation—odds ratios, we prefer to test for ordered local odds ratios; these tests are slightly easier to derive and compute, although the principles underlying the tests are the same.

When it comes to choosing between the proportional odds model and the cumulative probit model, we prefer the proportional odds model. For the proportional odds model, there are more methods to choose from, and they perform better than the methods for the cumulative probit model, both for testing effect and for estimating a confidence interval for the effect. The proportional odds model is also closely connected to standard binary logistic regression, with which most people working with applied statistics are familiar.

TABLE 6.16
Recommended tests and confidence intervals for ordered $2 \times c$ tables

Analysis	Recommended methods	Sample sizes
Tests for ordered local odds ratios	Exact conditional (Pearson)	all
	Exact conditional (LR)	all
Tests for ordered cumulative odds ratios*	Exact conditional (Pearson)	all
	Exact conditional (LR)	all
Testing the fit of a proportional odds model	Pearson goodness-of-fit	all
	Likelihood ratio (deviance)	all
	Brant	all
Tests for effect in the proportional odds model	Wald	all
	Score (Wilcoxon-Mann-Whitney)	all
	Mid-P linear rank	all
	Likelihood ratio	large
	Exact conditional linear rank	large
Confidence intervals for the proportional odds model	Wald	all
Testing the fit of a cumulative probit model[†]	Pearson goodness-of-fit	all
	Likelihood ratio (deviance)	all
Tests for effect in the cumulative probit model[†]	Wald	all
	Likelihood ratio	all
	Score	all
Confidence intervals for the cumulative probit model[†]	Wald	all

*These tests and recommendations also apply for testing ordered continuation odds ratios
[†]We recommend use of the proportional odds model ahead of the cumulative probit model

6.19.2 Testing the Fit of a Proportional Odds or Cumulative Probit Model

We have not evaluated the tests for fit and compared them to find which ones to recommend, simply because we recommend all of them. Testing for fit is like a screening test with no serious consequence of a false positive; if a test indicates lack of fit, we explore the data and the model in more detail before deciding how to proceed. A liberal test, a test with actual significance levels above the nominal level, is thus conservative for the task of selecting a suitable model. We are, therefore, not so concerned with how the tests for fit perform, except that we wish for excellent power to detect lack of fit. Calculating several tests for fit will increase power, at the expense of sometimes scrutinizing a model too much. Note, however, that for large sample sizes, the tests for fit may be significant even for quite small and inconsequential lacks of fit. In such cases, a cautious interpretation should be exercised.

6.19.3   Tests for Effect in the Proportional Odds Model

Three of the five tests that we have considered for testing the effect in the proportional odds model perform excellently, regardless of the sample size: the Wald, score (Wilcoxon-Mann-Whitney), and mid-P linear rank tests. The likelihood ratio test is too liberal for small sample sizes; however, it is acceptable when the marginal row sums are greater than 50. The exact conditional linear rank test is somewhat too conservative for small sample sizes, but it too performs very well for slightly larger sample sizes. When both the row sums are 100 or larger, all the five tests perform excellently.

6.19.4   Tests for Effect in the Cumulative Probit Model

The three tests that we have considered for the cumulative probit model—the Wald, likelihood ratio, and score tests—perform almost identically, except for very small sample sizes. Their actual significance levels exceed the nominal level, but not by much, usually below 5.5% for a nominal level of 5%. It may be the case that, for large sample sizes, this undesirable increase in significance levels may be inconsequentially small or non-existing, although we have no solid evidence for or against this. Our general recommendation is thus to use the proportional odds model—if it seems to fit the data well—instead of the cumulative probit model. This is not to say that we recommend that the cumulative probit model should be avoided, it just means that we recommend that the proportional odds model is the first choice of model for ordered $2 \times c$ tables.

6.19.5   Confidence Intervals for the Proportional Odds and Cumulative Probit Models

The excellent performance of the Wald interval for the proportional odds model is well illustrated in Figure 6.5. Quite remarkably, it has actual significance levels very close to the nominal level for all sample sizes $n_{1+} = n_{2+} \geq 10$, and this result was replicated for several other parameter settings. The coverage probability of the Wald interval for the cumulative probit model is usually at about 94.5% for a nominal level of 95%, which agrees with the results of the corresponding Wald test for effect.

The profile likelihood interval, for both models, will often provide similar confidence limits as the Wald interval, with a small distal shift in location.

Although we have not evaluated the profile likelihood interval in any detail, we expect it to perform similarly to the Wald interval. Because the Wald interval is so simple to calculate, with widespread availability in statistical software packages, whereas the profile likelihood interval is quite complicated to calculate, we do not include the profile likelihood interval in our recommendations for confidence interval estimation in the ordered $2 \times c$ table.

7

The $r \times c$ Table

7.1   Introduction

In this chapter, we consider unordered $r \times c$ tables, where r or c or both are greater than two, as well as ordered $r \times c$ tables, where both r and c are greater than two. Table 7.1 shows the notation for the observed counts of a general $r \times c$ table. As in previous chapters, we let—when appropriate—the rows comprise a grouping of the data, and the columns represent the outcomes.

TABLE 7.1
The observed counts of an $r \times c$ table

Variable 1	Variable 2				Total
	Category 1	Category 2	...	Category c	
Group 1	n_{11}	n_{12}	...	n_{1c}	n_{1+}
Group 2	n_{21}	n_{22}	...	n_{2c}	n_{2+}
⋮	⋮	⋮	⋱	⋮	⋮
Group r	n_{r1}	n_{r2}	...	n_{rc}	n_{r+}
Total	n_{+1}	n_{+2}	...	n_{+c}	N

Additional notation:
$\mathbf{n} = \{n_{11}, n_{12}, \ldots, n_{1c}, n_{21}, n_{22}, \ldots, n_{2c}, \ldots, n_{r1}, n_{r2}, \ldots, n_{rc}\}$: the observed table
$\mathbf{x} = \{x_{11}, x_{12}, \ldots, x_{1c}, x_{21}, x_{22}, \ldots, x_{2c}, \ldots, x_{r1}, x_{r2}, \ldots, x_{rc}\}$: any possible table

The choice of analysis of an $r \times c$ table depends on whether the categories in the rows and columns shall be regarded as unordered, i.e., corresponding to nominal variables, or ordered, i.e., corresponding to ordinal variables. For example, if three treatments represent increasing doses of a drug, it may be appropriate to consider the treatments as ordered. If, on the other hand, it is possible that the medium dose is the most beneficial, the outcome probabilities are not ordered, and we should regard the three treatments as unordered. An $r \times c$ table is said to be unordered, singly ordered, or doubly ordered, if none, one, or both variables are ordinal, respectively. Moreover, the choice of analysis depends on whether $r > 2$, $c > 2$, or both $r > 2$ and $c > 2$. A summary is given in Table 7.2.

The rest of this chapter is organized as follows. Section 7.2 introduces the examples, which will be used to illustrate how to use and interpret differ-

TABLE 7.2

Different $r \times c$ contingency tables, with examples of recommended methods

Size and type of table*	Examples of recommended methods	Described in
$r = c = 2$	Newcombe hybrid score interval Fisher mid-P test	Chapter 4
Ordered $r \times 2$	Cochran-Armitage test for trend	Chapter 5
Ordered $2 \times c$	Wilcoxon-Mann-Whitney test Ordinal logistic regression with binary covariate	Chapter 6
$r > 2$ and/or $c > 2$, both unordered	Pearson chi-squared test	Section 7.5
$r > 2$ and $c > 2$, singly ordered: ordered columns, unordered rows	Kruskal-Wallis test Ordinal logistic regression with column as outcome and row as nominal covariate	Section 7.6
$r > 2$ and $c > 2$, singly ordered: ordered rows, unordered columns	Kruskal-Wallis test *Nominal logistic regression with column as outcome and row as ordinal covariate*[†]	Section 7.6
$r > 2$ and $c > 2$, doubly ordered	Linear-by-linear test for association Jonckheere-Terpstra test Correlation measures Ordinal logistic regression	Section 7.8 / Section 7.9 / Section 7.10

*When appropriate, the rows comprise a grouping of the data and the columns represent the outcomes
[†]We regard this as an unrealistic situation, and we have yet to find an actual example of this setup

ent statistical methods for the analysis of $r \times c$ tables later in the chapter. We present notation and sampling distributions in Section 7.3 and different odds ratios as effect measures in Section 7.4. Unordered tables are covered in Section 7.5, and singly ordered tables are covered in Section 7.6 (tests for comparing r groups) and in Section 7.7 (ordinal logistic regression). Doubly ordered tables are the topic of three sections: Section 7.8 (tests for association), Section 7.9 (correlation measures), and Section 7.10 (ordinal logistic regression). Finally, we summarize our recommendations for $r \times c$ tables in Section 7.11.

7.2 Examples

7.2.1 Treatment for Ear Infection (Unordered 3 × 2 Table)

Van Balen et al. (2003) report a randomized, double-blind, controlled trial comparing three treatments for an ear infection. The numbers and proportions of patients reported cured and not cured after 21 days of treatment are summarized in Table 7.3. Because there is no ordering between the treatments, we regard Table 7.3 as an unordered 3 × 2 table.

TABLE 7.3
Status after 21 days treatment of the ear infection acute otitis externa (Van Balen et al., 2003)

Treatment	Cured	Not cured	Total
Acetic acid	40 (62%)	25 (38%)	65
Corticosteroid and acetic acid	54 (89%)	7 (12%)	61
Corticosteroid and antibiotic	63 (86%)	10 (14%)	73
Total	157 (79%)	42 (21%)	199

7.2.2 Psychiatric Diagnoses and Physical Activity (Unordered 6 × 2 Table)

Table 7.4 shows the number of subjects participating in team sports within each of six psychiatric diagnoses, based on data from a study of physical activity in adolescents aged 13 to 18 years who were referred to a child and adolescent psychiatric clinic from 2009 to 2001 (Mangerud et al., 2014). The psychiatric diagnoses are unordered, and we shall treat this as an unordered 6 × 2 table.

7.2.3 Psychiatric Diagnoses and BMI (Unordered or Singly Ordered 6 × 3 Table)

Table 7.5 shows the number of thin, normal weight, and overweight subjects within each of six psychiatric diagnoses, based on the same study as in Section 7.2.2 (Mangerud et al., 2014). Body mass index (BMI) is calculated as the weight in kg divided by the squared height in meters. In subjects aged 18 years or older, the cut-off points for being categorized as thin, normal weight, and overweight are BMI less than 18.5, BMI between 18.5 and 25, and BMI above 25, respectively. For younger subjects (below 18 years of age), the categorization was done following internationally adopted cut-off points for age and sex (Cole et al., 2000, 2007). For example, the cut-off point for being overweight at age 13 is 21.91 for males and 22.58 for females.

TABLE 7.4

Psychiatric diagnoses and participation in team
sports (Mangerud et al., 2014)

Main diagnosis	Participation in team sports		Total
	Yes	No	
Mood (affective) disorders	21 (25%)	62 (75%)	83
Anxiety disorders	48 (33%)	97 (67%)	145
Eating disorders	12 (55%)	10 (46%)	22
Autism spectrum disorders	7 (19%)	30 (81%)	37
Hyperkinetic disorders	78 (37%)	132 (63%)	210
Other disorders	17 (33%)	34 (67%)	51
Total	183 (33%)	365 (67%)	548

TABLE 7.5

Psychiatric diagnoses and weight categories based on age- and
sex-adjusted BMI (Mangerud et al., 2014)

Main diagnosis	Thin	Normal	Overweight	Total
Mood (affective) dis.	3 (3.7%)	55 (68%)	23 (28%)	81
Anxiety disorders	8 (5.5%)	102 (70%)	36 (25%)	146
Eating disorders	6 (29%)	14 (67%)	1 (4.8%)	21
Autism spectrum dis.	5 (13%)	21 (55%)	12 (32%)	38
Hyperkinetic dis.	19 (8.9%)	130 (61%)	64 (30%)	213
Other disorders	7 (14%)	26 (51%)	18 (35%)	51
Total	48 (8.7%)	348 (63%)	154 (28%)	550

The psychiatric diagnoses in Table 7.5 are unordered. But how shall we
regard the weight categories? In terms of increasing weight, the categories
thin, normal weight, and overweight are naturally ordered, so Table 7.5 may
be regarded as a singly ordered 6 × 3 table. Yet, in some situations, it can be
appropriate to regard weight category as an unordered variable. Being thin, as
well as being overweight, can have negative health effects compared to being
normal weight. Hence, in some contexts, it may be more appropriate to regard
Table 7.5 as an unordered 6 × 3 table.

7.2.4 Low Birth Weight and Psychiatric Morbidity (Singly Ordered 3 × 3 Table)

Lund et al. (2012) report psychiatric morbidity in young adulthood in two low
birth weight groups and a control group. The subjects were born between 1986
and 1988. The very low birth weight (VLBW) group consisted of babies born
preterm with birth weight ≤ 1500 grams. The small for gestational age at term

(SGA) group was born at term with birth weight below the 10th percentile adjusted for gestational age, sex, and parity. The control group was born at term, and was not small for gestational age. Table 7.6 shows the severity level of psychiatric problems at age 20 years. We shall regard the birth groups as unordered; however, the diagnostic groups are naturally ordered. Hence, Table 7.6 is a singly ordered 3 × 3 table with unordered rows and ordered columns.

TABLE 7.6

Categories of birth weight and psychiatric problems at age 20 years (Lund et al., 2012)

Birth weight	No diagnosis	Subthreshold diagnosis	Definite diagnosis	Total
VLBW*	22 (58%)	4 (11%)	12 (32%)	38
SGA[†]	24 (56%)	9 (21%)	10 (23%)	43
Control	51 (80%)	7 (11%)	6 (9.4%)	64
Total	97 (67%)	20 (14%)	28 (19%)	145

*Very low birth weight (< 1500 grams)
[†]Small for gestational age at term (birth weight below 10th percentile adjusted for age, sex, and parity)

7.2.5 Colorectal Cancer (Doubly Ordered 4 × 4 Table)

Early detection and treatment of colorectal cancer is beneficial, because advanced stages of colorectal cancer have poorer prognosis. Table 7.7 displays duration of symptoms (rows) versus tumor stage (columns) in a study of 784 patients treated for colorectal cancer at a regional hospital in Norway from 1980 to 2004 (Jullumstrø et al., 2009). The rows as well as the columns are ordered, and Table 7.7 can be regarded as a doubly ordered 4 × 4 table.

TABLE 7.7

Duration of symptoms and tumor stage for patients treated for colorectal cancer (Jullumstrø et al., 2009)

Duration	Tumor stage				Total
	T-1	T-2	T-3	T-4	
< 1 week	2 (3.7%)	4 (7.4%)	29 (54%)	19 (35%)	54
2–8 weeks	7 (3.9%)	6 (3.3%)	116 (64%)	51 (28%)	180
2–6 months	19 (5.9%)	27 (8.4%)	201 (62%)	76 (24%)	323
> 6 months	18 (7.9%)	22 (9.7%)	133 (59%)	54 (24%)	227
Total	46 (5.9%)	59 (7.5%)	479 (61%)	200 (26%)	784

7.2.6   Breast Tumor (Doubly Ordered 3 × 5 Table)

Bofin et al. (2004) studied associations between different findings in fine needle aspiration (FNA) smears from breast tumors and the final histological
diagnosis of tumor type in 133 patients. The aim of the study was to identify variables developed from FNA smears that could differentiate between
the different tumor diagnoses. Table 7.8 presents the cross-classification of the
FNA variable nuclear pleomorphism with tumor types. Both variables can be
considered as ordered, with tumor type ordered from benign (as in NPBD) to
most malign (as in IDC).

TABLE 7.8

Nuclear pleomorphism from fine needle aspiration smears and breast
tumor type (Bofin et al., 2004)

NP[†]	Tumor type[*]					Total
	NPBD	PBD	AIDH	DCIS	IDC	
None/mild	15 (21%)	35 (49%)	6 (8.5%)	9 (13%)	6 (8.5%)	71
Moderate	2 (6.7%)	4 (13%)	2 (6.7%)	11 (37%)	11 (37%)	30
Profound	0	0	1 (3.1%)	10 (31%)	21 (66%)	32
Total	17 (13%)	39 (29%)	9 (6.8%)	30 (23%)	38 (29%)	133

[*]NPBD = non-proliferative breast disease; PBD = proliferative breast disease; AIDH
= atypical intraductal hyperplasia; DCIS = ductal carcinoma in situ; IDC = invasive
ductal carcinoma
[†]Nuclear pleomorphism

7.2.7   Self-Rated Health (Doubly Ordered 4 × 4 Table)

In the HUNT study (Nord-Trøndelag county health survey), one of the questions is: "How is your overall health at the moment?" The outcome categories
are "Very good", "Good", "Not very good", and "Poor". Table 7.9 shows
the counts for the adolescents aged 12 to 17 years in 1995 to 1997 (Young-
HUNT 1), and for the same individuals four years later (Young-HUNT 2;
Breidablik et al. (2008)). Both the rows and the columns are ordered. In
this example, it may be appropriate to regard self-rated health as an unobserved (latent) continuous variable, where only a categorized version has been
observed. Then, it may be of interest to study polychoric correlation (see Section 7.9.5). In this chapter, we shall use this example to study the association
between Young-HUNT 1 and Young-HUNT 2.

Table 7.9 is actually an example of a paired $c \times c$ table with ordinal data.
A relevant question can be whether there is an increase or decrease in self-
rated health from Young-HUNT 1 to Young-HUNT 2. Relevant methods for
this problem are covered in Chapter 9. Also, an extension of this example is
discussed in Section 13.3 (on missing data).

TABLE 7.9

Self-rated health for 12 to 17 years old adolescents in Young-HUNT 1 and four years later in Young-HUNT 2 (Breidablik et al., 2008)

Young-HUNT 1	Young-HUNT 2				Total
	Poor	Not very good	Good	Very good	
Poor	2 (0.1%)	3 (0.1%)	3 (0.1%)	3 (0.1%)	11 (0.5%)
NVG*	2 (0.1%)	58 (2.5%)	98 (4.2%)	14 (0.6%)	172 (7.3%)
Good	8 (0.3%)	162 (6.9%)	949 (40%)	252 (11%)	1371 (58%)
Very good	4 (0.2%)	48 (2.0%)	373 (16%)	369 (16%)	794 (34%)
Total	16 (0.7%)	271 (12%)	1423 (61%)	638 (27%)	2348 (100%)

*Not very good

7.3 Notation and Sampling Distributions

An $r \times c$ table—much like the tables discussed in the previous chapters—may be the result of different sampling models or study designs. Two designs are of particular interest: the row margins fixed model and the total sum fixed model. In the row margins fixed model, we let the row sums $(n_{1+}, n_{2+}, \ldots, n_{r+})$ be fixed. In such a model, the rows represent a grouping of the data, for instance into exposure or treatment groups, while the columns represent an outcome variable. One example is the randomized, controlled trial in Table 7.3. The rows comprise three different treatments for ear infection, and the columns represent a dichotomous outcome (cured versus not cured). Table 7.3 is a 3×2 table, where the row sums are assumed fixed by design. Another example is the study of low birth weight and psychiatric morbidity shown in Table 7.6. Within each low birth weight group (the rows), the number of subjects is regarded as fixed; thus, Table 7.6 is a 3×3 table with fixed row sums.

In a total number fixed model, only the total count N is assumed fixed by design. This is typically the case with cross-sectional studies, where the observations are regarded as a random sample of size N from the target population. This is a reasonable model for most of the remaining examples in Section 7.2. One example is the data shown in Table 7.5, which comprise the cross-classification of psychiatric diagnosis (rows) with BMI weight categories (columns) for 550 adolescents referred to a psychiatric clinic. Neither the row sums nor the column sums can be assumed fixed by design.

In the total sum fixed model, we have $r \cdot c$ probabilities, which sum to one, as shown in Table 7.10. The sampling distribution is the multinomial distribution with parameters N and $\boldsymbol{\pi} = \{\pi_{11}, \pi_{12}, \ldots, \pi_{rc}\}$:

$$f(\mathbf{x} \mid \boldsymbol{\pi}, N) = \frac{N!}{x_{11}! \, x_{12}! \, \cdots \, x_{rc}!} \, \pi_{11}^{x_{11}} \cdot \pi_{12}^{x_{12}} \cdots \pi_{rc}^{x_{rc}}.$$

The null hypothesis of interest in the total sum fixed model is

$$H_0 : \pi_{ij} = \pi_{i+}\pi_{+j},$$

for all i, j, where $\pi_{i+} = \pi_{i1} + \pi_{i2} + \ldots + \pi_{ic}$ and $\pi_{+j} = \pi_{1j} + \pi_{2j} + \ldots + \pi_{rj}$, see Table 7.10.

TABLE 7.10

The probabilities of an $r \times c$ table under the total sum fixed model

	Variable 2				
Variable 1	Category 1	Category 2	...	Category c	Total
Group 1	π_{11}	π_{12}	...	π_{1c}	π_{1+}
Group 2	π_{21}	π_{22}	...	π_{2c}	π_{2+}
⋮	⋮	⋮	⋱	⋮	⋮
Group r	π_{r1}	π_{r2}	...	π_{rc}	π_{r+}
Total	π_{+1}	π_{+2}	...	π_{+c}	1

Additional notation:

$\boldsymbol{\pi} = \{\pi_{11}, \pi_{12}, \ldots, \pi_{1c}, \pi_{21}, \pi_{22}, \ldots, \pi_{2c}, \ldots, \pi_{r1}, \pi_{r2}, \ldots, \pi_{rc}\}$

Under the row margins fixed model, the counts in row number i are multinomially distributed with parameters n_{i+} and $\pi_{1|i}, \pi_{2|i}, \ldots, \pi_{c|i}$, where $\pi_{j|i} = \pi_{ij}/\pi_{i+}$ is the conditional probability of column category j given row category i, see Table 7.11. The sampling distribution for row number i under the row margins fixed model is

$$f(x_{i1}, \ldots, x_{ic} \mid \pi_{1|i}, \ldots, \pi_{c|i}; n_{i+}) = \frac{n_{i+}!}{x_{i1}! x_{i2}! \cdots x_{ic}!} \, \pi_{1|i}^{x_{i1}} \cdot \pi_{2|i}^{x_{i2}} \cdots \pi_{c|i}^{x_{ic}},$$

and the sampling distribution for the whole table is

$$f(\mathbf{x} \mid \boldsymbol{\pi}_{\mathrm{cond}}, \mathbf{n}_+) = \prod_{i=1}^{r} \frac{n_{i+}!}{x_{i1}! x_{i2}! \cdots x_{ic}!} \, \pi_{1|i}^{x_{i1}} \cdot \pi_{2|i}^{x_{i2}} \cdots \pi_{c|i}^{x_{ic}}, \qquad (7.1)$$

where $\boldsymbol{\pi}_{\mathrm{cond}}$ are the conditional probabilities in Table 7.11 and $\mathbf{n}_+ = \{n_{1+}, n_{2+}, \ldots, n_{r+}\}$ denotes the fixed row sums. The null hypothesis for the row margins fixed model is

$$H_0 : \pi_{j|1} = \pi_{j|2} = \ldots = \pi_{j|r}, \qquad (7.2)$$

for $j = 1, 2, \ldots, c$. If neither the rows nor the columns are ordered, the alternative hypothesis is simply "not H_0". In ordered tables, the alternative hypothesis is further specified, and the type of analysis depends on the specific alternative hypothesis.

One impediment to calculating exact tests of the null hypothesis in Equation 7.2 is the existence of several unknown nuisance parameters (the $\boldsymbol{\pi}_{\mathrm{cond}}$)

TABLE 7.11
Conditional probabilities within the rows of an $r \times c$ table

Variable 1	Variable 2				Total
	Category 1	Category 2	...	Category c	
Group 1	$\pi_{1\|1}$	$\pi_{2\|1}$	$\cdots$	$\pi_{c\|1}$	1
Group 2	$\pi_{1\|2}$	$\pi_{2\|2}$	$\cdots$	$\pi_{c\|2}$	1
$\vdots$	$\vdots$	$\vdots$	$\ddots$	$\vdots$	$\vdots$
Group r	$\pi_{1\|r}$	$\pi_{2\|r}$	$\cdots$	$\pi_{c\|r}$	1

Additional notation:
$\pi_{\text{cond}} = \{\pi_{1|1}, \pi_{2|1}, \ldots, \pi_{c|1}, \pi_{1|2}, \pi_{2|2}, \ldots, \pi_{c|2}, \ldots, \pi_{1|r}, \pi_{2|r}, \ldots, \pi_{c|r}\}$

in Equation 7.1. Under the null hypothesis in Equation 7.2, there are c unknown parameters. The standard remedy is to make an additional condition on the column totals, such that all marginals can be considered fixed. Under the null hypothesis in Equation 7.2, the resulting sampling distribution is the multiple hypergeometric distribution given by

$$f(\mathbf{x} \,|\, \mathbf{n}_{++}) = \frac{\left(\prod_{i=1}^{r} n_{i+}!\right)\left(\prod_{j=1}^{c} n_{+j}!\right)}{N! \prod_{i=1}^{r}\prod_{j=1}^{c} x_{ij}!}, \tag{7.3}$$

where $\mathbf{n}_{++} = \{n_{1+}, n_{2+}, \ldots, n_{r+}, n_{+1}, n_{+2}, \ldots, n_{+c}\}$ denotes the fixed row and column sums. Equation 7.3 does not contain any unknown parameters, and we will use it to derive the Fisher-Freeman-Halton exact test in Section 7.5.3 and the Kruskal-Wallis exact test in Section 7.6.3.

7.4   Odds Ratios in $r \times c$ Tables

In a 2×2 table, the degree of association can be summarized in one single quantity, such as the difference between the two success probabilities, the ratio of probabilities, or the odds ratio. This is not possible in an unordered $r \times c$ table without loss of information. In this section, we will describe different types of odds ratios and their ability to describe the association between rows and columns in $r \times c$ tables. To get the "full picture" with odds ratios, one could calculate all the $(r-1)(c-1)$ local odds ratios for 2×2 tables of neighbouring cells in the table:

$$\theta_{ij}^{\text{LO}} = \frac{\pi_{ij}\pi_{i+1,j+1}}{\pi_{i,j+1}\pi_{i+1,j}},$$

for $i = 1, 2, \ldots, r-1$ and $j = 1, 2, \ldots, c-1$. The local odds ratios are illustrated in Panel A of Figure 7.1. The natural estimate of this odds ratio is

$$\hat{\theta}_{ij}^{\text{LO}} = \frac{n_{ij} n_{i+1,j+1}}{n_{i,j+1} n_{i+1,j}}.$$

As already mentioned, there are more than one type of odds ratio that can be used to describe association in $r \times c$ tables. Instead of local odds ratios, one could calculate the $(r-1)(c-1)$ odds ratios relative to a reference cell, such as cell $(1, 1)$, as illustrated in Panel B of Figure 7.1.

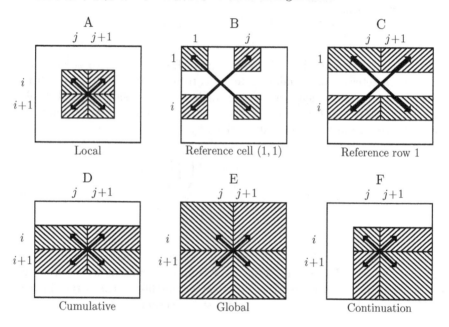

FIGURE 7.1
Different sets of $(r-1)(c-1)$ odds ratios in an $r \times c$ table: local odds ratios (A), odds ratio relative to a reference cell (B), cumulative odds ratios relative to a reference row (C), cumulative odds ratios (D), global odds ratios (E), and continuation odds ratios (F).

In singly ordered or doubly ordered $r \times c$ tables, it is sometimes useful to consider odds ratios for combination of cells in a row or in a column or both, such as illustrated in Panels C to F of Figure 7.1. Note that for ordered $2 \times c$ tables, the local odds ratio is called the *adjacent category odds ratio*. Moreover, the cumulative odds ratio relative to a reference row, cumulative odds ratio, and the continuation odds ratios coincide for the ordered $2 \times c$ table.

Without any restrictions on the odds ratios, there are $(r-1)(c-1)$ parameters defining the odds ratios in an $r \times c$ table. For singly or doubly ordered tables, it is usually pertinent to use a more parsimonious model with substantially fewer than $(r-1)(c-1)$ parameters, which we typically achieve with logistic regression models, as described in Sections 7.7 (singly ordered tables) and 7.10 (doubly ordered tables).

Tests for unspecific ordering, on the other hand, do not assume that the data follow any underlying parametric model. In Section 5.6, we considered tests for unspecific ordering in $r \times 2$ tables, which can also be used for testing unspecific ordering in $2 \times c$ tables. Generalizations of these tests to $r \times c$ tables are discussed in the seminal article by Agresti and Coull (2002). In the $r \times c$ case, this is a substantially more complex task than in $r \times 2$ and $2 \times c$ tables, and we do not consider this topic further in this chapter. We refer the interested reader to Agresti and Coull (2002) and the articles by Cohen et al. (2003) and Colombi and Forcina (2016).

7.5 Unordered $r \times c$ Tables

Most of the tests for association in 2×2 tables have straightforward generalizations to unordered $r \times c$ tables. These generalizations hold under the row margins fixed model, the total sum fixed model, and the no sums fixed model. Under the null hypothesis of no association, the estimated expected cell counts are

$$m_{ij} = n_{i+}n_{+j}/N,$$

for $i = 1, 2, \ldots, r$ and $j = 1, 2, \ldots, c$.

7.5.1 The Pearson Chi-Squared Test

The Pearson chi-squared statistic for the $r \times c$ table has the same expression as that of the 2×2 table:

$$T_{\text{Pearson}}(\mathbf{n}) = \sum_{i,j} \frac{(n_{ij} - m_{ij})^2}{m_{ij}}.$$

Asymptotically, T_{Pearson} is chi-squared distributed with $(r-1)(c-1)$ degrees of freedom. Hence, the asymptotic P-value for the Pearson chi-squared test is given by

$$P\text{-value} = \Pr\left[\chi^2_{(r-1)(c-1)} \geq T_{\text{Pearson}}(\mathbf{n})\right].$$

According to Cochran's criterion, the asymptotic P-value should be used only if at least 80% of the m_{ij} are greater than five and all m_{ij} are greater than one.

7.5.2   The Likelihood Ratio Test

The likelihood ratio statistic for an $r \times c$ table is

$$T_{\text{LR}}(\mathbf{n}) = 2 \sum_{i,j} n_{ij} \log \left(\frac{n_{ij}}{m_{ij}} \right),$$

where a term in the summation is zero if $n_{ij} = 0$. The likelihood ratio statistic, like the Pearson chi-squared statistic, has an asymptotic chi-squared distribution with $(r-1)(c-1)$ degrees of freedom. In small to moderate samples, the chi-squared approximation is generally reported to be slightly better for the Pearson statistic than for the likelihood ratio statistic (Larntz, 1978; Koehler, 1986; Cressie and Read, 1989).

7.5.3   The Fisher-Freeman-Halton and Related Exact Tests

In tables with small cell counts, asymptotic tests such as those described above are not recommended, because the chi-squared approximation does not hold. Here, we describe a generalization of the Fisher exact test (Section 4.4.5) that can be used for all unordered $r \times c$ tables, including tables with small cell counts. It is usually called the *Fisher-Freeman-Halton exact test*, and it is available in some general purpose statistical software packages. The Fisher-Freeman-Halton exact test, like the Fisher exact test, conditions on both the row and column sums. Under the null hypothesis, the probability distribution for an arbitrary table $\mathbf{x}$ is the multiple hypergeometric distribution in Equation 7.3, which is free of nuisance parameters. Now, let $T(\mathbf{n})$ be a test statistic, for which higher values provide increasing evidence against the null hypothesis. An exact conditional P-value is obtained as the sum of conditional probabilities for tables with the same marginals as the observed table—denoted by the set $\Omega(\mathbf{x}|\mathbf{n}_{++})$—and with test statistic at least as large as the observed table:

$$\text{exact cond. } P\text{-value} = \sum_{\Omega(\mathbf{x}|\mathbf{n}_{++})} I\big[T(\mathbf{x}) \geq T(\mathbf{n})\big] \cdot f(\mathbf{x}\,|\,\mathbf{n}_{++}), \qquad (7.4)$$

where $I()$ is the indicator function and $f(\mathbf{x}\,|\,\mathbf{n}_{++})$ is given in Equation 7.3.

The Fisher-Freeman-Halton exact test uses the point probability, $f(\mathbf{x}\,|\,\mathbf{n}_{++})$ as test statistic, for which smaller values provide evidence against the null hypothesis. Hence, the P-value equals the sum of probabilities for tables that are equally or less probable than the one actually observed:

$$\text{exact cond. } P\text{-value} = \sum_{\Omega(\mathbf{x}|\mathbf{n}_{++})} I\big[f(\mathbf{x}\,|\,\mathbf{n}_{++}) \leq f(\mathbf{n}\,|\,\mathbf{n}_{++})\big] \cdot f(\mathbf{x}\,|\,\mathbf{n}_{++}).$$

Exact conditional tests in $r \times c$ tables may also be defined with the Pearson statistic or the likelihood ratio statistic:

$$\text{exact cond. } P\text{-value} = \sum_{\Omega(\mathbf{x}|\mathbf{n}_{++})} I\big[T_{\text{Pearson}}(\mathbf{x}) \geq T_{\text{Pearson}}(\mathbf{n})\big] \cdot f(\mathbf{x}\,|\,\mathbf{n}_{++})$$

or

$$\text{exact cond. } P\text{-value} = \sum_{\Omega(\mathbf{x}|\mathbf{n}_{++})} I\big[T_{\text{LR}}(\mathbf{x}) \geq T_{\text{LR}}(\mathbf{n})\big] \cdot f(\mathbf{x}\,|\,\mathbf{n}_{++}).$$

7.5.4 Mid-*P* Tests

As discussed and illustrated in Chapter 4, exact conditional tests can be un-necessary conservative. This conservatism is due to discreteness, which comes about when the conditioning reduces the sample space considerably. Discrete-ness induced by conditioning is less of a problem in larger tables than in 2×2 tables; however, it still may be of interest to consider approaches that adjust for the discreteness of exact conditional tests. The mid-*P* value—see Section 1.10 for a general description—based on an exact conditional test for the $r \times c$ table is defined as

$$\text{mid-}P \text{ value} \;=\; \sum_{\Omega(\mathbf{x}|\mathbf{n}_{++})} I\big[T(\mathbf{x}) > T(\mathbf{n})\big] \cdot f(\mathbf{x}\,|\,\mathbf{n}_{++})$$

$$+ \;\; 0.5 \cdot \sum_{\Omega(\mathbf{x}|\mathbf{n}_{++})} I\big[T(\mathbf{x}) = T(\mathbf{n})\big] \cdot f(\mathbf{x}\,|\,\mathbf{n}_{++}). \quad (7.5)$$

Compared with the expression for the exact conditional test in Equation 7.4, the mid-*P* value is equal to the exact conditional *P*-value, except for a small downward adjustment to the probability weight of tables that agree equally with the null hypothesis as the observed table.

In principle, any test statistic can be used in Equation 7.5; however, the most likely candidates are the point probability, $f(\mathbf{x}\,|\,\mathbf{n}_{++})$, the Pearson statis-tic, and the likelihood ratio statistic. If the point probability is used, we denote the test by the Fisher-Freeman-Halton mid-*P* test.

7.5.5 The Fisher-Freeman-Halton Asymptotic Test

The Fisher-Freeman-Halton exact test can be quite computer-intensive, espe-cially if the $r \times c$ table has many rows or columns or both. An asymptotic version of the Fisher-Freeman-Halton test is based on a simple transformation of the point probability of the observed table:

$$T_{\text{FFH}}(\mathbf{n}) = -2\log\big[\gamma f(\mathbf{n}\,|\,\mathbf{n}_{++})\big],$$

where

$$\gamma = \left[(2\pi)^{(r-1)(c-1)} N^{-(rc-1)} \prod_{i=1}^{r} n_{i+}^{c-1} \prod_{j=1}^{c} n_{+j}^{r-1}\right]^{1/2}$$

is a normalizing constant that makes T_{FFH} asymptotically chi-squared dis-tributed with $(r-1)(c-1)$ degrees of freedom.

7.5.6   Exact Unconditional Tests

One can define exact unconditional tests for $r \times c$ tables, in a similar manner as for tables where r or c equals two; see Sections 5.7.12 and 6.14.6. In a row margins fixed design, the unconditional distribution under the null hypothesis of $\mathbf{x}$—an arbitrary table with row sums equal to $\mathbf{n}_+ = \{n_{1+}, n_{2+}, \ldots, n_{r+}\}$—is

$$
f(\mathbf{x} \,|\, \pi_1, \ldots, \pi_c; \mathbf{n}_+) \;=\; \prod_{i=1}^{r} \frac{n_{i+}!}{x_{i1}!\,x_{i2}! \cdots x_{ic}!} \; \pi_1^{x_{i1}} \cdot \pi_2^{x_{i2}} \cdots \pi_c^{x_{ic}}
$$

$$
=\; \pi_1^{x_{+1}} \cdot \pi_2^{x_{+2}} \cdots \pi_c^{x_{+c}} \prod_{i=1}^{r} \frac{n_{i+}!}{x_{i1}!\,x_{i2}! \cdots x_{ic}!} \;.
$$

This distribution follows from Equation 7.1, and the fact that under the null hypothesis, $\pi_{j|1} = \pi_{j|2} = \ldots = \pi_{j|r} = \pi_j$ for all $j = 1, 2, \ldots, c$. For a suitable test statistic $T(\mathbf{n})$, the exact unconditional P-value can then be expressed as

$$
P\text{-value} = \max_{\{\pi_1, \ldots, \pi_c\}} \left\{ \sum_{\Omega(\mathbf{x}|\mathbf{n}_+)} I\big[T(\mathbf{x}) \geq T(\mathbf{n})\big] \cdot f(\mathbf{x} \,|\, \pi_1, \ldots, \pi_c; \mathbf{n}_+) \right\},
$$

where $\Omega(\mathbf{x}|\mathbf{n}_+)$ denotes the set of tables $\mathbf{x}$ with row sums equal to $\mathbf{n}_+$.

Mehta and Hillton (1993) have studied the exact unconditional test with the Pearson statistic for comparing three binomials (3×2 tables, in our notation). They report some power gain compared to exact conditional tests in tables with spares counts; however, the difference in power diminishes rapidly with increasing sample size. Exact unconditional tests pose very difficult computational challenges in $r \times c$ tables, and the complexity increases rapidly with increasing r and c. We are not aware of any software packages that provide exact unconditional tests in $r \times c$ tables with $r > 2$ or $c > 2$, and we have not seen such tests used in applied publications.

7.5.7   Identifying Cells with Significantly Deviating Counts

A rejection of the null hypothesis for an unordered $r \times c$ table indicates that there is some association between rows and columns in the table. The overall strength of the association can be estimated by association measures proposed in the research literature (Theil, 1970; Goodman and Kruskal, 1979; Haberman, 1982); however, these are difficult to interpret and not much used. Summary measures of association seem to be more useful in ordered $r \times c$ tables, as will be described in subsequent sections in this chapter.

Rather than using one summary measure for the association in an unordered $r \times c$ table, one may look for cells with deviations from independence. Good (1956) proposed the quantities $\pi_{ij}/\pi_{i+}\pi_{+j}$ as measures of association in $r \times c$ tables. The quantities have been termed *association factors*, and they are estimated as the observed frequency divided by the expected frequency under the null hypothesis: n_{ij}/m_{ij}. Agresti (2013, p. 56) suggests that values below 1/2 or above 2 may be regarded as noteworthy deviations from independence. It may, however, be more informative to study residuals or standardized residuals. The Pearson residuals are defined as

$$r_{ij} = \frac{n_{ij} - m_{ij}}{\sqrt{m_{ij}}}.$$

Under the null hypothesis, these residuals are approximately normally distributed with mean zero and with variance slightly less than one. The standardized Pearson residuals are more relevant, because they are standardized with respect to the variances and are thus asymptotically standard normal distributed. For the $r \times c$ table, they are:

$$r_{ij}^* = \frac{n_{ij} - m_{ij}}{\sqrt{m_{ij}(1 - p_{i+})(1 - p_{+j})}},$$

where $p_{i+} = n_{i+}/N$ and $p_{+j} = n_{+j}/N$ (Haberman, 1973). Note that these residuals may be termed differently in some textbooks and statistical software packages, such as standardized residuals (for the Pearson residuals), and adjusted or adjusted standardized residuals (for the standardized Pearson residuals). Standardized Pearson residuals exceeding 2 or 3 in absolute value can be taken as indications of departure from the null hypothesis of independence in that cell. Larger threshold values are more relevant with larger degrees of freedom.

7.5.8 Multiple Comparisons of Binomial Probabilities

Now, we consider $r \times 2$ tables where the row sums are given, and the null hypothesis of independence is the one in Equation 7.2. Suppose that the P-value for testing independence is less than the nominal significance level. Then, we would like to identify the deviances from the null hypothesis.

Here, we consider three approaches to this question: first, we describe a method for simultaneous confidence interval estimation of linear combinations of the binomial probabilities. These intervals will be applied to the difference between binomial probabilities for the $r \times 2$ table. Second, we consider the closed testing procedure, which we use for the special case of multiple comparisons of three binomial probabilities. Finally, we describe the score studentized range procedure, which we also use to derive confidence intervals for the difference between binomial probabilities.

Simultaneous Confidence Intervals

Methods for deriving simultaneous confidence intervals for multinomial probabilities were derived in Section 3.6.1. Here, we extend the methods to simultaneous comparisons of linear contrasts of binomial probabilities. For an $r \times c$ table, a linear contrast in the probabilities is defined as a linear combination $\sum_{i,j} d_{ij} \pi_{j|i}$, where $\sum_i d_{ij} = 0$ for $j = 1, 2, \ldots, c$.

Simultaneous confidence intervals can be derived from the modified Pearson chi-squared statistic $\sum_{i,j}(n_{ij} - m_{ij})^2/n_{ij}$, which differs from the Pearson chi-squared statistic, which has m_{ij} in the denominator. Under the null hypothesis, the two test statistics are asymptotically equivalent. When the P-value for the modified Pearson chi-squared statistic is less than the nominal significance level α, there exists at least one statistically significant linear contrast, see Goodman (1964) or Miller (1981, p. 219), which derive Scheffé-type confidence intervals for linear contrasts with an overall coverage of $1 - \alpha$.

Suppose that C contrasts are of interest. Bonferroni confidence intervals can be derived, see Goodman (1964) or Hjort (1988). As pointed out by Hjort (1988), the Bonferroni-type intervals are shorter than the Scheffé-type intervals, see also Section 3.6.1.

In the simple case where we have only two columns, we may want to carry out pairwise comparisons between the binomial probabilities in the $r \times 2$ table. In some settings, it may be of interest to make comparisons between all pairs. There are $C = r(r-1)/2$ possible pairs. Consistent estimates of the conditional probabilities of success are $\hat{\pi}_{1|i} = n_{i1}/n_{i+}$ for $i = 1, 2, \ldots, r$. Scheffé-type simultaneous $1 - \alpha$ confidence intervals for the differences between the success probabilities in row i and row j are

$$\hat{\pi}_{1|i} - \hat{\pi}_{1|j} \pm \sqrt{\chi^2_{r-1}(\alpha)\left[\frac{\hat{\pi}_{1|i}\left(1 - \hat{\pi}_{1|i}\right)}{n_{i+}} + \frac{\hat{\pi}_{1|j}\left(1 - \hat{\pi}_{1|j}\right)}{n_{j+}}\right]}, \qquad (7.6)$$

for $i \neq j$ and $i, j = 1, 2, \ldots, r$. The Bonferroni-type intervals are obtained by substituting $\chi^2_1(\alpha/C)$ for $\chi^2_{r-1}(\alpha)$ in Equation 7.6. Both the Scheffé-type intervals and the Bonferroni-type intervals are Wald-type intervals. Table 7.13 contains an example of the Bonferroni-type intervals for pairwise comparisons in a 3×2 table.

In other settings, only some of these comparisons are of interest. For example, in a randomized, controlled trial of r treatments, out of which one is a placebo, the primary interest may be to compare each of the $r - 1$ active treatments with the placebo. Pairwise comparisons with Bonferroni-type intervals correspond to separate analyses of 2×2 tables, which can be carried out as described in Chapter 4. Because we make several pairwise comparisons, we must adjust the analysis if we want to preserve the *familywise error rate*. Preserving the familywise error rate means that the probability of rejecting at least one true null hypothesis shall not exceed the (pre-specified) nominal significance level. When the familywise error rate is α, the overall confidence level for the simultaneous confidence intervals is $1 - \alpha$.

The Closed Testing Procedure

If only three groups are involved, such as in the ear infection trial in Table 7.3, the familywise error rate can be preserved in a particularly simple and powerful way called the *closed testing procedure*. First, compute the global P-value,

P_0, for comparing the three groups. Second, compute the (unadjusted) local P-values, P_1, P_2, and P_3, corresponding to the three pairwise comparisons. To preserve the familywise error rate, simply compute the adjusted local P-values as $P_{k,\text{adjusted}} = \max(P_k, P_0)$, for $k = 1, 2, 3$. The fact that this procedure actually preserves the familywise error rate for the comparisons of three groups follows from the closed testing principle (Levin et al., 1994). It is also mentioned in Bender and Lange (2001, p. 345), but seems not to be widely known. We emphasize that this procedure only holds when three groups are involved.

Whereas adjusted confidence intervals can be constructed corresponding to some methods for multiplicity adjustment, such as the Bonferroni adjustment, this is not the case for the closed testing procedure and many other procedures. This is noted by Dmitrienko and D'Agostino (2013, p. 5205), who in such cases, find it acceptable to present unadjusted confidence intervals, with the understanding that the overall coverage probability of these intervals is not controlled in this setting.

The Score Studentized Range Procedure

With four or more groups, some other type of adjustment than the closed testing procedure is needed to preserve the familywise error rate. This is even the case if the only comparisons are between each of three alternative treatment groups and a control group, because that would involve four groups. A *Bonferroni adjustment* will always preserve the familywise error rate; however, it can be unnecessary conservative. A less conservative method uses a score statistic and the studentized range distribution (Agresti et al., 2008). If $x_1, x_2, \ldots, x_r$ are r independent, normally distributed variables with common mean μ and standard deviation σ, the *studentized range* is defined as

$$
\begin{aligned}
q_{r,\nu} &= \frac{\max(x_1, x_2, \ldots, x_r) - \min(x_1, x_2, \ldots, x_r)}{s} \\
&= \max_{i,j=1,\ldots,r} \frac{x_i - x_j}{s},
\end{aligned}
$$

where s^2 is an unbiased estimate of the variance σ^2, with ν degrees of freedom. Now, let $\pi_{1|i}$ and $\pi_{1|j}$ denote the success probabilities in rows i and j; let n_{i1}/n_{i+} and n_{j1}/n_{j+} denote the sample proportions; and let $\hat{\pi}_{1|i}$ and $\hat{\pi}_{1|j}$ denote the maximum likelihood estimate under the constraint $\Delta_{ij,0} = \pi_{1|i} - \pi_{1|j}$. The score statistic for testing the difference between the two success probabilities is

$$
z_{ij}(\Delta_{ij,0}) = \frac{(n_{i1}/n_{i+} - n_{j1}/n_{j+}) - \Delta_{ij,0}}{\sqrt{\dfrac{\hat{\pi}_{1|i}(1 - \hat{\pi}_{1|i})}{n_{i+}} + \dfrac{\hat{\pi}_{1|j}(1 - \hat{\pi}_{1|j})}{n_{j+}}}}.
$$

A confidence interval for the difference $\pi_{1|i} - \pi_{1|j}$ consists of the values of $\Delta_{ij,0}$ that satisfy

$$
\left| z_{ij}(\Delta_{ij,0}) \right| < Q_r(\alpha)/\sqrt{2},
$$

where $Q_r(\alpha)$ is the upper α percentile of the studentized range distribution for r observations and an infinite number of degrees of freedom. The set of all these $r(r-1)/2$ confidence intervals has a large sample simultaneous confidence level of approximately $(1-\alpha)$. Confidence intervals for odds ratios based on the same principle also work well (Agresti et al., 2008). The method is implemented in R (The R Project for Statistical Computing, https://www.r-project.org/), and the code is available at http://www.stat.ufl.edu/~aa/cda/R/multcomp/ryu-simultaneous.pdf.

7.5.9   Examples

Treatment of Ear Infection (Table 7.3)

The expected cell counts under the null hypothesis are readily computed as $m_{ij} = n_{i+}n_{+j}/N$. The smallest of the expected counts is 12.9, which is much higher than five, and by Cochran's criterion, we may safely use the asymptotic Pearson chi-squared test. The Pearson statistic is

$$T_{\text{Pearson}}(\mathbf{n}) = \frac{(40-51.3)^2}{51.3} + \frac{(54-48.1)^2}{48.1} + \ldots + \frac{(10-15.4)^2}{15.4} = 17.56.$$

Under the null hypothesis, T_{Pearson} is asymptotically chi-squared distributed with $(3-1)(2-1) = 2$ degrees of freedom. The corresponding P-value is $P = 0.000154$. Table 7.12 summarizes the observed test statistics for the Pearson, likelihood ratio, and Fisher-Freeman-Halton statistics, and the corresponding P-values for the asymptotic, exact conditional, and mid-P versions of the tests.

TABLE 7.12

Test statistics and P-values for an overall comparison of three treatments for ear infection (data from Table 7.3)

Test	Statistic	df*	P-value		
			Asymptotic	Exact cond.	Mid-P
Pearson	17.56	2	0.000154	0.000155	0.000151
LR[†]	16.70	2	0.000236	0.000303	0.000299
FFH[‡]	16.26	2	0.000295	0.000271	0.000267

*Degrees of freedom for the chi-squared distribution
[†]Likelihood ratio
[‡]Fisher-Freeman-Halton (point probability)

In Table 7.12, as in several other tables in this book, we have reported the P-values with more decimals than usual, to illustrate the differences between the tests. In practice, P-values such as those in Table 7.12, are usually reported as $P < 0.001$ or perhaps with only one significant digit, such as $P = 0.0002$. For this example, we observe that the Pearson asymptotic P-value is closer to its exact conditional and mid-P counterparts than is the case for the likelihood ratio P-value and the Fisher-Freeman-Halton P-value. This illustrates that the asymptotic approximation is usually better for the Pearson statistic than for the likelihood ratio statistic, as noted in Section 7.5.2.

Overall, there is a highly significant P-value for comparing the three treatments. Next, it may be of interest to investigate which treatment or treatments are better. A natural approach is to consider all three pairwise comparisons between two of the treatments. This corresponds to analyzing the three 2×2 subtables that can be formed from the original 3×2 table. Table 7.13 shows the results. The two treatments that include corticosteroid are each significantly better than acetic acid, $P < 0.001$ for both comparisons. The corresponding estimated difference between the probabilities are 27% and 25%, which would be considered a clinically relevant difference in practically any setting. There is no significant difference between the two treatments that include corticosteroid.

Because we only have three quantities to compare, we can preserve the familywise error rate with the simple approach explained in Section 7.5.8 of using the global P-value as a threshold. The global P-value from the original 3×2 table is $P = 0.00027$ (Fisher-Freeman-Halton mid-P test), and the adjusted P-values for the three pairwise comparisons are

$$\text{Acetic acid vs. CS + acetic acid: } P_{\text{adj}} = \max(0.00034, 0.00027) = 0.00034,$$
$$\text{Acetic acid vs. CS + antibiotic: } P_{\text{adj}} = \max(0.0012, 0.00027) = 0.0012,$$
$$\text{CS + acetic acid vs. CS + antibiotic: } P_{\text{adj}} = \max(0.70, 0.00027) = 0.70.$$

In this particular example, the adjusted local P-values are equal to the unadjusted local P-values. In general, this will not be the case.

The Newcombe hybrid score intervals with Bonferroni adjustment—which amounts to deriving $100(1 - \alpha/C)\%$ confidence intervals, where C is the number of comparisons—are quite similar to the Bonferroni-type simultaneous intervals, except for a slight shift in location. The Bonferroni-type intervals are Wald-type intervals, and they are likely to have inferior coverage properties compared with the Newcombe hybrid score intervals (see the evaluations in Section 4.5.8). The Newcombe hybrid score intervals are also easy to calculate, so we prefer them ahead of the Bonferroni-type simultaneous intervals.

Psychiatric Diagnoses and Physical Activity (Table 7.4)

Table 7.4 shows the number of subjects participating in team sports within each of six main psychiatric diagnoses. If we calculate the expected cell counts, we find that all counts are well above five, so Cochran's criterion for using the chi-squared approximation to the Pearson tests statistic is met. The observed value of the Pearson test statistics is 11.69, and with $(6 - 1)(2 - 1) = 5$ degrees of freedom, the asymptotic P-value is 0.039, indicating some differences between the six groups.

Now, we consider the pairwise comparisons of the six groups, which

TABLE 7.13

Pairwise comparisons of the probability of being cured for three
treatments for ear infection (data from Table 7.3)

Comparison	Cured	Not cured	P-value	$\hat{\Delta}$	Difference between probabilities 95% CI
Acetic acid	40 (62%)	25			-0.432 to -0.085[†]
versus			0.00034*	-0.270	
Corticosteroid and acetic acid	54 (89%)	7			-0.444 to -0.096[‡]
Acetic acid	40 (62%)	25			-0.412 to -0.069[†]
versus			0.0012*	-0.248	
Corticosteroid and antibiotic	63 (86%)	10			-0.421 to -0.074[‡]
Corticosteroid and acetic acid	54 (89%)	7			-0.127 to 0.162[†]
versus			0.701*	0.022	
Corticosteroid and antibiotic	63 (86%)	10			-0.115 to 0.159[‡]

*Fisher mid-P test (Section 4.4.6) adjusted with the closed testing procedure
[†]Newcombe hybrid score interval (Section 4.5.4) with Bonferroni adjustment
[‡]Bonferroni-type simultaneous intervals

amounts to comparing two binomial probabilities. Table 7.14 shows confidence intervals for the difference between the probabilities of participating in team sports for all pairs of psychiatric diagnoses, computed in two ways: (i) Newcombe hybrid score interval with Bonferroni adjustment; and (ii) method based on the score statistic and the studentized range distribution. With six groups, there are $6(6-1)/2 = 15$ pairwise comparisons. Hence, the confidence coefficient for the Bonferroni adjusted confidence intervals are $(1 - 0.05/15)\% = 99.667\%$. None of the adjusted confidence intervals exclude zero; however the two intervals for the difference between row 3 versus row 4 come close. These rows correspond to the diagnoses eating disorder (proportion that participates in team sports: $12/22 = 54.5\%$) and autism spectrum disorders (proportion that participates in team sports: $7/37 = 18.9\%$). The unadjusted P-value for comparing the two groups is 0.0062 (Fisher mid-P test; see Section 4.4.6).

It is noteworthy that the Bonferroni adjusted Newcombe hybrid score intervals and the score studentized range intervals are approximately equal. In most cases in this example, they differ only in the third decimal. The score studentized range intervals tend to be slightly more distal—and 0-2% narrower—than the Bonferroni adjusted intervals. This agrees well with Agresti et al.

(2008), who state that studentized range intervals are typically 2–3% narrower than Bonferroni adjusted intervals.

TABLE 7.14

Pairwise comparisons of the probability of participating in team sports for six psychiatric diagnoses (data from Table 7.4)

			Confidence interval for Δ	
Rows	Observed proportions	$\hat{\Delta}$	NHS* Bonferroni adjusted	Score studentized range
1 vs 2	25% vs 33%	-0.078	(-0.243 to 0.111)	(-0.243 to 0.106)
1 vs 3	25% vs 55%	-0.292	(-0.568 to 0.027)	(-0.575 to 0.021)
1 vs 4	25% vs 19%	0.064	(-0.200 to 0.263)	(-0.195 to 0.267)
1 vs 5	25% vs 37%	-0.118	(-0.269 to 0.064)	(-0.269 to 0.058)
1 vs 6	25% vs 33%	-0.080	(-0.314 to 0.143)	(-0.314 to 0.149)
2 vs 3	33% vs 55%	-0.214	(-0.486 to 0.088)	(-0.488 to 0.083)
2 vs 4	33% vs 19%	0.142	(-0.118 to 0.313)	(-0.110 to 0.318)
2 vs 5	33% vs 37%	-0.040	(-0.184 to 0.111)	(-0.182 to 0.107)
2 vs 6	33% vs 33%	-0.001	(-0.231 to 0.196)	(-0.227 to 0.195)
3 vs 4	55% vs 19%	0.365	(-0.010 to 0.636)	(-0.001 to 0.652)
3 vs 5	55% vs 37%	0.174	(-0.121 to 0.442)	(-0.116 to 0.441)
3 vs 6	55% vs 33%	0.212	(-0.133 to 0.509)	(-0.132 to 0.521)
4 vs 5	19% vs 37%	-0.182	(-0.340 to 0.073)	(-0.344 to 0.065)
4 vs 6	19% vs 33%	-0.144	(-0.382 to 0.141)	(-0.389 to 0.137)
5 vs 6	37% vs 33%	0.038	(-0.186 to 0.225)	(-0.181 to 0.223)

*Newcombe hybrid score interval (Section 4.5.4)

Psychiatric Diagnoses and BMI (Table 7.5)

After calculating the expected cell counts for the data in Table 7.5, we find that three of 18 cells—which is less than 20%—have expected cell counts below five. The smallest expected cell count is 1.83, which is above 1.0. Hence, Cochran's criterion for using the asymptotic Pearson chi-squared test is met. The observed test statistics and corresponding asymptotic P-values for the Pearson chi-squared, likelihood ratio, and Fisher-Freeman-Halton tests are shown in Table 7.15. We have not included the exact conditional and mid-P tests in Table 7.15, because the calculations are too computer-intensive.

In this example, the overall test of association is highly significant, with $P = 0.005$ for the Pearson chi-squared test. For the upper left cell in Table 7.5, the association factor is $3/7.1 = 0.42$, which can be regarded as a noteworthy deviation from independence. There are two other cells with factors below $1/2$ or above 2: thin patients with eating disorders ($6/1.8 = 3.3$), and overweight patients with eating disorders ($1/5.9 = 0.17$).

The standardized Pearson residuals are shown in Table 7.16. We observe that there is a clear over-prevalence of thin patients among those diagnosed

TABLE 7.15

Test statistics and P-values for an overall test of association
between psychiatric diagnoses and BMI (data from Table 7.5)

Test	Statistic	df*	Asymptotic P-value
Pearson	25.09	10	0.005174
Likelihood ratio	24.27	10	0.006910
Fisher-Freeman-Halton	23.77	10	0.008234

*Degrees of freedom for the chi-squared distribution

with eating disorders, with a residual equal to $r_{31} = 3.29$. We also see an
indication of under-prevalence of overweight patients in the eating disorders
group ($r_{33} = -2.42$). No other residuals exceed two in absolute value. Hence,
within the eating disorders group, there are deviations from the null hypothesis
that are noteworthy in size as well as being statistically significant. This is
not surprising for an association with body mass index.

TABLE 7.16

Standardized Pearson residuals for the study of psychiatric
diagnoses and BMI (data from Table 7.5)

Main diagnosis	Thin	Normal	Overweight
Mood (affective) disorders	-1.73	0.94	0.09
Anxiety disorders	-1.62	1.93	-1.05
Eating disorders	3.29	0.33	-2.42
Autism spectrum disorders	1.00	-1.06	0.51
Hyperkinetic disorders	0.13	-0.87	0.85
Other disorders	1.33	-1.91	1.22

We return to this example in Section 7.6.4, where we analyze the data
under the assumption that the BMI categories are ordered.

7.5.10   Evaluations of Tests

Evaluation Criteria

Suppose that we have sampling under the row margins fixed design, with
fixed row sums $\mathbf{n}_+ = \{n_{1+}, n_{2+}, \ldots, n_{r+}\}$, so that the sampling distribution
is given by the product of r multinomial distributions, see Equation 7.1. The
null hypothesis of interest is

$$H_0 : \pi_{j|1} = \pi_{j|2} = \ldots = \pi_{j|r},$$

for $j = 1, 2, \ldots, c$. The conditional probabilities $\boldsymbol{\pi}_{\text{cond}} = \{\pi_{1|1}, \pi_{2|1}, \ldots, \pi_{c|r}\}$
are defined in Table 7.11. The actual significance level can be calculated ex-

actly by

$$\text{ASL}(\boldsymbol{\pi}_{\text{cond}}, \mathbf{n}_+, \alpha \mid H_0) = \sum_{\Omega(\mathbf{x}|\mathbf{n}_+)} I\big[P(\mathbf{x}) \leq \alpha\big] \cdot f(\mathbf{x} \mid \boldsymbol{\pi}_{\text{cond}}, \mathbf{n}_+), \qquad (7.7)$$

where the conditional probabilities are subject to the constraints of the null hypothesis, α is the nominal significance level, $\Omega(\mathbf{x}|\mathbf{n}_+)$ is the set of possible tables $\mathbf{x}$ with row marginals equal to $\mathbf{n}_+$, $I()$ is the indicator function, $P(\mathbf{x})$ is the P-value for a test on the table $\mathbf{x}$, and $f()$ is the probability distribution for the one margin fixed model given in Equation 7.1.

Power is also calculated by Equation 7.7, only now, we assume that the probabilities $\boldsymbol{\pi}_{\text{cond}}$ agree with the alternative hypothesis H_{A} instead of the null hypothesis.

Evaluation of Actual Significance Level

We have three test statistics, the Pearson chi-squared, likelihood ratio, and Fisher-Freeman-Halton statistics, and each statistic gives rise to an asymptotic test, an exact test, and a mid-P test. We consider an unordered 3×2 table, that is, we assume that we have three (unordered) groups and a dichotomous outcome.

First, we calculate actual significance levels of the asymptotic tests. Figure 7.2 shows the results for a small sample size, with 15 observations in each of the three groups. The likelihood ratio test does not perform acceptably, with actual significance levels far above the nominal level for a wide range of parameter values. The Pearson chi-squared and the Fisher-Freeman-Halton tests perform similarly; however, the Pearson chi-squared test violates the nominal level slightly, whereas the Fisher-Freeman-Halton test does not. The Pearson chi-squared test is a little bit less conservative than the Fisher-Freeman-Halton test, particularly for small or large values of $\pi_{1|i}$. These performances are typical for other settings; however, the Fisher-Freeman-Halton test may also violate the nominal level but not by much and not very often.

Figure 7.3 shows an example of the actual significance levels of the asymptotic tests for a larger sample size, with 50 observations in each group. The likelihood ratio test still performs poorly, whereas the Pearson chi-squared test and the Fisher-Freeman-Halton test have actual significance levels quite close to the nominal level. Similar results are obtained for unequal sample sizes in the three groups, and for 3×3 tables.

Now, we turn to the three exact tests, and Figure 7.4 shows an example of their performance. There is not much difference between the tests, and they are—as usual for exact conditional tests—quite conservative, particularly for small sample sizes.

The actual significance levels of the three corresponding mid-P tests are shown in Figure 7.5. The mid-P tests all perform excellently for such a small sample size. There are some minor differences between the tests, but they are not consistent across different parameter settings. All three mid-P tests

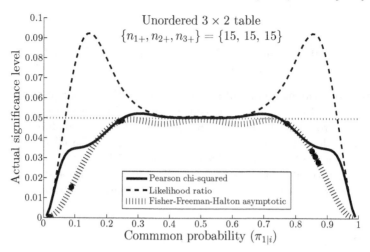

FIGURE 7.2
Actual significance levels of three asymptotic tests for unordered $r \times c$ tables as functions of the common probability under H_0

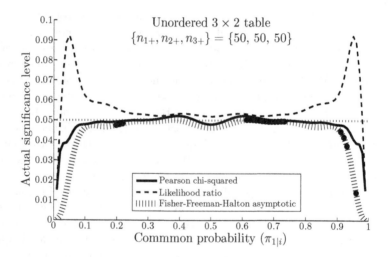

FIGURE 7.3
Actual significance levels of three asymptotic tests for unordered $r \times c$ tables as functions of the common probability under H_0

may violate the nominal level but not by much. The likelihood ratio mid-P test violates the nominal level slightly more often than the other two mid-

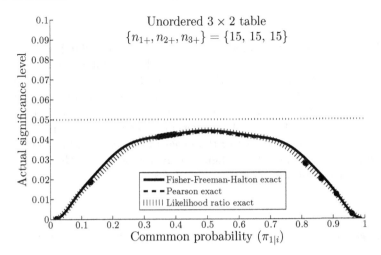

FIGURE 7.4
Actual significance levels of three exact tests for unordered $r \times c$ tables as functions of the common probability under H_0

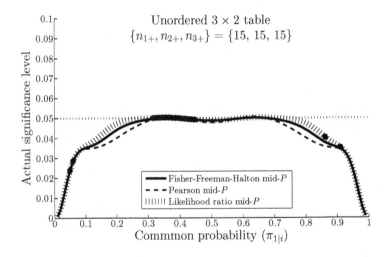

FIGURE 7.5
Actual significance levels of three mid-P tests for unordered $r \times c$ tables as functions of the common probability under H_0

P tests. Overall, the Pearson chi-squared and Fisher-Freeman-Halton mid-P tests perform equivalently.

Evaluation of Power

We refer to Lydersen et al. (2005) and Lydersen et al. (2007) for a detailed evaluation of the power of tests for unordered $r \times c$ tables, with particular emphasis on small sample sizes. Here, we briefly summarized the results of those two articles. The exact tests can be overly conservative and thereby have lower power than the mid-P tests. The Pearson chi-squared and Fisher-Freeman-Halton tests have similar power, and—averaged over many parameter settings—slightly higher power than that of the likelihood ratio tests. The likelihood ratio tests may have slightly higher power than the Pearson and Fisher-Freeman-Halton tests in poorly balanced designs.

7.6   Singly Ordered $r \times c$ Tables

7.6.1   Introduction and Null Hypothesis

In this section, we consider $r \times c$ tables in which r and c are greater than two, and one of the variables is regarded as ordered, whereas the other variables are regarded as unordered. In the study of low birth weight and psychiatric morbidity (Table 7.6), the rows represent low birth weight groups among which there is no ordering. The columns, however, are ordered and represent increasing degrees of psychiatric problems. Table 7.6 is thus a singly ordered 3×3 table with unordered rows and ordered columns.

Another example is the data in Table 7.5, in which the rows represent psychiatric diagnoses that are unordered, whereas the columns represent the BMI categories thin, normal, and overweight, which are ordered in terms of increasing weight. Table 7.5 can be regarded as a singly ordered 6×3 table with unordered rows and ordered columns.

In general, we let the rows represent an unordered grouping of the data, for instance into exposure or treatment groups, and we let the columns represent an ordered outcome variable. This is a generalization of the ordered $2 \times c$ table (Chapter 6) to the situation with more than two unordered categories.

In the reversed situation, where the grouping variable is ordered and the outcome is unordered, one could, in theory, use nominal logistic regression with an ordinal covariate to analyze the data. It is, however, difficult to envisage a practical situation where such a setup is a relevant representation of the data, and we do not consider this situation further.

The null hypothesis of interest for singly ordered $r \times c$ tables is that the probability of each outcome category is equal across the r groups. We assume sampling under the row margins fixed model, for which the appropriate null hypothesis was presented in Equation 7.2; however, we repeat it here for quick reference:

$$H_0 : \pi_{j|1} = \pi_{j|2} = \ldots = \pi_{j|r}, \tag{7.8}$$

for $j = 1, 2, \ldots, c$.

In the following two subsections, we describe the Kruskal-Wallis test for comparing r groups, both as an asymptotic test (Section 7.6.2) and as an exact test and as a mid-P test (Section 7.6.3). Section 7.7 will describe how to analyze the singly ordered $r \times c$ table with ordinal logistic regression.

7.6.2 The Kruskal-Wallis Asymptotic Test

The Kruskal-Wallis test is a straightforward generalization of the Wilcoxon-Mann-Whitney test for two samples described in Section 6.14.3. As noted in Section 6.14.3, the Wilcoxon-Mann-Whitney test is equivalent to a score test for the proportional odds model. Similarly, the Kruskal-Wallis test is equivalent to a score test for the proportional odds model with group as an unordered categorical covariate (Agresti, 2010, p. 81).

The Kruskal-Wallis test statistic is computed as follows. First, rank the observations from the smallest to the largest. Since the data are categorical, there will be many ties, and we use midranks. The n_{+1} smallest observations in column 1 are all equal, and each is assigned the midrank $m_1 = (1 + n_{+1})/2$, which is the average of the numbers from 1 to n_{+1}. In column number j, the midrank equals $m_j = n_{+1} + \ldots + n_{+(j-1)} + (1 + n_{+j})/2$.

The rank sum for row number i is $W_i = n_{i1}m_1 + \ldots + n_{ic}m_c$, and the average rank in row number i is W_i/n_{i+}. Under the null hypothesis, all the r average ranks have expectation $(N + 1)/2$, which is the average rank for all the N observations. High variation between these r average ranks is taken as evidence against the null hypothesis. The Kruskal-Wallis test statistic is

$$T_{\text{KW}}(\mathbf{n}) = \frac{12}{N(N+1)C_{\text{ties}}} \sum_{i=1}^{r} n_{i+} \left(\frac{W_i}{n_{i+}} - \frac{N+1}{2} \right)^2,$$

where C_{ties} is a correction term for ties, given by

$$C_{\text{ties}} = 1 - \sum_{j=1}^{c} \left(n_{+j}^3 - n_{+j} \right) / \left(N^3 - N \right).$$

Under the null hypothesis, T_{KW} is asymptotically chi-squared distributed with $r - 1$ degrees of freedom. The asymptotic Kruskal-Wallis P-value is thus

$$P\text{-value} = \Pr\left[\chi_{r-1}^2 \geq T_{\text{KW}}(\mathbf{n}) \right].$$

Alternative approximations to the Kruskal-Wallis test statistic, including beta and gamma approximations, are given in Meyer and Seaman (2013).

7.6.3 The Kruskal-Wallis Exact and Mid-P Tests

In small samples, the chi-squared approximation to the Kruskal-Wallis test statistic may not hold, and we should instead calculate an exact or mid-P test.

Under the null hypothesis in Equation 7.8, we have $c - 1$ unknown nuisance parameters, because the c column probabilities sum to one (see Table 7.11). By design, we regard the row sums as fixed. To eliminate the unknown nuisance parameters, we also condition on the column sums in the same manner as we did for the exact tests for unordered $r \times c$ tables (the Fisher-Freeman-Halton and related exact tests in Section 7.5.3). The Kruskal-Wallis exact conditional P-value is given by

$$\text{exact cond. } P\text{-value} = \sum_{\Omega(\mathbf{x}|\mathbf{n}_{++})} I\big[T_{\text{KW}}(\mathbf{x}) \geq T_{\text{KW}}(\mathbf{n})\big] \cdot f(\mathbf{x} \,|\, \mathbf{n}_{++}),$$

where $\Omega(\mathbf{x}|\mathbf{n}_{++})$ is the set of all possible tables $(\mathbf{x})$ with the same row and column sums $(\mathbf{n}_+)$ as the observed table $(\mathbf{n})$, $I()$ is the indicator function, and $f(\mathbf{x} \,|\, \mathbf{n}_{++})$ is the point probability in the multiple hypergeometric distribution, as shown in Equation 7.3.

As usual with exact conditional tests, there may be some conservatism in small samples that can be remedied with the mid-P approach—although this is less of a problem in $r \times c$ tables than in 2×2 tables. The Kruskal-Wallis mid-P test is given by

$$\text{mid-}P \text{ value} = \sum_{\Omega(\mathbf{x}|\mathbf{n}_{++})} I\big[T_{\text{KW}}(\mathbf{x}) > T_{\text{KW}}(\mathbf{n})\big] \cdot f(\mathbf{x} \,|\, \mathbf{n}_{++})$$

$$+ \; 0.5 \cdot \sum_{\Omega(\mathbf{x}|\mathbf{n}_{++})} I\big[T_{\text{KW}}(\mathbf{x}) = T_{\text{KW}}(\mathbf{n})\big] \cdot f(\mathbf{x} \,|\, \mathbf{n}_{++}).$$

7.6.4   Examples

Low Birth Weight and Psychiatric Morbidity (Table 7.6)

In Table 7.6, we have three unordered rows that represent birth weight groups and three ordered columns that represent increasing diagnostic severity, so this is a singly ordered 3×3 table. Here, we illustrate how to compute the Kruskal-Wallis test statistic. The average ranks in the VLBW, SGA, and control groups are $W_1 = 81.21$, $W_2 = 80.43$, and $W_3 = 63.13$, respectively. The average rank for all the $N = 145$ observations is $(145+1)/2 = 73$, and the correction for ties is $C_{\text{ties}} = 0.6908$. The observed value of the Kruskal-Wallis test statistic is thus $T_{\text{KW}}(\mathbf{n}) = 9.162$. We have $r - 1 = 2$ degrees of freedom, which gives an asymptotic P-value of 0.0102.

We may also calculate the Kruskal-Wallis exact P-value, which in this case is $P = 0.00955$. The mid-P version is not much different: $P = 0.00954$.

The Kruskal-Wallis tests indicate that there are differences between at least two of the birth weight groups. To identify which groups are different, we can carry out pairwise comparisons. Because we have used the Kruskal-Wallis test to test for an overall group difference, it is natural to use the Wilcoxon-Mann-Whitney test (Section 6.14.3) for the pairwise comparisons.

We obtain the results in Table 7.17. The asymptotic, exact, and mid-P values agree quite well, and they indicate that the distribution of diagnostic severity is different in the control group compared with both the VLBW and SGA groups. No difference is indicated between the VLBW and SGA groups.

TABLE 7.17

Pairwise comparisons of the diagnostic severity between three birth weight groups (data from Table 7.6)

	P-value		
Comparison	**Asymptotic***	**Exact[†]**	**Mid-P[‡]**
VLBW vs. SGA	0.8566	0.8651	0.8452
VLBW vs. Control	0.0099	0.0115	0.0104
SGA vs. Control	0.0080	0.0101	0.0086

*The Wilcoxon-Mann-Whitney test (Section 6.14.3)
[†]The exact conditional linear rank test (Section 6.14.4)
[‡]The mid-P linear rank test (Section 6.14.5)

If we would like to preserve the familywise error rate in the pairwise comparisons, we can use the closed testing procedure as described in Section 7.5.7, because only three groups are involved. We thus take the adjusted P-values to be the maximum of the local P-value and the overall P-value. The local P-values are given in Table 7.17, and the overall P-value is one of the P-values (asymptotic, exact, or mid-P) for the Kruskal-Wallis test. There is not much difference between the asymptotic, exact, and mid-P tests, so we used the asymptotic P-values. The adjusted asymptotic P-values are:

$$\text{VLBW vs. SGA: } P_{\text{adj}} = \max(0.8566, 0.0102) = 0.8566,$$

$$\text{VLBW vs. Control: } P_{\text{adj}} = \max(0.0099, 0.0102) = 0.0102,$$

$$\text{SGA vs. Control: } P_{\text{adj}} = \max(0.0080, 0.0102) = 0.0102.$$

We conclude that the VLBW and SGA groups have more severe psychiatric diagnoses than the control group, and that this conclusion holds under preservation of the familywise error rate at a level of 5%. We will return to this example in Section 7.7.6, where we analyze the data with ordinal logistic regression.

Psychiatric Diagnoses and BMI (Table 7.5)

We return to the data on the association between psychiatric diagnoses (in six unordered categories) and BMI (in three ordered or unordered categories), as shown in Table 7.5. In Section 7.5.9, we analyzed these data under the assumption that the BMI categories were unordered. As discussed in our introduction to this example in Section 7.2.3, we may argue for regarding the BMI categories as unordered in some settings and ordered in other settings, depending on the context. Here, we shall treat the categories as ordered.

The observed value of the Kruskal-Wallis test statistic is $T_{\text{KW}}(\mathbf{n}) = 11.97$.

There are $r = 6$ groups, so the degrees of freedom are $r - 1 = 5$, and the corresponding asymptotic P-value is 0.0353. Hence, there is some evidence of an association between psychiatric diagnoses and BMI. Next, we could carry out pairwise comparisons of the six diagnostic groups with, for instance, the Wilcoxon-Mann-Whitney test, possibly with a Bonferroni adjustment to the P-values—or some other approach—to account for multiple testing. With six groups, there are $6(6 - 1)/2 = 15$ pairwise comparisons. By just looking at the numbers in Table 7.5, we observe that some of the diagnostic groups have quite similar distributions of BMI categories, and that eating disorders stands out as the diagnostic group that is most different from the others. Hence, we only show the results for the comparisons between eating disorders and the five other diagnostic groups. When we adjust for multiple testing, however, we have to adjust for all possible 15 pairwise comparisons, not just the five that we have carried out, because we have selected the five comparisons after looking at the data. The results are shown in Table 7.18. After adjusting for multiple testing, such that the familywise error rate is preserved at 5%, we find that three psychiatric diagnoses—mood (affective) disorders, anxiety disorders, and hyperkinetic disorders—have significantly different distribution of BMI compared with eating disorders.

TABLE 7.18
Comparison of BMI between the diagnostic group eating disorders and five other diagnostic groups (data from Table 7.6)

| | P-value* | |
Comparison	Unadjusted	Bonferroni adjusted[†]
Eating dis. vs. Mood (affective) dis.	0.0005	0.0078
Eating dis. vs. Anxiety disorders	0.0009	0.0138
Eating dis. vs. Autism spectrum dis.	0.0152	0.2272
Eating dis. vs. Hyperkinetic dis.	0.0013	0.0192
Eating dis. vs. Other disorders	0.0078	0.1176

*The Wilcoxon-Mann-Whitney test (Section 6.14.3)
[†]Adjusted for 15 possible pairwise comparisons

We will return to this example in Section 7.7.6, where we analyze the data with ordinal logistic regression.

7.6.5 Evaluations of Tests

The criteria for evaluating tests for singly ordered $r \times c$ tables are the same as the ones for evaluating tests for unordered $r \times c$ tables, see Section 7.5.10.

Here, we consider a 2×3 table, and we have three tests to compare: the asymptotic, exact, and mid-P versions of the Kruskal-Wallis test. We use a simple 2×3 table to be able to calculate exact actual significance levels for the exact and mid-P tests, which would be very time consuming to perform for a table with more than two rows. A singly ordered 2×3 table is an example of an ordered $2 \times c$ table, which was the topic of Chapter 6, and the Kruskal-Wallis test is in this case equal to the Wilcoxon-Mann-Whitney test in Section 6.14.3.

The actual significance levels of the three tests for $n_{1+} = n_{2+} = 15$ are shown in Figure 7.6. The actual significance level is shown as a function of $\pi_{1|i}$, the common probability of $Y_2 = 1$. The probabilities of $Y_2 = 2$ and $Y_2 = 3$ are given as $\pi_{2|i} = (1 - \pi_{1|i})/3$ and $\pi_{3|i} = 2(1 - \pi_{1|i})/3$.

The Kruskal-Wallis asymptotic and the mid-P tests both perform excellently, with actual significance levels close to the nominal level, except for values of $\pi_{1|i} > 0.8$. That is not surprising, because the probabilities of $Y_2 = 2$ and $Y_2 = 3$ then are very small, and with a sample size of only 15 observations in each group, there is not much information in the data to provide precise inference for such small probabilities.

The Kruskal-Wallis exact test is rather conservative, and its actual significance level does not reach more than 3.9% in this example.

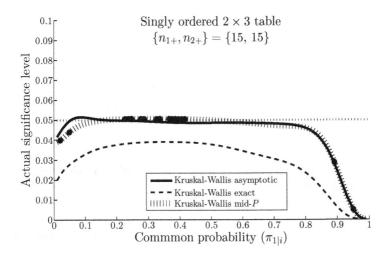

FIGURE 7.6
Actual significance levels of three Kruskal-Wallis tests for singly ordered $r \times c$ tables as functions of the common probability under H_0

7.7 Ordinal Logistic Regression with a Nominal Covariate for Singly Ordered $r \times c$ Tables

7.7.1 The Cumulative Logit Model

In Sections 6.10–6.15, we described ordinal logistic regression models for comparing two groups. These models are readily extended to the comparison of r groups. In this section, we limit our attention to the cumulative logit model. This is by far the most widely used ordinal model in practical application, see our comments on pages 241 and 255.

In singly ordered $r \times c$ tables, it is quite usual that one of the r groups is a control group, an unexposed group, or otherwise a natural reference group. In the study of low birth weight and psychiatric morbidity (Table 7.6), for example, we have an explicit control group; however, in Table 7.5, where six psychiatric diagnoses are cross-classified with three categories of BMI, the choice of reference group is not obvious.

Let Y_1 denote the row number, and let Y_2 denote the column number. The cumulative odds ratios for $Y_1 = i$ relative to $Y_1 = 1$, which we now take to be the reference category, is given by:

$$
\begin{aligned}
\theta_{ij}^{\text{CU}} &= \frac{\Pr(Y_2 \le j \,|\, Y_1 = i)/\Pr(Y_2 > j \,|\, Y_1 = i)}{\Pr(Y_2 \le j \,|\, Y_1 = 1)/\Pr(Y_2 > j \,|\, Y_1 = 1)} \\
&= \frac{(\pi_{1|i} + \ldots + \pi_{j|i})/(\pi_{j+1|i} + \ldots + \pi_{c|i})}{(\pi_{1|1} + \ldots + \pi_{j|1})/(\pi_{j+1|1} + \ldots + \pi_{c|1})},
\end{aligned}
$$

for $i = 2, 3, \ldots, r$ and $j = 1, 2, \ldots, c - 1$. This cumulative odds ratio is illustrated in Panel C of Figure 7.1.

To set up a logistic regression model, we define a binary indicator variable—i.e., a dummy variable—for each group except the reference group. The binary indicator variable, x_i, takes the value 1 if the observation is in group i and 0 otherwise. If the first group is the reference group, we have the indicator variables $x_2, x_3, \ldots, x_r$, where $x_i = 1$ if $Y_1 = i$, else $x_i = 0$. With a proportional odds logistic model, the cumulative logits are modelled as

$$
\begin{aligned}
\log \left[\frac{\Pr(Y_2 \le j \,|\, \mathbf{x})}{\Pr(Y_2 > j \,|\, \mathbf{x})} \right] &= \alpha_j - \boldsymbol{\beta}\mathbf{x} \\
&= \alpha_j - (\beta_1 x_1 + \beta_2 x_2 + \ldots + \beta_r x_r),
\end{aligned}
$$

which can be written

$$
\log \left[\frac{\Pr(Y_2 \le j \,|\, Y_1 = i)}{\Pr(Y_2 > j \,|\, Y_1 = i)} \right] = \alpha_j - \beta_i, \tag{7.9}
$$

for $i = 1, 2, \ldots, r$ and $j = 1, 2, \ldots, c - 1$, where $\beta_i = 0$ for the reference group, here $\beta_1 = 0$. The above notation with a minus sign for the β parameter is the same notation as we used in Chapter 6. As noted on page 245, and elsewhere, some textbooks and software packages define the proportional odds model with a positive sign for the β parameter.

7.7.2   Estimating Proportional Odds Models

The maximum likelihood principle is the standard procedure for estimating the parameters in the proportional odds model. The likelihood function is a straightforward generalization of Equation 6.28:

$$L(\alpha_1, \ldots, \alpha_{c-1}, \beta_2, \ldots, \beta_r)$$

$$= \prod_{i=1}^{r} \prod_{j=1}^{c} \left[\Pr(Y_2 \leq j \mid Y_1 = i) - \Pr(Y_2 \leq j - 1 \mid Y_1 = i) \right]^{n_{ij}}$$

$$= \prod_{i=1}^{r} \prod_{j=1}^{c} \left[\frac{\exp(\alpha_j - \beta_i)}{1 + \exp(\alpha_j - \beta_i)} - \frac{\exp(\alpha_{j-1} - \beta_i)}{1 + \exp(\alpha_{j-1} - \beta_i)} \right]^{n_{ij}},$$

where $\alpha_c = 0$ and $\beta_1 = 0$. We obtain estimates and standard errors with the methods described in Section 6.12.

7.7.3   Testing the Fit of a Proportional Odds Model

The proportional odds assumption made in Equation 7.9 implies that the coefficients β_i do not depend on the column number j. A general alternative is

$$\log \left[\frac{\Pr(Y_2 \leq j \mid Y_1 = i)}{\Pr(Y_2 > j \mid Y_1 = i)} \right] = \alpha_j - \beta_{ij},$$

for $i = 1, 2, \ldots, r$ and $j = 1, 2, \ldots, c-1$, where $\beta_{ij} = 0$ for the reference group, here $\beta_{1j} = 0$ for $j = 1, 2, \ldots, c - 1$ with row 1 as the reference group. Note that there are $c - 1$ parameters α_j and $(r - 1)(c - 1)$ parameters β_{ij}, because the subscripts of β_{ij} are $i = 2, 3, \ldots, r$ and $j = 1, 2, \ldots, c - 1$. Hence, the total number of free parameters in the model is $r(c - 1)$. The model is saturated in the sense that these $r(c - 1)$ parameters can be uniquely expressed by the conditional probabilities $\pi_{j|i}$. There are $r(c - 1)$ free conditional probabilities, because they sum to one in each of the rows.

The null hypothesis for testing the proportional odds assumption is now

$$H_0 : \beta_{i1} = \beta_{i2} = \ldots = \beta_{i,c-1} = \beta_i,$$

for all $i = 2, 3, \ldots, r$. The general alternative has the structural problem that the cumulative probabilities can be out of order for some values of the parameters (Agresti, 2010, p. 70). As a result, it is not always possible to maximize the likelihood function. A score test, however, maximizes the likelihood function under the null hypothesis, and is thus more widely applicable than a likelihood ratio test or a Wald test for testing the proportional odds assumption.

Unfortunately, this score test has drawbacks. It can perform poorly for sparse data (Peterson and Harrell, 1990), and it also tends to be too liberal when data are not sparse (Agresti, 2010, p. 71). When the general model can be fitted, the Wald tests, such as those proposed by Brant (1990), or a likelihood ratio test, can be used. In some software packages, a test for the proportional odds assumption is called a *test for parallel lines*.

Is a statistically significant deviation from the proportional odds assumption necessarily of practical importance? A proportional odds model with fewer parameters may more appropriately answer the research question than a fully saturated model. In our view, a test for the proportional odds assumption should not be emphasized too much, and sometimes it does not need to be calculated at all. Especially in large samples, a small P-value for this test does not always imply any meaningful deviations from the proportional odds assumption. This view is also expressed by Agresti (2010, p. 71).

7.7.4   Testing for Effect in a Proportional Odds Model

We now turn to the proportional odds model in Equation 7.9. Under the null hypothesis of no association between Y_1 (rows) and Y_2 (columns), all the regression coefficients are zero:

$$H_0 : \beta_2 = \beta_3 = \ldots = \beta_r = 0. \tag{7.10}$$

The general alternative hypothesis is $H_A : \beta_i \neq 0$ for at least one i. Common tests include the Wald test, the likelihood ratio test, and the score test, which are described in detail for the ordered $2 \times c$ table in Section 6.14. The generalizations to the singly ordered $r \times c$ table are rather straightforward.

In Chapter 6, we found that the Wald and score tests performed better than the likelihood ratio test, see our recommendations in Section 6.19.

7.7.5   Confidence Intervals for the Effect in a Proportional Odds Model

Let $\hat{\beta}_{i,\mathrm{ML}}$ denote the maximum likelihood estimate of β_i, and let $\widehat{\mathrm{SE}}(\hat{\beta}_{i,\mathrm{ML}})$ denote the corresponding standard error. A Wald confidence interval for β_i is given by

$$\hat{\beta}_{i,\mathrm{ML}} \pm z_{\alpha/2}\widehat{\mathrm{SE}}(\hat{\beta}_{i,\mathrm{ML}}),$$

where $z_{\alpha/2}$ is the upper $\alpha/2$ percentile of the standard normal distribution. Alternatively, a profile likelihood confidence interval can be constructed as described in Section 6.18.2.

7.7.6   Examples

Low Birth Weight and Psychiatric Morbidity (Table 7.6)

We return to the example of low birth weight and psychiatric morbidity. In Section 7.6.4, we tested for the overall association between the birth weight groups and the psychiatric diagnoses with the Kruskal-Wallis tests, and found that $P \approx 0.01$. Pairwise comparisons of the birth weight groups revealed that the VLBW and SGA groups were significantly different from the control group, but that there was no difference between the VLBW and SGA groups.

Now, we analyze the data in Table 7.6 with the proportional odds model. First, we test the assumption of proportional odds. The likelihood ratio test has an observed test statistic equal to 2.163. With $3 - 1 = 2$ degrees of freedom, we get $P = 0.339$. The Brant test gives similar results, with $P = 0.299$. Hence, we find no significant departure from the proportional odds assumption.

For the null hypothesis in Equation 7.10, the observed value of the likelihood ratio test statistic is 9.579. It is asymptotically chi-squared distributed with $r - 1$ degrees of freedom, here $3 - 1 = 2$ degrees of freedom, and we get $P = 0.0083$, which is slightly smaller than the P-values of the asymptotic, exact, and mid-P Kruskal-Wallis tests. We can now estimate the strength of the association with odds ratios from the proportional odds model. We use row 3 (the control group) as reference group, and we obtain the following results:

$$\text{VLBW vs. Control: } OR_1 = 3.22 \ (95\% \text{ CI } 1.35 \text{ to } 7.66), P = 0.0081,$$
$$\text{SGA vs. Control: } OR_2 = 2.97 \ (95\% \text{ CI } 1.28 \text{ to } 6.91), P = 0.0114.$$

The two low birth weight groups can be compared if we use SGA (or VLBW) as reference group:

$$\text{VLBW vs. SGA: } OR_3 = 1.08 \ (95\% \text{ CI } 0.465 \text{ to } 2.53), P = 0.852.$$

We may want to adjust these P-values to preserve the familywise error rate. Again, we can simply take the maximum of the local and global P-values, as explained in Section 7.5.7:

$$\text{VLBW vs. Control: } P_{\text{adj}} = \max(0.0081, 0.0083) = 0.0083,$$
$$\text{SGA vs. Control: } P_{\text{adj}} = \max(0.0114, 0.0083) = 0.0114,$$
$$\text{VLBW vs. SGA: } P_{\text{adj}} = \max(0.852, 0.0083) = 0.852.$$

Because the first two P-values are less than 5%, we conclude that the VLBW and SGA groups have more severe psychiatric diagnoses than the control group, while preserving the familywise error rate at a level of 5%.

Psychiatric Diagnoses and BMI (Table 7.5)

Now to the data in Table 7.5, which cross-classifies six psychiatric diagnoses with three categories of BMI. The observed value of the likelihood ratio test statistic for testing the proportional odds assumption is 11.29 with $6 - 1 = 5$ degrees of freedom. This gives a statistical significant P-value of 0.046; however, as discussed in Section 7.7.3, this does not necessarily mean that the proportional odds model is unsuitable to analyze these data. We continue this example under the assumption that the proportional odds model provides sensible results.

The likelihood ratio test for the null hypothesis in Equation 7.10 gives $P = 0.024$ ($T = 12.98$ with five degrees of freedom), and indicates that at least one diagnostic group is different from the others. But which group should be the reference group? In this example, we should not use the last row, "Other diagnoses", as reference group, because it contains several different diagnoses. There is no obvious choice; however, we may choose "hyperkinetic disorders" (row 5) as the reference group, partly because it is a large group. The results are shown in Table 7.19. The diagnostic group "eating disorders" has an OR equal to 0.19, which is statistically highly significant ($P = 0.0005$) and substantially lower than one. This means that the odds of being in a higher BMI category is substantially lower in the eating disorders group, compared with the reference group. All the other diagnostic groups have OR between 0.90 and 1.09, that is, rather close to one. If we use "eating disorders" as the reference group, we obtain the results in the third column of Table 7.19. The P-values for the comparisons with eating disorders range from 0.0005 to 0.0048. There are, however, 15 possible pairwise comparisons, and these P-values should be adjusted for multiple testing. One possible but somewhat conservative way to preserve the familywise error rate is to use Bonferroni adjustments, as was done when these pairwise comparisons were carried out with the Wilcoxon-Mann-Whitney test in Table 7.18. After Bonferroni adjustment—that is, we multiply each P-value with 15—the P-values range from 0.0073 to 0.071, and all would be regarded as statistically significant at the 5% level, except for autism spectrum disorders. In this context, we should note that autism spectrum disorders has an OR similar to the other diagnostic groups, except for eating disorders, and that it is one of the smallest groups, hence comparisons with this group have low power. Comparing any two diagnostic groups other than eating disorders shows similar OR and P-values far from significant. Hence, we will conclude that eating disorders is the only group deviating from the others in terms of BMI category.

In the two examples above, the P-values from the proportional odds model are similar but slightly smaller than the ones we obtained from the Kruskal-Wallis and pairwise Wilcoxon-Mann-Whitney tests in Section 7.6.4. One additional benefit with the proportional odds model is that we get estimated odds ratios as measures of effect size, with confidence intervals, in addition to the P-values.

TABLE 7.19
The results of fitting a proportional odds model of three categories of
birth weight on psychiatric diagnoses with two different reference groups,
based on the data in Table 7.5

	OR (95% CI), *P*-value	
Diagnostic group	**Hyperkinetic disorders as reference group**	**Eating disorders as reference group**
Mood (affective) dis.	1.09 (0.65 to 1.83), 0.74	5.71 (2.11 to 15.4), 0.0006
Anxiety disorders	0.90 (0.59 to 1.39), 0.64	4.72 (1.83 to 12.2), 0.0013
Eating disorders	0.19 (0.075 to 0.49), 0.0005	Reference
Autism spectrum dis.	0.93 (0.46, 1.89), 0.85	4.88 (1.62 to 14.7), 0.0048
Hyperkinetic dis.	Reference	5.23 (2.06 to 13.3), 0.0005
Other disorders	1.07 (0.50 to 1.89), 0.84	5.58 (1.95 to 15.9), 0.0013

7.8   Tests for Association in Doubly Ordered $r \times c$ Tables

We now turn to $r \times c$ tables where r and c are greater than two and both
variables are regarded as ordered. Tables 7.7, 7.8, and 7.9 are all examples
of doubly ordered tables. In some cases, such as the example in Table 7.8,
it is natural to consider one variable as an explanatory—or group—variable
(nuclear pleomorphism in the rows of Table 7.8) and the other variable as an
outcome variable (tumor type in the columns of Table 7.8). In these cases,
we assume that the table is arranged such that the explanatory variable is
in the rows and the outcome variable is in the columns. This setup can be
regarded as a generalization of the ordered $r \times 2$ table in Chapter 5, as well
as a generalization of the ordered $2 \times c$ table in Chapter 6.

In the following subsections, we present tests for association suitable for
the doubly ordered $r \times c$ table. Correlation measures for doubly ordered $r \times c$
tables are considered in Section 7.9, and ordinal logistic regression for doubly
ordered $r \times c$ tables is described in Section 7.10.

7.8.1   The Linear-by-Linear Test for Association

The linear-by-linear test for association is also called the *Mantel-Haenszel
test for trend* (Mantel, 1963)—not to be confused with the Cochran-Mantel-
Haenszel test for the common odds ratio in stratified 2×2 tables, as described
in Section 10.13.3. The Mantel-Haenszel test for trend has already been described for the ordered $r \times 2$ table in Section 5.7.6.

To calculate the linear-by-linear test statistic, we first assign scores to
the rows $(a_1, a_2, \ldots, a_r)$ and to the columns $(b_1, b_2, \ldots, b_c)$. As discussed in
Section 5.5, we will usually advocate use of equally spaced scores; however,
there are situations in which other choices should be considered. The linear-

by-linear test statistic is

$$T_{\text{linear}}(\mathbf{n}) = (N-1)\hat{r}_{\text{P}}^2,$$

where $\hat{r}_{\text{P}}$ is the sample Pearson correlation coefficient between the scores in the rows and the scores in the columns, defined in the usual way (see the upcoming Equation 7.12). Under the null hypothesis of no association, T_{linear} is approximately chi-squared distributed with one degree of freedom. An alternative version of the test statistic is

$$Z_{\text{linear}}(\mathbf{n}) = \sqrt{(N-1)}\hat{r}_{\text{P}},$$

which is asymptotically standard normally distributed under the null hypothesis, and we obtain P-values for the linear-by-linear test for association as

$$P\text{-value} = \Pr\!\Big[Z \geq \big|Z_{\text{linear}}(\mathbf{n})\big|\Big],$$

where Z is a standard normal variable. The sign of Z_{linear} shows the direction of the association.

7.8.2   The Jonckheere-Terpstra Test for Association

In an $r \times c$ table, there are $r(r-1)/2$ pairs of rows. The Jonckheere-Terpstra test statistic (Terpstra, 1952; Jonckheere, 1954) combines the Mann-Whitney U statistics (see Section 6.14.3) from all the $r(r-1)/2$ pairwise comparisons in the following way. For each pair of rows i_1 and i_2, where $i_1 < i_2$, compute the Mann-Whitney U statistic

$$U_{i_1,i_2} = \sum_{j=1}^{c}(m_{i_1,i_2,j}\, n_{i_1 j}) - n_{i_1+}(n_{i_1+}+1)/2,$$

where $m_{i_1,i_2,j}$ is the midrank in column number j in the $2 \times c$ table formed by the rows i_1 and i_2 of the $r \times c$ table. The midranks are calculated as shown in Section 6.8, culminating in Equation 6.19. The Jonckheere-Terpstra test statistic is formed as the sum of these $r(r-1)/2$ Mann-Whitney U statistics:

$$T_{\text{JT}}(\mathbf{n}) = \sum_{i_1=1}^{r-1}\sum_{i_2=i_1+1}^{r} U_{i_1,i_2}. \tag{7.11}$$

Under the null hypothesis, T_{JT} has expectation

$$\mathrm{E}\big[T_{\text{JT}}(\mathbf{n})\big] = \Big(N^2 - \sum_{i=1}^{r} n_{i+}^2\Big)\Big/ 4$$

and variance

$$\begin{aligned}
\mathrm{Var}\big[T_{\text{JT}}(\mathbf{n})\big] &= \frac{N(N-1)(2N+5) - A - B}{72} \\
&+ \frac{CD}{36N(N-1)(N-2)} + \frac{EF}{8N(N-1)},
\end{aligned}$$

where

$$A = \sum_{i=1}^{r} n_{i+}(n_{i+}-1)(2n_{i+}+5),$$

$$B = \sum_{j=1}^{c} n_{+j}(n_{+j}-1)(2n_{+j}+5),$$

$$C = \sum_{i=1}^{r} n_{i+}(n_{i+}-1)(n_{i+}-2),$$

$$D = \sum_{j=1}^{c} n_{+j}(n_{+j}-1)(n_{+j}-2),$$

$$E = \sum_{i=1}^{r} n_{i+}(n_{i+}-1),$$

$$F = \sum_{j=1}^{c} n_{+j}(n_{+j}-1),$$

see, for example, Hollander et al. (2014) or Neuhäuser (2012, p. 145). In the equations above, the terms involving B, D, and F account for the non-trivial number of ties. These are terms that include the column sums n_{+j}, which would all equal one if there were no ties. The normalized Jonckheere-Terpstra test statistic is

$$Z_{\mathrm{JT}}(\mathbf{n}) = \frac{T_{\mathrm{JT}}(\mathbf{n}) - \mathrm{E}\big[T_{\mathrm{JT}}(\mathbf{n})\big]}{\sqrt{\mathrm{Var}\big[T_{\mathrm{JT}}(\mathbf{n})\big]}}.$$

This statistic is asymptotically standard normally distributed under the null hypothesis. Equivalently, one can use $T_{\mathrm{JT}}^* = Z_{\mathrm{JT}}^2$ as a test statistic, with a chi-squared reference distribution with one degree of freedom. The sign of the test statistic Z_{JT} shows the direction of the association.

7.8.3 Exact and Mid-P Tests

In small samples, the chi-squared approximation to the linear-by-linear or Jonckheere-Terpstra test statistics may be inaccurate, and we may want to calculate an exact P-value (or a mid-P value) to get more reliable results. If we condition on the row and column sums, the unknown probability parameters (nuisance parameters) will be eliminated in the same manner as for the tests in unordered $r \times c$ tables, such as the Fisher-Freeman-Halton exact test in Section 7.5.3. The exact conditional P-value for the linear-by-linear test is given by

$$\text{exact cond. } P\text{-value} = \sum_{\Omega(\mathbf{x}|\mathbf{n}_{++})} I\big[T_{\mathrm{linear}}(\mathbf{x}) \geq T_{\mathrm{linear}}(\mathbf{n})\big] \cdot f(\mathbf{x} \mid \mathbf{n}_{++}),$$

where $\Omega(\mathbf{x}|\mathbf{n}_{++})$ is the set of all possible tables $(\mathbf{x})$ with the same row and column sums $(\mathbf{n}_{++})$ as the observed table $(\mathbf{n})$, $I()$ is the indicator function, and $f(\mathbf{x}|\mathbf{n}_{++})$ is the multiple hypergeometric probability distribution in Equation 7.3.

We may also derive an exact conditional test based on the Jonckheere-Terpstra test statistic in Equation 7.11:

$$\text{exact cond. } P\text{-value} = \sum_{\Omega(\mathbf{x}|\mathbf{n}_{++})} I\left[T_{\text{JT}}(\mathbf{x}) \geq T_{\text{JT}}(\mathbf{n})\right] \cdot f(\mathbf{x}|\mathbf{n}_{++}).$$

Mid-P versions of the exact conditional linear-by-linear test and the Jonckheere-Terpstra exact conditional test can be calculated with the usual modification to the exact conditional P-value of only including half the point probability of tables that agree equally with the null hypothesis as the observed table. We refer to Equation 7.5 for the expression of a mid-P value in $r \times c$ tables with a general test statistic. A simple substitution of T_{linear} or T_{JT} for T in Equation 7.5 will give the linear-by-linear mid-P value and the Jonckheere-Terpstra mid-P value, respectively.

7.8.4 Examples

In this section, we illustrate the use of the linear-by-linear and Jonckheere-Terpstra tests for association on the data from two examples of doubly ordered $r \times c$ tables: the colorectal cancer data in Table 7.7 and the data on breast tumor in Table 7.8.

Colorectal Cancer (Table 7.7)

For the colorectal cancer data, we carry out two analyses of the linear-by-linear asymptotic test: one with equally spaced scores $\{1, 2, 3, 4\}$ and one with mid-interval scores $\{0.5, 5, 17, 39\}$ assigned to the rows. The mid-interval scores are based on the number of weeks of symptoms for each row category. In both cases, we assign the equally spaced scores $\{1, 2, 3, 4\}$ to the columns, which represent the tumor stages T-1, T-2, T-3, and T-4. Table 7.20 shows the results, which display a highly statistically significant association between duration of symptoms and colorectal tumor stage. The observed value of the test statistics are negative, thus shorter duration of symptoms is associated with more severe tumor stage. There are some minor differences between the three test statistics, also between the two linear-by-linear tests with different choices of scores assigned to the rows; however, the conclusions based on the results of the three tests remain the same.

The strength of the association between duration of symptoms and colorectal tumor stage can be measured by the Pearson correlation coefficient, which we need to compute as an intermediate step in the calculation of the linear-by-linear test statistic. With the scores $\{a_1, a_2, a_3, a_4\} = \{1, 2, 3, 4\}$, the Pearson correlation coefficient is $\hat{r}_P = -0.102$, and with the mid-interval scores

TABLE 7.20

Test results for the association between duration of
symptoms and colorectal tumor stage, based on the
data in Table 7.7

Test	Statistic	*P*-value
Linear-by-linear asymptotic	-2.85	0.0043
Linear-by-linear asymptotic with mid-interval scores	-2.67	0.0077
Jonckheere-Terpstra asymptotic	-2.71	0.0067

$\{a_1, a_2, a_3, a_4\} = \{0.5, 5, 17, 39\}$, it is $\hat{r}_P = -0.095$. Although the association
is highly significant, the strength of the association is weak. The association
is slightly weaker when mid-interval scores are used instead of equally spaced
scores.

We have not included any exact conditional test for the colorectal cancer
data, because the computational burden is too heavy to carry out, at least for
an ordinary laptop computer.

Breast Tumor (Table 7.8)

For the breast tumor data in Table 7.8, we have a highly significant association
between the degree of nuclear pleomorphism found in fine needle aspiration
(the rows) and tumor type (the columns), see Table 7.21. The observed values
of the test statistics are quite large, and the *P*-values are in the order of
10^{-14} for the two asymptotic tests, and in the order of 10^{-16} for the two
exact conditional tests. Because the observed value of the test statistics are
positive, the association is positive, which means that higher degrees of nuclear
pleomorphism are associated with more malign tumor types.

TABLE 7.21

Test results for the association between nuclear pleomorphism
and breast tumor type, based on the data in Table 7.8

Test	Statistic	*P*-value
Linear-by-linear asymptotic	7.57	< 0.0001
Linear-by linear exact conditional		< 0.0001
Jonckheere-Terpstra asymptotic	7.66	< 0.0001
Jonckheere-Terpstra exact conditional		< 0.0001

The estimated Pearson correlation coefficient for the association between
nuclear pleomorphism and tumor type is $\hat{r}_P = 0.659$, which indicates a rather
strong association.

7.8.5 Evaluations of Tests

The criteria for evaluating tests for doubly ordered $r \times c$ tables are the same as the ones for evaluating tests for unordered $r \times c$ tables, see Section 7.5.10.

Here, we consider a 3×3 table, and we calculate actual significance levels as functions of $\pi_{1|i}$, the common probability of $Y_2 = 1$. The probabilities of $Y_2 = 2$ and $Y_2 = 3$ are given as $\pi_{2|i} = (1 - \pi_{1|i})/3$ and $\pi_{3|i} = 2(1 - \pi_{1|i})/3$. Figure 7.7 shows the actual significance levels of the (asymptotic) linear-by-linear and Jonckheere-Terpstra tests for $n_{1+} = n_{2+} = n_{3+} = 10$. The two tests perform quite similarly, with actual significance levels remarkably close to the nominal level. The exception is when $\pi_{1|i} > 0.8$, for which the probabilities of $Y_2 = 2$ and $Y_2 = 3$ are very small; we do not expect any test to perform well for such parameters with such a small sample size.

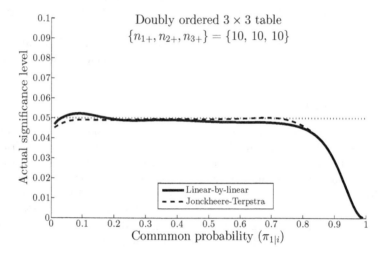

FIGURE 7.7
Actual significance levels of the asymptotic linear-by-linear and Jonckheere-Terpstra tests for doubly ordered $r \times c$ tables as functions of the common probability under H_0

7.9 Correlation Measures in Doubly Ordered $r \times c$ Tables

One way to quantify the association between row and columns in a doubly ordered $r \times c$ table is to use a correlation coefficient. The most commonly used correlation measures are the Pearson correlation, Spearman correlation,

The $r \times c$ Table 317

Kendall's tau-b, and polychoric correlation, and these are described in the following subsections. In addition, we describe the gamma coefficient. The gamma coefficient is not a correlation measure; however, it is closely related to Kendall's tau-b.

7.9.1 The Pearson Correlation Coefficient

Assume that we have assigned scores to the rows $(a_1, a_2, \ldots, a_r)$ and to the columns $(b_1, b_2, \ldots, b_c)$. The Pearson correlation coefficient, r_P, is defined by means of the expectations, variances, and covariance of Y_1 (the variable defining the rows) and Y_2 (the variable defining the columns):

$$E(Y_1) = \sum_{i=1}^{r} a_i \pi_{i+}, \quad E(Y_2) = \sum_{j=1}^{c} b_j \pi_{+j},$$

$$\text{Var}(Y_1) = \sum_{i=1}^{r} \left[a_i - E(Y_1)\right]^2 \pi_{i+}, \quad \text{Var}(Y_2) = \sum_{j=1}^{c} \left[b_j - E(Y_2)\right]^2 \pi_{+j},$$

$$\text{Cov}(Y_1, Y_2) = \sum_{i=1}^{r} \sum_{j=1}^{c} \left[a_i - E(Y_1)\right]\left[b_j - E(Y_2)\right] \pi_{ij},$$

$$r_P = \frac{\text{Cov}(Y_1, Y_2)}{\sqrt{\text{Var}(Y_1)\text{Var}(Y_2)}}, \tag{7.12}$$

where the probabilities π_{i+}, π_{+j}, and π_{ij} are given in Table 7.10.

The Pearson correlation coefficient quantifies the degree of linear association between the row scores Y_1 and the column scores Y_2. The possible values of r_P range from -1 to 1, where -1 and 1 correspond to perfect (deterministic) linear relation with negative and positive slopes, respectively. We obtain an estimate of the Pearson correlation coefficient, $\hat{r}_P$, by inserting the sample proportions for the probabilities in the equations above, i.e.,

$$\pi_{i+} \rightarrow n_{i+}/N,$$
$$\pi_{+j} \rightarrow n_{+j}/N,$$
$$\pi_{ij} \rightarrow n_{ij}/N.$$

If $\{Y_1, Y_2\}$ is bivariate normal distributed, the following procedure can be used for inference about the Pearson correlation coefficient. First, apply the inverse hyperbolic tangent transformation—also called the *Fisher Z transformation*—to $\hat{r}_P$:

$$\hat{z} = \tanh^{-1}(\hat{r}_P) = \frac{1}{2} \log\left(\frac{1 + \hat{r}_P}{1 - \hat{r}_P}\right). \tag{7.13}$$

If the bivariate normality assumption holds, $\hat{z}$ is approximately normally distributed with expectation equal to the true population correlation, and standard deviation equal to $1/\sqrt{N-3}$. To construct a $1 - \alpha$ confidence interval

(L, U) for r_{P}, first calculate the limits (l, u) for a $1 - \alpha$ confidence interval for z:

$$(l, u) = \hat{z} \mp z_{\alpha/2} \frac{1}{\sqrt{N - 3}}.$$

Next, transform (l, u) back to the original scale to obtain an interval for r_{P}:

$$L = \frac{\exp(2l) - 1}{\exp(2l) + 1}$$

and

$$U = \frac{\exp(2u) - 1}{\exp(2u) + 1}.$$

With a limited number of row and column categories, we cannot rely on the normality assumption. It is thus questionable whether this procedure can be recommended for making inference about the Pearson correlation coefficient in $r \times c$ tables. Yet, it seems to work well in the examples we present in Section 7.9.6. The Fisher Z transformation can, however, be recommended for making inference about the Spearman correlation and Kendall's tau correlation coefficients, as we shall see in Sections 7.9.2 and 7.9.4.

A procedure that can be recommended for the Pearson correlation coefficient is to compute a bootstrap confidence interval for r_{P} based on the sample correlation coefficient. Our general recommendation is to use the *bias-corrected and accelerated bootstrap interval*—see, for instance, Storvik (2012)—which is available in many software packages.

7.9.2 The Spearman Correlation Coefficient

The Spearman correlation coefficient ρ_{S}—also called the *Spearman rank correlation coefficient* or simply *Spearman rho*—is based on midranks (see Section 7.6.2) as scores. The midrank in row number i is

$$a_i = n_{1+} + n_{2+} + \ldots + n_{i-1,+} + (1 + n_{i+})/2,$$

for $i = 1, 2, \ldots, r$, and the midrank in column number j is

$$b_j = n_{+1} + n_{+2} + \ldots + n_{+,j-1} + (1 + n_{+j})/2,$$

for $j = 1, 2, \ldots, c$. The Spearman correlation coefficient is defined as the sample Pearson correlation for these midranks:

$$\hat{\rho}_{\mathrm{S}} = \frac{\sum\limits_{i=1}^{r} \sum\limits_{j=1}^{c} \left[a_i - (N+1)/2 \right] \left[b_j - (N+1)/2 \right] \pi_{ij}}{\sqrt{\left\{ \sum\limits_{i=1}^{r} \left[a_i - (N+1)/2 \right]^2 \pi_{i+} \right\} \left\{ \sum\limits_{j=1}^{c} \left[b_j - (N+1)/2 \right]^2 \pi_{+j} \right\}}}.$$

This also equals the sample Pearson correlation for the ridits, see Agresti (2010, p. 192). The Spearman correlation coefficient equals -1 or 1 if there is a strictly monotone negative or positive (not necessarily linear) relationship between the row scores Y_1 and the column scores Y_2.

To carry out inference for the Spearman correlation coefficient, one can use the Fisher Z transformation (Equation 7.13), as described in Section 7.9.1, see also (Altman et al., 2000, pp. 90–91); however, for the Spearman correlation, the standard deviation of $\hat{z}$ is somewhat higher than $1/\sqrt{N-3}$. Fieller et al. (1957) recommend use of $1.06/\sqrt{N-3}$, whereas Bonett and Wright (2000) report that the factor $\sqrt{(1+\hat{\rho}_S^2/2)/(N-3)}$ gives a better approximation.

Note that the Spearman correlation coefficient is a sample statistic and does not necessarily represent an estimate of a population value. Hence, it is not always relevant to compute a confidence interval for the Spearman correlation coefficient.

7.9.3 The Gamma Coefficient

Several measures of association in doubly ordered $r \times c$ tables are based on the concept of concordant and discordant pairs. Consider two pairs of observations, A and B, with observed values of Y_1 (the variable defining the rows) and Y_2 (the variable defining the columns) equal to $\{Y_1^A, Y_2^A\}$ and $\{Y_1^B, Y_2^B\}$. The pairs are *concordant* if $Y_1^A < Y_2^A$ and $Y_1^B < Y_2^B$, or if $Y_1^A > Y_2^A$ and $Y_1^B > Y_2^B$. Conversely, the pairs are *discordant* if $Y_1^A < Y_2^A$ and $Y_1^B > Y_2^B$, or if $Y_1^A > Y_2^A$ and $Y_1^B < Y_2^B$. If $Y_1^A = Y_2^A$ or $Y_1^B = Y_2^B$, the pairs are neither concordant nor discordant, and we consider them as *tied*. For a given table, the numbers of concordant and discordant pairs are given as

$$C = \sum_{i<k}\sum\sum_{j<l}\sum n_{ij}n_{kl} \quad \text{and} \quad D = \sum_{i<k}\sum\sum_{j>l}\sum n_{ij}n_{kl},$$

respectively. The association is said to be positive if $C > D$, and negative if $C < D$. Among the $C+D$ pairs of observations that are concordant or discordant, the proportion of concordant pairs is $C/(C+D)$, and the proportion of discordant pairs is $D/(C+D)$. The difference between these two proportions is called the *gamma coefficient* and was proposed by Goodman and Kruskal (1954):

$$\hat{\gamma} = \frac{C-D}{C+D}.$$

The population analog of the gamma coefficient is

$$\gamma = \frac{\Pi_C - \Pi_D}{\Pi_C + \Pi_D},$$

where

$$\Pi_C = 2\sum_{i<k}\sum\sum_{j<l}\sum \pi_{ij}\pi_{kl} \quad \text{and} \quad \Pi_D = 2\sum_{i<k}\sum\sum_{j>l}\sum \pi_{ij}\pi_{kl}$$

are the probabilities of concordance and discordance, respectively, for a randomly selected pair of observations. The factor 2 is included in these expressions because the first observation could be in cell (i, j) and the second observation in cell (k, l), or vice versa.

We will use the bias-corrected and accelerated bootstrap interval to construct a confidence interval for γ.

7.9.4   The Kendall Rank Correlation Coefficient

The Kendall rank correlation coefficient (Kendall, 1945) is also commonly referred to as Kendall's tau coefficient, and several versions of it exist. In ordered $r \times c$ tables, the number of ties cannot be ignored, and we regard the version referred to as *Kendall's tau-b* as the appropriate one in this context. Kendall's tau-b is based on the number of concordant and discordant pairs, as well as the number of tied pairs. The number of pairs tied on the rows is

$$T_1 = \sum_{i=1}^{r} \frac{n_{i+}(n_{i+} - 1)}{2};$$

the number of pairs tied on the columns is

$$T_2 = \sum_{j=1}^{c} \frac{n_{+j}(n_{+j} - 1)}{2};$$

and the number of pairs tied on both rows and columns is

$$T_{12} = \sum_{i=1}^{r} \sum_{j=1}^{c} \frac{n_{ij}(n_{ij} - 1)}{2}.$$

For N observations, the total number of pairs is

$$\frac{N(N-1)}{2} = C + D + T_1 + T_2 - T_{12}.$$

In the equation above, T_{12} is subtracted because the pairs that are tied in both rows and columns are already counted in T_1 and T_2. The sample version of Kendall's tau-b is defined as

$$\hat{\tau}_b = \frac{C - D}{\sqrt{\left[\dfrac{N(N-1)}{2} - T_1\right]\left[\dfrac{N(N-1)}{2} - T_2\right]}}.$$

The population version is

$$\tau_b = \frac{\Pi_C - \Pi_D}{\sqrt{\left(1 - \sum_{i=1}^{r} \pi_{i+}^2\right)\left(1 - \sum_{j=1}^{c} \pi_{+j}^2\right)}}.$$

For 2 × 2 tables, Kendall's tau-b equals the Pearson correlation coefficient. In fact, Kendall's tau-b is also a type of correlation for ordered $r \times c$ tables. Consider any two pairs of observations, A and B, with observed values of Y_1 (the variable defining the rows) and Y_2 (the variable defining the columns) equal to $\{Y_1^A, Y_2^A\}$ and $\{Y_1^B, Y_2^B\}$. Let $S_1 = \text{sign}(Y_1^B - Y_1^A)$ and $S_2 = \text{sign}(Y_2^B - Y_2^A)$, where sign() is 1, -1, or 0, when the argument is positive, negative, or zero, respectively. It can be shown that the Pearson correlation between S_1 and S_2 equals Kendall's tau-b, see Agresti (2010, pp. 188–189).

Inference about Kendall's tau-b can be carried out with the Fisher Z transformation (Equation 7.13), in a similar manner as for Spearman rho, but with $\sqrt{0.437/(N - 4)}$ as the approximate standard deviation for $\hat{z}$ (Fieller et al., 1957).

Note that Kendall's tau-b is a sample statistic—as is the case for the Spearman correlation coefficient—and it does not necessarily represent an estimate of a population value. Calculation of a confidence interval for Kendall's tau-b is thus not always relevant.

7.9.5 Polychoric Correlation

For some doubly ordered $r \times c$ tables, it may be appropriate to regard the rows and columns as categorizations of a pair of unobserved, continuous latent variables. As is customary, we assume that the underlying probability distribution to be bivariate normal. The situation is illustrated in Figure 7.8. The row (Y_1) and column (Y_2) variables are related to the underlying continuous variables (Z_1, Z_2) by

$$Y_1 = a_i \quad \text{if} \quad \gamma_{i-1} \le Z_1 < \gamma_i,$$

for $i = 1, 2, \ldots r$, and

$$Y_2 = b_j \quad \text{if} \quad \tau_{j-1} \le Z_2 < \tau_j,$$

for $j = 1, 2, \ldots c$. The correlation in the underlying bivariate distribution of (Z_1, Z_2) is called the *polychoric correlation*. The same idea can be applied to 2×2 tables, for which the underlying bivariate normal distribution is called the *tetrachoric correlation*. The names polychoric and tetrachoric were introduced by Karl Pearson in 1904. Pearson, as well as others, has published approximation formulas for the underlying correlation. Today, such approximation formulas are not needed, because inference can be made with maximum likelihood methods. The parameters γ_i and τ_j are called *thresholds*. If we want to identify the thresholds, the expectation and variance of Z_1 and Z_2 must be fixed, usually at 0 and 1.

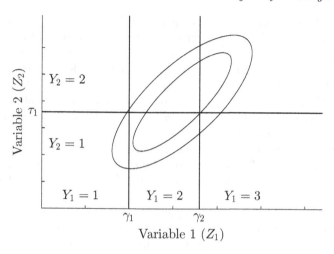

FIGURE 7.8

Example of two ordered categorical variables Y_1 and Y_2 formed by categorizing bivariate normally distributed variables Z_1 and Z_2

7.9.6   Examples

Colorectal Cancer (Table 7.7)

Table 7.22 shows the different correlation measures and sample statistics computed for the colorectal cancer data in Table 7.7. The Pearson correlation coefficient is calculated with two choices of scores for the rows: (i) equally spaced scores $\{a_1, a_2, a_3, a_4\} = \{1, 2, 3, 4\}$; and (ii) mid-interval scores $\{a_1, a_2, a_3, a_4\} = \{0.5, 5, 17, 39\}$, representing the number of weeks of symptoms. In both cases, the column scores are $\{b_1, b_2, b_3, b_4\} = \{1, 2, 3, 4\}$, which represent the tumor stages T-1, T-2, T-3, and T-4. The other correlation measures—Spearman, Kendall's tau-b, and polychoric correlation—depend only on the ordering of the observations and are not based on scores for the rows or columns.

The numerical values of the Pearson and Spearman correlation coefficients agree quite well. Kendall's tau-b is slightly closer to zero (no correlation), and the polychoric correlation is further from zero. We note that for the Pearson and Spearman correlation coefficients, the confidence intervals obtained by different methods are typically equal within two decimals. This is even the case with the Fisher Z interval, which assumes normally distributed variables. This may indicate that the Fisher Z interval is fairly robust against deviations from normality; however, we state this with caution; we are not aware of research confirming this theory. For Kendall's tau-b, we note that the bootstrap interval is wider than the Fieller interval based on the Fisher Z transformation. This

TABLE 7.22
Estimated correlation measures and sample correlation statistics
with confidence intervals for the colorectal cancer data in Table 7.7

Correlation measure	Estimate	95% CI	Width
Pearson correlation coefficient with equally spaced scores			
Fisher Z interval	$\hat{r}_P = -0.102$	-0.171 to -0.032	0.139
Bootstrap interval*		-0.171 to -0.031	0.140
Pearson correlation coefficient with mid-interval scores[†]			
Fisher Z interval	$\hat{r}_P = -0.095$	-0.164 to -0.025	0.139
Bootstrap interval*		-0.166 to -0.023	0.143
Spearman correlation coefficient			
Fieller interval		-0.169 to -0.022	0.147
Bonett-Wright interval	$\hat{\rho}_S = -0.096$	-0.165 to -0.026	0.139
Bootstrap interval*		-0.167 to -0.024	0.143
The gamma coefficient			
Bootstrap interval*	$\hat{\gamma} = -0.139$	-0.239 to -0.036	0.203
Kendall's tau-b			
Fieller interval		-0.132 to -0.040	0.092
Bootstrap interval*	$\hat{\tau}_b = -0.086$	-0.149 to -0.022	0.127
Polychoric correlation[‡]			
ML-based interval	-0.120	-0.206 to -0.034	0.172

*The bias-corrected and accelerated bootstrap interval with 50 000 bootstrap
samples
[†] $\{a_1, a_2, a_3, a_4\} = \{0.5, 5, 17, 39\}$
[‡] Calculated in Stata (StataCorp LP) with the `polychoric` package, written by
Stas Kolenikov

makes us question whether the coverage probability of the Fieller interval is
too low; however, we have no evidence to back this up.

   The example we have just considered has a fairly large sample size of
$N = 784$. We have two other examples of doubly ordered $r \times c$ tables in
this book: the breast cancer example in Table 7.8, which has a small sample
size of $N = 133$, and the self-rated health example in Table 7.9, which has a
large sample size of $N = 2238$. We do not show the results of the correlation
measures for these two data examples; however, we note that the between-
method differences are very similar to the ones in Table 7.22, and that our
discussion of the results above holds for these two examples as well.

7.10   Ordinal Logistic Regression with an Ordinal Covariate for Doubly Ordered $r \times c$ Tables

7.10.1   The Cumulative Logit Model

Ordinal logistic regression with a dichotomous covariate for the analysis of ordered $2 \times c$ tables was covered in Sections 6.10–6.15, and ordinal logistic regression with a nominal covariate was the topic of Section 7.7. In a doubly ordered $r \times c$ table, we can use an ordinal logistic regression model with columns representing the levels of the dependent (outcome) variable, and row number, or row number score, as a covariate (explanatory variable). As noted in Section 6.10, we have several different types of ordinal logistic regression models available; however, in this section, as in Section 7.7, we shall confine our attention to the cumulative logit model. The cumulative odds ratio for comparing $Y_1 = i$ relative to $Y_1 = 1$ is given by

$$
\begin{aligned}
\theta_{ij}^{\mathrm{CU}} &= \frac{\Pr(Y_2 \le j \,|\, Y_1 = i+1) / \Pr(Y_2 > j \,|\, Y_1 = i+1)}{\Pr(Y_2 \le j \,|\, Y_1 = i) / \Pr(Y_2 > j \,|\, Y_1 = i)} \\[2mm]
&= \frac{(\pi_{1|i+1} + \ldots + \pi_{j|i+1}) / (\pi_{j+1|i+1} + \ldots + \pi_{c|i+1})}{(\pi_{1|i} + \ldots + \pi_{j|i}) / (\pi_{j+1|i} + \ldots + \pi_{c|i})},
\end{aligned} \tag{7.14}
$$

for $i = 1, 2, \ldots, r-1$ and $j = 1, 2, \ldots, c-1$. This cumulative odds ratio is illustrated in Panel D of Figure 7.1. The proportional odds assumption means that the odds ratio is the same for all $j = 1, 2, \ldots, c-1$. The cumulative logits are modelled as

$$
\log\left[\frac{\Pr(Y_2 \le j \,|\, Y_1 = i)}{\Pr(Y_2 > j \,|\, Y_1 = i)} \right] = \alpha_j - \beta x_i. \tag{7.15}
$$

In this model, the association structure in the $r \times c$ table is modelled in terms of just one parameter β. If the row numbers are used as scores, such that $x_i = i$ for all $i = 1, 2, \ldots, r-1$, the odds ratio in Equation 7.14 reduces to simply $\theta_{ij}^{\mathrm{CU}} = \exp(\beta)$. Alternatively, one can use more general scores $\{a_1, a_2, \ldots, a_r\}$ to represent the rows, such that the x_i in Equation 7.15 is $x_i = a_i$.

The linear-by-linear and Jonckheere-Terpstra tests for association and the correlation measures described earlier are symmetric in rows and columns, that is, they give exactly the same results if the rows and columns are interchanged. The ordinal logistic regression model described here, however, is not symmetric in this sense.

7.10.2   Estimating Proportional Odds Models

As noted earlier, the maximum likelihood principle is the standard procedure for estimating the parameters in the proportional odds model. The likelihood function is a straightforward generalization of Equation 6.28:

$$L(\alpha_1, \ldots, \alpha_{c-1}, \beta)$$

$$= \prod_{i=1}^{r} \prod_{j=1}^{c} \left[\Pr(Y_2 \le j \,|\, Y_1 = i) - \Pr(Y_2 \le j - 1 \,|\, Y_1 = i) \right]^{n_{ij}}$$

$$= \prod_{i=1}^{r} \prod_{j=1}^{c} \left[\frac{\exp(\alpha_j - \beta x_i)}{1 + \exp(\alpha_j - \beta x_i)} - \frac{\exp(\alpha_{j-1} - \beta x_i)}{1 + \exp(\alpha_{j-1} - \beta x_i)} \right]^{n_{ij}},$$

where $\alpha_c = 0$. We obtain estimates and standard errors with the methods described in Section 6.12.

7.10.3 Testing the Fit of a Proportional Odds Model

The proportional odds assumption made in Equation 7.9 implies that the coefficients β_i do not depend on the column number j. A general alternative is

$$\log \left[\frac{\Pr(Y_2 \le j \,|\, Y_1 = i)}{\Pr(Y_2 > j \,|\, Y_1 = i)} \right] = \alpha_j - \beta_j x_i,$$

for $j = 1, 2, \ldots, c - 1$. The null hypothesis for testing the proportional odds assumption is now

$$H_0 : \beta_1 = \beta_2 = \ldots = \beta_{c-1} = \beta.$$

The comments about alternative tests for singly ordered $r \times c$ tables noted in Section 7.7.3 are also valid for doubly ordered $r \times c$ tables. When the general model can be fitted, the Wald tests such as those proposed by Brant (1990) can be used. We reiterate our comments from Section 7.7.3 that a statistically significant deviation from the proportional odds assumption is not always of practical importance (see page 308). Note again that in some software packages, a test for the proportional odds assumption is called a *test for parallel lines*.

7.10.4 Testing for Effect in a Proportional Odds Model

We now return to the proportional odds model in Equation 7.15. Under the null hypothesis of no association between Y_1 (rows) and Y_2 (columns), the regression coefficient is zero:

$$H_0 : \beta = 0.$$

The alternative hypothesis is $H_A : \beta \ne 0$. Common tests include the Wald test, the likelihood ratio test, and the score test, which are described in detail for the ordered $2 \times c$ table in Section 6.14. The generalizations to the doubly ordered $r \times c$ table are rather straightforward.

7.10.5 Confidence Intervals for the Effect in a Proportional Odds Model

Let $\hat{\beta}_{\text{ML}}$ denote the maximum likelihood estimate of β, and let $\widehat{\text{SE}}(\hat{\beta}_{\text{ML}})$ denote the corresponding standard error. A Wald confidence interval for β is given by

$$\hat{\beta}_{\text{ML}} \pm z_{\alpha/2}\widehat{\text{SE}}(\hat{\beta}_{\text{ML}}),$$

where $z_{\alpha/2}$ is the upper $\alpha/2$ percentile of the standard normal distribution. Alternatively, a profile likelihood confidence interval can be constructed as described in Section 6.18.2.

7.10.6 Examples

Colorectal Cancer (Table 7.7), Breast Cancer (Table 7.8), and Self-Rated Health (Table 7.9)

Table 7.23 shows the results of applying the proportional odds logistic regression model to the data on colorectal cancer in Table 7.7, the data on breast tumor in Table 7.8, and the data on self-rated health in Table 7.9. The P-values are of the same size of order as those obtained from the linear-by-linear and Jonckheere-Terpstra tests for association (see Tables 7.20 and 7.21). The colorectal cancer data merits some comments. The group or explanatory variable was duration of symptoms in four categories, namely < 1 week, 2–8 weeks, 2–6 months, and > 6 months. The first row in Table 7.23 shows the results when equally spaced scores $\{1, 2, 3, 4\}$ are used for the explanatory variable. The second row shows the results when the mid-interval scores $\{0.5, 5, 17, 39\}$, based on the number of weeks, are used. The P-values are in the same size of order in these two analyses ($P = 0.0058$ and $P = 0.0125$). This is not surprising, because the P-value seldom depends heavily on the selection of scores in this context. The estimated odds ratios, however, differ substantially. This is also to be expected, because the odds ratio 0.987 represent the odds ratio for a one-week increase in symptom duration, whereas the odds ratio 0.801 represents the odds ratio for an increase of one category in the original categorization, which in turn represents an increase in symptom duration of between 4.5 and 22 weeks.

7.11 Recommendations

7.11.1 Summary

Our main recommendations for analyzing unordered, singly ordered, and doubly ordered $r \times c$ tables are shown in Table 7.24. More details on each type of analysis are given in Sections 7.11.2–7.11.4.

TABLE 7.23
Results of applying the proportional odds model to the data in
Tables 7.7, 7.8, and 7.9

Example	Estimate	95% CI	*P*-value
Colorectal cancer (Table 7.7)	0.801	0.684 to 0.938	0.0058
Colorectal cancer with mid-interval scores (Table 7.7)	0.987	0.977 to 0.997	0.0125
Breast tumor (Table 7.8)	6.12	3.74 to 10.0	< 0.0001
Self-rated health (Table 7.9)	3.26	2.82 to 3.76	< 0.0001

TABLE 7.24
Recommended tests and confidence intervals for *r × c* tables

Analysis	Recommended methods	Sample sizes
Unordered tables	Pearson chi-squared mid-*P* test	small/medium
	FFH* mid-*P* test	small/medium
	Pearson chi-squared exact test	small/medium
	FFH* exact test	small/medium
	Pearson chi-squared asymptotic test	large
Singly ordered tables	Proportional odds model	all
	Kruskal-Wallis mid-*P* test	small
	Kruskal-Wallis exact test	small
	Kruskal-Wallis asymptotic test	medium/large
Doubly ordered tables	Proportional odds model	all
	Spearman correlation coefficient	all
	Linear-by-linear mid-*P* test	small
	Jonckheere-Terpstra mid-*P* test	small
	Linear-by-linear exact test	small
	Jonckheere-Terpstra exact test	small
	Linear-by-linear asymptotic test	medium/large
	Jonckheere-Terpstra asymptotic test	medium/large

*FFH = Fisher-Freeman-Halton

7.11.2 Unordered Tables

The three test statistics for unordered *r × c* tables—Pearson chi-squared, likelihood ratio, and Fisher-Freeman-Halton—are asymptotically equivalent and give approximately the same results in large samples. In small to moderate samples, on the other hand, the Pearson chi-squared and Fisher-Freeman-Halton statistics perform better than the likelihood ratio statistic in most settings. This result is in accordance with the general result that the chi-squared approximation is slightly better for the Pearson statistic than for the likelihood ratio statistic in small to moderate samples (Larntz, 1978; Koehler, 1986; Cressie and Read, 1989).

The exact conditional tests are conservative, although not as much as exact conditional tests for other types of tables, such as the 2×2 table. The mid-P versions of the Pearson chi-squared and Fisher-Freeman-Halton tests perform excellently. In large samples, the exact conditional and mid-P tests can be too computer-intensive for practical use.

The Pearson chi-squared statistic represents a type of metric that is simple, intuitively understandable, and can be used in a wide range of settings. Because its properties are at least as good as, or approximately as good as, the alternative statistics in most settings, we generally prefer the Pearson statistic for testing association in unordered $r \times c$ tables. In small samples, we recommend the mid-P version, or—if the actual significance level needs to be bounded by the nominal level—the exact conditional test. In samples large enough to satisfy the Cochran criterion (see page 285), and when exact computations are not feasible, we recommend the asymptotic Pearson chi-squared test. As a secondary recommendation, the three tests based on the Fisher-Freeman-Halton statistic can also be recommended, although the asymptotic Fisher-Freeman-Halton test does not look particularly appealing compared with the simplicity of the Pearson chi-squared asymptotic test.

We refer to Lydersen et al. (2005) and Lydersen et al. (2007) for a detailed examination of the actual significance level and power of the asymptotic, exact, and mid-P tests based on the Pearson chi-squared, likelihood ratio, and Fisher-Freeman-Halton statistics.

Further, we may want to identify which parts of the table that contribute to the deviation from the null hypothesis. The association factors are defined as n_{ij}/m_{ij}, where n_{ij} are the observed and m_{ij} the estimated expected cell counts. Values below $1/2$ or above 2 may be regarded as noteworthy deviations from independence. Statistical significance may be studied with the standardized Pearson residuals.

In the special case of $c = 2$, it can be relevant to compare some of, or all of, the $r(r-1)/2$ pairs of binomial probabilities, possibly with adjustment for multiple comparisons, such as Bonferroni adjustment. If only three groups are involved ($r = 3$), we recommend the closed testing procedure for multiplicity adjustment: for each of the pairwise comparisons, take the maximum of the local P-value and the global P-value.

7.11.3 Singly Ordered Tables

In this situation, we assume that the rows represent an unordered grouping of the data—such as into exposure or treatment groups—and that the columns represent an ordered outcome variable. The non-parametric Kruskal-Wallis test can be recommended for this situation. It comes in an asymptotic version, an exact conditional version, and a mid-P version. In small samples, we recommend the mid-P test if minor violations of the nominal significance level are acceptable, or else the exact test. Note, however, that the exact test can be quite conservative. In moderate to large samples, the asymptotic version can be used.

An alternative way of analyzing singly ordered tables is to use a parametric model. The most commonly used ordinal regression model in this setting is the proportional odds logistic model, with the column variable as the outcome variable and the row variable as a nominal covariate (explanatory variable). Some researchers will want to test the proportional odds assumption before carrying out the actual modeling; however, we regard this to be of limited importance, because the model can be fit for purpose even with some violations of the proportional odds assumption. An important advantage of a logistic regression model over a non-parametric method such as the Kruskal-Wallis test is that the regression model provides you with an effect estimate and a confidence interval in addition to a *P*-value, whereas the non-parametric test only gives you a *P*-value.

After obtaining a significant overall test, we may want to investigate which of the groups (rows) differ significantly from each other. These pairwise comparisons can be carried out with the corresponding methods for two groups, i.e., the Wilcoxon-Mann-Whitney test or ordinal logistic regression with a dichotomous covariate, possibly with an adjustment for multiple comparisons. In particular, if only three groups are involved in the comparisons, we recommend the closed testing procedure.

7.11.4   Doubly Ordered Tables

Tests for Association

Both the linear-by-linear test and the Jonckheere-Terpstra test are appropriate for testing association in doubly ordered $r \times c$ tables. The two tests are not asymptotically equivalent; however, we are not aware of research that compares the power of the tests. The linear-by-linear test originates from analysis in a log-linear model, while the Jonckheere-Terpstra test was developed as a non-parametric test for comparing r groups of continuous variables. The tests differ in that the linear-by-linear test uses scores for the rows and columns (see Section 5.5 for a general discussion about how to assign scores), whereas the Jonckheere-Terpstra test uses the ranking of the observations and no scores.

In small samples, both the linear-by-linear test and the Jonckheere-Terpstra test have an exact conditional version and a mid-P version that can be used.

Correlation Measures

All the four correlation measures we have described—The Pearson and Spearman correlation coefficients, Kendall's tau-b, and polychoric correlation—equal zero if the row variable Y_1 and the column variable Y_2 are independent. A positive correlation indicates a positive association between rows and columns, in the sense that higher values of Y_1 tend to co-occur with higher values of Y_2. A negative correlation indicates a negative association between rows and columns, which means that higher values of Y_1 tend to co-occur with lower values of Y_2. Moreover, all four correlation measures are bounded by one in absolute value.

The Pearson correlation coefficient equals one if there is a perfect linear relationship between the two variables, and this is also the case for the Spearman correlation coefficient and Kendall's tau-b. If, on the other hand, there is a strictly increasing but non-linear relationship between the variables, for example, if $Y_2 = Y_1^3$, the Pearson correlation coefficient will be less than one, whereas the Spearman correlation and Kendall's tau-b will still equal one. In many cases, non-linear but monotone relationships should be considered just as interesting as a linear relationship, which would favor the Spearman correlation coefficient and Kendall's tau-b over the Pearson correlation coefficient. To quote Altman (1991, p. 288): *"Rank correlation should be used more often. It is the only non-parametric method which gives as much information as its parametric equivalent (rather than just a P value), and it is of wider validity"*.

Among the two non-parametric measures, the Spearman correlation coefficient seems to be the most used. This may be due to ease of computation, which was an important issue before personal computers were omnipresent, or it may be because the definition of the Spearman correlation coefficient is more straightforward. Kraemer (2006) finds both the Spearman correlation coefficient and Kendall's tau-b to be adequate measures for doubly ordered $r \times c$ tables.

The polychoric and tetrachoric correlation coefficients rely heavily on the underlying distribution of bivariate normality. When this assumption does not hold, the correlation estimated from observed data may be misleading. Moreover, large sample approximations may perform poorly in small samples, even if the bivariate normality assumption holds. These unfavorable properties may explain why these correlation measures are not in much use for the analysis of contingency tables (Kraemer, 2006). For factor analysis of ordinal variables, however, Holgado-Tello et al. (2010) recommend use of polychoric correlation ahead of the Pearson correlation coefficient.

Ordinal Logistic Regression

Section 7.11.3, on recommendations for the singly ordered table, provided arguments for analyzing $r \times c$ tables with the proportional odds logistic regression model, and for not being too strict about violations of the proportional odds assumption. The same arguments also apply to doubly ordered $r \times c$ tables.

8

The Paired 2×2 Table

8.1 Introduction

This chapter considers tests for association, effect measures, and confidence intervals for paired binomial probabilities. Paired binomial probabilities arise in study designs such as matched and cross-over clinical trials, longitudinal studies, and matched case-control studies. The data consist of two samples of dichotomous events: Event A and Event B. Each observation of Event A is matched with one observation of Event B. The two observations in a matched pair may come from the same subject, such as in cross-over clinical trials, where each subject is measured twice (treatment A and treatment B). In matched case-control studies, however, each matched pair refers to two different subjects: one case (Event A) and one matching control (Event B). The purpose of a case-control study is to compare the exposure history between cases and controls. In both situations, the outcomes in the two samples are dependent. The paired 2×2 table may also the result of the measurements of two raters, and if inter-rater agreement is of interest, the methods in Section 13.2 should be used.

The results of studies of paired binomial probabilities can be summarized in a paired 2×2 table, as shown in Table 8.1. The possible outcomes for each event is either success or failure. As usual, success does not necessarily indicate a favorable outcome but rather the outcome of interest, which may, for instance, be the presence of a certain disease. The paired 2×2 table may look like the unpaired 2×2 table in Chapter 4 (see Table 4.1), but the statistical methods used to analyze unpaired and paired 2×2 tables are not the same. Because the two samples of observations in a paired 2×2 table are matched, the statistical methods used to analyze paired 2×2 tables must account for dependent data. We also note that Table 8.1 (unlike Table 4.1) contains $2N$ observations, because each count consists of a pair of observations.

Section 8.2 gives examples of published studies with paired 2×2 table data that illustrate different study designs, and Section 8.3 introduces the notation and the relevant sampling distribution. Two main categories of statistical models (marginal and subject-specific models) are described in Section 8.4. Tests for association are described in Section 8.5. The next four sections present confidence intervals for the difference between probabilities (Section 8.6), the number needed to treat (Section 8.7), the ratio of probabilities (Section 8.8),

TABLE 8.1

The observed counts of a paired 2×2 table

	Event B		
Event A	Success	Failure	Total
Success	n_{11}	n_{12}	n_{1+}
Failure	n_{21}	n_{22}	n_{2+}
Total	n_{+1}	n_{+2}	N

Additional notation:

$\mathbf{n} = \{n_{11}, n_{12}, n_{21}, n_{22}\}$: the observed table
$\mathbf{x} = \{x_{11}, x_{12}, x_{21}, x_{22}\}$: any possible table

and the odds ratio (Section 8.9). Section 8.10 gives recommendations for the practical use of the methods in Sections 8.5–8.9. This chapter is partly based on Fagerland et al. (2013) and Fagerland et al. (2014).

8.2 Examples

8.2.1 Airway Hyper-Responsiveness Status before and after Stem Cell Transplantation

Stem cell transplantation (SCT) is a recognized treatment option for patients with hematological (and various other) malignancies (Bentur et al., 2009). SCT is, however, associated with pulmonary complications. In a prospective longitudinal study, Bentur et al. (2009) measured the airway hyper-responsiveness (AHR) status of 21 children before and after SCT. The purpose of the study was to investigate whether the prevalence of AHR increases following SCT. The results of the study are summarized in Table 8.2. Two children (9.5%) had AHR before SCT and eight (38%) children had AHR after SCT.

TABLE 8.2

Airway hyper-responsiveness (AHR) status before and after stem cell transplantation (SCT) in 21 children (Bentur et al., 2009)

	After SCT		
Before SCT	AHR	No AHR	Total
AHR	1	1	2 (9.5%)
No AHR	7	12	19 (91%)
Total	8 (38%)	13 (62%)	21 (100%)

The two measurements of AHR that constitute a matched pair come from the same patient, and the matching is on the exposure variable (SCT). We can analyze Table 8.2 in several ways. We can formulate a null hypothesis that the probabilities of AHR before and after SCT are equal. Section 8.5 considers tests for association that can be used to test this hypothesis. We can also estimate the strength of the relationship between SCT and AHR status with four different effect measures: the difference between probabilities (Section 8.6), the number needed to treat (Section 8.7), the ratio of probabilities (Section 8.8), and the odds ratio (Section 8.9). We shall return to this example when we illustrate the statistical methods later in this chapter.

8.2.2 Complete Response before and after Consolidation Therapy

The study in the previous example had a small sample size with only 21 pairs of observations. We now consider a similar but larger study with 161 pairs of observations. Cavo et al. (2012) report the results of a randomized clinical trial of two induction therapy treatments before autologous stem cell transplantation for patients with multiple myeloma. A secondary endpoint of the trial was to assess the efficacy and safety of subsequent consolidation therapy. The results for one of the treatment arms are shown in Table 8.3. The outcome was complete response (CR), confirmed from bone marrow biopsy samples, and each patient was measured before and after consolidation therapy. The study design (longitudinal) is the same as in the previous example, and each matched pair consists of two measurements from one patient. An increase in the proportion of patients with CR following consolidation therapy can be observed: sixty-five (40%) patients had CR before consolidation therapy, and 75 (47%) patients had CR after consolidation therapy. Table 8.3 can be analyzed with tests for association and—because the matching is on the exposure variable (consolidation therapy)—with the same effect measures as the previous example.

TABLE 8.3

Complete response (CR) before and after consolidation therapy (Cavo et al., 2012)

Before consolidation	After consolidation		Total
	CR	**No CR**	
CR	59	6	65 (40%)
No CR	16	80	96 (60%)
Total	75 (47%)	86 (53%)	161 (100%)

8.2.3   The Association between Floppy Eyelid Syndrome and Obstructive Sleep Apnea-Hypopnea Syndrome

We now turn to a different study design: the matched case-control study. In a study reported by Ezra et al. (2010), 102 patients with floppy eyelid syndrome (FES, the cases) were 1:1 matched to 102 patients without FES (the controls). The patients were matched according to age, gender, and body mass index. One of the aims of the study was to investigate the association between FES (the disease) and obstructive sleep apnea-hypopnea syndrome (OSAHS, the exposure). Table 8.4 shows the results. Each pair of observations now consists of the OSAHS status of one case and the OSAHS status of one matching control. Thirty-two (31%) of the 102 cases had OSAHS, whereas only nine (8.8%) of the 102 controls had OSAHS.

TABLE 8.4

The observed association between floppy eyelid syndrome (FES, the disease) and obstructive sleep apnea-hypopnea syndrome (OSAHS, the exposure) in a matched case-control study (Ezra et al., 2010)

Cases (FES)	Controls (no FES)		Total
	OSAHS	No OSAHS	
OSAHS	7	25	32 (31%)
No OSAHS	2	68	70 (69%)
Total	9 (8.8%)	93 (91%)	102 (100%)

In this example, matching is on the outcome (disease) variable (FES) and not on the exposure variable, as in the previous examples. We thus have information on the distribution of the exposure given the disease but not the other way around. The odds ratio is an appropriate effect measure, because the odds ratio for the association of disease given exposure is equal to the odds ratio of the association of exposure given disease. It is, however, noteworthy that the ordinary unconditional maximum likelihood is inconsistent as an estimate of the odds ratio, and it is crucial to use the conditional maximum likelihood estimate, see Section 8.9.1.

Table 8.4 can also be analyzed with tests for association, and we shall revisit this example in Section 8.5.6 (tests for association) and in Section 8.9.5 (estimation of the conditional odds ratio).

8.3 Notation and Sampling Distribution

Suppose that we have observed N pairs of dichotomous events (A and B), and let Y_1 denote the outcome of Event A, with $Y_1 = 1$ for a success and $Y_1 = 0$ for a failure. Likewise, let Y_2 denote the outcome of Event B, with $Y_2 = 1$ for a success and $Y_2 = 0$ for a failure. Each n_{ij} in Table 8.1 ($i, j = 1, 2$) corresponds to the number of pairs with outcomes $Y_1 = 2 - i$ and $Y_2 = 2 - j$. The $n_{11} + n_{22}$ pairs with identical outcomes are referred to as *concordant pairs*, whereas the $n_{12} + n_{21}$ pairs with unequal outcomes are referred to as *discordant pairs*. Sometimes, we use "subject" to mean a pair of observations, independent of whether the two observations originate from the same study participant or two matched participants.

Let π_{ij} denote the joint probability that $Y_1 = 2 - i$ and $Y_2 = 2 - j$, for $i, j = 1, 2$, such that we have the probability structure in Table 8.5. The joint sampling distribution for the paired 2×2 table is the multinomial distribution with probabilities $\boldsymbol{\pi} = \{\pi_{11}, \pi_{12}, \pi_{21}, \pi_{22}\}$ and N:

$$f(\mathbf{x} \mid \boldsymbol{\pi}, N) = \frac{N!}{x_{11}! x_{12}! x_{21}! x_{22}!} \pi_{11}^{x_{11}} \pi_{12}^{x_{12}} \pi_{21}^{x_{21}} \pi_{22}^{x_{22}}. \tag{8.1}$$

TABLE 8.5

The joint probabilities of a paired 2×2 table

Event A	Event B		Total
	Success	Failure	
Success	π_{11}	π_{12}	π_{1+}
Failure	π_{21}	π_{22}	π_{2+}
Total	π_{+1}	π_{+2}	1

Additional notation: $\boldsymbol{\pi} = \{\pi_{11}, \pi_{12}, \pi_{21}, \pi_{22}\}$

In problems with paired 2×2 data, we are usually interested in the marginal success probabilities π_{1+} and π_{+1}, that is the success probabilities for Event A and Event B. To study π_{1+} and π_{+1} is equivalent to studying π_{12} and π_{21}. The joint distribution of $\{x_{11}, x_{12}, x_{21}, x_{22}\}$ is given in Equation 8.1. As in Section 1.6, we can use the conditional approach to eliminate the nuisance parameters by conditioning on the sufficient statistics for them. When we condition on n_{11} and the total number of discordant pairs, $n_d = n_{12} + n_{21}$, only x_{12} and $n_d - x_{12}$ remain as variables. The conditional distribution is the binomial probability distribution given by

$$f(x_{12} \mid \mu, n_{11}, n_d) = \binom{n_d}{x_{12}} \mu^{x_{12}} (1 - \mu)^{n_d - x_{12}}, \tag{8.2}$$

where $\mu = \pi_{12}/(\pi_{12}+\pi_{21})$. As we shall see in Section 8.5.3, the distribution in Equation 8.2, under the null hypothesis, will be free of unknown parameters.

The unconditional approach is to consider all possible tables with N pairs of observations. The full likelihood for the unknown parameters is given in Equation 8.1. This likelihood can be factorized into three binomial probabilities, see Lloyd (2008). One of these factors depends solely on the binomial distribution of n_{11}; however, n_{11} is sufficient for π_{11}/π_{22}. It contains no information about the discordant pairs, and we can ignore it without losing information about the association. The distribution of the discordant pairs x_{12} and x_{21} is given by the trinomial probability distribution

$$f(x_{12}, x_{21} \,|\, \pi_{12}, \pi_{21}, N) = \tag{8.3}$$

$$\frac{N!}{x_{12}!x_{21}!(N - x_{12} - x_{21})!} \pi_{12}^{x_{12}} \pi_{21}^{x_{21}} (1 - \pi_{12} - \pi_{21})^{N-x_{12}-x_{21}}.$$

8.4   Statistical Models

8.4.1   Marginal Models

If we assume that the probability of a specific realization of the kth pair, $k = 1, 2, \ldots, N$ is independent of k (the subject), we have a marginal (or population-averaged) model. The probability of success for Event A (π_{1+}) and the probability of success for Event B (π_{+1}) are the marginal probabilities that $Y_1 = 1$ and $Y_2 = 1$, respectively. A marginal probability model for the relationship between the success probabilities and the events can be formulated as the generalized linear model

$$\text{link}\big[\Pr(Y_t = 1 \,|\, x_t)\big] = \alpha + \beta x_t,$$

where $t = 1, 2$ indexes the events, with $x_1 = 1$ for Event A and $x_2 = 0$ for Event B. Interest is on the parameter β, and the choice of link function determines how β is interpreted. We use the identity link to study the difference between probabilities (Section 8.6), the log link for the ratio of probabilities (Section 8.8), and the logit link for the odds ratio (Section 8.9).

8.4.2   Subject-Specific Models

In the previous section, we assumed that the probabilities were independent of the subject. When we have matched pairs data, it is often more realistic to assume that the π_{ij} vary by subject, such that the probabilities are subject specific. Interest is then on the association within the pair, conditional on the subject. We may view the data from N matched pairs as N 2×2 tables, one for each pair (Table 8.6). Collapsing over the subjects results in Table 8.1. A subject-specific model includes a subject-specific parameter (α_k):

$$\text{link}\big[\Pr(Y_t = 1 \mid x_{kt})\big] = \alpha_k + \beta x_{kt}, \qquad (8.4)$$

where $t = 1, 2$ indexes the events and $k = 1, 2, \ldots, N$ indexes the subjects. For subject number k, we have that $x_{k1} = 1$ for Event A and $x_{k2} = 0$ for Event B. The effect of event (β) on the probability of success is now conditional on the subject. Equation 8.4 is a *conditional model*, and β is a measure of the within-subject association, which is generally of greater interest than the marginal association. The practical consequences of assuming either a marginal or a subject-specific model will be explained when we consider tests for association (Section 8.5), confidence intervals for the difference between probabilities (Section 8.6), confidence intervals for the ratio of probabilities (Section 8.8), and confidence intervals for the odds ratio (Section 8.9). The subject-specific model is of special interest for the odds ratio.

TABLE 8.6

Matched pairs data displayed as N 2 × 2 tables, where the first four subjects (matched pairs) represent each of the four possible outcomes

	Event B		
Event A	**Success**	**Failure**	**Subject (pair)**
Success	1	0	1
Failure	0	0	
Success	0	1	2
Failure	0	0	
Success	0	0	3
Failure	1	0	
Success	0	0	4
Failure	0	1	
$\vdots$	$\vdots$		$\vdots$
Success	n_{11k}	n_{12k}	k
Failure	n_{21k}	n_{22k}	
$\vdots$	$\vdots$		$\vdots$
Success	n_{11N}	n_{12N}	N
Failure	n_{21N}	n_{22N}	

8.5   Tests for Association

8.5.1   The Null and Alternative Hypotheses

In studies of paired binomial probabilities, interest is on the marginal success probabilities π_{1+} and π_{+1}. When $\pi_{1+} = \pi_{+1}$, we also have that $\pi_{2+} = \pi_{+2}$. A test for H_0: $\pi_{1+} = \pi_{+1}$ is thus a test for *marginal homogeneity*. If we assume a subject-specific model, interest is on the *conditional independence* between Y_1 and Y_2 in the three-way $2 \times 2 \times N$ table. Testing for conditional independence (controlling for subject) is equivalent to testing for marginal homogeneity, and we shall treat the two situations as one. The following sets of hypotheses are equivalent:

$$H_0 : \pi_{1+} = \pi_{+1} \quad \text{versus} \quad H_A : \pi_{1+} \neq \pi_{+1}$$
$$\Updownarrow$$
$$H_0 : \pi_{2+} = \pi_{+2} \quad \text{versus} \quad H_A : \pi_{2+} \neq \pi_{+2}$$
$$\Updownarrow$$
$$H_0 : \pi_{12} = \pi_{21} \quad \text{versus} \quad H_A : \pi_{12} \neq \pi_{21}$$

8.5.2   The McNemar Asymptotic Test

Under the null hypothesis, the expected number of success-failure pairs is equal to the expected number of failure-success pairs. Conditional on n_{11} and the total number of discordant pairs ($n_d = n_{12} + n_{21}$), n_{12} is binomially distributed with parameters n_d and μ, see Section 8.3.

Under H_0, $\mu = 1/2$, and the standard error estimate of n_{12} is

$$\widehat{\mathrm{SE}}_0(n_{12}) = \sqrt{n_d \mu (1 - \mu)} = \frac{1}{2}\sqrt{n_{12} + n_{21}}.$$

The McNemar asymptotic test is based on the McNemar (1947) test statistic:

$$Z_{\mathrm{McNemar}}(\mathbf{n}) = \frac{n_{12} - \frac{1}{2}(n_{12} + n_{21})}{\widehat{\mathrm{SE}}_0(n_{12})} = \frac{n_{12} - n_{21}}{\sqrt{n_{12} + n_{21}}}, \qquad (8.5)$$

which, under H_0, has an asymptotic standard normal distribution. Because we have estimated the standard error under the null hypothesis, Z_{McNemar} is a score statistic (see Section 1.7). We obtain P-values for the McNemar asymptotic test as

$$P\text{-value} = \Pr\left[Z \geq \left|Z_{\mathrm{McNemar}}(\mathbf{n})\right|\right],$$

where Z is a standard normal variable.

The concordant pairs of observations (n_{11} and n_{22}) do not contribute to the test statistic in Equation 8.5 because the statistic is derived under the condition that the total number of discordant pairs is fixed. This might seem like a disadvantage of the method because, intuitively, the evidence of a true difference between the events should decrease when the number of identical outcomes (success-success and failure-failure) increases. It turns out, however, that the concordant pairs have negligible effect on tests of association, but they may affect measures of effect size, both in terms of estimates and precision (Agresti and Min, 2004).

Edwards (1948) proposed a continuity corrected version of the McNemar asymptotic test. The purpose of the continuity correction was to approximate the McNemar exact conditional test (see Section 8.5.3). The continuity corrected test statistic is

$$Z_{\text{McNemarCC}}(\mathbf{n}) = \frac{|n_{12} - n_{21}| - 1}{\sqrt{n_{12} + n_{21}}},$$

and its approximate distribution is the standard normal distribution.

Both versions of the McNemar asymptotic test (with and without continuity correction) are undefined when $n_{12} = n_{21} = 0$.

8.5.3 The McNemar Exact Conditional Test

Recall from Section 1.9 that an exact test derives P-values by summing the (exact) probabilities of all possible tables ($\mathbf{x}$) that agree less than or equally with the null hypothesis than does the observed table ($\mathbf{n}$):

$$\text{exact } P\text{-value} = \Pr\big[T(\mathbf{x}) \geq T(\mathbf{n}) \,|\, H_0\big].$$

Here, $T()$ denotes an arbitrary test statistic, defined such that large values indicate less agreement with the null hypothesis than do small values. Under H_0, we have an unknown common success probability $\pi = \pi_{1+} = \pi_{+1}$, and this is a nuisance parameter. As explained in Section 8.3, we can eliminate the nuisance parameter by conditioning on n_{11} and the total number of discordant pairs, $n_{\text{d}} = n_{12} + n_{21}$. The McNemar test statistic in Equation 8.5 can then be reduced to

$$T_{\text{McNemar}}(\mathbf{n} \,|\, n_{\text{d}}) = n_{12},$$

and the probability of observing x_{12}, which now completely characterizes the entire 2 × 2 table, is given by the binomial probability distribution in Equation 8.2. Under the null hypothesis, we have that $\mu = \pi_{12}/(\pi_{12} + \pi_{21}) = 1/2$, and we may simplify Equation 8.2 to

$$f(x_{12} \,|\, n_{11}, n_{\text{d}}) = \binom{n_{\text{d}}}{x_{12}} \left(\frac{1}{2}\right)^{n_{\text{d}}}.$$

The one-sided McNemar exact conditional P-value is

$$\text{one-sided } P\text{-value} = \sum_{x_{12}=0}^{\min(n_{12}, n_{21})} f(x_{12} \,|\, n_{11}, n_{\text{d}}), \tag{8.6}$$

which we multiply by two to obtain the two-sided P-value. If $n_{12} = n_{21}$, let the two-sided P-value be 1.0. The McNemar exact conditional test is sometimes called the *exact conditional binomial test*.

The McNemar exact conditional test is the uniformly most powerful unbiased test for testing H_0, see Section 1.6.

8.5.4   The McNemar Mid-P Test

Section 1.10 presented the mid-P approach as a way to reduce the conservatism of exact conditional methods. Here, we use the mid-P approach on the McNemar exact conditional test. To obtain the mid-P value, we subtract half the probability of the observed outcome (n_{12}) from the one-sided exact conditional P-value in Equation 8.6 and double the results:

$$
\begin{aligned}
\text{mid-}P \text{ value} \;&=\; 2 \cdot \left[\text{one-sided } P\text{-value} - \frac{1}{2} f(n_{12} \mid n_{11}, n_{\mathrm{d}})\right] \\
&=\; \text{two-sided } P\text{-value} - f(n_{12} \mid n_{11}, n_{\mathrm{d}}). \qquad (8.7)
\end{aligned}
$$

When $n_{12} = n_{21}$, the McNemar mid-P value is

$$
\text{mid-}P \text{ value} = 1 - \frac{1}{2} f(n_{12} \mid n_{11}, n_{\mathrm{d}}).
$$

8.5.5   The McNemar Exact Unconditional Test

In Section 8.5.3, we eliminated the nuisance parameter by conditioning on n_{11} and n_d to obtain an exact conditional test. The unconditional test, on the other hand, uses information from both types of discordant pairs, x_{12} and x_{21}. The distribution of the discordant pairs is given by the trinomial probability distribution in Equation 8.3, which under the null hypothesis $\pi_{12} = \pi_{21}$ reduces to

$$
f(x_{12}, x_{21} \mid \pi, N) \;=\;
$$

$$
\frac{N!}{x_{12}! x_{21}! (N - x_{12} - x_{21})!} \left(\frac{\pi}{2}\right)^{x_{12}+x_{21}} (1 - \pi)^{N - x_{12} - x_{21}},
$$

where $\pi = \pi_{12} + \pi_{21}$ is the probability of a discordant pair (the nuisance parameter). The exact unconditional approach is to eliminate the nuisance parameter by maximization over the domain of π:

$$
P\text{-value} = \max_{0 \leq \pi \leq 1} \left\{ \sum_{\Omega(\mathbf{x}\mid N)} I\big[T(\mathbf{x}) \geq T(\mathbf{n})\big] \cdot f(x_{12}, x_{21} \mid \pi, N) \right\}, \qquad (8.8)
$$

where $\Omega(\mathbf{x}\mid N)$ denotes the set of all tables with N observations, $I()$ is the indicator function, and $T()$ is a test statistic, defined such that tables with large values of T agree less with the null hypothesis than do tables with small values of T.

The Berger and Boos procedure (see Section 4.4.7) can be used to reduce the nuisance parameter space:

$$P\text{-value} = \max_{\pi \in C_\gamma} \left\{ \sum_{\Omega(\mathbf{x}|N)} I\big[T(\mathbf{x}) \geq T(\mathbf{n})\big] \cdot f(x_{12}, x_{21} \mid \pi, N) \right\} + \gamma,$$

where C_γ is a $100(1 - \gamma)\%$ confidence interval for π, and γ is a small value, for instance $\gamma = 0.0001$.

The first to propose an exact unconditional test for the paired 2×2 table was Suissa and Shuster (1991) who used the McNemar test statistic

$$T_{\text{McNemar}}(\mathbf{n}) = \frac{n_{12} - n_{21}}{\sqrt{n_{12} + n_{21}}}$$

in Equation 8.8. We shall refer to this test as the McNemar exact unconditional test.

A reasonable alternative to the McNemar statistic is the likelihood ratio statistic. Lloyd (2008) compared exact unconditional tests based on the two statistics and found no practical differences between them.

8.5.6 Examples

Airway Hyper-Responsiveness Status before and after Stem Cell Transplantation (Table 8.2)

The null hypothesis of interest is that the probability of AHR before SCT is equal to the probability of AHR after SCT: H_0: $\pi_{1+} = \pi_{+1}$. We test this against the alternative hypothesis: H_A: $\pi_{1+} \neq \pi_{+1}$. The observed proportions of AHR are $\hat{\pi}_{1+} = 2/21 = 0.095$ (before SCT) and $\hat{\pi}_{1+} = 8/21 = 0.38$ (after SCT). The observed value of the McNemar test statistic (Equation 8.5) is

$$Z_{\text{McNemar}}(\mathbf{n}) = \frac{1 - 7}{\sqrt{1 + 7}} = -2.12.$$

To obtain the P-value for the asymptotic McNemar test, we can refer -2.12 to the standard normal distribution or we may take the square of the observed value, $-2.12^2 = 4.50$, and refer that to the chi-squared distribution with one degree of freedom. The resulting P-value will be the same no matter the method. Here, we use the chi-squared distribution, which is consistent with the way we calculated the Pearson chi-squared test for the unpaired 2×2 table in Section 4.4. Thus, the P-value for the asymptotic McNemar test is

$$P\text{-value} = \Pr(\chi_1^2 \geq 4.50) = 0.0339.$$

If we use the Edwards continuity correction, the test statistic is

$Z_{\text{McNemarCC}}(\mathbf{n}) = 1.77$. The P-value then is $P = 0.0771$, which is quite a bit higher than the P-value of the uncorrected test.

The sample size is small in this example, and we may question whether it is appropriate to use an asymptotic test. For the unpaired 2×2 table in Chapter 4, we used Cochran's criterion (see page 100) as a rule of thumb to decide if it was safe to use asymptotic tests. There is no Cochran's criterion for the paired 2×2 table, and there are no other obvious criteria for deciding when the sample size is sufficiently large to allow for asymptotic tests. An evaluation of the tests will be carried out in Section 8.5.7, and we shall gain more insight into the performances and scopes of application of the tests. Here, we proceed with the calculation of the tests for the data in Table 8.2 and leave the recommendations of which test to use in which situation to Section 8.10.

The McNemar exact conditional test reduces the sample space to tables that have the same number of discordant pairs as the observed table, $n_{\text{d}} = 1 + 7 = 8$. Thus, nine tables are possible; however, because the minimum value of n_{12} and n_{21} is 1, only two probabilities are needed to calculate the one-sided P-value according to Equation 8.6. The calculations are shown in Table 8.7. To obtain the two-sided P-value, we double the one-sided P-value and get $P = 0.0703$. This value is similar to the P-value for the asymptotic McNemar test with continuity correction. As this example illustrates, the McNemar exact conditional test is afflicted by discreteness; only two probabilities went into the calculations of the P-value. As with other exact conditional methods, the result is conservative inference.

TABLE 8.7

Quantities involved in the calculation of the one-sided P-value of the McNemar exact conditional test on the data in Table 8.2

x_{12}	$f(x_{12} \mid n_{\text{d}})$	Cumulative probability
0	0.0039	0.0039
1	0.0313	0.0352

We now turn to the McNemar mid-P test. To calculate it (see Equation 8.7), we need the P-value from the McNemar exact conditional test ($P = 0.0703$) and the probability of the observed outcome. The latter is shown in the second column of the last row in Table 8.7. The McNemar mid-P value then is mid-$P = 0.0703 - 0.0313 = 0.0391$.

The McNemar exact unconditional test includes all possible tables with $N = 21$ pairs. There are 253 ways of distributing the 21 pairs to the cell counts x_{12} and x_{21}, and for 112 of these tables, the McNemar test statistic is equal to or greater than that for the observed table. So, for each value of π (the nuisance parameter), a P-value is obtained as the sum of 112 probabilities (Equation 8.8). Figure 8.1 shows the P-value as a function of π. The exact unconditional P-value without the Berger and Boos procedure is taken as the maximum of this function across the entire nuisance parameter space, which results in $P = 0.0353$. To apply the Berger and Boos procedure with $\gamma = 0.0001$, we calculate a 99.99% confidence interval for π with the Clopper-Pearson exact interval: $C_\gamma = (0.070, 0.79)$. The exact unconditional P-value is now the maximum P-value over C_γ, to which we add the value of γ. The C_γ interval is indicated as the shaded area in Figure 8.8. The maximum P-value over C_γ is the same as the maximum P-value over $(0, 1)$. The exact unconditional P-value with Berger and Boos procedure is therefore $P = 0.0353 + 0.0001 = 0.0354$.

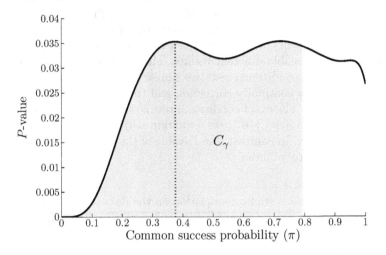

FIGURE 8.1
P-value as a function of the common success probability (π) for the McNemar exact unconditional test on the data in Table 8.2. The dotted vertical line shows the maximum P-value and its corresponding value of π. The shaded area indicates the C_γ interval.

Table 8.8 summarizes the results.

TABLE 8.8
Results of six tests for association on the data in Table 8.2

Test	P-value
McNemar asymptotic	0.0339
McNemar asymptotic with continuity correction	0.0771
McNemar exact conditional	0.0703
McNemar mid-P	0.0391
McNemar exact unconditional	0.0353
McNemar exact unconditional*	0.0354

*Calculated with Berger and Boos procedure ($\gamma = 0.0001$)

Complete Response before and after Consolidation Therapy (Table 8.3)

The previous example showed that quite different results were obtained for the six tests on a table with 21 pairs of observations. The study by Cavo et al. (2012)—with results shown in Table 8.3—is similar, in that each patient in the study is measured before and after treatment; however, the sample size is considerably larger, with 161 pairs of observations (patients). The null hypothesis is that the probability of complete response is the same before (π_{1+}) and after (π_{+1}) consolidation therapy: H_0: $\pi_{1+} = \pi_{+1}$. The two-sided alternative is H_A: $\pi_{1+} \neq \pi_{+1}$. We do not give the details of the computations of the tests but show the results in Table 8.9. Even with this medium-to-large sample size, we obtain noticeable different results. The P-values of the asymptotic, mid-P, and exact unconditional tests are similar, whereas the P-values of the asymptotic test with continuity correction and the exact conditional test are considerably higher. The exact conditional test is still a victim of discreteness: only seven ($= \min(n_{12}, n_{21}) + 1$, see Equation 8.6) probabilities are used to compute the P-value. In contrast, the P-value of the exact unconditional test is a sum of 10 290 probabilities.

TABLE 8.9
Results of six tests for association on the data in Table 8.3

Test	P-value
McNemar asymptotic	0.0330
McNemar asymptotic with continuity correction	0.0550
McNemar exact conditional	0.0525
McNemar mid-P	0.0347
McNemar exact unconditional	0.0342
McNemar exact unconditional*	0.0341

*Calculated with Berger and Boos procedure ($\gamma = 0.0001$)

The Association between Floppy Eyelid Syndrome and Obstructive Sleep Apnea-Hypopnea Syndrome (Table 8.4)

The study by Ezra et al. (2010), summarized in Table 8.4, is a matched case-control study. Each pair of outcomes consists of the exposure status (OSAHS) of one case (a patient with FES) and the exposure status of one matching control (a patient without FES). The null hypothesis is that the proportion of exposed cases is equal to the proportion of exposed controls: H_0: $\pi_{1+} = \pi_{+1}$ versus H_A: $\pi_{1+} \neq \pi_{+1}$. This hypothesis setup is the same as in the other two examples. Testing for association in matched case-control studies is thus identical to testing for association in cohort studies where each participant is measured twice.

The observed proportion of exposed cases is $\hat{\pi}_{1+} = 32/102 = 0.31$, and the observed proportion of exposed controls is $\hat{\pi}_{1+} = 9/102 = 0.088$. All the six tests for association give $P < 0.00011$. A strong association between FES and OSAHS is indicated; however, in a matched case-control study, it is more appropriate to study the within-subject association, for which the subject-specific model in Equation 8.4 can be used. In Section 8.9.5, we estimate the conditional odds ratio and its confidence interval to quantify the within-subject association.

8.5.7 Evaluation of Tests

Evaluation Criteria

We evaluate tests for association by calculating their actual significance levels and power. The actual significance level and power depend on the probabilities π_{11}, π_{12}, π_{21}, and π_{22}, the number of pairs (N), and the nominal significance level α. Because the parameters of interest are the probabilities of success for Event A (π_{1+}) and Event B (π_{+1}), we reparameterize $\{\pi_{11}, \pi_{12}, \pi_{21}, \pi_{22}\}$ into the equivalent parameter set $\{\pi_{1+}, \pi_{+1}, \theta\}$, where $\theta = \pi_{11}\pi_{22}/\pi_{12}\pi_{21}$. For each parameter space point—any realization of $\{\pi_{1+}, \pi_{+1}, \theta, N, \alpha\}$—we use complete enumeration to calculate the actual significance level (ASL) if $\pi_{1+} = \pi_{+1} = \pi$, or power if $\pi_{1+} \neq \pi_{+1}$. That is, we perform the tests on all possible tables with N pairs and add the probability of all tables with P-values less than the nominal significance level:

$$\text{ASL}(\pi, \theta, N, \alpha) =$$

$$\sum_{x_{11}=0}^{N} \sum_{x_{12}=0}^{N-x_{11}} \sum_{x_{21}=0}^{N-x_{11}-x_{12}} I\big[P(\mathbf{x}) \leq \alpha\big] \cdot f(\mathbf{x} \,|\, \pi, \theta, N)$$

and

$$\text{Power}(\pi_{1+}, \pi_{+1}, \theta, N, \alpha) =$$

$$\sum_{x_{11}=0}^{N} \sum_{x_{12}=0}^{N-x_{11}} \sum_{x_{21}=0}^{N-x_{11}-x_{12}} I\big[P(\mathbf{x}) \leq \alpha\big] \cdot f(\mathbf{x} \,|\, \pi_{1+}, \pi_{+1}, \theta, N),$$

where $I()$ is the indicator function, $P(\mathbf{x})$ is the P-value for a test on $\mathbf{x} = \{x_{11}, x_{12}, x_{21}, N - x_{11} - x_{12} - x_{21}\}$, and $f()$ is the multinomial probability distribution (Equation 8.1).

Evaluation of Actual Significance Level

By fixing the number of matched pairs (N) and the parameter θ, and setting $\alpha = 0.05$, we can plot the actual significance level as a function of the common success probability (π). Figure 8.2 shows the three non-exact McNemar tests for $N = 50$ and $\theta = 2.0$. The McNemar asymptotic test violates the nominal significance level for nearly half the range of π; however, the violations are small: the maximum actual significance level is 5.3%. The McNemar mid-P test has actual significance levels close to but below the nominal level. The McNemar asymptotic test with continuity correction, on the other hand, is very conservative: it has significance levels below 3% for all the shown parameter space points. Later in this section, we shall see that the McNemar exact conditional test performs similarly.

The results in Figure 8.2 are typical for a wide range of situations. In an evaluation study covering almost 10 000 scenarios (Fagerland et al., 2013), the McNemar asymptotic test frequently violated the nominal level, but its actual significance level was never above 5.37%. The McNemar mid-P test did not violate the nominal level in any of the almost 10 000 scenarios. This latter result is unusual: mid-P tests (and confidence intervals) usually exhibit occasional but small infringements on the nominal level.

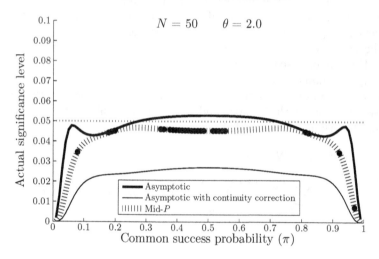

FIGURE 8.2
Actual significance levels of three McNemar tests

We now turn to three exact tests: the McNemar exact conditional test and the McNemar exact unconditional tests with ($\gamma = 0.0001$) and without ($\gamma = 0$) the Berger and Boos procedure. The situation in Figure 8.3, which shows the actual significance levels of the exact tests for $N = 20$ and $\theta = 2.0$, is both typical and atypical. The typical results are that the exact conditional test is overly conservative, here with an actual significance level below 2%, and that the exact unconditional tests perform much better. The atypical result is the large difference between the exact unconditional tests with and without the Berger and Boos procedure. In most of the situations we consider in this book, the Berger and Boos procedure may have a noticeable impact on P-values and confidence intervals for particular data (see, for instance, Table 4.15); however, we rarely see such an obvious improvement as in Figure 8.3. This large improvement in performance for the McNemar exact unconditional test seems to be confined to small sample sizes ($N < 25$). For larger sample sizes, there is no noticeable difference in actual significance levels between the tests with and without the Berger and Boos procedure. Figure 8.4 shows an example with $N = 50$. Note that the exact conditional test is still very conservative. It performs similarly to the McNemar asymptotic test with continuity correction, which can be seen in Figure 8.2.

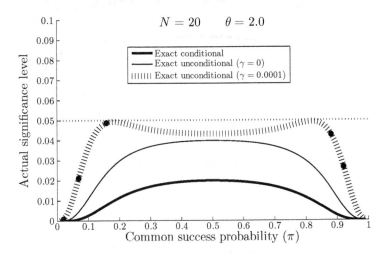

FIGURE 8.3
Actual significance levels of three exact McNemar tests

The McNemar asymptotic test (without continuity correction) also performs well for small sample sizes. Figure 8.5 shows the actual significance levels of the asymptotic test, the mid-P test, and the exact unconditional test with $\gamma = 0.0001$ for a total of only 15 matched pairs. The maximum actual significance level of the asymptotic test in this case is 5.03%. This performance of the standard asymptotic test is excellent and surprising: we are used to the fact that simple asymptotic tests (and confidence intervals) produce substantial violations of the nominal level in small samples. This is certainly the case with the Pearson chi-squared test for the unpaired 2 × 2 table; see, for instance, Figure 4.4 and the discussions in Section 4.4.9.

Evaluation of Power

In the preceding evaluations of actual significance level, we plotted the actual significance level as a function of the common success probability and kept the sample size fixed. To evaluate power, it is more instructive to treat the success probabilities π_{1+} and π_{+1} as fixed, and consider power as a function of the sample size. In Figure 8.6, we have fixed $\pi_{1+} = 0.25$, $\pi_{+1} = 0.5$, and $\theta = 2.0$.

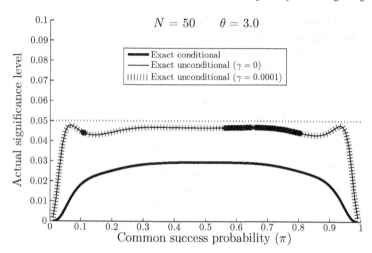

FIGURE 8.4
Actual significance levels of three exact McNemar tests

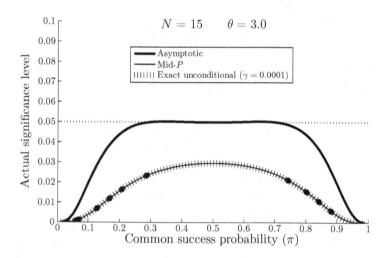

FIGURE 8.5
Actual significance levels of three McNemar tests

The plot shows how the probability (power) to detect a difference in success probabilities of 25% versus 50% depends on the number of matched pairs (N). We have restricted N to values between 40 and 80 so that the power of most of the tests is between 65% and 95%, which should be the most interesting range of power for most practical situations. We observe several interesting differences between the tests. The asymptotic test is clearly the most powerful test, followed by the mid-P test and the exact unconditional test (without the Berger and Boos procedure). The powers of the exact conditional test and the asymptotic test with continuity correction trail that of the other tests considerably. If we were to design a study with an 80% chance of detecting $\pi_{1+} = 0.25$ versus $\pi_{+1} = 0.5$, a plan to use the asymptotic test would require nine or ten fewer matched pairs than a plan to use the exact conditional test.

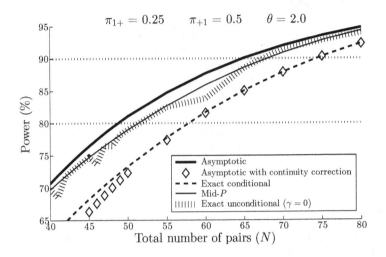

FIGURE 8.6
Power of five McNemar tests

We observed a noticeable improvement in actual significance levels for the McNemar exact unconditional test when the Berger and Boos procedure with $\gamma = 0.0001$ was used. Figure 8.7 shows that the Berger and Boos procedure also evokes a small benefit in power. This benefit is related to the paradoxical result that the exact unconditional test without Berger and Boos procedure sometimes loses power when the number of matched pairs is increased by one. We have yet to experience this unwanted behavior when the Berger and Boos procedure is in use.

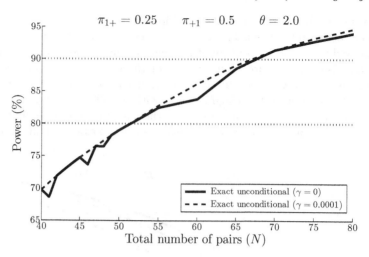

FIGURE 8.7
Power of the McNemar exact unconditional test with ($\gamma = 0.0001$) and without ($\gamma = 0$) the Berger and Boos procedure

8.6 Confidence Intervals for the Difference between Probabilities

8.6.1 Introduction and Estimation

The difference between the marginal probabilities (the success probabilities) is a natural effect measure for paired randomized trials and paired longitudinal studies. The canonical link function for the generalized linear models in Section 8.4 is the linear link. The subject-specific model is given by

$$\Pr(Y_t = 1 \mid x_{kt}) = \alpha_k + \beta x_{kt},$$

for $t = 1, 2$ and $k = 1, 2, \ldots, N$. For the kth subject, $x_{k1} = 1$ for Event A and $x_{k2} = 0$ for Event B. We have that

$$\Pr(Y_1 = 1 \mid x_{k1}) - \Pr(Y_2 = 1 \mid x_{k2}) = \beta.$$

For each subject, β is the difference between the probabilities of Event A and Event B. By summation, we see that β is the difference between the marginal probabilities. If we assume a marginal model instead of a subject-specific model, we drop the subscript k from x and α in the preceding equations and obtain the same result; thus, the marginal association is the same as the within-subject association.

We define the difference between probabilities as

$$\Delta = \pi_{1+} - \pi_{+1}.$$

The maximum likelihood estimate of Δ is given by the sample proportions:

$$\hat{\Delta} = \hat{\pi}_{1+} - \hat{\pi}_{+1} = \frac{n_{1+} - n_{+1}}{N} = \frac{n_{12} - n_{21}}{N}.$$

Sections 8.6.2–8.6.5 present different confidence interval methods for Δ. In Section 8.6.6, we apply the methods to the examples presented in Section 8.2. The methods are evaluated in Section 8.6.7, and Section 8.10 provides recommendations.

8.6.2 Wald Intervals

The (asymptotic) *Wald interval* for Δ is the most used interval for paired binomial probabilities. It is defined as:

$$\hat{\Delta} \pm \frac{z_{\alpha/2}}{N} \sqrt{n_{12} + n_{21} - \frac{(n_{12} - n_{21})^2}{N}}.$$

When $n_{12} = n_{21} = 0$, the zero-width interval $(0,0)$ is produced.

A continuity correction—similar to the one for the asymptotic McNemar test in Section 8.5.2—can be applied to the Wald interval. We call the resulting interval the *Wald interval with continuity correction*:

$$\hat{\Delta} \pm \frac{z_{\alpha/2}}{N} \sqrt{n_{12} + n_{21} - \frac{(|n_{12} - n_{21}| - 1)^2}{N}}.$$

As with the Wald interval, the Wald interval with continuity correction gives the interval $(0,0)$ when $n_{12} = n_{21} = 0$.

Agresti and Min (2005b) investigate the effects of adding pseudo-frequencies to the observed cells in Table 8.1 before calculating the Wald interval. They find that adding $1/2$ to each cell improves performance:

$$\frac{\tilde{n}_{12} - \tilde{n}_{21}}{\tilde{N}} \pm \frac{z_{\alpha/2}}{\tilde{N}} \sqrt{\tilde{n}_{12} + \tilde{n}_{21} - \frac{(\tilde{n}_{12} - \tilde{n}_{21})^2}{\tilde{N}}},$$

where $\tilde{n}_{12} = n_{12} + 1/2$, $\tilde{n}_{21} = n_{12} + 1/2$, and $\tilde{N} = N + 2$. We refer to this interval as the *Wald interval with Agresti-Min adjustment*.

Another simple adjustment to the Wald interval was proposed by Bonett and Price (2012). First, calculate the *Laplace estimates* $\tilde{\pi}_{12} = (n_{12}+1)/(N+2)$ and $\tilde{\pi}_{21} = (n_{21}+1)/(N+2)$. Then, calculate a confidence interval for Δ as

$$\tilde{\pi}_{12} - \tilde{\pi}_{21} \pm z_{\alpha/2} \sqrt{\frac{\tilde{\pi}_{12} + \tilde{\pi}_{21} - (\tilde{\pi}_{12} - \tilde{\pi}_{21})^2}{N+2}}.$$

We refer to this interval as the *Wald interval with Bonett-Price adjustment*.

Neither of the four versions of the Wald interval is guaranteed to respect the $[-1, 1]$ boundary of Δ. When overshoot happens, the usual approach is to truncate the overshooting limit to 1 or -1. The disadvantage of this approach is that the interval can be artificially narrow, thus underestimating the uncertainty in the data.

8.6.3 The Newcombe Square-And-Add Interval (MOVER Wilson Score)

In Chapter 4, we encountered several applications of the square-and-add approach—also called the *method of variance estimates recovery (MOVER)*—for the construction of confidence intervals for different effect measures for the unpaired 2×2 table. Recall that MOVER is a general method that constructs a confidence interval for the difference of two parameters, $\theta_1 - \theta_2$, by combining two separate confidence intervals for θ_1 and θ_2. Let (l_1, u_1) denote the interval for θ_1, and let (l_2, u_2) denote the interval for θ_2. For paired binomial data, the confidence limits for $\theta_1 - \theta_2$ are

$$L^* = \hat{\theta}_1 - \hat{\theta}_2 - \sqrt{\left(\hat{\theta}_1 - l_1\right)^2 + \left(u_2 - \hat{\theta}_2\right)^2 - 2\psi\left(\hat{\theta}_1 - l_1\right)\left(u_2 - \hat{\theta}_2\right)} \qquad (8.9)$$

and

$$U^* = \hat{\theta}_1 - \hat{\theta}_2 + \sqrt{\left(u_1 - \hat{\theta}_1\right)^2 + \left(\hat{\theta}_2 - l_2\right)^2 - 2\psi\left(u_1 - \hat{\theta}_1\right)\left(\hat{\theta}_2 - l_2\right)}, \qquad (8.10)$$

where $\hat{\theta}_1$ and $\hat{\theta}_2$ are estimates of θ_1 and θ_2, and $\psi = \widehat{\text{corr}}(\hat{\theta}_1, \hat{\theta}_2)$ is an estimate of the correlation coefficient between $\hat{\theta}_1$ and $\hat{\theta}_2$. A derivation of and motivation for Equations 8.9 and 8.10 can be found in Newcombe (1998a) and Tang et al. (2010). Tang et al. also provide examples of early applications of the method.

Equations 8.9 and 8.10 give rise to many different confidence intervals. Each choice of confidence interval method for the binomial parameter—to calculate (l_1, u_1) and (l_2, u_2)—and each choice of estimate for ψ leads to a distinct method. Newcombe (1998a) proposed and evaluated several different square-and-add intervals for $\Delta = \pi_{1+} - \pi_{+1}$. Here, we consider the best performing of these intervals, which is based on Wilson score intervals (see Section 2.4.3) for the binomial probability π_{1+}:

$$(l_1, u_1) = \frac{2n_{1+} + z_{\alpha/2}^2 \mp z_{\alpha/2}\sqrt{z_{\alpha/2}^2 + 4n_{1+}\left(1 - \frac{n_{1+}}{N}\right)}}{2\left(N + z_{\alpha/2}^2\right)} \qquad (8.11)$$

and π_{+1}:

$$(l_2, u_2) = \frac{2n_{+1} + z_{\alpha/2}^2 \mp z_{\alpha/2}\sqrt{z_{\alpha/2}^2 + 4n_{+1}\left(1 - \frac{n_{+1}}{N}\right)}}{2\left(N + z_{\alpha/2}^2\right)}. \qquad (8.12)$$

If any of the marginal sums $(n_{1+}, n_{2+}, n_{+1}, n_{+2})$ is zero, set $\psi = 0$. Otherwise, let $A = n_{11}n_{22} - n_{12}n_{21}$ and compute ψ as

$$
\psi = \begin{cases}
(A - N/2)/\sqrt{n_{1+}n_{2+}n_{+1}n_{+2}} & \text{if } A > N/2, \\
0 & \text{if } 0 \le A \le N/2, \\
A/\sqrt{n_{1+}n_{2+}n_{+1}n_{+2}} & \text{if } A < 0.
\end{cases}
$$

The lower (L) and upper (U) limits of the Newcombe square-and-add interval for Δ are given by

$$
L = \hat{\Delta} - \sqrt{\left(\hat{\pi}_{1+} - l_1\right)^2 + \left(u_2 - \hat{\pi}_{+1}\right)^2 - 2\psi\left(\hat{\pi}_{1+} - l_1\right)\left(u_2 - \hat{\pi}_{+1}\right)} \quad (8.13)
$$

and

$$
U = \hat{\Delta} + \sqrt{\left(\hat{\pi}_{+1} - l_2\right)^2 + \left(u_1 - \hat{\pi}_{1+}\right)^2 - 2\psi\left(\hat{\pi}_{+1} - l_2\right)\left(u_1 - \hat{\pi}_{1+}\right)}, \quad (8.14)
$$

where $\hat{\pi}_{1+} = n_{1+}/N$ and $\hat{\pi}_{+1} = n_{+1}/N$.

8.6.4 The Tango Asymptotic Score Interval

Tango (1998) developed an asymptotic score interval for the difference between paired probabilities based on inverting two asymptotic $\alpha/2$ level score tests (the tail method). For a specified value $\Delta_0 \in [-1, 1]$, the score statistic is

$$
T_{\text{score}}(\mathbf{n} \mid \Delta_0) = \frac{n_{12} - n_{21} - N\Delta_0}{\sqrt{N\left[2\tilde{p}_{21} + \Delta_0(1 - \Delta_0)\right]}}, \quad (8.15)
$$

where $\mathbf{n} = \{n_{11}, n_{12}, n_{21}, n_{22}\}$, as usual, denotes the observed table and $\tilde{p}_{21}$ is the maximum likelihood estimate of π_{21}, constrained to $\pi_{1+} - \pi_{+1} = \Delta_0$, given as

$$
\tilde{p}_{21} = \frac{\sqrt{B^2 - 4AC} - B}{2A},
$$

where $A = 2N$, $B = -n_{12} - n_{21} + (2N - n_{12} + n_{21})\Delta_0$, and $C = -n_{21}\Delta_0(1 - \Delta_0)$. The Tango asymptotic score interval (L, U) for Δ is obtained by solving

$$
T_{\text{score}}(\mathbf{n} \mid L) = z_{\alpha/2}
$$

and

$$
T_{\text{score}}(\mathbf{n} \mid U) = -z_{\alpha/2}
$$

iteratively, for instance, with the secant or bisection method. It is possible—although tricky—to derive closed-form expressions for L and U (Newcombe, 2013, Chapter 8). An Excel implementation is given as web-based supplementary material to Newcombe (2013).

8.6.5   The Sidik Exact Unconditional Interval

The score statistic in Equation 8.15 can also be used to derive exact unconditional tests, which in turn may be inverted to obtain exact unconditional
confidence intervals for Δ. There are two main approaches: we can invert two
one-sided $\alpha/2$ level tests or one two-sided α level test. The first approach (the
tail method) ensures that the non-coverage in each tail does not exceed $\alpha/2$.
The limits from such an interval are thereby consistent with the results of
the corresponding exact unconditional one-sided test. An interval based on
inverting one two-sided test, on the other hand, guarantees that the overall
non-coverage does not exceed α but makes no claims about the left and right
tails. It is consistent with the results of the corresponding exact unconditional
two-sided test.

Here, we consider an interval first proposed by Hsueh et al. (2001), which
inverts two one-sided exact score tests. We have two nuisance parameters:
π_{12} and π_{21}. The version described in the following is due to Sidik (2003),
who showed how to simplify the computations of the interval by reducing the
dimensions of the nuisance parameter space from two to one.

Let $\mathbf{x} = \{x_{11}, x_{12}, x_{21}, x_{22}\}$ denote an arbitrary outcome with N pairs. The
probability of observing $\mathbf{x}$ is given by the trinomial probability distribution:

$$f(x_{12}, x_{21} \,|\, \pi_{12}, \Delta_0, N) =$$

$$\frac{N!}{x_{12}! x_{21}! (N - x_{12} - x_{21})!} \pi_{12}^{x_{12}} (\pi_{12} - \Delta_0)^{x_{21}} (1 - 2\pi_{12} + \Delta_0)^{N - x_{12} - x_{21}},$$

where $\Delta_0 = \pi_{12} - \pi_{21}$. This is a reparameterized version of Equation 8.3.

As shown in Sidik (2003), we can eliminate the remaining nuisance param-
eter (π_{12}) by taking the maximum value over the domain $D(\Delta_0) : \{0 \leq \pi_{12} \leq
(1 + \Delta_0)/2\}$. The lower ($L$) and upper ($U$) confidence limits of the Sidik exact
unconditional interval for Δ are the solutions—calculated iteratively—of the
two equations:

$$\max_{\pi_{12} \in D(\Delta_0)} \left\{ \sum_{\Omega(\mathbf{x}|\Delta_0, N)} I\big[T(\mathbf{x}\,|\,L) \geq T(\mathbf{n}\,|\,L)\big] \cdot f(x_{12}, x_{21}\,|\,\pi_{12}, L, N) \right\} = \alpha/2 \tag{8.16}$$

and

$$\max_{\pi_{12} \in D(\Delta_0)} \left\{ \sum_{\Omega(\mathbf{x}|\Delta_0, N)} I\big[T(\mathbf{x}\,|\,U) \leq T(\mathbf{n}\,|\,U)\big] \cdot f(x_{12}, x_{21}\,|\,\pi_{12}, U, N) \right\} = \alpha/2, \tag{8.17}$$

where $T()$ is the score statistic in Equation 8.15.

The Berger and Boos procedure (Section 4.4.7) may be used to reduce
the domain of π_{12} for the maximizations in Equations 8.16 and 8.17. This
is not as straightforward as in the previous cases in this book, because the
Berger and Boos procedure must be applied to the two-dimensional nuisance

parameter space defined by π_{12} and π_{21}. Sidik (2003) has shown how to define a confidence interval, C_γ, for π_{12}, and that taking the maximum value over C_γ is equivalent to taking the maximum value over the two-dimensional confidence set for π_{12} and π_{21}. Let L_{CP} and U_{CP} denote a $100(1-\gamma)\%$ Clopper-Pearson exact interval (see Section 2.4.7) for $2\pi_{12} - \Delta_0$ based on the assumption that $x_{12} + x_{21}$ is binomially distributed with parameters N and $2\pi_{12} - \Delta_0$. Then, the lower limit of C_γ is $(L_{\text{CP}} + \Delta_0)/2$, and the upper limit of C_γ is $(U_{\text{CP}} + \Delta_0)/2$. The Sidik exact unconditional interval for Δ with Berger and Boos procedure is obtained by substituting Equations 8.16 and 8.17 with

$$\max_{\pi_{12} \in C_\gamma} \left\{ \sum_{\Omega(\mathbf{x}|\Delta_0, N)} I\big[T(\mathbf{x}\,|\,L) \geq T(\mathbf{n}\,|\,L)\big] \cdot f(x_{12}, x_{21}\,|\,\pi_{12}, L, N) \right\} + \gamma = \alpha/2$$

and

$$\max_{\pi_{12} \in C_\gamma} \left\{ \sum_{\Omega(\mathbf{x}|\Delta_0, N)} I\big[T(\mathbf{x}\,|\,U) \leq T(\mathbf{n}\,|\,U)\big] \cdot f(x_{12}, x_{21}\,|\,\pi_{12}, U, N) \right\} + \gamma = \alpha/2.$$

We suggest that $\gamma = 0.0001$ is used.

8.6.6 Examples

Airway Hyper-Responsiveness Status before and after Stem Cell Transplantation (Table 8.2)

The two parameters of interest are the probability of AHR before SCT (π_{1+}) and the probability of AHR after SCT (π_{+1}). The estimated probabilities are $\hat{\pi}_{1+} = 1/21 = 0.095$ and $\hat{\pi}_{+1} = 7/21 = 0.38$. The maximum likelihood estimate of the difference between the probabilities is

$$\hat{\Delta} = \frac{n_{12} - n_{21}}{N} = \frac{1 - 7}{21} = -0.286.$$

Table 8.10 gives eight different 95% confidence intervals for Δ. We do not go into the computational details of the methods here but refer the reader to Section 4.5.7, where we show how to calculate some similar confidence intervals for the difference between independent probabilities. The sample size is small in this example—the total number of pairs is only 21—and we would expect the different interval methods to vary considerably, as we observed with the tests for association in Table 8.8. The intervals in Table 8.10 are, however, quite similar, although the Sidik exact unconditional interval is slightly wider than the others. Interestingly, neither of the intervals contains zero, the null value. There is thus no interval for the difference between probabilities that gives results that agree with the McNemar exact conditional test ($P = 0.070$) or the McNemar asymptotic test with continuity correction ($P = 0.077$) for these data. All the intervals in Table 8.10 agree well with the McNemar asymptotic ($P = 0.034$), McNemar mid-P ($P = 0.039$), and McNemar exact unconditional ($P = 0.035$) tests.

TABLE 8.10

95% confidence intervals for the difference between probabilities
($\hat{\Delta} = -0.286$) based on the data in Table 8.2

	Confidence limits		
Interval	Lower	Upper	Width
Wald	-0.520	-0.052	0.468
Wald with continuity correction	-0.529	-0.042	0.487
Wald with Agresti-Min adjustment	-0.493	-0.029	0.465
Wald with Bonett-Price adjustment	-0.508	-0.013	0.495
Newcombe square-and-add	-0.507	-0.026	0.481
Tango asymptotic score	-0.517	-0.026	0.491
Sidik exact unconditional	-0.537	-0.020	0.517
Sidik exact unconditional*	-0.532	-0.020	0.512

*Calculated with Berger and Boos procedure ($\gamma = 0.0001$)

Complete Response before and after Consolidation Therapy (Table 8.3)

The aim of this example is to estimate the difference between the probabilities of complete response before and after consolidation therapy for patients with multiple myeloma. The sample proportion of patients with complete response before consolidation therapy is $\hat{\pi}_{1+} = 65/161 = 0.404$. After consolidation therapy, the sample proportion is $\hat{\pi}_{1+} = 75/161 = 0.466$. We estimate the difference between probabilities as

$$\hat{\Delta} = \frac{n_{12} - n_{21}}{N} = \frac{6 - 16}{161} = -0.0621.$$

Table 8.11 shows eight different 95% confidence intervals for Δ. Only minor differences between the methods can be observed, whereas the tests for association in Table 8.9 gave considerably larger variation in results for these data.

8.6.7 Evaluation of Intervals

Evaluation Criteria

We use three indices of performance to evaluate confidence intervals: coverage probability, width, and location (see Section 1.4). In the following, we show how coverage, width, and location for the difference between paired probabilities can be calculated exactly with complete enumeration.

The coverage probability, width, and location depend on the probabilities π_{11}, π_{12}, π_{21}, and π_{22}, and the number of pairs (N). Because the parameters of interest are the probabilities of success for Event A (π_{1+}) and Event B (π_{+1}), we reparameterize $\{\pi_{11}, \pi_{12}, \pi_{21}, \pi_{22}\}$ into the equivalent parameter set $\{\pi_{1+}, \pi_{+1}, \theta\}$, where $\theta = \pi_{11}\pi_{22}/\pi_{12}\pi_{21}$. The exact coverage probability

TABLE 8.11
95% confidence intervals for the difference between probabilities
($\hat{\Delta} = -0.0621$) based on the data in Table 8.3

| | Confidence limits | | |
Interval	Lower	Upper	Width
Wald	-0.118	-0.006	0.113
Wald with continuity correction	-0.119	-0.006	0.113
Wald with Agresti-Min adjustment	-0.118	-0.005	0.114
Wald with Bonett-Price adjustment	-0.120	-0.003	0.116
Newcombe square-and-add	-0.119	-0.005	0.114
Tango asymptotic score	-0.124	-0.005	0.119
Sidik exact unconditional	-0.126	-0.005	0.121
Sidik exact unconditional*	-0.124	-0.005	0.118

*Calculated with Berger and Boos procedure ($\gamma = 0.0001$)

for the difference between probabilities is

$$\mathrm{CP}(\pi_{1+}, \pi_{+1}, \theta, N, \alpha) =$$

$$\sum_{x_{11}=0}^{N} \sum_{x_{12}=0}^{N-x_{11}} \sum_{x_{21}=0}^{N-x_{11}-x_{12}} I(L \le \Delta \le U) \cdot f(\mathbf{x} \mid \pi_{1+}, \pi_{+1}, \theta, N), \quad (8.18)$$

where $I()$ is the indicator function, $L = L(\mathbf{x}, \alpha)$ and $U = U(\mathbf{x}, \alpha)$ are the lower and upper $100(1 - \alpha)\%$ confidence limits of an interval for the table $\mathbf{x} = \{x_{11}, x_{12}, x_{21}, N - x_{11} - x_{12} - x_{21}\}$, and $f()$ is the multinomial probability distribution (Equation 8.1). The exact expected interval width is defined as

$$\mathrm{Width}(\pi_{1+}, \pi_{+1}, \theta, N, \alpha) =$$

$$\sum_{x_{11}=0}^{N} \sum_{x_{12}=0}^{N-x_{11}} \sum_{x_{21}=0}^{N-x_{11}-x_{12}} (U - L) \cdot f(\mathbf{x} \mid \pi_{1+}, \pi_{+1}, \theta, N).$$

Location is measured by the MNCP/NCP index. The total non-coverage probability (NCP) is computed as $1 - \mathrm{CP}$, where CP is defined in Equation 8.18. The mesial non-coverage probability (MNCP) is defined as

$$\mathrm{MNCP}(\pi_{1+}, \pi_{+1}, \theta, N, \alpha) =$$

$$\sum_{x_{11}=0}^{N} \sum_{x_{12}=0}^{N-x_{11}} \sum_{x_{21}=0}^{N-x_{11}-x_{12}} I(L > \Delta \ge 0 \text{ or } U < \Delta \le 0) \cdot f(\mathbf{x} \mid \pi_{1+}, \pi_{+1}, \theta, N).$$

Evaluation of Coverage Probability

Figure 8.8 illustrates the coverage probability of the four Wald intervals. Here, $\alpha = 0.05$, such that 95% confidence intervals are calculated. We have a small sample size (25 pairs of observations) and the intervals perform quite differently. The standard Wald interval has unacceptable low coverage. An improvement is obtained with the Wald interval with continuity correction, although its coverage is still quite low. The Wald interval with Agresti-Min adjustment is a greater improvement with coverage probabilities mostly between 94% and 95%. The only interval with coverage above 95% in Figure 8.8 is the Wald interval with Bonett-Price adjustment. It is conservative in almost all situations and performs much like an exact interval, although it cannot guarantee coverage at least to the nominal level. As with other conservative intervals, it may produce too wide intervals.

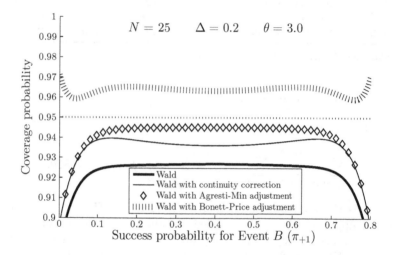

FIGURE 8.8
Coverage probabilities of four Wald intervals for the difference between probabilities

Figure 8.9 shows the same four Wald intervals for a sample size of $N = 100$ matched pairs. All intervals now have coverage closer to the nominal level compared with Figure 8.8, although the low coverage of the standard Wald interval may still cause concern. Note the excellent performance of the Wald interval with Bonett-Price adjustment: it has coverage slightly above the nominal level for all values of π_{+1}.

An example of the coverage properties of the Newcombe square-and-add, Tango asymptotic score, and Sidik exact unconditional intervals is shown in Figure 8.10. We include two versions of the Sidik exact unconditional interval: one with ($\gamma = 0.0001$) and one without ($\gamma = 0$) the Berger and Boos procedure. As noted in Section 8.5.7, when we evaluated the McNemar tests, we rarely see large effects of the Berger and Boos procedure on actual significance levels (tests) or coverage probabilities (confidence intervals). The paired 2×2 table seems to be an exception: the McNemar exact unconditional test with

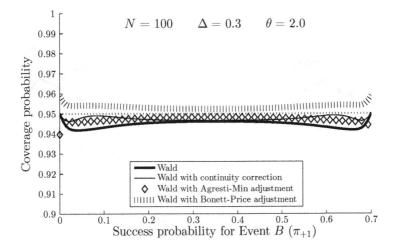

FIGURE 8.9
Coverage probabilities of four Wald intervals for the difference between probabilities

$\gamma = 0.0001$ has actual significance levels closer to the nominal level than the test with $\gamma = 0$, and the Sidik exact unconditional interval with $\gamma = 0.0001$ has coverage probabilities closer to the nominal level than the interval with $\gamma = 0$. For the exact unconditional test, this benefit is confined to $N < 25$, whereas for the exact unconditional interval, the benefit persists for many other combinations of N-, Δ-, and θ-values. The two other intervals in Figure 8.10 have coverage probabilities closer to the nominal level than the exact unconditional intervals. The Tango asymptotic score interval is particularly good in this example with only minor deviations from the nominal 95% coverage for all values of π_{+1}.

Unfortunately, the excellent performance of the Tango asymptotic score interval in Figure 8.10 does not continue for all choices of parameter values. Figure 8.11 shows that the coverage probability of the Tango asymptotic score interval can be quite low, even with as much as 40 matched pairs. In this example, the Newcombe square-and-add interval has the best coverage properties, although the Wald interval with Bonett-Price adjustment and the Sidik exact unconditional interval also perform quite well.

Evaluation of Width

Figure 8.12 shows an example of the expected widths of six confidence intervals for the difference between probabilities. The Wald interval and the Wald interval with continuity correction are not included because of their poor coverage properties. The situation in Figure 8.12 is representative for most other

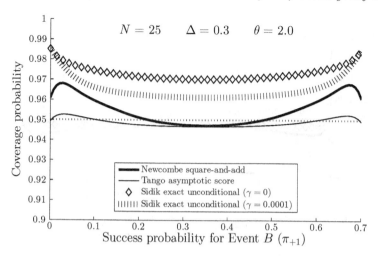

FIGURE 8.10
Coverage probabilities of four confidence intervals for the difference between
probabilities

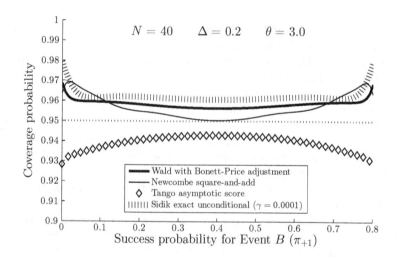

FIGURE 8.11
Coverage probabilities of four confidence intervals for the difference between
probabilities

choices of parameters: the Wald interval with Agresti-Min adjustment, the
Newcombe square-and-add interval, and the Tango asymptotic score interval
are the shortest intervals followed by the Wald interval with Bonett-Price ad-

justment. The exact unconditional intervals are wider than the other intervals, and the interval with Berger and Boos procedure ($\gamma = 0.0001$) is slightly more narrow than the interval without Berger and Boos procedure ($\gamma = 0$). It may seem from Figure 8.12 that the differences in interval widths are considerable; however, the range of the y-axis (the width) is limited to 0.1, which may trick the eye and exaggerate the differences. Tables 8.10 and 8.11 show two examples where the practical differences in interval widths are mostly small.

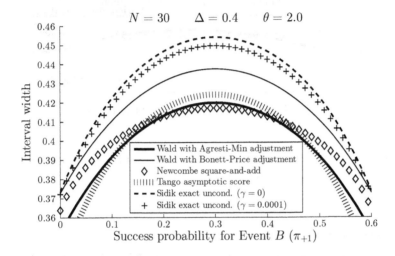

FIGURE 8.12
Expected width of six confidence intervals for the difference between probabilities

Evaluation of Location

Figure 8.13 shows a typical example of the location index MNCP/NCP for four of the confidence intervals for the difference between probabilities. The location of the Sidik exact unconditional interval (with and without Berger and Boos procedure) is usually in the satisfactory range ($0.4 \leq \text{MNCP/NCP} \leq 0.6$), as in Figure 8.13, although it can be slightly mesially located for other parameter values. The Wald interval with Bonett-Price adjustment and the Newcombe square-and-add interval are either slightly too mesially located (Figure 8.13) or with location just inside the satisfactory range. Four intervals are not shown: the Wald interval, the Wald interval with continuity correction, and the Tango asymptotic score interval have mostly satisfactory location, whereas the Wald interval with Agresti-Min adjustment has location similar to the Wald interval with Bonett-Price adjustment. Neither of the eight intervals is too distally located.

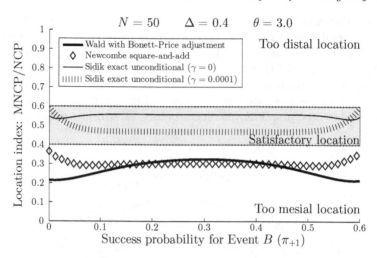

FIGURE 8.13
Location, as measured by the MNCP/NCP index, of four confidence intervals
for the difference between probabilities

8.7 Confidence Intervals for the Number Needed to Treat

8.7.1 Introduction and Estimation

The number needed to treat was introduced in Section 4.6 for the unpaired
2×2 table. We can also calculate a number needed to treat for the paired 2×2
table, and the underlying concepts and ideas are the same. We therefore refer
the reader to Section 4.6 for a general description and background material
for the number needed to treat, including a brief discussion of the practical
utility of the effect measure, and references to opposing views on whether and
how the number needed to treat should be used.

As with the number needed to treat for unpaired data in Section 4.6.2, we
estimate the number needed to treat for paired data as the reciprocal of the
difference between probabilities (Walter, 2001):

$$\mathrm{NNT} = \frac{1}{\hat{\pi}_{1+} - \hat{\pi}_{+1}}.$$

With this notation, we assume that the "treatment" in the number needed to
treat is associated with the binary event with success probability π_{+1} (Event B
in Table 8.1), and that success indicates an unfavorable outcome, such as the
presence of a certain disease. That is, when the difference between probabilities
is positive, NNT is also positive, and Event B represents a beneficial event

compared with Event A. If instead success represents a favorable outcome, a positive value of the difference between probabilities—and thereby a positive value of NNT—indicates a benefit for Event A compared with Event B. One may simply reverse the order of π_{1+} and π_{+1}, and define the difference between probabilities as $\pi_{+1} - \pi_{1+}$, to obtain the desired sign of NNT, if necessary.

As explained in Section 4.6.2, we may also—as suggested by Altman (1998)—denote positive values of NNT by NNTB: the number of patients needed to be treated for one additional patient to benefit; and negative values of NNT can be made positive and denoted by NNTH: the number of patients needed to be treated for one additional patient to be harmed. A proper interpretation of the number needed to treat thus is dependent on a careful definition of the items involved and their direction.

8.7.2 Confidence Intervals

A confidence interval for the number needed to treat is obtained by first calculating a confidence interval for the associated difference between probabilities. One of the methods in Section 8.6 should be used, and we denote the lower and upper confidence limits by L and U, respectively. If the confidence interval for the difference between probabilities does not include zero, the confidence interval for NNTB and NNTH can be obtained by taking the reciprocals of the absolute values of L and U and reversing their order:

$$1/|U| \text{ to } 1/|L|. \tag{8.19}$$

If, on the other hand, the interval (L, U) contains zero, the confidence interval for the number needed to treat should be denoted by (Altman, 1998):

$$\text{NNTH } 1/|L| \text{ to } \infty \text{ to NNTB } 1/U.$$

Figure 4.24 on page 133 illustrates the correspondence between the scales of the difference between probabilities and the number needed to treat, which may help deciphering the above expression.

8.7.3 Examples

Airway Hyper-Responsiveness Status before and after Stem Cell Transplantation (Table 8.2)

In this example, interest is on the probabilities of AHR before (π_{1+}) and after (π_{+1}) SCT. Here, the treatment is SCT, and the outcome denoted by "success" is an unfavorable event. A positive value of the difference between probabilities ($\Delta = \pi_{1+} - \pi_{+1}$) thereby indicates a beneficial effect of SCT, and vice versa for a negative value of Δ. With the data in Table 8.2, we get that

$$\hat{\Delta} = \frac{n_{12} - n_{21}}{N} = \frac{1 - 7}{21} = -0.286.$$

We estimate that SCT increases the probability of AHR by about 29 percentage points. Because AHR is a harmful event, we change the sign of $\hat{\Delta}$ and rephrase the number needed to treat in terms of the number needed to harm:

$$\text{NNTH} = \frac{1}{0.286} = 3.5.$$

We estimate that for every 3.5th patient treated with SCT, one additional patient will experience AHR.

To estimate a 95% confidence interval for the NNTH, we first calculate a 95% confidence interval for the corresponding Δ. This was done in Section 8.6.6 (see Table 8.10), where we observed quite similar results for the eight different interval methods. Here, we use the Wald interval with Bonett-Price adjustment, which is very easy to calculate and performs well in most situations. For the data in Table 8.2 (with Δ defined as $\pi_{+1} - \pi_{1+}$), the Wald interval with Bonett-Price adjustment is (0.013 to 0.508). Because this interval does not include zero, we use Equation 8.19 to find the corresponding 95% confidence interval for NNTH:

$$\left(\frac{1}{0.508} \text{ to } \frac{1}{0.013} \right) = (1.97 \text{ to } 76.9).$$

The frequency with which one additional patient will experience AHR may be as high as every 2nd patient or as low as every 77th patient treated with SCT.

Complete Response before and after Consolidation Therapy (Table 8.3)

When we defined the number needed to treat in Section 8.7.1, we assumed that "success" indicated an unfavorable outcome. Now, the outcome of interest is a beneficial one: complete response. To obtain a proper interpretation of the number needed to treat, we therefore reverse the sign of the estimate of the difference between probabilities. The sample proportions of patients with complete response before and after consolidation therapy are $\hat{\pi}_{1+} = 65/161 = 0.404$ and $\hat{\pi}_{+1} = 75/161 = 0.466$, respectively. The (reversed) estimate of the difference between probabilities then is $\hat{\Delta} = \hat{\pi}_{+1} - \hat{\pi}_{1+} = 0.0621$. The consolidation treatment seems to increase the probability of complete response, and we rephrase the number needed to treat as the number needed to benefit:

$$\text{NNTB} = \frac{1}{0.0621} = 16.1.$$

We estimate that for every 16 patients treated with consolidation therapy, one additional patient will have complete response.

A 95% Wald interval with Bonett-Price adjustment for Δ is (0.003 to 0.120), see Table 8.11. Because this interval does not contain zero, we use Equation 8.19 to find the corresponding 95% confidence interval for NNTB:

$$\left(\frac{1}{0.120} \text{ to } \frac{1}{0.003} \right) = (8.33 \text{ to } 333).$$

We state, with 95% confidence, that as few as 8.3 or as many as 333 patients need to be treated for one additional patient to benefit.

8.8 Confidence Intervals for the Ratio of Probabilities

8.8.1 Introduction and Estimation

In this section, we assume that the link function for the generalized linear models in Section 8.4 is the log link. The subject-specific model is given by

$$\log\left[\Pr(Y_t = 1 \mid x_{kt})\right] = \alpha_k + \beta x_{kt},$$

for $t = 1, 2$ and $k = 1, 2, \ldots, N$. For the kth subject, $x_{k1} = 1$ for Event A and $x_{k2} = 0$ for Event B. We have that

$$\Pr(Y_t = 1 \mid x_{kt}) = \exp(\alpha_k + \beta x_{kt})$$

and

$$\frac{\Pr(Y_1 = 1 \mid x_{k1})}{\Pr(Y_2 = 1 \mid x_{k2})} = \frac{\exp(\alpha_k + \beta \cdot 1)}{\exp(\alpha_k + \beta \cdot 0)} = \exp(\beta).$$

For each subject, the probability of Event A is $\exp(\beta)$ times the probability of Event B. By summation, we see that $\exp(\beta)$ is the ratio of the marginal probabilities. If we assume a marginal model, we drop the subject-specific subscript k from x and α in the preceding equations and obtain the same result. As for the difference between probabilities in Section 8.6.1, the marginal and the within-subject associations are the same.

We define the ratio of paired probabilities as the probability of success for Event A divided by the probability of success for Event B:

$$\phi = \frac{\pi_{1+}}{\pi_{+1}}.$$

The ratio of probabilities may be a more informative effect measure than the difference between probabilities in several situations, particularly when one or both probabilities are close to zero. We use the sample proportions to estimate ϕ:

$$\hat{\phi} = \frac{\hat{\pi}_{1+}}{\hat{\pi}_{+1}} = \frac{n_{1+}/N}{n_{+1}/N} = \frac{n_{11} + n_{12}}{n_{11} + n_{21}}.$$

Sections 8.8.2–8.8.5 present different confidence interval methods for ϕ. In Section 8.8.6, we apply the methods to the examples presented in Section 8.2. The methods are evaluated in Section 8.8.7, and Section 8.10 provides recommendations.

8.8.2 The Wald Interval

The Wald confidence interval for ϕ (Desu and Raghavarao, 2004, pp. 184–185) is obtained by exponentiating the endpoints of

$$\log \hat{\phi} \pm z_{\alpha/2} \sqrt{\frac{n_{12} + n_{21}}{n_{1+} \cdot n_{+1}}}. \tag{8.20}$$

When $n_{12} = n_{21} = 0$, the standard error estimate in (8.20) is zero, and the Wald interval produces the zero-width interval $(1, 1)$. If $n_{1+} = 0$, the estimate is $\hat{\phi} = 0$ and no upper limit is calculated. Similarly, if $n_{+1} = 0$, the estimate is infinite and no lower limit is calculated.

8.8.3 The Tang Asymptotic Score Interval

Under the constraint $\phi = \phi_0$, the score statistic for the ratio of paired binomial probabilities is (Tang et al., 2003, 2012)

$$T_{\text{score}}(\mathbf{n} \mid \phi_0) = \frac{n_{1+} - n_{+1}\phi_0}{\sqrt{N(1 + \phi_0)\tilde{p}_{21} + (n_{11} + n_{12} + n_{21})(\phi_0 - 1)}},$$

where

$$\tilde{p}_{21} = \frac{-B + \sqrt{B^2 - 4AC}}{2A},$$

and

$$
\begin{aligned}
A &= N(1 + \phi_0), \\
B &= (n_{11} + n_{21})\phi_0^2 - (n_{11} + n_{12} + 2n_{21}), \\
C &= n_{21}(1 - \phi_0)(n_{11} + n_{12} + n_{21})/N.
\end{aligned}
$$

The Tang asymptotic score interval (L, U) for ϕ is obtained by solving the equations

$$T_{\text{score}}(\mathbf{n} \mid L) = z_{\alpha/2}$$

and

$$T_{\text{score}}(\mathbf{n} \mid U) = -z_{\alpha/2}.$$

An iterative algorithm is needed to solve the equations.

8.8.4 The Bonett-Price Hybrid Wilson Score Interval

Bonett and Price (2006) proposed a closed-form confidence interval for ϕ based on combining two Wilson score intervals (see Section 2.4.3) for the binomial parameters π_{1+} and π_{+1}. Let $n^* = n_{11} + n_{12} + n_{21}$, and define

$$A = \sqrt{\frac{n_{12} + n_{21} + 2}{(n_{1+} + 1)(n_{+1} + 1)}}, \quad B = \sqrt{\frac{1 - \frac{n_{1+}+1}{n^*+2}}{n_{1+} + 1}}, \quad C = \sqrt{\frac{1 - \frac{n_{+1}+1}{n^*+2}}{n_{+1} + 1}},$$

and

$$z = \frac{A}{B+C} z_{\alpha/2}.$$

The Wilson score interval for π_{1+} is

$$(l_1, u_1) = \frac{2n_{1+} + z^2 \mp z\sqrt{z^2 + 4n_{1+}\left(1 - \frac{n_{1+}}{n^*}\right)}}{2(n^* + z^2)}, \tag{8.21}$$

and for π_{+1}, it is

$$(l_2, u_2) = \frac{2n_{+1} + z^2 \mp z\sqrt{z^2 + 4n_{+1}\left(1 - \frac{n_{+1}}{n^*}\right)}}{2(n^* + z^2)}. \tag{8.22}$$

The Bonett-Price hybrid Wilson score interval for ϕ is

$$\left(\frac{l_1}{u_2} \text{ to } \frac{u_1}{l_2}\right).$$

A continuity corrected version is obtained with the following adjustments to Equations 8.21 and 8.22:

$$l_1 = \frac{2n_{1+} + z^2 - 1 - z\sqrt{z^2 - 2 - \frac{1}{n^*} + 4n_{1+}\left(1 - \frac{n_{1+}+1}{n^*}\right)}}{2(n^* + z^2)},$$

$$u_1 = \frac{2n_{1+} + z^2 + 1 + z\sqrt{z^2 + 2 - \frac{1}{n^*} + 4n_{1+}\left(1 - \frac{n_{1+}-1}{n^*}\right)}}{2(n^* + z^2)},$$

and

$$l_2 = \frac{2n_{+1} + z^2 - 1 - z\sqrt{z^2 - 2 - \frac{1}{n^*} + 4n_{1+}\left(1 - \frac{n_{1+}+1}{n^*}\right)}}{2(n^* + z^2)},$$

$$u_2 = \frac{2n_{+1} + z^2 + 1 + z\sqrt{z^2 + 2 - \frac{1}{n^*} + 4n_{1+}\left(1 - \frac{n_{1+}-1}{n^*}\right)}}{2(n^* + z^2)}.$$

If $n_{1+} = 0$ ($n_{+1} = 0$), set $l_1 = 0$ ($l_2 = 0$). If $n_{1+} = n^*$ ($n_{+1} = n^*$), set $u_1 = 1$ ($u_2 = 1$). The *Bonett-Price hybrid Wilson score interval with continuity correction* provides more conservative confidence intervals than the uncorrected interval.

8.8.5   The MOVER Wilson Score Interval

Section 8.6.3 introduced a MOVER confidence interval for the difference between paired binomial probabilities. That approach can also be used to construct confidence intervals for the ratio of paired probabilities. To find the lower confidence limit L for $\phi = \pi_{1+}/\pi_{+1}$, let $\theta_1 = \pi_{1+}$ and $\theta_2 = L\pi_{+1}$. As shown in Tang et al. (2012), we can use Equation 8.9 and the fact that

$$\Pr\big(\pi_{1+}/\pi_{+1} \le L\big) = \Pr\big(\pi_{1+} - L\pi_{+1} \le 0\big) = \alpha/2 \quad \Rightarrow \quad L^* = 0$$

to obtain

$$L = \frac{A - \hat{\pi}_{1+}\hat{\pi}_{+1} + \sqrt{\big(A - \hat{\pi}_{1+}\hat{\pi}_{+1}\big)^2 - l_1\big(2\hat{\pi}_{1+} - l_1\big)u_2\big(2\hat{\pi}_{+1} - u_2\big)}}{u_2\big(u_2 - 2\hat{\pi}_{+1}\big)}, \quad (8.23)$$

where $A = (\hat{\pi}_{1+} - l_1)(u_2 - \hat{\pi}_{+1})\widehat{\mathrm{corr}}(\hat{\pi}_{1+}, \hat{\pi}_{+1})$. The upper confidence limit U for ϕ is found in a similar manner:

$$U = \frac{B - \hat{\pi}_{1+}\hat{\pi}_{+1} - \sqrt{\big(B - \hat{\pi}_{1+}\hat{\pi}_{+1}\big)^2 - u_1\big(2\hat{\pi}_{1+} - u_1\big)l_2\big(2\hat{\pi}_{+1} - l_2\big)}}{l_2\big(l_2 - 2\hat{\pi}_{+1}\big)},$$

$$(8.24)$$

where $B = (u_1 - \hat{\pi}_{1+})(\hat{\pi}_{+1} - l_2)\widehat{\mathrm{corr}}(\hat{\pi}_{1+}, \hat{\pi}_{+1})$. As in Newcombe (1998a) and Tang et al. (2012), we can use the phi coefficient, which in this case, also is the Pearson correlation coefficient, given by

$$\widehat{\mathrm{corr}}(\hat{\pi}_{1+}, \hat{\pi}_{+1}) = \frac{n_{11}n_{22} - n_{12}n_{21}}{\sqrt{n_{1+}n_{2+}n_{+1}n_{+2}}}.$$

If the denominator is 0, set $\widehat{\mathrm{corr}}(\hat{\pi}_{1+}, \hat{\pi}_{+1}) = 0$.

The confidence limits L and U in Equations 8.23 and 8.24 depend on the particular confidence interval used to obtain (l_1, u_1) and (l_2, u_2). Tang et al. (2012) consider several different interval methods for the binomial parameter and their corresponding MOVER intervals, and recommend using the Wilson score interval (see Section 2.4.3). In that case, the appropriate expressions for (l_1, u_1) and (l_2, u_2) are given in Equations 8.11 and 8.12.

The MOVER Wilson score interval produces the zero-width interval $(1, 1)$ when $n_{11} = n_{22}$ and $n_{12} = n_{21} = 0$.

8.8.6   Examples

Airway Hyper-Responsiveness Status before and after Stem Cell Transplantation (Table 8.2)

The proportion of patients with AHR before SCT is $\hat{\pi}_{1+} = 2/21 = 0.096$, and the proportion of patients with AHR after SCT is $\hat{\pi}_{+1} = 8/21 = 0.38$. We estimate the ratio of probabilities as

$$\hat{\phi} = \frac{n_{11} + n_{12}}{n_{11} + n_{21}} = \frac{1 + 1}{1 + 7} = 0.25.$$

The probability of AHR after SCT is estimated to be four times the probability of AHR before SCT. Table 8.12 shows five different 95% confidence intervals for ϕ. The MOVER Wilson score interval is the shortest interval, followed by the Bonett-Price hybrid Wilson score and Tang asymptotic score intervals. Neither of these three intervals contains the null value ($\phi = 1.0$). The Wald interval is slightly wider and has 1.0 as the upper limit. The Bonett-Price hybrid Wilson score interval with continuity correction is considerably wider than the other intervals. Overall, there is less agreement between the intervals for the ratio of probabilities than was the case for the difference between probabilities (Table 8.10), for which none of the seven intervals contained the null value ($\Delta = 0$).

TABLE 8.12
95% confidence intervals for the ratio of probabilities ($\hat{\phi} = 0.25$) based on the data in Table 8.2

	Confidence limits		
Interval	Lower	Upper	Log width
Wald	0.063	1.000	2.77
Tang asymptotic score	0.065	0.907	2.63
Bonett-Price hybrid Wilson score	0.068	0.923	2.61
Bonett-Price hybrid Wilson score CC*	0.042	1.127	3.29
MOVER Wilson score	0.069	0.869	2.54

*CC = continuity correction

Complete Response before and after Consolidation Therapy (Table 8.3)

An estimate of the ratio of the probabilities for the data in Table 8.3 is

$$\hat{\phi} = \frac{n_{11} + n_{12}}{n_{11} + n_{21}} = \frac{59 + 6}{59 + 16} = 0.867.$$

We estimate the probability of complete response before consolidation therapy to be 13% smaller than the probability of complete response after consolidation therapy. Table 8.13 provides 95% confidence intervals for ϕ. All five intervals give similar confidence limits, although the Bonett-Price hybrid score interval with continuity correction is slightly wider than the other intervals. It is the only interval that includes the null value ($\phi = 1.0$); however, the other four intervals have upper limits that are marginally below the null value ($U \approx 0.99$ for all four intervals).

Statistical Analysis of Contingency Tables

TABLE 8.13

95% confidence intervals for the ratio of probabilities ($\hat{\phi} = 0.867$) based on the data in Table 8.3

Interval	Confidence limits		Log width
	Lower	Upper	
Wald	0.760	0.989	0.263
Tang asymptotic score	0.748	0.988	0.278
Bonett-Price hybrid Wilson score	0.758	0.991	0.268
Bonett-Price hybrid Wilson score CC*	0.747	1.006	0.297
MOVER Wilson score	0.759	0.987	0.262

*CC = continuity correction

8.8.7 Evaluation of Intervals

Evaluation Criteria

As usual, we use three indices of performance to evaluate confidence intervals: coverage probability, width, and location (see Section 1.4 for general descriptions). In the following, we show how coverage, width, and location for the ratio of probabilities can be calculated exactly with complete enumeration. The succeeding expressions are simple modifications of the formulas in Section 8.6.7. The exact coverage probability for the ratio of probabilities is defined as

$$\mathrm{CP}(\pi_{1+}, \pi_{+1}, \theta, N, \alpha) =$$

$$\sum_{x_{11}=0}^{N} \sum_{x_{12}=0}^{N-x_{11}} \sum_{x_{21}=0}^{N-x_{11}-x_{12}} I(L \leq \phi \leq U) \cdot f(\mathbf{x} \mid \pi_{1+}, \pi_{+1}, \theta, N), \quad (8.25)$$

where $\theta = \pi_{11}\pi_{22}/\pi_{12}\pi_{21}$, $I()$ is the indicator function, $L = L(\mathbf{x}, \alpha)$ and $U = U(\mathbf{x}, \alpha)$ are the lower and upper $100(1 - \alpha)\%$ confidence limits of an interval for the table $\mathbf{x} = \{x_{11}, x_{12}, x_{21}, N - x_{11} - x_{12} - x_{21}\}$, and $f()$ is the multinomial probability distribution (Equation 8.1). The exact expected interval width (on the logarithmic scale) is defined as

$$\mathrm{Width}(\pi_{1+}, \pi_{+1}, \theta, N, \alpha) =$$

$$\sum_{x_{11}=0}^{N} \sum_{x_{12}=0}^{N-x_{11}} \sum_{x_{21}=0}^{N-x_{11}-x_{12}} \left[\log(U) - \log(L)\right] \cdot f(\mathbf{x} \mid \pi_{1+}, \pi_{+1}, \theta, N).$$

To calculate the location index MNCP/NCP, we compute NCP $= 1 - $ CP, where CP is defined in Equation 8.25 and

$$\mathrm{MNCP}(\pi_{1+}, \pi_{+1}, \theta, N, \alpha) =$$

$$\sum_{x_{11}=0}^{N} \sum_{x_{12}=0}^{N-x_{11}} \sum_{x_{21}=0}^{N-x_{11}-x_{12}} I(L, U, \phi) \cdot f(\mathbf{x} \mid \pi_{1+}, \pi_{+1}, \theta, N),$$

where $I(L, U, \phi) = I\big[\log(L) > \log(\phi) \geq 0$ or $\log(U) < \log(\phi) \leq 0\big]$.

Evaluation of Coverage Probability

We illustrate the coverage properties of the five confidence intervals for the ratio of probabilities by plotting the coverage probability against the probability of success for Event A (π_{1+}). That means that we hold N, ϕ (and thereby π_{+1}), and θ fixed. Two examples with small sample sizes are shown in Figures 8.14 and 8.15. These figures show that each of the intervals is associated with drawbacks; neither interval always performs well. The standard Wald interval often performs adequately, such as in Figure 8.14; however, it can have coverage probabilities considerably lower than the nominal level, usually when $\pi_{1+} > 0.7$ and the number of matched pairs is fairly low, say, $N \leq 40$. An example can be seen in Figure 8.15. The Tang asymptotic score and Bonett-Price hybrid Wilson score intervals often have similar coverage probabilities. The coverage probabilities of the Bonett-Price interval are often slightly closer to the nominal level than those of the asymptotic score interval. Both intervals may have coverage considerably lower than the nominal level for small values of π_{1+} and moderately large values of ϕ (Figure 8.14). For an interval with closed-form expression, the Bonett-Price interval performs excellently. The MOVER Wilson score interval, also a closed-form method, performs well; however—although not shown here—it has lower and more frequent dips in coverage below the nominal level than do the asymptotic score and Bonett-Price intervals. The Bonett-Price interval with continuity correction is very conservative: it has coverage above 98% for more than half the parameter space points in Figures 8.14 and 8.15.

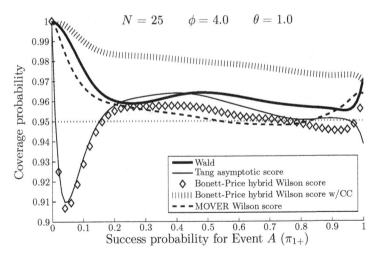

FIGURE 8.14
Coverage probabilities of five confidence intervals for the ratio of probabilities

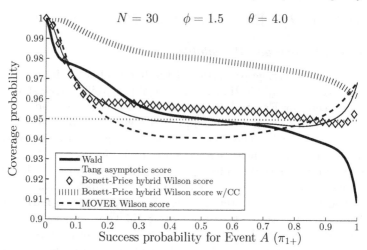

FIGURE 8.15

Coverage probabilities of five confidence intervals for the ratio of probabilities

Figure 8.16 illustrates how the intervals perform when we increase the sample size to 60 matched pairs. The Bonett-Price hybrid score interval with continuity correction is still very conservative; its minimum coverage probability is just below 97%. The Tang asymptotic score, Bonett-Price hybrid Wilson score, and MOVER Wilson score intervals all perform excellently, while the Wald interval is a bit too conservative.

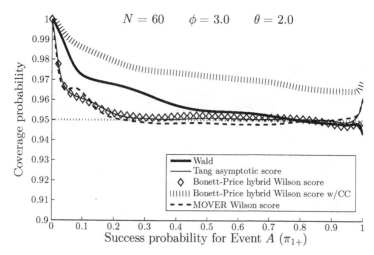

FIGURE 8.16

Coverage probabilities of five confidence intervals for the ratio of probabilities

Evaluation of Width

Figure 8.17 gives an example of the expected width of the intervals. The intervals can be ordered from the widest to the narrowest as follows: Bonett-Price hybrid score with continuity correction, Wald, Tang asymptotic score, Bonett-Price hybrid score, and MOVER Wilson score. In most cases, there is little to distinguish the widths of the latter three intervals.

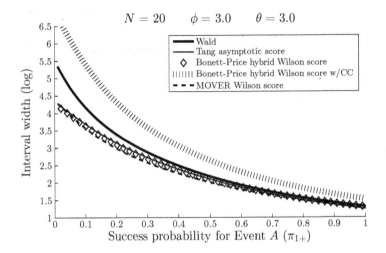

FIGURE 8.17
Expected width of six confidence intervals for the ratio of probabilities

Evaluation of Location

All intervals are too mesially located for most choices of parameter values (Figure 8.18). The MNCP/NCP values of the Tang asymptotic score and MOVER Wilson score intervals sometimes reach the satisfactory range (0.4, 0.6), but only for values of ϕ not too far from 1.0. The Wald interval and the Bonett-Price hybrid score interval with continuity correction have the worst location indices.

8.9 Confidence Intervals for the Odds Ratio

8.9.1 Introduction and Estimation

In matched cohort studies or clinical trials, we have exposure- (or treatment-) matching, in which exposed subjects are paired with unexposed subjects. In

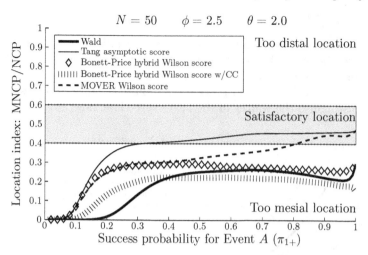

FIGURE 8.18
Location, as measured by the MNCP/NCP index, of four confidence intervals
for the ratio of probabilities

matched case-control studies, on the other hand, matching is of diseased to
non-diseased.

When the link function for the generalized linear models in Section 8.4 is
the logit link, β is a log odds ratio. The subject-specific model is

$$\text{logit}\big[\text{Pr}(Y_t = 1 \mid x_{kt})\big] = \alpha_k + \beta x_{kt},$$

for $t = 1, 2$ and $k = 1, 2, \ldots, N$. For the kth subject, $x_{k1} = 1$ for Event A and
$x_{k2} = 0$ for Event B. The odds for $Y_1 = 1$ is $\exp(\alpha_k + \beta)$ and the odds for
$Y_2 = 1$ is $\exp(\alpha_k)$. Hence, for each subject, the odds of success for Event A is
$\exp(\beta)$ times the odds for Event B. Averaging over the subjects will not give
us the same interpretation of β as in the marginal model. For the marginal
model, β equals the log odds ratio of the marginal probabilities in Table 8.1:

$$\beta_{\text{marginal}} = \log \left[\frac{\pi_{1+}/(1 - \pi_{1+})}{\pi_{+1}/(1 - \pi_{+1})} \right],$$

with maximum likelihood estimate $\hat{\beta}_{\text{marginal}} = \log[(n_{1+}/n_{2+})/(n_{+1}/n_{+2})]$.

For the subject-specific model, ordinary maximum likelihood estimation
of β does not work because the number of α_k is proportional to N (Andersen,
1970; Agresti and Min, 2004). As shown in Andersen (1970), the unconditional
maximum likelihood estimate of β converges to 2β. The solution is to use the
conditional maximum likelihood estimate, obtained by conditioning on the
number of discordant pairs ($n_d = n_{12} + n_{21}$), which is a sufficient statistic for

α_k. The conditional distribution is given in Equation 8.2, where

$$\mu = \frac{\pi_{12}}{\pi_{12} + \pi_{21}} = \frac{\theta_{\text{cond}}}{1 + \theta_{\text{cond}}}.$$

The conditional maximum likelihood estimate of θ_{cond} is

$$\hat{\beta}_{\text{cond}} = \log\left(\frac{n_{12}}{n_{21}}\right).$$

Note that $\hat{\beta}_{\text{cond}}$ equals the Mantel-Haenszel estimate (Breslow and Day, 1980, p. 165) of the common log odds ratio across N strata of matched case-control pairs (as in Table 8.6). We estimate the *conditional odds ratio* by

$$\hat{\theta}_{\text{cond}} = \frac{n_{12}}{n_{21}}.$$

We use the subscript "cond" to separate the paired-data conditional odds ratio ($\hat{\theta}_{\text{cond}}$) from the ordinary unconditional odds ratio ($\hat{\theta}$) used in several places throughout the book. The conditional odds ratio is the within pairs association, which generally is of more interest than the marginal association.

In a case-control study, the estimated within pairs association is the number of pairs with exposed cases and unexposed controls, divided by the number of pairs with unexposed cases and exposed controls.

Sections 8.9.2–8.9.4 present different confidence interval methods for θ_{cond}. In Section 8.9.5, we apply the methods to the examples presented in Section 8.2. The methods are evaluated in Section 8.9.6, and Section 8.10 provides recommendations.

8.9.2 The Wald Interval

An estimate of the asymptotic variance of $\hat{\beta}_{\text{cond}}$ is given by $1/n_{12}+1/n_{21}$. This is the standard Taylor series variance estimate of $\log(n_{12}/n_{21})$ and equals the Mantel-Haenszel variance estimate (Robins et al., 1986). To obtain the Wald interval for θ_{cond}, exponentiate the endpoints of

$$\log\left(\hat{\theta}_{\text{cond}}\right) \pm z_{\alpha/2}\sqrt{\frac{1}{n_{12}} + \frac{1}{n_{21}}}.$$

An equivalent expression is given by

$$\left(\hat{\theta}_{\text{cond}}/\text{EF} \text{ to } \hat{\theta}_{\text{cond}} \cdot \text{EF}\right),$$

where EF is the error factor:

$$\text{EF} = \exp\left(z_{\alpha/2}\sqrt{\frac{1}{n_{12}} + \frac{1}{n_{21}}}\right).$$

The Wald interval is undefined if $n_{12} = 0$ or $n_{21} = 0$.

8.9.3   The Wald Interval with Laplace Adjustment

Greenland (2000) evaluated different bias-corrections for the odds ratio. We consider the simple *Laplace adjustment* obtained by adding 1 to each of n_{12} and n_{21} before calculating the Wald interval. The Wald interval with Laplace adjustment is given by exponentiating the endpoints of

$$\log\left(\tilde{\theta}_{\text{cond}}\right) \pm z_{\alpha/2}\sqrt{\frac{1}{\tilde{n}_{12}} + \frac{1}{\tilde{n}_{21}}},$$

where $\tilde{\theta}_{\text{cond}} = \tilde{n}_{12}/\tilde{n}_{21}$ and $\tilde{n}_{12} = n_{12} + 1$ and $\tilde{n}_{21} = n_{21} + 1$. The adjusted interval copes with $n_{12} = 0$ or $n_{21} = 0$ or both.

8.9.4   Intervals Obtained by Transforming Intervals for $\pi_{12}/(\pi_{12} + \pi_{21})$

In this section, we consider two asymptotic and two exact confidence intervals based on an approach described in Agresti and Min (2005b). Let (L_μ, U_μ) denote a confidence interval for the binomial parameter

$$\mu = \frac{\pi_{12}}{\pi_{12} + \pi_{21}}.$$

Because $\theta_{\text{cond}} = \mu/(1 - \mu)$, a confidence interval for θ_{cond} is obtained as

$$(L \text{ to } U) = \left(\frac{L_\mu}{1 - L_\mu} \text{ to } \frac{U_\mu}{1 - U_\mu}\right). \tag{8.26}$$

In principle, any interval for μ can be used, and the confidence interval for θ_{cond} inherits the properties of the single binomial interval. In the following, let $n_{\text{d}} = n_{12} + n_{21}$.

Transforming the Wilson Score Interval

The Wilson (1927) score confidence interval for μ is given as

$$(L_\mu \text{ to } U_\mu) = \frac{2n_{12} + z_{\alpha/2}^2 \mp z_{\alpha/2}\sqrt{z_{\alpha/2}^2 + 4n_{12}\left(1 - \frac{n_{12}}{n_{\text{d}}}\right)}}{2\left(n_{\text{d}} + z_{\alpha/2}^2\right)}.$$

(See also Section 2.4.3). The transformation in Equation 8.26 gives the corresponding confidence interval for θ_{cond}.

The transformed Wilson score interval is equal to the approximate interval in Breslow and Day (1980, p. 166) without continuity correction. The interval with continuity correction is overly conservative (Agresti and Min, 2005b).

Transforming the Clopper-Pearson Exact Interval

Section 2.4.7 introduced the Clopper-Pearson exact interval for the binomial parameter and showed that the interval could be expressed with a beta distribution. Here, we repeat the expressions in terms of L_μ and U_μ, the lower and upper confidence limits for μ:

$$L_\mu = B(\alpha/2;\ n_{12},\ n_{21} + 1)$$

and

$$U_\mu = B(1 - \alpha/2;\ n_{12} + 1,\ n_{21}).$$

$B(z; a, b)$ is the lower z-quantile of the beta distribution with parameters a and b. The transformation in Equation 8.26 yields an exact confidence interval for θ_{cond}.

Transforming the Clopper-Pearson Mid-P Interval

The Clopper-Pearson mid-P interval was introduced in Section 2.4.8. A mid-P interval (L_μ to U_μ) for μ can be obtained by iteratively solving

$$\sum_{i=n_{12}}^{n_d} \binom{n_d}{i} L_\mu^{\,i}(1 - L_\mu)^{n_d - i} - \frac{1}{2}\binom{n_d}{n_{12}} L_\mu^{\,n_{12}}(1 - L_\mu)^{n_d - n_{12}} = \alpha/2$$

and

$$\sum_{i=0}^{n_{12}} \binom{n_d}{i} U_\mu^{\,i}(1 - U_\mu)^{n_d - i} - \frac{1}{2}\binom{n_d}{n_{12}} U_\mu^{\,n_{12}}(1 - U_\mu)^{n_d - n_{12}} = \alpha/2.$$

No simplification using the beta distribution is available for the Clopper-Pearson mid-P interval. An interval for θ_{cond} is obtained with the transformation in Equation 8.26.

Transforming the Blaker Exact Interval

Section 2.4.7 also included a description of the Blaker exact interval, for which the evaluations in Section 2.4.10 revealed some beneficial properties as compared with the Clopper-Pearson exact interval. For convenience, we repeat the expressions for the Blaker exact interval here, with notation appropriate for the problem of computing a confidence interval for θ_{cond}.

For $k = 0, 1, \ldots, n_d$, define the function

$$\gamma(k, \pi_*) = \min\left[\sum_{i=k}^{n_d} \binom{n_d}{i} \pi_*^{\,i}(1 - \pi_*)^{n_d - i},\ \sum_{i=0}^{k} \binom{n_d}{i} \pi_*^{\,i}(1 - \pi_*)^{n_d - i} \right],$$

where π_* denotes an arbitrary confidence limit for μ. Let $\gamma(n_{12}, \pi_*)$ denote the value of γ for the observed data. The confidence limits of the Blaker exact interval for μ are the two solutions of π_* that satisfy the equation

$$\sum_{k=0}^{n_\mathrm{d}} I\left[\gamma(k, \pi_*) \leq \gamma(n_{12}, \pi_*)\right] \cdot \binom{n_\mathrm{d}}{k} \pi_*^k (1 - \pi_*)^{n_\mathrm{d}-k} = \alpha,$$

where $I()$ is the indicator function. The transformation in Equation 8.26 gives the corresponding exact interval for θ_cond.

8.9.5   Examples

The Association between Floppy Eyelid Syndrome and Obstructive Sleep Apnea-Hypopnea Syndrome (Table 8.4)

Previous sections in this chapter have shown how to estimate the difference between probabilities (Section 8.6.6) and the ratio of probabilities (Section 8.8.6)—with confidence intervals—for the data in Tables 8.2 and 8.3. Here, we do not estimate the odds ratio for these examples but turn our attention to the matched case-control study of the association between floppy eyelid syndrome (FES) and obstructive sleep apnea-hypopnea syndrome (OS-AHS), for which the observed data is shown in Table 8.4.

Because this is a case-control study, we are unable to use the difference between probabilities and the ratio of probabilities as effect measures. In Section 8.5.6, we calculated five tests for association for these data and observed a strong association between FES and OSAHS ($P < 0.00011$ for all tests). Now, we use the odds ratio to estimate the size of this association:

$$\hat{\theta}_\mathrm{cond} = \frac{n_{12}}{n_{21}} = \frac{25}{2} = 12.5.$$

The odds of OSAHS among the patients with FES is estimated to be 12.5 times the odds of OSAHS among the patients without FES. Alternatively—because of the interchangeable nature of the odds ratio—the odds of FES for patients with OSAHS is estimated to be 12.5 times the odds of FES for patients without OSAHS.

Table 8.14 shows 95% confidence intervals for θ_cond. All six intervals have lower limits well above the null value ($\theta_\mathrm{cond} = 1.0$). Still, there is considerable variation in the upper limits and the interval widths. Note the close agreement between the transformed Clopper-Pearson mid-P and the transformed Blaker exact intervals.

8.9.6   Evaluation of Intervals

Evaluation Criteria

Again, we use three indices of performance to evaluate confidence intervals: coverage probability, width, and location (see Section 1.4 for general descriptions). The calculations of coverage probability, width, and location for the

TABLE 8.14
95% confidence intervals for the odds ratio ($\hat{\theta}_{\mathrm{cond}} = 12.5$) based on the data in Table 8.4

	Confidence limits		
Interval	Lower	Upper	Log width
Wald	2.96	52.8	2.88
Wald with Laplace adjustment	2.62	28.6	2.39
Transformed Wilson score	3.28	47.7	2.68
Transformed Clopper-Pearson exact	3.12	109	3.55
Transformed Clopper-Pearson mid-P	3.47	78.3	3.12
Transformed Blaker exact	3.30	74.1	3.11

odds ratio differ from those for the difference between probabilities and the ratio of probabilities. As shown in Section 8.9.1, the odds ratio is defined conditional on the discordant pairs. Under this condition, the sample space is one-dimensional: any one possible table is completely characterized by the count of one cell (x_{12}). Because of the conditional nature of the odds ratio, the coverage probability, width, and location are also defined conditional on the discordant pairs. The exact coverage probability for the odds ratio is defined as

$$\mathrm{CP}(\pi_{12}, n_{\mathrm{d}}, \alpha) = \sum_{x_{12}=0}^{n_{\mathrm{d}}} I(L \leq \theta_{\mathrm{cond}} \leq U) \cdot f(x_{12} \mid n_{\mathrm{d}}, \pi_{12}), \qquad (8.27)$$

where $n_{\mathrm{d}} = n_{12} + n_{21}$, $I()$ is the indicator function, $L = L(x_{12}, \alpha)$ and $U = U(x_{12}, \alpha)$ are the lower and upper $100(1-\alpha)\%$ confidence limits of an interval for any table with x_{12} and $x_{21} = n_{\mathrm{d}} - x_{12}$ discordant pairs, and $f()$ is the binomial probability distribution with parameters n_{d} and π_{12} evaluated at x_{12}:

$$f(x_{12} \mid n_{\mathrm{d}}, \pi_{12}) = \binom{n_{\mathrm{d}}}{x_{12}} \pi_{12}^{x_{12}} (1 - \pi_{12})^{n_{\mathrm{d}} - x_{12}}.$$

The exact expected interval width (on the logarithmic scale) is defined as

$$\mathrm{Width}(\pi_{12}, n_{\mathrm{d}}, \alpha) = \sum_{x_{12}=0}^{n_{\mathrm{d}}} \left[\log(U) - \log(L) \right] \cdot f(x_{12} \mid n_{\mathrm{d}}, \pi_{12}).$$

To calculate the location index MNCP/NCP, we compute $\mathrm{NCP} = 1 - \mathrm{CP}$, where CP is defined in Equation 8.27 and

$$\mathrm{MNCP}(\pi_{12}, n_{\mathrm{d}}, \alpha) = \sum_{x_{12}=0}^{n_{\mathrm{d}}} I(L, U, \theta_{\mathrm{cond}}) \cdot f(x_{12} \mid n_{\mathrm{d}}, \pi_{12}).$$

where

$$I(L, U, \theta_{\mathrm{cond}}) = I\left[\log(L) > \log(\theta_{\mathrm{cond}}) \geq 0 \ \text{or} \ \log(U) < \log(\theta_{\mathrm{cond}}) \leq 0 \right].$$

Evaluation of Coverage Probability

We fix the number of discordant pairs and plot the coverage probability as a function of π_{12}, the probability of success for Event A and failure for Event B. The coverage probability is a highly discontinuous function of π_{12}, as was the case for the confidence intervals for the binomial parameter in Chapter 2. Newcombe and Nurminen (2011) argue that in these cases, it is more informative to consider the moving average of the coverage probabilities, because this smoothed curve provides a realistic assessment of the coverage achieved in practice. An example with 30 discordant pairs is shown in Figure 8.19, where the moving average curves of the two Wald intervals are superimposed on their coverage probabilities. The Wald interval tends to be conservative, although it has coverage probabilities quite close to the nominal level for values of π_{12} close to 0.5. The Wald interval with Laplace adjustment has good average coverage for parts of the parameter space; however, coverage can be very low for small and large values of π_{12}.

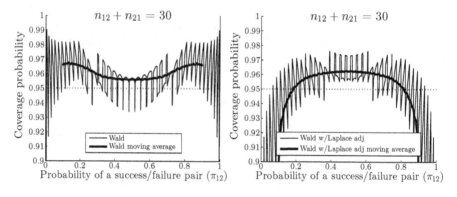

FIGURE 8.19

Coverage probabilities (with moving averages over the range $[\pi_{12} - 0.1, \pi_{12} + 0.1]$) of two Wald intervals for the odds ratio

Figure 8.20 shows an example of the coverage probabilities of the transformed Wilson score and transformed Clopper-Pearson mid-P intervals. Both intervals have excellent average coverage for most values of π_{12}. The Wilson score interval tends to fluctuate slightly more and dip slightly lower below the nominal level than do the mid-P interval. These performance traits persist for larger values of $n_{12} + n_{21}$, at least up to 100.

The coverage probabilities of the two exact intervals, the transformed Clopper-Pearson and Blaker intervals, are illustrated in Figure 8.21. Because these are exact intervals, their coverage probabilities are bounded below by the nominal level. The Blaker interval is considerably less conservative than the Clopper-Pearson interval. This difference is still clearly visible when the number of discordant pairs is increased to 100.

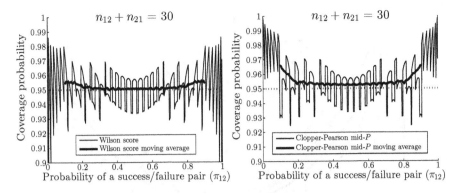

FIGURE 8.20
Coverage probabilities (with moving averages over the range $[\pi_{12} - 0.1, \pi_{12} + 0.1]$) of the transformed Wilson score and transformed Clopper-Pearson mid-P intervals for the odds ratio

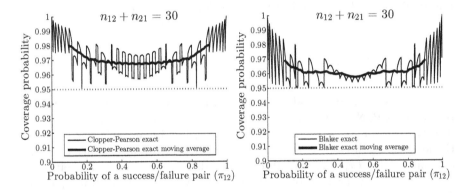

FIGURE 8.21
Coverage probabilities (with moving averages over the range $[\pi_{12} - 0.1, \pi_{12} + 0.1]$) of the transformed Clopper-Pearson exact and transformed Blaker exact intervals for the odds ratio

Evaluation of Width

The widths of the intervals can be ordered from the widest to the narrowest as follows: the transformed Clopper-Pearson exact interval, the transformed Blaker exact and the transformed Clopper-Pearson mid-P intervals (these two intervals have almost identical widths), the Wald interval, the transformed Wilson score interval, and the Wald interval with Laplace adjustment. Figure 8.22 gives an example for a small sample size ($n_{12} + n_{21} = 15$), where the between-interval differences are clearly seen. When the number of discordant pairs is greater than 50, the widths of all six intervals are similar.

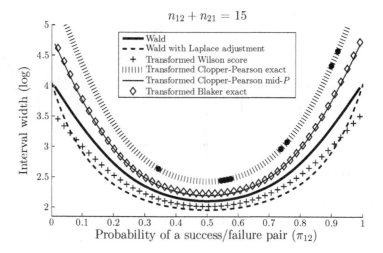

FIGURE 8.22

Expected width of six confidence intervals for the odds ratio

Evaluation of Location

The locations of the transformed Clopper-Pearson exact, transformed Clopper-Pearson mid-P, and transformed Blaker exact intervals, as measured by the MNCP/NCP index, are satisfactory for most combinations of parameters. An example is given in the right panel of Figure 8.23, which shows the location of the mid-P interval as a function of the probability of a success/failure pair (π_{12}) for a fixed total of 40 discordant pairs. In the left panel of Figure 8.23, the location of the transformed Wilson score interval is plotted. This interval has location mostly in the satisfactory range, except for small and large values of π_{12}, for which it is too mesially located. The Wald interval is slightly more mesially located than the transformed Wilson score interval, whereas the Wald interval with Laplace adjustment is too mesially located for all parameters, except when π_{12} is between 0.4 and 0.6.

8.10   Recommendations

8.10.1   Summary

Section 4.9 (recommendations for the unpaired 2×2 table) described the properties of the ideal method and made several general observations about how the different types of methods perform. Most of these observations also apply for the paired 2×2 table; however, we make two adaptations:

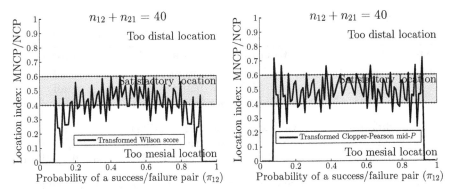

FIGURE 8.23
Location, as measured by the MNCP/NCP index, of the transformed Wilson score and transformed Clopper-Pearson mid-P intervals for the odds ratio

- The McNemar asymptotic test perform well even for small sample sizes

- Asymptotic score intervals do not perform as well for the paired 2 × 2 table as for the unpaired 2 × 2 table

Table 8.15 provides a summary of the recommended tests and confidence intervals, and gives the sample sizes for which the recommended methods are appropriate. The labels small, medium, and large cannot be given precise definitions, they will vary from one analysis to the other, and some subjectivity needs to be applied. As a rule of thumb, small may be taken as less than 50 number of pairs, medium as between 50 and 200 number of pairs, and large as more than 200 number of pairs. Sections 8.10.2–8.10.6 discuss the recommendations in more detail and summarize the merits of the different methods.

8.10.2 Tests for Association

Contrary to expectations, the simple McNemar asymptotic test performs well for all sample sizes. Its actual significance levels are close to the nominal level for almost any situation, except when the total number of matched pairs is very low ($N < 15$), in which case it is still better than the other tests in Section 8.5. The power of the McNemar asymptotic test is equal to or greater than that of the other tests for all situations. It frequently violates the nominal significance level, but not by much. The maximum actual significance level of the McNemar asymptotic test we have observed is 5.37% for a 5% nominal level. If that amount of infringement on the nominal level is acceptable—and we are of the opinion that it is—the asymptotic McNemar test can be considered the best test for association for the paired 2 × 2 table.

TABLE 8.15
Recommended tests and confidence intervals (CIs) for paired 2×2 tables

Analysis	Recommended methods	Sample sizes
Tests for association	McNemar asymptotic*	all
	McNemar mid-P*	all
	McNemar exact unconditional[†]	small/medium
CIs for difference between probabilities	Wald with Bonett-Price adjust.*	all
	Newcombe square-and-add*	small/medium
	Sidik exact unconditional[†]	small/medium
CIs for number needed to treat	The reciprocals of the limits of the recommended intervals for the difference between probabilities	
CIs for ratio of probabilities	Bonett-Price hybrid Wilson score*	all
	Tang asymptotic score	all
	MOVER Wilson score*	medium/large
	Wald*	large
CIs for odds ratio	Transformed Wilson score*	all
	Trans. Clopper-Pearson mid-P	all
	Transformed Blaker exact	small/medium
	Wald*	large

*These methods have closed-form expression
[†]Preferably with the Berger and Boos procedure ($\gamma = 0.0001$)

If an exact test is required, we recommend the McNemar exact unconditional test, preferable with the Berger and Boos procedure ($\gamma = 0.0001$), which is particularly beneficial in small sample sizes ($N < 25$). The commonly used McNemar exact conditional test is very conservative, and we do not recommend its use. Nor yet do we recommend use of the McNemar asymptotic test with continuity correction, which is at least as conservative as the exact conditional test. An easy-to-calculate alternative to the exact unconditional test is the McNemar mid-P test. Although the mid-P test cannot guarantee that the actual significance level does not exceed the nominal level, Fagerland et al. (2013) did not observe any violations of the nominal level in almost 10 000 considered scenarios. The supplementary materials to Fagerland et al. (2013) show how to calculate the mid-P test in eight commonly used software packages.

8.10.3   Confidence Intervals for the Difference between Probabilities

The Wald interval with Bonett-Price adjustment is a very good all-round interval, particularly considering how easy it is to calculate, and we recommend it for all sample sizes. Usually, we would recommend a standard (unadjusted)

Wald interval for large sample sizes; however, in this case, the Wald interval with Bonett-Price adjustment is so easy to calculate that there is no reason to resort to the standard Wald interval, which can have coverage slightly below the nominal level even for quite large sample sizes. The Wald interval with Bonett-Price adjustment can be a little bit conservative for small sample sizes, for which the Newcombe square-and-add interval often provides more narrow intervals. The Newcombe interval requires slightly more elaborate calculations than the Wald intervals; however, it has a closed-form expression and does not require dedicated software resources. The Tango asymptotic score interval usually performs quite well; however, it sometimes has coverage too far below the nominal level to recommend it for general use.

The Sidik exact unconditional interval also performs well. It is based on inverting two one-sided exact tests (the tail method) and sometimes provide overly conservative inference. One remedy, which seems to be far more effective for the paired 2 × 2 table than the unpaired 2 × 2 table, is to use the Berger and Boos procedure with $\gamma = 0.0001$. Another alternative is to use an exact unconditional interval that inverts one two-sided exact test (Tang et al., 2005). These intervals usually provide less conservative inference than the exact interval considered in Section 8.6.5; however, they have not yet found broad usage and are not available in standard software packages. The Sidik exact unconditional interval based on the tail method is available in the software StatXact (Cytel Inc., Cambridge, MA).

8.10.4   Confidence Intervals for the Number Needed to Treat

The calculation of a confidence interval for the number needed to treat is based on the confidence limits for the associated difference between probabilities. Hence, the recommended intervals for the difference between probabilities (Table 8.15) apply for the number needed to treat as well. The Wald interval with Bonett-Price adjustment deserves particular attention: it is very easy to calculate and performs well for most sample sizes and parameter values.

8.10.5   Confidence Intervals for the Ratio of Probabilities

The Bonett-Price hybrid Wilson score and the Tang asymptotic score intervals usually perform well, although low coverage can occur for small values of π_{1+} combined with moderately large values of ϕ. The Bonett-Price interval is particularly useful because it has a closed-form expression and thereby can be calculated without dedicated software resources. The Tang asymptotic score interval, on the other hand, requires iterative calculations. One advantage of the Tang interval is that it belongs to the family of score intervals, a well-known and general approach for constructing tests and confidence intervals for categorical data. According to our evaluations, the Bonett-Price hybrid Wilson score interval has coverage probabilities slightly closer to the nominal level than do the Tang asymptotic score interval. Both intervals can be recommended for general use. We also recommend the MOVER Wilson score interval, which, like the Bonett-Price hybrid Wilson score interval, can be calculated with simple arithmetics. Because of a high probability of low coverage, we do not recommend that the MOVER Wilson score interval is used for very small sample sizes, say, when the total number of matched pairs is less than 25.

No exact interval is available for the ratio of probabilities; however, the Bonett-Price hybrid Wilson score interval with continuity correction is conservative to the extent that it has a very low probability of coverage below the nominal level.

The standard Wald interval needs a large sample size to perform as well as the Bonett-Price hybrid Wilson score interval. When $N = 200$, the two intervals have similar coverage probabilities for most parameter values; however, the Bonett-Price interval has coverage probabilities slightly closer to the nominal level than the Wald interval for $\pi_{1+} < 0.3$. We believe the simplicity of the Wald interval makes up for this small difference in coverage probabilities and recommend the Wald interval when $N \geq 200$.

8.10.6   Confidence Intervals for the Odds Ratio

The transformed Wilson score and transformed Clopper-Pearson mid-P intervals have excellent average coverage probabilities for small as well as large sample sizes. The coverage probabilities of both intervals, however, fluctuate above and below the nominal level, and the smaller the sample size, the greater the fluctuations. Thus, for small sample sizes, the minimum coverage can be low. Nevertheless, we agree with Newcombe and Nurminen (2011), who argue for aligning the mean coverage—and not the minimum coverage—with the nominal $1 - \alpha$, and we recommend both intervals for all sample sizes. One advantage with the transformed Wilson score interval is that it has a closed-form expression, whereas the transformed Clopper-Pearson mid-P interval requires iterative calculations.

If the coverage probability is required to be at least to the nominal level, the transformed Blaker exact interval is superior to the transformed Clopper-Pearson exact interval. The Blaker interval has coverage closer to the nominal level, and it is shorter, than the Clopper-Pearson interval; however, it is also more complex to calculate and not well supported in software packages. The Clopper-Pearson exact interval (for the binomial parameter), on the other hand, is widely available in standard software packages.

When the sample size is large, say, with 100 or more discordant pairs, the Wald interval (without adjustment) performs about as well as the transformed Wilson score and transformed Clopper-Pearson mid-P intervals.

9

The Paired $c \times c$ Table

9.1 Introduction

Chapter 8 considered the paired 2×2 table, which consists of N pairs of dichotomous outcomes. In this chapter, we consider a similar situation, in which the two outcome variables (Variable 1 and Variable 2) have $c > 2$ categories. The categories may be either nominal or ordinal. This distinction is important, because the statistical methods we use for the two situations are not the same. As in Chapter 8, the data come in N pairs and arise from study designs such as matched and cross-over clinical trials, longitudinal studies, and matched case-control studies. The two observations in a matched pair may come from the same subject, such as in cross-over clinical trials, where each subject is measured twice (Variable 1 and Variable 2). In matched case-control studies, however, each matched pair refers to two different subjects: one case (Variable 1) and one matching control (Variable 2). If, however, the paired $c \times c$ table is the result of the measurements of two raters and inter-rater agreement is of interest, the methods in Section 13.2 should be used.

The observed data can be summarized in a $c \times c$ contingency table, such as Table 9.1. The rows form the c categories of Variable 1 and the columns form the c categories of Variable 2. Each cell consists of the number of pairs, n_{ij}, with outcome equal to i for Variable 1 and outcome equal to j for Variable 2. The total number of observations is thus $2N$, because each count in Table 9.1 consists of a pair of observations. It might be tempting to form a $2 \times c$ table with one row for each variable and one column for each outcome category; however, the analysis of such a table would ignore the pairing and is not appropriate.

The rest of this chapter is organized as follows. Section 9.2 introduces two examples of paired $c \times c$ tables, one example with nominal outcome categories and one example with ordinal outcome categories. Section 9.3 contains notation, the sampling distribution, and relevant null and alternative hypotheses for both nominal and ordinal categories. Statistical models and methods for tables with nominal categories are described in Section 9.4, while Section 9.5 considers tables with ordinal categories. The final section of this chapter provides recommendations for practical analysis of paired $c \times c$ tables (Section 9.6).

TABLE 9.1

The observed counts of a paired $c \times c$ table, where $c > 2$

Variable 1	Variable 2				Total
	Category 1	Category 2	...	Category c	
Category 1	n_{11}	n_{12}	...	n_{1c}	n_{1+}
Category 2	n_{21}	n_{22}	...	n_{2c}	n_{2+}
$\vdots$	$\vdots$	$\vdots$	$\ddots$	$\vdots$	$\vdots$
Category c	n_{c1}	n_{c2}	...	n_{cc}	n_{c+}
Total	n_{+1}	n_{+2}	...	n_{+c}	N

Additional notation:

$\mathbf{n} = \{n_{11}, \dots, n_{1c}, n_{21}, \dots, n_{2c}, \dots, n_{c1}, \dots, n_{cc}\}$: the observed table

$\mathbf{x} = \{x_{11}, \dots, x_{1c}, x_{21}, \dots, x_{2c}, \dots, x_{c1}, \dots, x_{cc}\}$: any possible table

9.2 Examples

9.2.1 Pretherapy Susceptibility of Pathogens in Patients with Complicated Urinary Tract Infection

Peterson et al. (2007) report the results of a post hoc analysis of a multicenter, randomized, controlled trial of levofloxacin versus ciprofloxacin for patients with complicated urinary tract infection or acute pyelonephritis. The purpose of the study was to assess the pretherapy susceptibility of pathogens from the patients enrolled in the clinical trial. The authors categorized the susceptibility of 680 isolated pathogens to levofloxacin and ciprofloxacin as susceptible, intermediately resistant, resistant, or not applicable. The last category was used for isolates for which no performance standards existed (per January 2005). The cross-classification of results is shown in Table 9.2. We will treat the categories in Table 9.2 as nominal, although one might consider the three-level category variable obtained by excluding the not applicable category as ordinal. The research question of interest in this example is whether the distribution of observations to the categories for levofloxacin is different from the distribution of observations to the categories for ciprofloxacin.

9.2.2 A Comparison between Serial and Retrospective Measurements of Change in Patients' Health Status

Clinical trials commonly use different methods of measuring patients' change in health status than do physicians and nurses in everyday clinical practice. Clinicians often ask the patients for a retrospective assessment, whereas clinical trials often use serial measurements of the patients at two time points. Fischer et al. (1999) sought to compare retrospective and serial measurements of changes in pain and disability for patients starting a new therapy for chronic

TABLE 9.2

The paired susceptibility of pathogens to levofloxacin and ciprofloxacin (Peterson et al., 2007)

	Ciprofloxacin				
Levofloxacin	Susceptible	Intermediately resistant	Resistant	N/A*	Total
Susceptible	596	18	6	5	625
Intermediately resistant	0	2	0	0	2
Resistant	0	0	42	0	42
N/A*	11	0	0	0	11
Total	607	20	48	5	680

*Not applicable

arthritis. Here, we consider the pain measurements only. The serial measurements of pain used a 0–10 cm visual analog scale. Two assessments were done: one at baseline (before start of treatment) and another after six weeks of treatment. The serial measurement of change was taken as the difference between the baseline and six weeks scores, categorized into five categories, which were interpreted to range from "much worse" to "much better". The retrospective measurement of change in pain was assessed six weeks after treatment. The patients reported their perception of the change in pain on a seven-point Likert scale that ranged from "very much worse" to "very much better". The first two categories (1 and 2) were grouped, as were the last two categories (6 and 7). The results are shown in Table 9.3. The categories are clearly ordinal, and we are interested in assessing whether patients report greater change with one of the measurements as compared with the other.

TABLE 9.3

Serial versus retrospective measurements of change in pain (Fischer et al., 1999)

	Retrospective measurements*					
Serial measurements†	1 or 2	3	4	5	6 or 7	Total
10 to 6 (Much worse)	1	0	1	0	0	2
5 to 2	0	2	8	4	4	18
1 to −1 (No change)	1	1	31	14	11	58
−2 to −5	1	0	15	9	12	37
−6 to −10 (Much better)	0	0	2	1	3	6
Total	3	3	57	28	30	121

*Likert values of change
†Visual analog scale of change

9.3   Notation, Sampling Distribution, and Hypotheses

Let Y_1 and Y_2 denote the outcomes of Variables 1 and 2. We assume a marginal (population-averaged) model, in which the joint probabilities

$$\pi_{ij} = \Pr(Y_1 = i, Y_2 = j)$$

are independent of the subject. We will briefly discuss a subject-specific model for nominal categories in Section 9.4.2 and a subject-specific model for ordinal categories in Section 9.5.2. In a marginal model, we regard the cell counts $\mathbf{x} = \{x_{11}, x_{12}, \ldots, x_{cc}\}$ as multinomially distributed with parameters N and $\boldsymbol{\pi} = \{\pi_{11}, \pi_{12}, \ldots, \pi_{cc}\}$:

$$f(\mathbf{x} \mid \boldsymbol{\pi}, N) = N! \prod_{i=1}^{c} \prod_{j=1}^{c} \frac{\pi_{ij}^{x_{ij}}}{x_{ij}!}, \tag{9.1}$$

where $\sum_{ij} \pi_{ij} = 1$. One hypothesis of interest is *marginal homogeneity*, which is defined as

$$H_0 : \pi_{i+} = \pi_{+i}, \tag{9.2}$$

for $i = 1, 2, \ldots, c$, where $\pi_{i+} = \Pr(Y_1 = i)$ and $\pi_{+i} = \Pr(Y_2 = i)$ are the marginal probabilities of outcome equal to category i for Variables 1 and 2, respectively. The two-sided alternative hypothesis is $H_A : \pi_{i+} \neq \pi_{+i}$ for some $i = 1, 2, \ldots, c$. Another hypothesis of interest is that of *symmetry*, defined as

$$H_0 : \pi_{ij} = \pi_{ji}, \tag{9.3}$$

for all $i \neq j$, with two-sided alternative hypothesis equal to $H_A : \pi_{ij} \neq \pi_{ji}$ for at least one pair of $i \neq j$. For the paired 2×2 table (Chapter 8), symmetry is equivalent to marginal homogeneity; however, for the paired $c \times c$ table with $c > 2$, marginal homogeneity may be present without symmetry, whereas symmetry always implies marginal homogeneity. For many practical problems, we consider symmetry to be too restrictive and consider marginal homogeneity to be of primary interest.

9.4   Nominal Categories

In this section, we assume that the categories of Variables 1 and 2 are nominal (unordered). Examples of nominal categorical variables include blood type, classified as O, A, B, or AB, health insurance plan A, B, or C, and the WHO classification of cardiovascular diseases into coronary heart disease, cerebrovascular disease, peripheral arterial disease, rheumatic heart disease, congenital heart disease, or deep vein thrombosis and pulmonary embolism.

9.4.1 A Marginal Model

A marginal model for paired nominal outcomes is the baseline-category logit model:

$$\log\left[\frac{\Pr(Y_t = j \mid x_t)}{\Pr(Y_t = 1 \mid x_t)}\right] = \alpha_j + \beta_j x_t, \tag{9.4}$$

for $t = 1, 2$ and $j = 2, 3, \ldots, c$, where $x_1 = 1$ denotes Variable 1 and $x_2 = 0$ denotes Variable 2. Marginal homogeneity is given by $\beta_2 = \ldots = \beta_c = 0$.

9.4.2 A Subject-Specific Model

The baseline-category logit model in Equation 9.4 is a marginal (population-averaged) model, in which the probabilities π_{ij} are independent of the subject. If we allow the probabilities to be subject specific by adding a subject-specific intercept, we may consider the following subject-specific baseline-category logit model:

$$\log\left[\frac{\Pr(Y_t = j \mid x_{kt})}{\Pr(Y_t = 1 \mid x_{kt})}\right] = \alpha_{kj} + \beta_j x_{kt},$$

for $t = 1, 2$, $j = 2, 3, \ldots, c$, where $k = 1, 2, \ldots, N$ indexes the subjects. In a subject-specific model, interest is on the association within the pair, conditional on the subject, and we may view the data as N $c \times c$ tables, one for each pair. Collapsing over the subjects results in Table 9.1. Estimating subject-specific models with unconditional maximum likelihood estimation is difficult, because of the large number of parameters (the α_{kj}), see Section 8.9.1 for the paired 2×2 table. One solution is to treat the α_{kj} as *random effects*. Then, the generalized linear models (GLM) are extended to generalized linear mixed models (GLMM) to incorporate random effects. See, for instance, Rabe-Hesketh and Skrondal (2012, Chapter 12) or Agresti (2013, p. 514).

The subject-specific baseline-category model is closely connected to the concept of *quasi-symmetry*, which is a more flexible model assumption than symmetry. We refer the reader to Agresti (2013, Chapter 11) for an in-depth discussion on marginal homogeneity, symmetry, quasi-symmetry, and quasi-independence.

9.4.3 The Bhapkar Test for Marginal Homogeneity

Bhapkar (1966) suggested a Wald test for the null hypothesis of marginal homogeneity (Equation 9.2). Let d_i denote the differences between the marginal sums:

$$d_i = n_{i+} - n_{+i}, \tag{9.5}$$

for $i = 1, 2, \ldots, c$. Define the vector $\mathbf{d}^T = \{d_1, d_2, \ldots, d_{c-1}\}$. Note that we omit category c, because $\sum_i d_i = 0$. We could as well omit one of the other categories; the forthcoming Bhapkar statistic is invariant to the choice of which of the $c - 1$ categories to use. Form the sample covariance matrix $\widehat{\boldsymbol{\Sigma}}$ with

$$n_{i+} + n_{+i} - 2n_{ii} - \frac{(n_{+i} - n_{i+})^2}{N} \tag{9.6}$$

as its diagonal entries and

$$-(n_{ij} + n_{ji}) - \frac{(n_{+i} - n_{i+})(n_{+j} - n_{j+})}{N} \tag{9.7}$$

as its off-diagonal entries. Under marginal homogeneity, the Bhapkar test statistic

$$T_{\text{Bhapkar}}(\mathbf{n}) = \mathbf{d}^T \widehat{\boldsymbol{\Sigma}}^{-1} \mathbf{d}$$

is asymptotically chi-squared distributed with $c - 1$ degrees of freedom. We obtain P-values for the Bhapkar test as

$$P\text{-value} = \Pr\left[\chi^2_{c-1} \geq T_{\text{Bhapkar}}(\mathbf{n})\right].$$

9.4.4   The Stuart Test for Marginal Homogeneity

In the previous section, Bhapkar (1966) used the sample covariance matrix, $\widehat{\boldsymbol{\Sigma}}$, to construct a Wald test for marginal homogeneity. Stuart (1955) tested marginal homogeneity using the null covariance matrix, $\widehat{\boldsymbol{\Sigma}}_0$, which results in a score test. $\widehat{\boldsymbol{\Sigma}}_0$ is obtained by omitting the last term in Equation 9.6 and in Equation 9.7, such that $\widehat{\boldsymbol{\Sigma}}_0$ has diagonal entries equal to $n_{i+} + n_{+i} - 2n_{ii}$ and off-diagonal entries equal to $-(n_{ij} + n_{ji})$. The Stuart test statistic is thus

$$T_{\text{Stuart}}(\mathbf{n}) = \mathbf{d}^T \widehat{\boldsymbol{\Sigma}}_0^{-1} \mathbf{d}, \tag{9.8}$$

which is asymptotically chi-squared distributed with $c - 1$ degrees of freedom.

The Stuart test statistic in Equation 9.8 is equal to the square of the McNemar test statistic in Equation 8.5 when $c = 2$.

For large values of N, the difference between the Bhapkar and Stuart test statistics is negligible, due to the diminishing size of the last terms in Equations 9.6 and 9.7. This may also be illustrated by the result shown in Ireland et al. (1969) that

$$T_{\text{Bhapkar}} = \frac{T_{\text{Stuart}}}{1 - (T_{\text{Stuart}}/N)}.$$

The Stuart test is sometimes called the Stuart-Maxwell test, after Maxwell (1970) published the same test, apparently independent of Stuart, 15 years later. Here, we refer to it as the *Stuart test*.

9.4.5   Closed-Form Expression of the Stuart Test

Both the Bhapkar and Stuart tests require inversion of a covariance matrix and are thus not of closed-form expression. Fleiss and Everitt (1971) derived explicit formulas for the Stuart test statistic when $c = 3$ and $c = 4$. Here, we show the formula for $c = 3$, which is simple and concise. The formula for $c = 4$ is also simple but quite elaborate, and we refer the reader to the original article by Fleiss and Everitt (1971) for the expression. The Fleiss-Everitt test statistic is

$$T_{\text{Fleiss-Everitt}}(\mathbf{n}) = \frac{\bar{n}_{23}d_1^2 + \bar{n}_{13}d_2^2 + \bar{n}_{12}d_3^2}{2(\bar{n}_{12}\bar{n}_{23} + \bar{n}_{12}\bar{n}_{13} + \bar{n}_{13}\bar{n}_{23})},$$

where $\bar{n}_{ij} = (n_{ij} + n_{ji})/2$, and d_1, d_2, and d_3 are defined in Equation 9.5. $T_{\text{Fleiss-Everitt}}$ is asymptotically chi-squared distributed with two degrees of freedom.

9.4.6 Tests and Confidence Intervals for Individual Categories

When the null hypothesis of marginal homogeneity can be rejected, it is usually of interest to identify the categories with unequal marginal probabilities. One simple method is to collapse the paired $c \times c$ table into c paired 2×2 tables, one for each category, and to use the methods in Chapter 8 on each 2×2 table. For Category k, the collapsed 2×2 table is obtained as

$$\begin{bmatrix} n_{kk} & \sum_{i \neq k} n_{ki} \\ \sum_{j \neq k} n_{jk} & \sum_{i \neq k}\sum_{j \neq k} n_{ij} \end{bmatrix}. \tag{9.9}$$

The marginal probabilities for Category k in the $c \times c$ table can be compared with a test for the null hypothesis $H_0 : \pi_{1+} = \pi_{+1}$ in the 2×2 table. One of the McNemar tests in Section 8.5 can be used for this purpose. We can also calculate a confidence interval for $\pi_{1+} - \pi_{+1}$ by the methods in Section 8.6.

The preceding approach does not take into account the increased probability of false positive findings as a result of multiple comparisons. One simple solution for testing equality of each pair of marginal probabilities by the collapsing table approach—while maintaining control over the *familywise error rate* (Benjamini and Hochberg, 1995)—is to increase the degrees of freedom for the chi-squared version of the McNemar asymptotic test (Z_{McNemar}^2) from one to $c - 1$ (Fleiss et al., 2003, p. 382). For simultaneous confidence interval estimation, we present two methods suggested by Fleiss and Everitt (1971).

Scheffé-Type Simultaneous Confidence Intervals for $\pi_{i+} - \pi_{+i}$

Let

$$\Delta_i = \pi_{i+} - \pi_{+i},$$

for $i = 1, 2, \ldots, c$, denote the differences between the marginal probabilities, with unbiased estimate

$$\widehat{\Delta}_i = \frac{n_{i+} - n_{+i}}{N}.$$

An estimate of the squared standard error of $\widehat{\Delta}_i$ is given by (Fleiss and Everitt, 1971):

$$\widehat{SE}^2(\widehat{\Delta}_i) = \frac{1}{N}\left(\frac{n_{i+}}{N} + \frac{n_{+i}}{N} - \frac{2n_{ii}}{N} - \widehat{\Delta}_i^2\right).$$

Let $\chi_{c-1}^2(\alpha)$ denote the upper α percentile of the chi-squared distribution with $c - 1$ degrees of freedom. Then, we have that

$$\frac{\left(\widehat{\Delta}_i - \Delta_i\right)^2}{\widehat{SE}^2(\widehat{\Delta}_i)} \leq \chi_{c-1}^2(\alpha) \tag{9.10}$$

with probability approximately equal to $1 - \alpha$. Equation 9.10 is a quadratic inequality in Δ_i, and we obtain the lower and upper confidence limits as the two solutions of Δ_i to the equation

$$A\Delta_i^2 + B\Delta_i + C = 0, \tag{9.11}$$

where

$$
\begin{aligned}
A &= 1 + \frac{\chi_{c-1}^2(\alpha)}{N} \\
B &= -2\widehat{\Delta}_i \\
C &= \widehat{\Delta}_i^2 - \frac{\chi_{c-1}^2(\alpha)}{N^2}(n_{i+} + n_{+i} - 2n_{ii}).
\end{aligned} \tag{9.12}
$$

The Scheffé-type simultaneous confidence intervals have an overall confidence level of $1 - \alpha$.

Bonferroni-Type Simultaneous Confidence Intervals for $\pi_{i+} - \pi_{+i}$

The Bonferroni confidence intervals for $\pi_{i+} - \pi_{+i}$ are equal to the Scheffé confidence intervals for $\pi_{i+} - \pi_{+i}$, except that we substitute $\chi_1^2(\alpha/c)$ for $\chi_{c-1}^2(\alpha)$ in Equations 9.10–9.12. Because $\chi_1^2(\alpha/c) < \chi_{c-1}^2(\alpha)$ for $c \geq 3$, the Bonferroni intervals are shorter than the Scheffé intervals. This difference increases with increasing c.

A Note on the General Scheffé and Bonferroni Adjustments

The Scheffé and the Bonferroni simultaneous confidence intervals described here are specific realizations of the general Scheffé and Bonferroni adjustments, which are examples of *multiple-comparison adjustments* that preserve the familywise error rate. The aim of such a multiple-comparison adjustment is to make sure that the probability is $1 - \alpha$ that a set of confidence intervals simultaneously contain the true parameter value. Or—for hypothesis testing—that the probability is α that at least one null hypothesis is falsely rejected among several simultaneous hypothesis tests. The Scheffé adjustment is, in general, more conservative than the Bonferroni adjustment because it accounts for all possible *contrasts* (i.e., linear combinations of the parameters) that can be defined, whereas the Bonferroni adjustment accounts for only the pairwise comparisons of the parameters. The general recommendation is therefore to use the Bonferroni adjustment when the comparisons have been specified a priori, and the Scheffé adjustment for unplanned exploratory analyses. We refer to Kirk (2013, Chapter 5) and Dmitrienko and D'Agostino (2013) for more information on multiple-comparison adjustment.

9.4.7 The McNemar-Bowker Test for Symmetry

The Bhapkar, Stuart, and Fleiss-Everitt tests in the preceding sections are tests of the hypothesis of marginal homogeneity (Equation 9.2). We now consider a test for the stronger hypothesis of symmetry (Equation 9.3). The following test, due to Bowker (1948), is a generalization of the McNemar test (Section 8.5.2), and it is usually referred to as the *McNemar-Bowker test*. The test statistic

$$T_{\text{McNemar-Bowker}}(\mathbf{n}) = \sum_{i>j} \frac{(n_{ij} - n_{ji})^2}{n_{ij} + n_{ji}} \tag{9.13}$$

is asymptotically chi-squared distributed with $c(c-1)/2$ degrees of freedom. If, for any $i > j$, the denominator is zero, set that fraction equal to zero.

9.4.8 Examples

Pretherapy Susceptibility of Pathogens in Patients with Complicated Urinary Tract Infection (Table 9.2)

To investigate the research question of whether the distribution of observations to the categories for levofloxacin is different from the distribution of observations to the categories of ciprofloxacin, we test the null hypothesis of marginal homogeneity:

$$H_0 : \pi_{1+} = \pi_{+1}, \quad \pi_{2+} = \pi_{+2}, \quad \pi_{3+} = \pi_{+3}, \quad \text{and} \quad \pi_{4+} = \pi_{+4}.$$

Estimates of the marginal probabilities are

$$\hat{\pi}_{1+} = 0.919 \quad \text{versus} \quad \hat{\pi}_{+1} = 0.893,$$
$$\hat{\pi}_{2+} = 0.003 \quad \text{versus} \quad \hat{\pi}_{+2} = 0.029,$$
$$\hat{\pi}_{3+} = 0.062 \quad \text{versus} \quad \hat{\pi}_{+3} = 0.071,$$
$$\hat{\pi}_{4+} = 0.016 \quad \text{versus} \quad \hat{\pi}_{+4} = 0.007.$$

The vector of the difference between the marginal sums is $\mathbf{d}^T = \{18, -18, -6\}$, and the sample and null covariance matrices are

$$\widehat{\Sigma} = \begin{bmatrix} 39.52 & -17.52 & -5.84 \\ -17.52 & 17.52 & -0.16 \\ -5.84 & -0.16 & 5.95 \end{bmatrix}$$

and

$$\widehat{\Sigma}_0 = \begin{bmatrix} 40 & -18 & -6 \\ -18 & 18 & 0 \\ -6 & 0 & 6 \end{bmatrix},$$

respectively. The Bhapkar test statistic is $T_{\text{Bhapkar}} = 27.30$, and the Stuart test statistic is $T_{\text{Stuart}} = 26.25$, with $P < 0.0001$ for both tests (the 95th percentile of the chi-squared distribution with $c - 1 = 3$ degrees of freedom is 7.81). We conclude that the distributions of observations to the different categories of susceptibility are different for levofloxacin and ciprofloxacin.

We now turn to an analysis of the individual categories to identify for which categories levofloxacin and ciprofloxacin differ, and how large these differences are. We will only consider methods that adjust for multiple analyses. To obtain a test for the equality of each of the individual categories, we collapse the 4×4 table into four 2×2 tables using Equation 9.9:

Category 1   Category 2   Category 3   Category 4

$$\begin{bmatrix} 596 & 29 \\ 11 & 44 \end{bmatrix} \quad \begin{bmatrix} 2 & 0 \\ 18 & 660 \end{bmatrix} \quad \begin{bmatrix} 42 & 0 \\ 6 & 632 \end{bmatrix} \quad \begin{bmatrix} 0 & 11 \\ 5 & 664 \end{bmatrix}$$

The results for the adjusted asymptotic McNemar test with $c - 1 = 3$ degrees of freedom are shown in Table 9.4. Also shown are estimates of the differences between the marginal probabilities with 95% Scheffé and Bonferroni confidence intervals. The main differences between levofloxacin and ciprofloxacin stem from the first two categories: susceptibility and intermediately resistant. The probability of susceptible pathogens seems to be greater with levofloxacin, whereas intermediately resistant pathogens are more common with ciprofloxacin.

TABLE 9.4

Adjusted analyses (for multiple comparisons) of the individual categories of Table 9.2

Category	Estimate	McNemar* P-value	Scheffé 95% CI	Bonferroni 95% CI
$\pi_{1+} - \pi_{+1}$	0.0265	0.044	0.001 to 0.052	0.003 to 0.049
$\pi_{2+} - \pi_{+2}$	-0.0265	0.0004	-0.043 to -0.009	-0.042 to -0.011
$\pi_{3+} - \pi_{+3}$	-0.0088	0.11	-0.019 to 0.001	-0.018 to 0.0002
$\pi_{4+} - \pi_{+4}$	0.0088	0.52	-0.008 to 0.025	-0.006 to 0.023

*The asymptotic McNemar test for $H_0 : \pi_{i+} = \pi_{+i}$ with $c - 1 = 3$ degrees of freedom

We end this example by also testing for symmetry, that is $H_0 : \pi_{ij} = \pi_{ji}$, for all $i \neq j$. Testing for symmetry in this example is admittedly superfluous, because we have already rejected marginal homogeneity, and marginal homogeneity is always present under symmetry, so a rejection of marginal homogeneity is also a rejection of symmetry. For the sake of illustration, we carry out the exercise. The McNemar-Bowker test statistic (Equation 9.13) is

$$T = \frac{(18-0)^2}{18+0} + \frac{(6-0)^2}{68+0} + \frac{(5-11)^2}{5+11} = 18 + 6 + 2.25 = 26.25.$$

With $c(c-1)/2 = 6$ degrees of freedom, we obtain $P = 0.0002$. We observe that the greatest contribution to the McNemar-Bowker test statistic comes from the cells $n_{12} = 18$ and $n_{21} = 0$, which strengthens the previous conclusion that intermediately resistant pathogens are more common with ciprofloxacin than with levofloxacin.

9.4.9 Evaluation of Tests

Evaluation Criteria

We evaluate tests for marginal homogeneity and symmetry by calculating their actual significance levels (type I error rate) and power. The actual significance level and power depend on the probabilities $\boldsymbol{\pi} = \{\pi_{11}, \pi_{12}, \ldots, \pi_{cc}\}$, for which $\sum_{ij} \pi_{ij} = 1$, the number of pairs (N), and the nominal significance level α. To calculate actual significance levels, we assume that the null hypothesis (H_0) is true, that is $H_0 : \pi_{i+} = \pi_{+i}$ for all $i = 1, 2, \ldots, c$ (tests for marginal homogeneity), and $H_0 : \pi_{ij} = \pi_{ji}$ for all $i \neq j$ (tests for symmetry). To calculate power, we assume that the alternative hypothesis (H_A) is true, that is $H_A : \pi_{i+} \neq \pi_{+i}$ for at least one $i \in \{1, 2, \ldots, c\}$ (tests for marginal homogeneity), and $\pi_{ij} \neq \pi_{ji}$ for at least one pair of $i \neq j$ (tests for symmetry). The calculations are carried out with complete enumeration: for a given realization of $\{\boldsymbol{\pi}, N, \alpha\}$, we perform the tests on all possible tables with N pairs and add the probability of all tables with P-values less than the nominal significance level:

$$\text{ASL}(\boldsymbol{\pi}, N, \alpha \,|\, H_0) = \sum_{x_{11}=0}^{N} \sum_{x_{12}=0}^{N-x_{11}} \cdots \sum_{x_{c,c-1}=0}^{N_{c-2}} I\big[P(\mathbf{x}) \leq \alpha\big] \cdot f(\mathbf{x} \,|\, \boldsymbol{\pi}, N)$$

and

$$\text{Power}(\boldsymbol{\pi}, N, \alpha \,|\, H_A) = \sum_{x_{11}=0}^{N} \sum_{x_{12}=0}^{N-x_{11}} \cdots \sum_{x_{c,c-1}=0}^{N_{c-2}} I\big[P(\mathbf{x}) \leq \alpha\big] \cdot f(\mathbf{x} \,|\, \boldsymbol{\pi}, N),$$

where $N_{c-2} = N - x_{11} - x_{12} - \ldots - x_{c,c-2}$, $I()$ is the indicator function, $P(\mathbf{x})$ is the P-value for a test on $\mathbf{x} = \{x_{11}, x_{12}, \ldots, x_{c,c-1}, N_{c-2} - x_{c,c-1}\}$, and $f()$ is the multinomial probability distribution in Equation 9.1.

Evaluation of Actual Significance Level

To avoid the tedious details involved in handling a large number of parameters—and the increased computational effort required as c increases—we only consider the paired 3×3 table in this evaluation, for which the cell and marginal probabilities are shown in Table 9.5.

TABLE 9.5

The cell and marginal probabilities of the paired 3×3 table

Variable 1	Variable 2			Total
	Category 1	Category 2	Category 3	
Category 1	π_{11}	π_{12}	π_{13}	π_{1+}
Category 2	π_{21}	π_{22}	π_{23}	π_{2+}
Category 3	π_{31}	π_{32}	π_{33}	π_{3+}
Total	π_{+1}	π_{+2}	π_{+3}	1

Consider the probability values in Table 9.6, which we refer to as Evaluation Scenario #1. This scenario complies with the hypothesis of marginal homogeneity but not with the hypothesis of symmetry. We evaluate the actual significance levels of the tests for marginal homogeneity (and postpone the evaluation of the McNemar-Bowker test for symmetry to later in this section) by setting $\alpha = 0.05$ and varying the number of pairs from $N = 5$ to $N = 30$. We augment the exact evaluations with simulations for $N > 30$. The results are shown in Figure 9.1. The Bhapkar test violates the nominal significance level by a large margin, even for $N > 50$, whereas the actual significance level of the Stuart test approaches—without exceeding—the nominal significance level from below quite rapidly with increasing N. For $N \geq 18$, the actual significance level of the Stuart test is between 4% and 5%, which we regard as a highly satisfactory performance for such small sample sizes in a table with nine cells.

TABLE 9.6

Probability values for Evaluation Scenario #1, which satisfies marginal homogeneity but not symmetry

Variable 1	Variable 2			Total
	Category 1	Category 2	Category 3	
Category 1	0.30	0.15	0.05	0.50
Category 2	0.10	0.10	0.10	0.30
Category 3	0.10	0.05	0.05	0.20
Total	0.50	0.30	0.20	1

The distribution of cell probabilities in Evaluation Scenario #1 has center of mass in the upper left corner of the table. In Evaluation Scenario #2 (Table 9.7), the cell probabilities are quite evenly distributed, as are the marginal probabilities. Evaluation Scenario #2 complies with both the hypothesis of marginal homogeneity and the hypothesis of symmetry, so we may evaluate the actual significance levels of the McNemar-Bowker test for symmetry together with the Bhapkar and Stuart tests for marginal homogeneity (Figure 9.2, where the computations are exact for $N \leq 30$ and simulated for

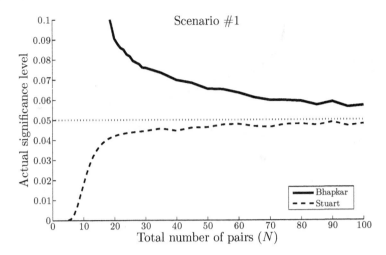

FIGURE 9.1
Actual significance levels of the Bhapkar and Stuart tests for marginal homogeneity

$N > 30$). The actual significance levels of the Bhapkar and Stuart tests for Evaluation Scenario #2 are quite similar to those in Evaluation Scenario #1. The McNemar-Bowker test is rather conservative for small sample sizes, with actual significance levels at about 2.4% for $N = 20$ and about 3.5% for $N = 30$. For $N \geq 50$, the actual significance levels of the McNemar-Bowker test are above 4% and very similar to those of the Stuart test.

TABLE 9.7
Probability values for Evaluation Scenario #2, which satisfies both marginal homogeneity and symmetry

| | Variable 2 | | | |
Variable 1	Category 1	Category 2	Category 3	Total
Category 1	0.15	0.10	0.10	0.35
Category 2	0.10	0.15	0.10	0.35
Category 3	0.10	0.10	0.10	0.30
Total	0.35	0.35	0.30	1

Finally, we consider Evaluation Scenario #3 (Table 9.8), in which the distribution of cell probabilities has center of mass in the center of the table. Evaluation Scenario #3, as was the case for Evaluation Scenario #2, satisfies both the hypothesis of marginal homogeneity and the hypothesis of symmetry. The results (Figure 9.3, where the computations are exact for $N \leq 30$ and simulated for $N > 30$) are not too different from the previous two evaluation

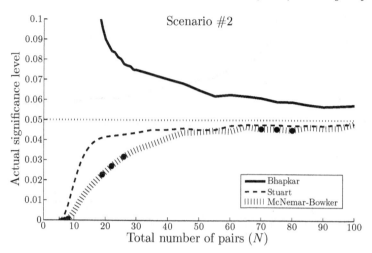

FIGURE 9.2
Actual significance levels of the Bhapkar and Stuart tests for marginal homogeneity and the McNemar-Bowker test for symmetry

scenarios, and confirm the overall impression that the Bhapkar test violates
the nominal significance level by an unacceptable amount, the Stuart test
performs quite well, and the McNemar-Bowker test is quite conservative for
$N \leq 40$.

TABLE 9.8
Probability values for Evaluation Scenario #3, which satisfies both
marginal homogeneity and symmetry

		Variable 2		
Variable 1	Category 1	Category 2	Category 3	Total
Category 1	0.05	0.10	0.05	0.20
Category 2	0.10	0.45	0.05	0.60
Category 3	0.05	0.05	0.10	0.20
Total	0.20	0.60	0.20	1

Evaluation of Power

In this section, we will not compare the power of the Bhapkar and Stuart
tests because the two tests have widely different actual significance levels. The
Bhapkar test has consistent and unacceptable high actual significance levels,
even for moderate to large sample sizes, and, as a consequence, its power is
not a relevant quantity to describe its performance. The actual significance

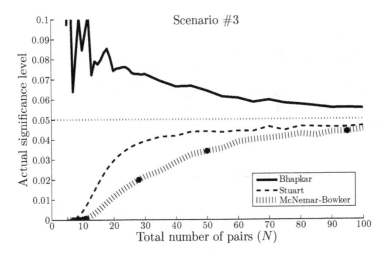

FIGURE 9.3
Actual significance levels of the Bhapkar and Stuart tests for marginal homogeneity and the McNemar-Bowker test for symmetry

levels of the Stuart test, on the other hand, are quite satisfactory. The aim of this section is thus to illustrate the effect sizes needed to achieve different power levels with the Stuart test for marginal homogeneity and the McNemar-Bowker test for symmetry.

Consider the probability values of Power Scenario #1 (Table 9.9). Observations of Variable 1 are most likely to fall in Category 1, whereas observations of Variable 2 are most likely to fall in Category 2. Power Scenario #1 complies with neither the hypothesis of marginal homogeneity nor the hypothesis of symmetry. To illustrate the size of the departures from the two null hypotheses, we note that the marginal differences are 0.35, −0.25, and −0.10, and the differences between the symmetric cell pairs are 0.20, 0.15, and −0.05. Figure 9.4 shows the power of the Stuart and McNemar-Bowker tests for N (the total number of pairs) in the range 5–100. Exact calculations are used for $N \leq 40$, and simulations are used for $N > 40$. The power of the two tests are quite similar, and we need about 50 pairs of observations to achieve 80% power and about 65 pairs of observations to achieve 90% power to detect that the scenario in Table 9.9 does not comply with marginal homogeneity or symmetry.

TABLE 9.9

Probability values for Power Scenario #1, in which neither marginal homogeneity nor symmetry is satisfied

	Variable 2			
Variable 1	Category 1	Category 2	Category 3	Total
Category 1	0.10	0.30	0.20	0.60
Category 2	0.10	0.05	0.05	0.20
Category 3	0.05	0.10	0.05	0.20
Total	0.25	0.45	0.30	1

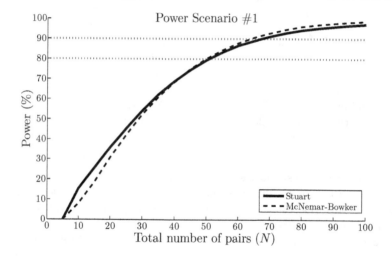

FIGURE 9.4

Power of the Stuart test for marginal homogeneity and the McNemar-Bowker test for symmetry

9.4.10   Evaluation of Intervals

Evaluation Criteria

Because we already know that Bonferroni intervals are shorter than Scheffé intervals (see page 394), we do not evaluate the width of the two intervals; however, we will consider coverage probability and interval location (see Section 1.4 for general descriptions). Both indices depend on the probabilities $\boldsymbol{\pi} = \{\pi_{11}, \pi_{12}, \ldots, \pi_{cc}\}$, for which $\sum_{ij} \pi_{ij} = 1$, the number of pairs (N), and the nominal significance level α. Because the Scheffé and Bonferroni interval methods are designed to account for multiple estimation, we define the *overall coverage probability* as the probability that all the c intervals contains the true parameter value. Overall non-coverage occurs when at least one of the c

intervals does not contain the true parameter value. Similarly, overall mesial non-coverage occurs when at least one of the c intervals is mesially located. The exact overall coverage probability is defined as

$$CP(\boldsymbol{\pi}, N, \alpha) =$$

$$\sum_{x_{11}=0}^{N} \sum_{x_{12}=0}^{N-x_{11}} \cdots \sum_{x_{c,c-1}=0}^{N_{c-2}} I(L_i \le \Delta_i \le U_i; \ i = 1, \ldots, c) \cdot f(\mathbf{x} \mid \boldsymbol{\pi}, N),$$

where $N_{c-2} = N - x_{11} - x_{12} - \ldots - x_{c,c-2}$, $I()$ is the indicator function, $L_i = L_i(\mathbf{x}, \alpha)$ and $U_i = U_i(\mathbf{x}, \alpha)$ are the lower and upper $100(1 - \alpha)\%$ confidence limits of an interval for $\Delta_i = \pi_{i+} - \pi_{+i}$ on the table $\mathbf{x} = \{x_{11}, x_{12}, \ldots, x_{c,c-1}, N_{c-2} - x_{c,c-1}\}$, and $f()$ is the multinomial probability distribution in Equation 9.1.

Location is measured by the MNCP/NCP index. The overall non-coverage probability (NCP) is computed as $1 - CP$. The overall mesial non-coverage probability (MNCP) is defined as

$$MNCP(\boldsymbol{\pi}, N, \alpha) = \sum_{x_{11}=0}^{N} \sum_{x_{12}=0}^{N-x_{11}} \cdots \sum_{x_{c,c-1}=0}^{N_{c-2}} I(L_i, \Delta_i, U_i) \cdot f(\mathbf{x} \mid \boldsymbol{\pi}, N).$$

where $I(L_i, \Delta_i, U_i) = I(L_i > \Delta_i \ge 0 \text{ or } U_i < \Delta_i \le 0; \ i = 1, 2, \ldots, c)$.

Evaluation of Coverage Probability

For simplicity, we use the probability values in Power Scenario #1 (Table 9.9). The marginal differences are $\Delta_1 = 0.35$, $\Delta_2 = -0.25$, and $\Delta_3 = -0.10$. We let $N = 5, 10, \ldots, 100$ and calculate the coverage probability for each value of N, using exact calculations for $N \le 40$, and simulations for $N > 40$. The results are shown in Figure 9.5. Except for the situation with $N = 5$, both the Scheffé and Bonferroni intervals have coverage probabilities above the nominal level. The Bonferroni intervals perform excellently, with coverage probabilities consistently close to the nominal level. Other choices of probability values (not shown here) give similar results, although coverage probabilities may be below 95% for $N \le 20$.

Evaluation of Location

We use the same probability values as above (Power Scenario #1; Table 9.9) to evaluate the location properties of the Scheffé and Bonferroni intervals. For each value of $N = 5, 10, \ldots, 100$, we calculate the index MNCP/NCP, exactly for $N \le 40$, and approximately with simulations for $N > 40$. The results are shown in Figure 9.6. Both intervals are too mesially located, although their MNCP/NCP values approach the satisfactory range with increasing N. The Bonferroni intervals have slightly higher MNCP/NCP values than the Scheffé intervals. Other choices of probability values (not shown here) give similar results.

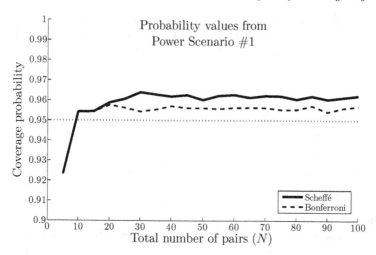

FIGURE 9.5
Overall coverage probabilities of three Scheffé and three Bonferroni confidence intervals for the difference between marginal probabilities

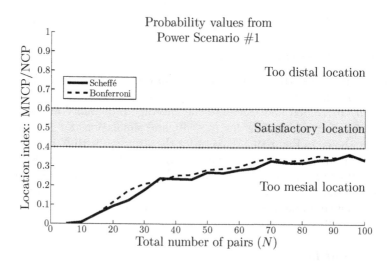

FIGURE 9.6
Overall location, as measured by the MNCP/NCP index, of three Scheffé and three Bonferroni confidence intervals for the difference between marginal probabilities

9.5   Ordinal Categories

Ordinal outcome categories are arguably more common than nominal outcome categories. It may be helpful to regard an ordinal variable to belong to one of two distinct types. *Grouped continuous variables* are explicitly derived from the categorization of a continuous variable, whereas *assessed variables* may be related to an underlying continuous variable but only measured on an ordinal categorical scale (Anderson, 1984). Body mass index (BMI) with categories underweight (less than 18.5 kg/m^2), normal range (between 18.5 and 25.0 kg/m^2), overweight (between 25.0 and 30.0 kg/m^2), and obese (over 30.0 kg/m^2) is an example of a grouped continuous variable. Level of pain, measured as none, mild, moderate, or severe, is an example of an assessed variable. In the following, we do not emphasize the difference between grouped continuous and assessed variables, and we will treat them as equal. Sometimes, however, we need to give scores to the categories of an ordinal variable, and to do that properly, we must consider its origins.

9.5.1 A Marginal Model

A marginal model for paired ordinal outcomes is the proportional-odds logit model:

$$\log \left[\frac{\Pr(Y_t \leq j \mid x_t)}{\Pr(Y_t > j \mid x_t)} \right] = \alpha_j - \beta x_t, \tag{9.14}$$

for $t = 1, 2$ and $j = 2, 3, \ldots, c$, where $x_1 = 1$ denotes Variable 1 and $x_2 = 0$ denotes Variable 2. The odds ratio for comparing $Y_2 \leq j$ with $Y_1 \leq j$ is $\exp(\alpha_j)/\exp(\alpha_j - \beta) = \exp(\beta)$. Marginal homogeneity is given by $\beta = 0$. This model has only a single parameter of interest (β), whereas the marginal model for nominal outcomes given in Equation 9.4 has $c - 1$ parameters of interest $(\beta_2, \beta_3, \ldots, \beta_c)$.

The simplicity of the proportional odds logit model comes with the price of a restrictive assumption: the odds ratio for comparing $Y_2 \leq j$ with $Y_1 \leq j$ is independent on the category j. Thus, if Y_1 and Y_2 have four categories, we have the following equalities:

$$\text{OR}(1,2\text{--}4) = \text{OR}(1\text{--}2,3\text{--}4) = \text{OR}(1\text{--}3,4) = \exp(\beta),$$

where OR(a, b) denotes the odds ratio comparing category a with category b. The proportional odds logit model is also called the *(constrained) cumulative logit model* or the *parallel lines logit model*. The proportional odds model was considered in more detail in Chapter 6, see in particular Sections 6.10–6.12.

When the proportional odds assumption does not hold, several other models can be used. Alternative models that use the logit link include the adjacent category and the constrained and unconstrained continuation ratio models (Fagerland, 2014). The adjacent category and constrained continuation ratio models are similar to the proportional odds model in that they only have one parameter of interest, whereas the unconstrained continuation ratio model has $c - 1$ parameters of interest. Section 6.10 contains more information on the adjacent category and continuation ratio models.

Another model within the logistic family is the stereotype logistic model (Anderson, 1984), which is a compromise between the restrictive one-parameter models and the fully flexible model for nominal outcomes. Other link functions besides the logit can also be used, such as the standard normal cumulative distribution function, which leads to the cumulative probit model (see Section 6.15).

In Sections 9.5.3–9.5.7, we present some simple methods that take advantage of the inherent information in the ordinal scale of ordinal variables. These methods are more powerful than the ones presented for nominal categories; however, they are considerably less used and not so well supported by standard statistical software packages. The main interest is on the marginal probabilities (π_{i+} and π_{+i}), but first, we will briefly consider a subject-specific model in Section 9.5.2.

9.5.2   A Subject-Specific Model

The proportional odds model in Equation 9.14 was a marginal (or population-averaged) model, in which interest is on the two marginal distributions. One assumption of marginal models is that the probabilities π_{ij} are independent of the subject. If we allow each subject to have its own probability distribution, we can define a subject-specific model by including subject-specific intercepts. Then, the proportional odds model is:

$$\log \left[\frac{\Pr(Y_t \geq j \mid x_{kt})}{\Pr(Y_t < j \mid x_{kt})} \right] = \alpha_{kj} + \beta x_{kt}, \tag{9.15}$$

for $t = 1, 2$ and $j = 2, 3, \ldots, c$. Here, $k = 1, 2, \ldots, N$ indexes the subjects, and the subject-specific effects are reflected in the α_{kj}. Note that the model still assumes one common effect, given by β. In a subject-specific model, interest is on the association within the pair, conditional on the subject, and subject-specific models are also referred to as *conditional models*. In a subject-specific model formulation, it may be helpful to think of the data as consisting of N $c \times c$ tables, one for each pair. Each table contains zeros in each cell, except for a single "1" in the cell corresponding to that subject's response for the two variables. If we collapse the tables over the subjects, we obtain the summary data of Table 9.1.

As for the subject-specific model for nominal categories, see Section 9.4.2, estimation of the common effect is difficult because of the subject-specific parameters. Instead, random effects are often assumed, see Rabe-Hesketh and Skrondal (2012, Chapter 11) or Agresti (2010, p. 282).

9.5.3   Simple Test for Three-Level Outcomes

When we have paired ordinal outcomes, we are often interested in whether one of the variables tends to have more observations in higher categories and fewer observations in lower categories than the other variable. If we only have three categories, this problem can be summarized by the difference $d_1 - d_3$, where the ds are the differences between the marginal sums (Equation 9.5). If $d_1 - d_3$ is positive, Variable 1 tends to produce responses in higher categories than Variable 2, and vice versa for negative $d_1 - d_3$. A test statistic for the null hypothesis that the two variables do not differ in their responses for low and high categories is (Fleiss et al., 2003):

$$T_{\text{Fleiss-Levin-Paik}}(\mathbf{n}) = \frac{(d_1 - d_3)^2}{2(\bar{n}_{12} + 4\bar{n}_{13} + \bar{n}_{23})},$$

where $\bar{n}_{ij} = (n_{ij} + n_{ji})/2$. $T_{\text{Fleiss-Levin-Paik}}$ is asymptotically chi-squared distributed with one degree of freedom, and P-values are obtained as

$$P\text{-value} = \Pr\left[\chi_1^2 \geq T_{\text{Fleiss-Levin-Paik}}(\mathbf{n})\right].$$

If the comparison was suggested by the data, Fleiss et al. (2003) state that the degrees of freedom should be two.

9.5.4 Wald Test and Interval for Marginal Mean Scores

Assume that we are able to assign scores to the outcome categories, such that the scores reflect the "distance" between the categories. The scores must be either monotonically increasing $a_1 \leq a_2 \leq \ldots \leq a_c$ or monotonically decreasing $a_1 \geq a_2 \geq \ldots \geq a_c$. If the outcome variables represent a categorization of a continuous variable (grouped continuous variables), we may take the midpoints of the corresponding intervals as the scores. As discussed in Section 5.5, our general advice is to use equally spaced integer scores, unless there is information about the categories that strongly advises us not to.

With the scores $\mathbf{a} = \{a_1, a_2, \ldots, a_c\}$ in place, we want to make inference about the marginal mean scores

$$E(Y_1) = \sum_{i=1}^{c} a_i \pi_{i+} \quad \text{and} \quad E(Y_2) = \sum_{i=1}^{c} a_i \pi_{+i},$$

which, under marginal homogeneity, are equal. Specifically, we want to test the null hypothesis

$$H_0 : \Delta = E(Y_1) - E(Y_2) = 0 \tag{9.16}$$

against the two-sided alternative $H_1 : \Delta \neq 0$, and to estimate a confidence interval for Δ. The sample mean scores for the two variables are

$$\bar{Y}_1 = \sum_{i=1}^{c} a_i \hat{\pi}_{i+}$$

and

$$\bar{Y}_2 = \sum_{i=1}^{c} a_i \hat{\pi}_{+i},$$

where, as usual, $\hat{\pi}_{i+} = n_{i+}/N$ and $\hat{\pi}_{+i} = n_{+i}/N$ are the maximum likelihood estimates of π_{i+} and π_{+i}. Thus,

$$\hat{\Delta} = \bar{Y}_1 - \bar{Y}_2$$

is the maximum likelihood estimate of Δ, and the standard error estimate of $\hat{\Delta}$ is

$$\widehat{SE}(\hat{\Delta}) = \sqrt{\frac{\sum_i \sum_j (a_i - a_j)^2 \hat{\pi}_{ij} - \hat{\Delta}^2}{N}},$$

where $\hat{\pi}_{ij} = n_{ij}/N$. A Wald test statistic for testing the null hypothesis in Equation 9.16 is given by

$$Z_{\text{Wald},\Delta}(\mathbf{n}) = \frac{\hat{\Delta}}{\widehat{SE}(\hat{\Delta})}.$$

$Z_{\text{Wald},\Delta}$ has an asymptotic standard normal distribution (Bhapkar, 1965), and we obtain P-values for the Wald test for marginal mean scores as

$$P\text{-value} = \Pr\left[Z \geq \left|Z_{\text{Wald},\Delta}(\mathbf{n})\right|\right],$$

where Z is a standard normal variable.

A Wald confidence interval for the difference between the marginal mean scores (Δ) is given by

$$\hat{\Delta} \pm z_{\alpha/2} \cdot \widehat{SE}(\hat{\Delta}),$$

where $z_{\alpha/2}$ is the upper $\alpha/2$ percentile of the standard normal distribution.

9.5.5 Score Test and Interval for Marginal Mean Scores

The Wald test and confidence interval in the preceding section used the standard error of the maximum likelihood estimate. A score test and score confidence interval can be obtained by substituting $\widehat{SE}(\hat{\Delta})$ with the standard error under the null hypothesis. Under the null hypothesis, $\Delta = 0$, and we obtain

$$\widehat{SE}_0(\hat{\Delta}) = \sqrt{\frac{\sum_i \sum_j (a_i - a_j)^2 \hat{\pi}_{ij}}{N}}.$$

The score test statistic then is

$$Z_{\text{score},\Delta}(\mathbf{n}) = \frac{\hat{\Delta}}{\widehat{SE}_0(\hat{\Delta})},$$

which has an asymptotic standard normal distribution. The score confidence interval for Δ is given by

$$\hat{\Delta} \pm z_{\alpha/2} \cdot \widehat{SE}_0(\hat{\Delta}).$$

9.5.6 Wald Test and Interval for Marginal Mean Ranks/Ridits

Instead of assigning scores to the categories and comparing the marginal mean scores, we can compare the marginal mean ranks using midranks. The midrank for category i is

$$r_i = \sum_{j=1}^{i-1}(n_{j+} + n_{+j}) + \frac{1}{2}(1 + n_{i+} + n_{+i}),$$

and the marginal mean ranks for Variable 1 and Variable 2 are

$$\bar{R}_1 = \sum_{i=1}^{c} r_i \hat{\pi}_{i+}$$

and

$$\bar{R}_2 = \sum_{i=1}^{c} r_i \hat{\pi}_{+i},$$

respectively. The mean ranks themselves are not that interesting, because they increase with increasing sample size, and thereby do not have a simple, intuitive interpretation. A better approach is to use two summary measures, τ and α, which are based on the *ridits* (Bross, 1958). The ridits are the average cumulative proportion scores, and they are linearly related to the midranks. The ridit for category i with the sample marginal distribution of Variable 1 is

$$a_{1i} = \hat{\pi}_{1+} + \ldots + \hat{\pi}_{(i-1)+} + \frac{1}{2}\hat{\pi}_{i+},$$

and the ridit for category i with the sample marginal distribution of Variable 2 is

$$a_{2i} = \hat{\pi}_{+1} + \ldots + \hat{\pi}_{+(i-1)} + \frac{1}{2}\hat{\pi}_{+i}.$$

The relationship between the ridits and the midranks is given by

$$a_{1i} + a_{2i} = \frac{r_i - 0.5}{N}.$$

Agresti (1983) and Ryu and Agresti (2008) considered two equivalent measures, τ and α, which both can be expressed by the ridits above. The two measures describe the extent to which the marginal distribution of Y_2 is stochastically higher than the marginal distribution of Y_1:

$$\tau = \Pr(Y_2 > Y_1) - \Pr(Y_1 > Y_2)$$

and

$$\alpha = \Pr(Y_2 > Y_1) + \frac{1}{2}\Pr(Y_1 = Y_2).$$

The relationship between τ and α is given by $\alpha = (\tau + 1)/2$. Under marginal homogeneity, $\tau = 0$ and $\alpha = 1/2$. We can estimate τ and α by

$$\hat{\tau} = \sum_{i<j} \hat{\pi}_{i+}\hat{\pi}_{+j} - \sum_{i>j} \hat{\pi}_{i+}\hat{\pi}_{+j},$$

and

$$\hat{\alpha} = \sum_{i<j} \hat{\pi}_{i+}\hat{\pi}_{+j} + \frac{1}{2}\sum_{i=j} \hat{\pi}_{i+}\hat{\pi}_{+j},$$

where the summations are over all pairs of (i,j) that satisfy $i < j$, $i > j$, or $i = j$. The relationships between $\hat{\tau}$ and the ridits and between $\hat{\alpha}$ and the ridits are given by (Vigderhous, 1979)

$$\hat{\tau} = \bar{A}_{Y_1}(Y_2) - \bar{A}_{Y_2}(Y_1)$$

and

$$\hat{\alpha} = \bar{A}_{Y_1}(Y_2),$$

where $\bar{A}_{Y_1}(Y_2)$ is the mean ridit for the distribution of Y_2 with ridits defined by the marginal distribution of Y_1, and similarly, $\bar{A}_{Y_2}(Y_1)$ is the mean ridit for the distribution of Y_1 with ridits defined by the marginal distribution of Y_2:

$$\bar{A}_{Y_1}(Y_2) = \sum_{i=1}^{c} a_{1i}\hat{\pi}_{+i} \quad \text{and} \quad \bar{A}_{Y_2}(Y_1) = \sum_{i=1}^{c} a_{2i}\hat{\pi}_{i+}.$$

An estimate of the standard error of $\hat{\tau}$ is

$$\widehat{\text{SE}}(\hat{\tau}) = \sqrt{\frac{\sum_i \sum_j \hat{\phi}_{ij}^2 \hat{\pi}_{ij} - \left(\sum_i \sum_j \hat{\phi}_{ij}\hat{\pi}_{ij}\right)^2}{N}},$$

where $\hat{\phi}_{ij} = 2(a_{2j} - a_{1i})$. The corresponding standard error estimate of $\hat{\alpha}$ is $\widehat{\text{SE}}(\hat{\alpha}) = \widehat{\text{SE}}(\hat{\tau})/2$.

A Wald test statistic for the null hypothesis of marginal homogeneity, which we may formulate as $H_0 : \tau = 0$, or equivalently, as $H_0 : \alpha = 1/2$, is

$$Z_{\text{Wald},\tau,\alpha}(\mathbf{n}) = \frac{\hat{\tau}}{\widehat{\text{SE}}(\hat{\tau})} = \frac{\hat{\alpha} - 1/2}{\widehat{\text{SE}}(\hat{\alpha})}.$$

$Z_{\text{Wald},\tau,\alpha}$ has an asymptotic standard normal distribution (Agresti, 1983).

Wald confidence intervals for τ and α are given by

$$\hat{\tau} \pm z_{\alpha/2} \cdot \widehat{\text{SE}}(\hat{\tau}),$$

and

$$\hat{\alpha} \pm z_{\alpha/2} \cdot \widehat{\text{SE}}(\hat{\alpha}).$$

9.5.7 Wald Logit Test and Interval for Marginal Mean Ranks/Ridits

Ryu and Agresti (2008) compared several different confidence interval methods for α (and thereby also for τ), including Wald, Wald logit, likelihood ratio test-based, score, and pseudo-score intervals, and found that the Wald logit interval performed better than the more sophisticated intervals like the score interval. The Wald logit interval constructs an interval on the logit(α) scale, and then inverts the confidence limits to the α scale. On the logit scale, the Wald confidence limits are

$$(l, u) = \text{logit}(\hat{\alpha}) \mp \frac{z_{\alpha/2} \cdot \widehat{\text{SE}}(\hat{\alpha})}{\hat{\alpha}(1 - \hat{\alpha})}.$$

The confidence limits (L_α, U_α) on the α scale are obtained as

$$L_\alpha = \frac{\exp(l)}{1 + \exp(l)}, \quad U_\alpha = \frac{\exp(u)}{1 + \exp(u)},$$

and a simple transformation produces the corresponding confidence limits (L_τ, U_τ) for τ:

$$L_\tau = 2L_\alpha - 1, \quad U_\tau = 2U_\alpha - 1.$$

Similarly, a Wald test statistic for the null hypothesis of marginal homogeneity may also be constructed on the logit(α) scale:

$$Z_{\text{Wald logit}}(\mathbf{n}) = \frac{\hat{\alpha}(1 - \hat{\alpha}) \cdot \text{logit}(\hat{\alpha})}{\widehat{\text{SE}}(\hat{\alpha})}.$$

9.5.8 Examples

Pretherapy Susceptibility of Pathogens in Patients with Complicated Urinary Tract Infection (Table 9.2)

Section 9.4.8 considered tests and confidence intervals for the marginal probabilities in Table 9.2 under the assumption that the categories were unordered. If we exclude the not applicable (N/A) category, the remaining three categories—susceptible, intermediately resistant, and resistant—can be treated as ordinal. The row and column sums for the susceptible category are now 620 and 596, respectively, with all other marginal sums unchanged. The total sample size is 664.

Because we now have a three-level ordinal outcome, we can use the Fleiss-Levin-Paik test of the null hypothesis that levofloxacin and ciprofloxacin have equal probabilities of low (susceptible) and high (resistant) outcomes. We have that $d_1 = 620 - 596 = 24$, $d_3 = 42 - 48 = -6$, and $d_1 - d_3 = 30$. The test statistic is $T_{\text{Fleiss-Levin-Paik}} = 21.43$, and $P < 0.0001$ (with either one or two degrees of freedom). More pathogens seem to be susceptible to levofloxacin than to ciprofloxacin.

A Comparison between Serial and Retrospective Measurements of Change in Patients' Health Status (Table 9.3)

For this example, there is no doubt about the ordinal nature of the two variables. They are, however, measured with different instruments on different scales. Yet, their purpose is the same: to evaluate patients' change in health status. In the following, we analyze the data in Table 9.3 with the methods for marginal mean scores and the methods for marginal mean ranks.

To use the methods for marginal mean scores, we must first decide upon a scoring of the outcome categories. We may either assign scores based on the serial measurements (the rows in Table 9.3), which originate from a visual analog scale, or the retrospective measurements (the columns in Table 9.3), which originate from a seven point Likert scale. With the first approach, we use the midpoints of the category intervals as scores, and obtain $a_{serial} = \{8, 3.5, 0, -3.5, -8\}$. With the second approach, we take each score as the actual category value, except the first and last scores, which have been set to the midpoints of their corresponding category intervals. Thus we obtain $a_{retrospective} = \{1.5, 3, 4, 5, 6.5\}$. Even though the two set of scores may look quite different—one is decreasing, the other increasing—the spacings between the categories are quite similar. As we will see, we obtain very similar results with the two sets of scores.

First, we use the scores based on the visual analog scale (a_{serial}). The sample mean scores are

$$\bar{y}_1 = 8 \cdot \frac{2}{121} + 3.5 \cdot \frac{18}{121} + 0 \cdot \frac{58}{121} - 3.5 \cdot \frac{37}{121} - 8 \cdot \frac{6}{121} = -0.814$$

and

$$\bar{y}_2 = 8 \cdot \frac{3}{121} + 3.5 \cdot \frac{3}{121} + 0 \cdot \frac{57}{121} - 3.5 \cdot \frac{28}{121} - 8 \cdot \frac{30}{121} = -2.508,$$

and we estimate the difference between the marginal mean scores as

$$\hat{\Delta} = \bar{y}_1 - \bar{y}_2 = 1.694.$$

On the retrospective measurements, the patients reported an improvement in pain of about 2.5 points on a visual analog scale, whereas on the serial measurements, the patients reported a more modest improvement of about 0.8 points, which is within what the authors define as "no change". The difference between the two measurement scores ($\hat{\Delta} = 1.694$) is about 1/2 of a category. The Wald standard error estimate of $\hat{\Delta}$ is $\widehat{SE}(\hat{\Delta}) = 0.3875$ and the score standard error estimate is $\widehat{SE}_0(\hat{\Delta}) = 0.4170$. The results for the Wald and score confidence intervals for Δ, and the Wald and score tests for the null hypothesis that $\Delta = 0$, are shown in Table 9.10. The score interval is a bit wider than the Wald interval, although both intervals roughly state that the true difference between the retrospective and serial measurements is likely to be between 0.9 and 2.5 points on a visual analog scale. The difference is statistically highly significant.

TABLE 9.10
Results of the comparisons of serial and retrospective
measurements of changes in pain (data from Table 9.3)

$\hat{\Delta} = 1.694$	
95% Wald confidence interval for Δ	0.935 to 2.454
95% score confidence interval for Δ	0.877 to 2.512
Wald test for $H_0 : \Delta = 0$	$P < 0.0001$
Score test for $H_0 : \Delta = 0$	$P < 0.0001$
$\hat{\tau} = 0.239$	
95% Wald confidence interval for τ	0.118 to 0.360
95% Wald logit confidence interval for τ	0.115 to 0.356
Wald test for $H_0 : \tau = 0$	$P = 0.0001$
Wald logit test for $H_0 : \tau = 0$	$P = 0.0002$
$\hat{\alpha} = 0.620$	
95% Wald confidence interval for α	0.559 to 0.680
95% Wald logit confidence interval for α	0.558 to 0.678
Wald test for $H_0 : \alpha = 1/2$	$P = 0.0001$
Wald logit test for $H_0 : \alpha = 1/2$	$P = 0.0002$

If we use $\mathbf{a}_{\text{retrospective}}$ instead of $\mathbf{a}_{\text{serial}}$ in our calculations of the marginal mean scores, we obtain a mean score of $\bar{y}_1 = 4.240$ for the serial measurements and $\bar{y}_2 = 4.764$ for the retrospective measurements. Although we now interpret the results with respect to the Likert value of change categories instead of the visual analog scale, we get similar results. The patients report more improvement on the retrospective measurements than on the serial measurements, and the difference between the two measurement scores ($\hat{\Delta} = -0.525$) is about the size of $1/2$ category. The 95% Wald and score confidence intervals for Δ is now (-0.757 to -0.293) and (-0.775 to -0.275), respectively. The P-value for the null hypothesis of $\Delta = 0$ is $P < 0.0001$ with both the Wald and score tests.

We can also analyze the data in Table 9.3 with the methods for marginal mean ranks (or ridits). τ and α are measures of the extent to which the marginal distribution of the retrospective measurements are stochastically higher than the marginal distribution of the serial measurements. We obtain the estimates $\hat{\tau} = 0.239$ and $\hat{\alpha} = 0.620$. Positive values of τ, and values of α larger than 0.5, indicate that patients tend to report greater changes of pain with retrospective measurements as compared with serial measurements. Confidence intervals for τ and α, and tests for the null hypothesis that $\tau = 0$ and $\alpha = 1/2$, are shown in Table 9.10, and indicate a very similar conclusion as that for the marginal mean scores.

9.5.9   Evaluation of Tests

Evaluation Criteria

The evaluation criteria for ordinal categories are the same as those for nominal categories, and we refer the reader to the first paragraph of Section 9.4.9.

Evaluation of Actual Significance Level

We start by noting that the Fleiss-Levin-Paik test (for three-level outcomes) will be equal to the score test for Δ if the scores are equally spaced, but not otherwise. In this section, we consider the evaluation scenarios from Section 9.4.9, all with a three-level outcome. Moreover, we will use scores $\mathbf{a} = \{1, 2, 3\}$; therefore, we do not explicitly evaluate the Fleiss-Levin-Paik test because it performs equally with the score test for Δ under these conditions.

Consider the Wald and score tests for Δ and Evaluation Scenario #1 (Table 9.6), which satisfies marginal homogeneity. We calculate the actual significance level exactly for $N \leq 30$ and with simulations for $N > 30$, and show the results in Figure 9.7. The Wald test violates the nominal level considerably for small sample sizes, and also quite markedly for medium sample sizes. It performs much like the Bhapkar test for marginal homogeneity, which was evaluated in Figures 9.1–9.3. The score test, on the other hand, has actual significance levels very close to the nominal level, even for quite small sample sizes. The results for Evaluation Scenario #2 (Table 9.7) and Evaluation Scenario #3 (Table 9.8)—both of which comply with marginal homogeneity—are similar to those shown in Figure 9.7, and we do not show them here.

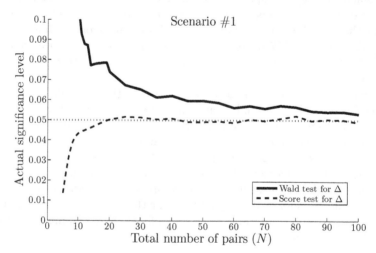

FIGURE 9.7
Actual significance levels of the Wald and score tests for $H_0 : \Delta = 0$

We use the same three evaluation scenarios on the Wald and Wald logit tests for τ. Because α is linearly related to τ by $\alpha = (\tau + 1)/2$, the Wald and Wald logit tests will give equivalent results for τ and α. The performances of the two tests are therefore equal for $H_0 : \tau = 0$ and $H_0 : \alpha = 1/2$. Figure 9.8 shows the actual significance levels of the Wald and Wald logit tests on Evaluation Scenario #2. Exact calculations are used for $N \leq 30$ and simulations are used for $N > 30$. The Wald test for τ and α performs similarly to the Wald test for Δ, with too high actual significance levels for small and moderate sample sizes. The performance of the Wald logit test is excellent in Figure 9.8. Its actual significance levels are very close to the nominal level for $N \geq 10$. The Wald logit test does not perform equally well on all other scenarios; however, it is still quite good and much better than the Wald test.

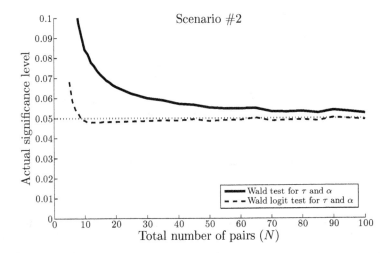

FIGURE 9.8
Actual significance levels of the Wald and Wald logit tests for $H_0 : \tau = 0$ and $H_0 : \alpha = 1/2$

Evaluation of Power

In the above, we observed that the actual significance levels of the Wald test for Δ and the Wald test for τ and α were much higher than the nominal level for small and moderate sample sizes. We thereby do not find it informative to calculate the power of these two tests. Instead, we will compare the power of the score test for Δ with that of the Wald logit test for τ and α. The results for Power Scenario #1 (Table 9.9) are shown in Figure 9.9, where the calculations are done exactly for $N \leq 30$ and with simulations for $N > 30$. The Wald logit test has the greatest power. When power is in the range 50%–90%, the Wald logit test has 3–3.5 percentage points more power than the score test. To achieve 80% power, 51 pairs of observations are needed with the Wald logit test, whereas 56 pairs of observations are needed with the score test. To achieve 90% power, the numbers of pairs are 68 (Wald logit test) and 75 (score test).

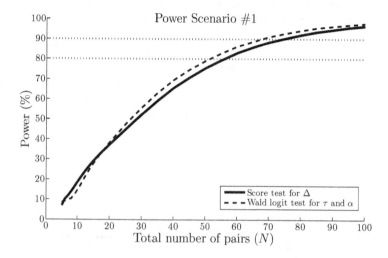

FIGURE 9.9
Power of the score test for $H_0 : \Delta = 0$ and the Wald logit test for $H_0 : \tau = 0$ and $H_0 : \alpha = 1/2$

In Section 9.4.9 (see Figure 9.4), we calculated the power of the Stuart and McNemar-Bowker tests for the same evaluation scenario as in Figure 9.9. If we were to superimpose one of the plots on the other, we would observe that the power of the Stuart test for marginal homogeneity is quite similar to the power of the Wald logit test for τ and α. This is contrary to the expectation that ordinal tests, such as the score and Wald logit tests, have higher power than nominal tests, such as the Stuart test, when the categories are ordered. The marginal probabilities in Power Scenario #1, however, do not have an ordered structure, and the alternative hypothesis in this case is not ordered. Consider now the probabilities in Power Scenario #2, given in Table 9.11, in which there is a clear ordinal structure to the marginal probabilities. The calculated powers for this scenario are shown in Figure 9.10. We now observe that the power of the ordinal tests (score and Wald logit) are considerably higher than the nominal test (Stuart), as expected. The overall higher power for Power Scenario #1 compared with Power Scenario #2 reflects that the sizes of the marginal differences are greater in the former.

We have added the McNemar-Bowker test to the plot in Figure 9.10, even though this is a test for symmetry, and as such, is not directly comparable to the other tests, which are tests of marginal homogeneity. The differences between the symmetric cell pairs are small (0, -0.10, and -0.10) in Table 9.11, and the power for the McNemar-Bowker test is also small.

TABLE 9.11
Probability values for Power Scenario #2, in which neither marginal homogeneity nor symmetry is satisfied

	Variable 2			
Variable 1	Category 1	Category 2	Category 3	Total
Category 1	0.05	0.05	0.05	0.15
Category 2	0.05	0.10	0.10	0.25
Category 3	0.15	0.20	0.25	0.60
Total	0.25	0.35	0.40	1

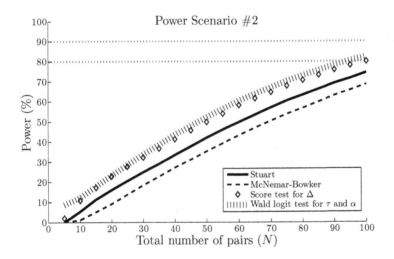

FIGURE 9.10
Power of three tests for marginal homogeneity (Stuart, score, and Wald logit) and one test for symmetry (McNemar-Bowker)

9.5.10 Evaluation of Intervals

Evaluation Criteria

The evaluation criteria for ordinal categories are the same as those for nominal categories; however, in our evaluation of the Bonferroni and Scheffé intervals in Section 9.4.10, we defined the overall coverage probability and the overall mesial non-coverage probability (and no width), whereas here, we define the coverage probability, width, and mesial non-coverage probability for a single confidence interval method. Moreover, the effect measure is different for the nominal and ordinal situations, although the calculations are, in principle, equal; only the notation is slightly different.

In the following, let $\boldsymbol{\pi} = \{\pi_{11}, \pi_{12}, \ldots, \pi_{cc}\}$ denote the cell probabilities of the $c \times c$ table, for which $\sum_{ij} \pi_{ij} = 1$. Let N denote the number of pairs, and let α denote the nominal significance level. We have also used α to denote one of the effect measures for marginal mean ranks. Because the effect measure α is linearly related to the effect measure τ by $\alpha = (\tau + 1)/2$, the performances of the two effect measures with respect to coverage probability, width, and location will be equivalent. We thereby only evaluate τ in this section, and let α denote the nominal significance level, although we sometimes refer to the effect measure α, but only when the context makes it clear that we are considering the effect measure and not the nominal significance level.

The exact coverage probability of an interval for Δ (the difference between the marginal mean scores) is

$$\text{CP}(\boldsymbol{\pi}, N, \alpha) = \sum_{x_{11}=0}^{N} \sum_{x_{12}=0}^{N-x_{11}} \cdots \sum_{x_{c,c-1}=0}^{N_{c-2}} I(L \leq \Delta \leq U) \cdot f(\mathbf{x} \,|\, \boldsymbol{\pi}, N), \quad (9.17)$$

where $N_{c-2} = N - x_{11} - x_{12} - \ldots - x_{c,c-2}$, $I()$ is the indicator function, $L = L(\mathbf{x}, \alpha)$ and $U = U(\mathbf{x}, \alpha)$ are the lower and upper $100(1 - \alpha)\%$ confidence limits of an interval for Δ on the table $\mathbf{x} = \{x_{11}, x_{12}, \ldots, x_{c,c-1}, N_{c-2} - x_{c,c-1}\}$, and $f()$ is the multinomial probability distribution in Equation 9.1.

The formula for the coverage probability of intervals for τ is obtained by substituting τ for Δ in Equation 9.17.

The exact expected interval width is defined as

$$\text{Width}(\boldsymbol{\pi}, N, \alpha) = \sum_{x_{11}=0}^{N} \sum_{x_{12}=0}^{N-x_{11}} \cdots \sum_{x_{c,c-1}=0}^{N_{c-2}} (U - L) \cdot f(\mathbf{x} \,|\, \boldsymbol{\pi}, N),$$

and the formula is equal for Δ and τ.

Location is measured by the MNCP/NCP index. The non-coverage probability (NCP) is computed as $1 - \text{CP}$. The mesial non-coverage probability (MNCP) of an interval for Δ is defined as

$$\text{MNCP}(\boldsymbol{\pi}, N, \alpha) =$$

$$\sum_{x_{11}=0}^{N} \sum_{x_{12}=0}^{N-x_{11}} \cdots \sum_{x_{c,c-1}=0}^{N_{c-2}} I(L > \Delta \geq 0 \text{ or } U < \Delta \leq 0) \cdot f(\mathbf{x} \,|\, \boldsymbol{\pi}, N).$$

The corresponding formula for intervals for τ is obtained by substituting τ for Δ.

Evaluation of Coverage Probability

First, we consider the two intervals for Δ. The coverage probabilities of the Wald and score intervals under Power Scenario #1 (Table 9.9) can be seen in

Figure 9.11. Exact calculations have been used for $N \leq 30$, and simulations have been used for $N > 30$. The two intervals perform very differently. The Wald interval has coverage probabilities markedly below the nominal level, particularly for small sample sizes, but also quite noticeably for $N > 70$. The score interval is conservative, with coverage mostly in the range of 96%–96.5%. The performance of the Wald interval is similar under other evaluation scenarios; however, the score interval sometimes performs much better, as seen in Figure 9.12, in which the score interval has coverage probabilities very close to the nominal level for $N > 15$. Evaluation Scenario #2, which is used in Figure 9.12, is defined in Table 9.7.

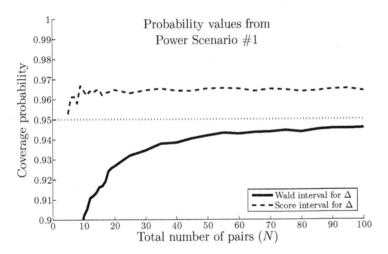

FIGURE 9.11
Coverage probabilities of the Wald and score confidence intervals for Δ

Next, we consider the Wald and Wald logit intervals for τ and α. A typical example of their coverage probabilities is shown in Figure 9.13. As usual in this chapter, the Wald interval performs poorly, with low coverage probabilities for small sample sizes. The Wald logit interval is somewhat conservative in Figure 9.13, although to a lesser extent than the score interval for Δ in Figure 9.11. Just like the score interval for Δ, the Wald logit interval for τ and α may also perform excellently under other scenarios, such as Evaluation Scenario #2.

Evaluation of Width

We do not present any results of calculations of the interval widths for the ordinal effect measures. In the evaluation of coverage probability, we observed that the Wald interval for Δ and the Wald interval for τ and α had, in general, coverage probabilities lower than the nominal level. For small sample

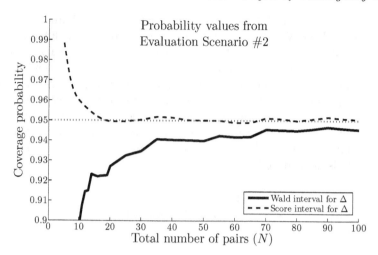

FIGURE 9.12
Coverage probabilities of the Wald and score confidence intervals for Δ

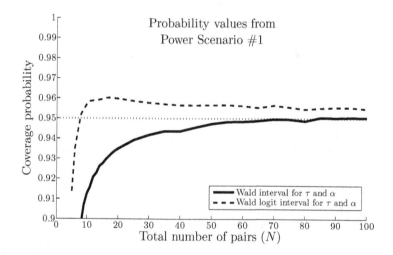

FIGURE 9.13
Coverage probabilities of the Wald and Wald logit confidence intervals for τ and α

sizes, the coverage of the Wald intervals was unacceptably low. The coverage probabilities of the score interval for Δ and the Wald logit interval for τ and α, on the other hand, are slightly to moderately greater than the nominal level. A comparison of the widths of the Wald and the score intervals for Δ, or a comparison of the widths of the Wald and the Wald logit intervals for τ and α, thereby is not particularly interesting, both because we would expect the liberal intervals (Wald) to be more narrow than the conservative intervals (score and Wald logit), and—most important—because we would not recommend using the Wald interval at all, no matter the width. Moreover, we do not compare the score interval for Δ with the Wald logit interval for τ and α simply because the two effect measures are defined on different scales, and they are not comparable.

Evaluation of Location

Figure 9.14 shows an example of the MNCP/NCP index of location for the two intervals with satisfactory coverage probabilities: the score interval for Δ and the Wald logit interval for τ and α. Both intervals are too mesially located,

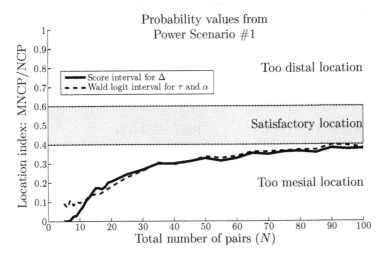

FIGURE 9.14
Location, as measured by the MNCP/NCP index, of the score interval for Δ and the Wald logit interval for τ and α

and there is not much that sets them apart. When the sample size increases toward 100 pairs of observations, the location of both intervals approaches the satisfactory range of the MNCP/NCP index. The performances of the two intervals under other evaluation scenarios are quite similar to those in Figure 9.14.

9.6 Recommendations

9.6.1 Summary

Table 9.12 provides a summary of the recommended tests and confidence intervals for paired $c \times c$ tables. The Wald tests and confidence intervals for ordinal categories are only recommended for large sample sizes, which we may take as 100 or more number of pairs. Sections 9.6.2 and 9.6.3 discuss the recommendations in more detail and summarize the merits of the different methods.

TABLE 9.12

Recommended tests and confidence intervals (CIs) for paired $c \times c$ tables

Analysis	Recommended methods	Sample sizes
Nominal categories		
Tests for marginal homogeneity	Stuart[*]	all
Tests for equality of marginal probabilities	McNemar asymptotic[†‡]	all
CIs for difference between marginal probabilities	Bonferroni-type[†]	all
Tests for symmetry	McNemar-Bowker[†]	all
Ordinal categories		
Tests and CIs for marginal mean scores	Score[†] Wald[†]	all large
Tests and CIs for marginal mean ranks	Wald logit[†] Wald[†]	all large

[*]Closed-form version exists for $c = 3$ (the Fleiss-Everitt test)
[†]These methods have closed-form expression
[‡]Adjusted with $c - 1$ degrees of freedom for the chi-squared statistic Z^2_{McNemar}

9.6.2 Nominal Categories

With nominal categories, we are mostly concerned with testing for marginal homogeneity or symmetry or both. Symmetry is the strongest of the two hypotheses and includes marginal homogeneity. In most practical applications, marginal homogeneity will be the hypothesis of interest. Under marginal homogeneity, the marginal probabilities of each category is the same for Variable 1 and Variable 2. The Stuart test for marginal homogeneity performs very well. It is slightly conservative, with actual significance levels between

4% and 5%, except when the sample size is very small, in which the actual significance levels are below 4%. One disadvantage with the Stuart test is the need to invert a covariance matrix to calculate it. A simple closed-form expression of the Stuart test, called the *Fleiss-Everitt test*, exists for $c = 3$ outcome categories. There is also a rather elaborate closed-form expression for the $c = 4$ case, for which we refer the reader to Fleiss and Everitt (1971).

If the Stuart test indicates that the marginal probabilities are unequal, it may be of interest to consider each of the marginal probability pairs, (π_{i+}, π_{+i}), $i = 1, 2, \ldots, c$. We can test the c null hypotheses that $\pi_{i+} = \pi_{+i}$ with the adjusted McNemar asymptotic test. The adjusted test is equal to the ordinary McNemar asymptotic test (see Section 8.5.2), except that we use the chi-squared version of the test statistic (Z_{McNemar}^2), and a chi-squared reference distribution with $c - 1$ degrees of freedom. Further, we can estimate the differences between the marginal probabilities with a confidence interval. The Bonferroni-type intervals are shorter than the Scheffé-type intervals, and they have coverage probabilities closer to the nominal level. Both the adjusted McNemar test and the Bonferroni-type intervals compensate for the increased probability of false positive findings as a result of multiple comparisons.

To test the null hypothesis of symmetry, we recommend the McNemar-Bowker test. The McNemar-Bowker test is a natural generalization of the McNemar test, and it has a very simple closed-form expression. It can be quite conservative for small sample sizes, and we do not recommend to test for symmetry unless the sample size is at least moderately large, say more than 40 pairs.

9.6.3 Ordinal Categories

When the outcome categories are ordinal, we may assign scores to the categories and analyze the distribution of the scores, or we may consider the ranks. With scores assigned, we can estimate the difference between the expected scores (Δ) with a confidence interval, and test for $\Delta = 0$. For this purpose, we recommend the score test and confidence interval, which always perform well. The Wald test and confidence interval are acceptable only for large sample sizes, such as $N \geq 100$.

The Fleiss-Levin-Paik test for three-level outcomes can be recommended for situations where it is reasonable to assume that the categories are evenly spaced; that is, if you can assign evenly spaced scores to the categories (although you do not need the scores to calculate the test). In that case, the Fleiss-Levin-Paik test is equal to the score test for marginal means with equally spaced scores, and it may be used as a simple workaround for the slightly more elaborate score test. If, however, you have the score test available, you might as well use it for all situations.

Instead of assigning scores to the categories, which sometimes may be difficult, we may analyze the ranks (or ridits) instead. Two equivalent effect measures are available (τ and α), and the statistical methods to analyze them are the same. The recommended methods are the Wald logit test and confidence interval, which are Wald tests and confidence intervals constructed on the logit scale. For large sample sizes ($N \geq 100$), the (ordinary) Wald test and confidence interval can also be used.

Finally, we note that methods for nominal categories can be used for ordinal data, with an expected loss of power. The loss of power increases with increasing number of categories.

10

Stratified 2×2 Tables and Meta-Analysis

10.1 Introduction

In Chapter 4, we considered tests and confidence intervals for the association between two dichotomous variables, summarized in a 2×2 table. Here, we extend those analyses to the situation where we have a third categorical variable, with K categories. The data can then be summarized in K 2×2 tables. The third variable that creates the K tables is a stratification variable, which we assume is unordered. The objective of the analysis is usually to study the association between an outcome variable and a grouping (or exposure) variable over K strata. The stratification variable may influence the association between the grouping and the outcome, and a standard procedure to account for the stratification variable is to consider the association in a stratified 2×2 table. Note that we in this chapter consider unpaired data. The analysis of stratified paired 2×2 tables is dealt with in Chapter 11.

An alternative way of summarizing the association between two variables stratified on a third variable is to consider the three variables symmetrically in a three-way table, that is, a $2 \times 2 \times K$ table. For the logit link function, the modeling of stratified two-way tables can be done with the logistic function, whereas the analysis of symmetrical three-way tables can be done with log-linear models. We refer to Section 1.5 for more information on the differences and analogies between the two approaches.

There are several study designs that lead to stratified 2×2 tables: randomized clinical trials, cohort studies, unmatched case-control studies, and cross-sectional studies. In *systematic reviews*, the summary information for categorical data is often presented in several 2×2 tables, where the stratification into the K tables is based on the included studies. The statistical methods that are used to analyze systematic reviews are usually labeled *meta-analysis*; however, the methods that fall under the meta-analysis appellation are not limited to systematic reviews, and can be used in many different settings, including for the analysis of the study designs mentioned above.

For randomized clinical trials, it is common to consider stratification by means of a third variable, for instance, by sex or age group. In multicenter trials, study center is usually a stratification variable. In observational studies, such as cohort studies, case-control studies, and cross-sectional studies, interest is on the exposure-disease relationship, which is commonly studied in a stratified 2×2 table, where the stratification variable may be a confounder or effect modifier.

Table 10.1 shows the general setup for a stratified 2×2 table, which is a straightforward generalization of the general 2×2 table in Table 4.1. The rows within each 2×2 table are denoted by groups, which may be treatment or exposure groups. The columns consist of the two outcomes, which we label Success and Failure. The data are summarized as the observed cell counts for each category of the outcome variables, cross-classified by the grouping variable and the stratification variable.

TABLE 10.1
The observed counts of a stratified 2×2 table

| Variable 1 | Variable 2 | | Total | Variable 3 |
	Success	Failure		
Group 1	n_{111}	n_{121}	n_{1+1}	
Group 2	n_{211}	n_{221}	n_{2+1}	Stratum 1
Total	n_{+11}	n_{+21}	n_{++1}	
Group 1	n_{112}	n_{122}	n_{1+2}	
Group 2	n_{212}	n_{222}	n_{2+2}	Stratum 2
Total	n_{+12}	n_{+22}	n_{++2}	
$\vdots$	$\vdots$	$\vdots$	$\vdots$	$\vdots$
Group 1	n_{11K}	n_{12K}	n_{1+K}	
Group 2	n_{21K}	n_{22K}	n_{2+K}	Stratum K
Total	n_{+1K}	n_{+2K}	n_{++K}	

Additional notation:

$\mathbf{n} = \{n_{111}, n_{121}, n_{211}, n_{221}, \ldots, n_{11K}, n_{12K}, n_{21K}, n_{22K}\}$: the observed table

$\mathbf{x} = \{x_{111}, x_{121}, x_{211}, x_{221}, \ldots, x_{11K}, x_{12K}, x_{21K}, x_{22K}\}$: any possible table

The total number of observations: $\sum_{k=1}^{K} n_{++k} = N$

This chapter is organized as follows. Section 10.2 introduces two examples of stratified 2×2 tables (one of which is a meta-analysis) that will be used to illustrate the methods later in the chapter. The general notation and the most common sampling designs are presented in Section 10.3. Section 10.4 presents statistical models, effect measures, and effect estimates for stratified 2×2 tables. Relevant sampling distributions are given in Section 10.5. An overview of the different steps in the analyses of stratified tables is provided in Section 10.6. A brief explanation on how to estimate the stratum specific effects with a confidence interval is given in Section 10.7. Section 10.8 considers the Mantel-Haenszel, inverse variance, maximum likelihood, and Peto odds ratio estimates. The next two sections deal with tests of homogeneity of effects over the K strata, first for a general link function and effect measure (Section 10.9), then for the logit link and the odds ratio (Section 10.10). If homogeneity of effects over the strata can be assumed, we can test whether the common effect is zero. Sections 10.11, 10.12, and 10.13 present tests for a common difference between probabilities, ratio of probabilities, and odds ratio, respectively. Confidence intervals for the common effect are considered in Section 10.14. Section 10.15 turns our attention from fixed effects models to random effects models and presents estimates and confidence intervals for the common effect parameter in two different random effects models. We analyze the two examples in Section 10.16, and briefly mention some additional methods for meta-analysis in Section 10.17. Finally, Section 10.18 provides our recommendations for the analysis of stratified 2 × 2 tables.

10.2 Examples

10.2.1 Smoking and Lung Cancer

The article by Doll and Hill (1950) is regarded as the path-breaking report on the association between smoking and lung cancer. Doll and Hill investigated the association between smoking and carcinoma of the lung in a case-control study of 709 patients with carcinoma in the lung (cases) and 709 patients with other diseases (controls). The purpose was to study the association between smoking and lung cancer. When this study was performed, smoking was a less common habit for women than for men, so the authors presented the data stratified according to sex. Table 10.2 shows the results.

10.2.2 Prophylactic Use of Lidocaine in Myocardial Infarction (Meta-Analysis)

The data in Table 10.3 originate from a study by Hine et al. (1989), and were reanalyzed in a tutorial on meta-analysis by Normand (1999). The purpose of the study was to investigate whether prophylactic lidocaine given to patients with proved or suspected myocardial infarction has any effect on mortality. The authors used a systematic review to identify six randomized clinical trials with available data on patients with proved myocardial infarction. The six trials form the strata in Table 10.3.

As noted by Normand (1999), the sample size in each trial is too small to detect any effect of prophylactic lidocaine if the trials are analyzed individually; however, by considering all trials in a meta-analysis, a useful conclusion may be obtained. After pooling the results of the six trials, 37 deaths (6.6%)

TABLE 10.2

The cross-classification of lung cancer and smoking, stratified by gender (Doll and Hill, 1950)

| Exposure | Lung cancer | | Total | Sex |
	Cases	Controls		
Smokers	647 (99.7%)	622 (95.8%)	1269 (97.8%)	
Non-smokers	2 (0.3%)	27 (4.2%)	29 (2.2%)	Males
Total	649	649	1298	
Smokers	41 (68%)	28 (47%)	69 (58%)	
Non-smokers	19 (32%)	32 (53%)	51 (43%)	Females
Total	60	60	120	

The total number of observations: $N = 1418$

were observed among the patients treated with lidocaine, and 21 deaths (3.8%) were observed among the control patients.

10.3 Notation and Sampling Designs

The presentation we use for data in stratified 2×2 tables is in accordance with the general setup in Table 10.1. Variable 1 (Y_1) is the grouping variable, and Variable 2 (Y_2) represents the outcome. Finally, Variable 3 (Y_3) is a stratification variable for the 2×2 tables. If Y_3 has K categories, stratification on Y_3 produces K 2×2 tables.

The grouping variable Y_1 and the outcome variable Y_2 take the values 1 or 2. The stratification variable Y_3 takes value $Y_3 = k$ for observations in stratum k, where $k = 1, 2, \ldots, K$. The total number of observations in stratum k is n_{++k}. The total number of observations in Group i in stratum k is n_{i+k}. The number of successes in Group i in stratum k is n_{i1k}, and the number of failures in Group i in stratum k is $n_{i2k} = n_{i+k} - n_{i1k}$.

The probability of a success in Group i and stratum k is $\Pr(Y_{i1k} = 1) = \pi_{i1k}$. The cell probabilities in stratum k are $\{\pi_{11k}, \pi_{12k}, \pi_{21k}, \pi_{22k}\}$, which add up to 1, and is a generalization of the notation in Chapter 4. The probabilities of the kth table in a stratified 2×2 table are shown in Table 10.4.

The stratified 2×2 table is produced by stratification on Y_3, which may be either by pre-specification or post-specification. Pre-specification is made by the design, for instance, by stratification into age groups, with predefined sample sizes for each stratum. In a randomized clinical trial and in a meta-analysis, the stratification is typically fixed by design, such as when the strata are made

TABLE 10.3

The cross-classification of mortality and treatment with prophylactic lidocaine, stratified by six randomized controlled trials (Hine et al., 1989)

Treatment	Mortality		Total	Trial
	Dead	Alive		
Lidocaine	2 (5.1%)	37 (95%)	39	
Control	1 (2.3%)	42 (98%)	43	Chopra et al.
Total	3 (3.7%)	79 (96%)	82	
Lidocaine	4 (9.1%)	40 (91%)	44	
Control	4 (9.1%)	40 (91%)	44	Mogensen
Total	8 (9.1%)	80 (91%)	88	
Lidocaine	6 (5.6%)	101 (94%)	107	
Control	4 (3.6%)	106 (96%)	110	Pitt et al.
Total	10 (4.6%)	207 (95%)	217	
Lidocaine	7 (6.8%)	96 (93%)	103	
Control	5 (5.0%)	95 (95%)	100	Darby et al.
Total	12 (5.9%)	191 (94%)	203	
Lidocaine	7 (6.4%)	103 (94%)	110	
Control	3 (2.8%)	103 (97%)	106	Bennett et al.
Total	10 (4.6%)	206 (95%)	216	
Lidocaine	11 (7.1%)	143 (93%)	154	
Control	4 (2.7%)	142 (97%)	146	O'Brian et al.
Total	15 (5.0%)	285 (95%)	300	

The total number of observations: $N = 1106$

up by centers or by studies. Post-stratification is done by conditioning on Y_3 in the data analysis. Then, the strata totals, $\{n_{++1}, n_{++2}, \ldots, n_{++K}\}$, are regarded as fixed in the analysis.

The grouping variable (Y_1) may also be fixed by design, or be fixed in the analysis. In a randomized clinical trial, for instance, the treatments are given by the design, with pre-specified row totals. In an observational study,

TABLE 10.4

The joint probabilities of the kth table in a stratified 2×2 table

	Variable 2		
Variable 1	Success	Failure	Total
Group 1	π_{11k}	π_{12k}	π_{1+k}
Group 2	π_{21k}	π_{22k}	π_{2+k}
Total	π_{+1k}	π_{+2k}	1

the grouping of the data may be specified in the analysis, for instance into exposure groups that were not specified in the design.

For the stratified 2×2 table, there are three sampling models of particular interest: the row margins fixed model (with fixed group sizes, see also Section 4.2.2), the column margins fixed model (with fixed numbers of successes and failures, see also Section 4.2.3), and the total sum fixed model (random samples, see also Section 4.2.4). Table 10.5 shows a summary.

TABLE 10.5

Designs in stratified 2×2 tables

Design	Specification
Fixed group sizes or stratification on Y_1	n_{1+k}, n_{2+k} fixed for $k = 1, 2, \ldots, K$
Case-control or stratification on Y_2	n_{+1k}, n_{+2k} fixed for $k = 1, 2, \ldots, K$
Random samples within strata	n_{++k} fixed for $k = 1, 2, \ldots, K$

Designs with random samples within strata occur, for instance, in cross-sectional studies. Fixed group sizes occur if either the group sizes are set a priori, like in clinical studies with two treatment groups, or by post-stratification with respect to the variable Y_1. In the latter case, we may have random samples within strata; however, the statistical analysis is done for the groups, formed by post-stratification on Y_1. The case with n_{+1k}, n_{+2k} fixed occurs in case-control studies. The cases are selected and compared with the controls with respect to the grouping variable, which usually is exposure status.

Because the outcome variable is Y_2, our main interest lies in the conditional probability

$$\pi_{1|ik} = \Pr(Y_2 = 1 \mid Y_1 = i, Y_3 = k), \tag{10.1}$$

for $i = 1, 2$ and $k = 1, 2, \ldots, K$. We assume that $\Pr(Y_3 = k)$ contains no information about the association between the outcome Y_2 and the grouping Y_1, given the stratum Y_3. Then, there is no loss of information if we make the assumption that the strata totals, $\{n_{++1}, n_{++2}, \ldots, n_{++K}\}$ are fixed, regardless of whether this is based on pre- or post-stratification.

In the row margins fixed model (with fixed group sizes), we have that the conditional probabilities in each row sum to one. We thereby simplify the notation for the conditional probabilities and use

$$\pi_{1k} = \Pr(Y_2 = 1 \,|\, Y_1 = 1, Y_3 = k) \tag{10.2}$$

and

$$\pi_{2k} = \Pr(Y_2 = 1 \,|\, Y_1 = 2, Y_3 = k), \tag{10.3}$$

for the conditional probabilities of success for Group 1 and Group 2 in the kth table, see also Table 10.6.

TABLE 10.6
Conditional probabilities within the rows of
the kth table

	Variable 2		
Variable 1	Success	Failure	Total
Group 1	π_{1k}	$1 - \pi_{1k}$	1
Group 2	π_{2k}	$1 - \pi_{2k}$	1

10.4 Statistical Models, Effect Measures, and Effect Estimates

Here, as in Chapter 4, we will study three effect measures: the difference between probabilities, the ratio of probabilities, and the odds ratio, all based on the conditional probabilities in Equations 10.2 and 10.3. We are interested in the parameters in the model for the K 2×2 tables, and the effect measures are associated with three link functions. The general link function with only main effects included is given by

$$\text{link}(\pi_{ik}) = \alpha + \beta_i + \gamma_k, \tag{10.4}$$

for $i = 1, 2$ and $k = 1, 2, \ldots, K$, where we for simplicity assume that $\beta_2 = \gamma_1 = 0$, i.e., Group 2 and stratum 1 are reference categories. Note that in Equation 10.4, there is only one main effect of the grouping, i.e., β_1 is the effect of being in Group 1, adjusted for the strata. The parameter vector $\{\gamma_2, \gamma_3, \ldots, \gamma_K\}$ contains the effects of the strata.

When the effect of the grouping varies with the strata, we reformulate Equation 10.4 to a model with interaction:

$$\text{link}(\pi_{ik}) = \alpha + \beta_i + \gamma_k + \delta_{ik}, \tag{10.5}$$

where $\beta_2 = \gamma_1 = \delta_{11} = 0$, and $\delta_{2k} = 0$. In Equation 10.5, β_1 is the effect of the grouping, adjusted for the strata, γ_k is the effect of being in stratum k, and the δ_{ik}s are the interaction effects between the grouping and the stratification. Note that there are $2K$ parameters in Equation 10.5; thus, a model with interactions is saturated.

Alternatively, we can express Equation 10.5 as

$$\mathrm{link}(\pi_{ik}) = \alpha + \beta_{ik} + \gamma_k, \tag{10.6}$$

where $\beta_{ik} = \beta_i + \delta_{ik}$. Because $\beta_{2k} = 0$ for $k = 1, 2, \ldots, K$, we have K parameters (the β_{ik}s) for the associations in the 2×2 tables, and $K - 1$ parameters (the γ_ks) for the stratum effects. Under the assumption of no interaction between the groups and the strata, we impose $K - 1$ restrictions on the β_{1k} in Equation 10.6.

As we discussed in Chapter 4, the association in a 2×2 table can be expressed by one parameter. In a stratified 2×2 table, the associations can be expressed by K association measures, one for each 2×2 table. When $\beta_{11} = \beta_{12} = \ldots = \beta_{1K} = \beta$ in Equation 10.6, there is no interaction in the association in the K tables, and the link function has only main effects. Note that we denote the common parameter for the grouping effect by β. Note also that "no interaction" refers to the specific link function; for instance, a positive interaction for a linear link does not necessarily give a positive interaction for a log link or a logit link, see VanderWeele and Knol (2014).

There are three link functions of particular interest: the linear link, the log link, and the logit link. For the general model with interaction (Equation 10.6), applying the linear link gives

$$\pi_{ik} = \alpha_{\mathrm{linear}} + \beta_{ik,\mathrm{linear}} + \gamma_{k,\mathrm{linear}}. \tag{10.7}$$

Applying the log link results in

$$\log(\pi_{ik}) = \alpha_{\log} + \beta_{ik,\log} + \gamma_{k,\log}, \tag{10.8}$$

and applying the logit link gives

$$\log\left(\frac{\pi_{ik}}{1 - \pi_{ik}}\right) = \alpha_{\mathrm{logit}} + \beta_{ik,\mathrm{logit}} + \gamma_{k,\mathrm{logit}}. \tag{10.9}$$

We note that in Equations 10.7–10.9, there is one parameter that describes the association between the outcome and the grouping for each stratum, namely β_{1k}. We apply different interpretations to β_{1k} depending on the link function. Table 10.7 summarizes the effect measures, and their estimates, that arise from the linear, log, and logit links.

Given any of the link functions above, we are interested in the overall effect of the grouping, and there are two roads to follow:

(i) We can assume a model without interaction (Equation 10.4) and obtain a model-based estimate of the grouping effect, for instance, with the principle of maximum likelihood estimation.

TABLE 10.7
Link functions, effect measures, and maximum likelihood (ML) estimates
(under the assumption of a saturated model) for stratified 2 × 2 tables

Link function	Notation	Effect measure	ML estimate
Linear link	$\Delta_k = \beta_{1k}$	Difference between probabilities	$\hat{\Delta}_k = \dfrac{n_{11k}}{n_{1+k}} - \dfrac{n_{21k}}{n_{2+k}}$
Log link	$\phi_k = \exp(\beta_{1k})$	Ratio of probabilities	$\hat{\phi}_k = \dfrac{n_{11k}/n_{1+k}}{n_{21k}/n_{2+k}}$
Logit link	$\theta_k = \exp(\beta_{1k})$	Odds ratio	$\hat{\theta}_k = \dfrac{n_{11k}/n_{12k}}{n_{21k}/n_{22k}}$

(ii) We obtain an estimate of the overall association as some sort of weighted average over the strata.

Option (ii) is particularly appealing for the logit link, for which the Mantel-Haenszel estimate—which is a weighted average of the stratum specific odds ratios—is often preferred to the maximum likelihood estimate (see also the upcoming Sections 10.6 and 10.8).

In a case-control study, the column margins are fixed. The notation for the 2 × 2 table in Section 4.3.1 can be extended to the case with K strata. The parameters of interest are π_{1k} and π_{2k} for $k = 1, 2, \ldots, K$, see Equations 10.2 and 10.3, with effect measures derived from these. In a case-control study, we are able to estimate the parameters $\gamma_{jk} = \Pr(Y_1 = 1 \mid Y_2 = j, Y_3 = k)$ for $j = 1, 2$ and $k = 1, 2, \ldots, K$, see Table 4.8 for the 2 × 2 table. If, however, γ_{jk} are given, it is not possible to derive π_{ik}, for $i = 1, 2$ and $k = 1, 2, \ldots, K$, without further information about the margins. Fortunately, for each 2 × 2 table, the odds ratio is the same, regardless of whether the data originate from a fixed row design or a fixed column design, see also Section 4.8

10.5 Sampling Distributions

The sampling distributions for the stratified 2 × 2 table are straightforward extensions of the ones for the 2 × 2 table in Section 4.3. The most relevant model is the one with fixed row sums (the row margins fixed model). The rows are either fixed by design or by conditioning at the analysis stage. Recall that $\mathbf{n} = \{n_{111}, n_{121}, n_{211}, n_{221}, \ldots, n_{11K}, n_{12K}, n_{21K}, n_{22K}\}$ denotes the observed table and $\mathbf{x} = \{x_{111}, x_{121}, x_{211}, x_{221}, \ldots, x_{11K}, x_{12K}, x_{21K}, x_{22K}\}$ denotes any possible table. The joint probability for the numbers of successes in Group 1 and Group 2 in stratum $k = 1, 2, \ldots, K$ is then

$$f(x_{111}, x_{211}, \ldots, x_{11K}, x_{21K} \,|\, \pi_{11}, \pi_{21}, \ldots, \pi_{1K}, \pi_{2K}; \mathbf{n}_+) = \qquad (10.10)$$

$$\prod_{k=1}^{K} \binom{n_{1+k}}{x_{11k}} \pi_{1k}^{x_{11k}} \left(1 - \pi_{1k}\right)^{n_{1+k} - x_{11k}} \cdot \binom{n_{2+k}}{x_{21k}} \pi_{2k}^{x_{21k}} \left(1 - \pi_{2k}\right)^{n_{2+k} - x_{21k}},$$

where $\mathbf{n}_+ = \{n_{1+1}, n_{2+1}, \ldots, n_{1+K}, n_{2+K}\}$ denotes the fixed row sums. The conditional probabilities of success, π_{1k} and π_{2k}, are given in Equations 10.2 and 10.3, and we regard them here as nuisance parameters.

If we, in addition to assuming that the row sums are fixed, also fix the column sums in each stratum (the both margins fixed model), we obtain the following sampling distribution for the numbers of successes in Group 1:

$$f(x_{111}, x_{112}, \ldots, x_{11K} \,|\, \theta_1, \theta_2, \ldots, \theta_K; \mathbf{n}_{++}) =$$

$$\prod_{k=1}^{K} \frac{\binom{n_{1+k}}{x_{11k}} \binom{n_{2+k}}{n_{+1k} - x_{11k}} \theta_k^{x_{11k}}}{\binom{n_{++k}}{x_{+1k}}}, \qquad (10.11)$$

where $\mathbf{n}_{++} = \{n_{1+1}, n_{2+1}, \ldots, n_{1+K}, n_{2+K}, n_{+11}, n_{+21}, \ldots, n_{+1K}, n_{+2K}\}$ denotes the fixed row and column sums, and $\theta_k = \exp(\beta_k) = \pi_{11k}\pi_{22k}/\pi_{12k}\pi_{21k}$ is the odds ratio in stratum k. In Equation 10.11, we have reduced the number of nuisance parameters to one (θ_k) for each stratum.

If $\theta_1 = \theta_2 = \ldots = \theta_K = \theta$, we have a homogeneous association between the rows (Y_1) and the columns (Y_2). When $\theta = 1$, we say that we have conditional independence between Y_1 and Y_2 given Y_3. For the case of $\theta_1 = \theta_2 = \ldots = \theta_K = \theta$, Gart (1970) has shown that the conditional distribution of the numbers of successes in Group 1 given the marginals $\mathbf{n}_{++}$ equals

$$f(x_{111}, x_{112}, \ldots, x_{11K} \,|\, \theta, \mathbf{n}_{++}) =$$

$$\frac{\displaystyle\prod_{k=1}^{K} \binom{n_{1+k}}{x_{11k}} \binom{n_{2+k}}{n_{+1k} - x_{11k}} \theta^{x_{11k}}}{\displaystyle\sum_{\Omega(\mathbf{x}|\mathbf{n}_{++})} \prod_{k=1}^{K} \binom{n_{1+k}}{x_{11k}} \binom{n_{2+k}}{n_{+1k} - x_{11k}} \theta^{x_{11k}}}, \qquad (10.12)$$

where $\Omega(\mathbf{x}|\mathbf{n}_{++})$ denotes the set of all samples $\{x_{111}, x_{112}, \ldots, x_{11K}\}$ with given marginals $\mathbf{n}_{++}$. When $\theta = 1$, the exact conditional distribution is the product of K hypergeometric distributions, given by inserting $\theta = 1$ in Equation 10.11, see also Cox (1966).

Here, $T(\mathbf{n}) = \sum_{k=1}^{K} n_{11k}$ is a sufficient statistic for θ. The conditional distribution of T given the marginals $\mathbf{n}_{++}$ equals

$$f(x_{111}, x_{112}, \ldots, x_{11K} \mid \theta, \mathbf{n}_{++}, t) = \frac{c(t)\theta^t}{\sum\limits_{t_{\min}}^{t_{\max}} c(u)\theta^u}, \qquad (10.13)$$

where

$$c(t) = \sum_{\Omega(\mathbf{x}|\mathbf{n}_{++}, t)} \prod_{k=1}^{K} \binom{n_{1+k}}{x_{11k}} \binom{n_{2+k}}{n_{+1k} - x_{11k}}$$

and $\Omega(\mathbf{x}|\mathbf{n}_{++}, t)$ is the set of all tables with $t = \sum_{k=1}^{K} n_{11k}$ and given marginals $\mathbf{n}_{++}$, $t_{\min} = \sum_{k=1}^{K} \max(0, n_{+1k} - n_{1+k})$, and $t_{\max} = \sum_{k=1}^{K} \min(n_{1+k}, n_{+1k})$.

10.6 A Guide to the Analyses of Stratified 2×2 Tables

When the aim of a study is to estimate the association between the grouping (Y_1) and the outcome (Y_2), the association estimated from a 2×2 table of Y_1 and Y_2 is called a crude, or unadjusted, effect estimate. The crude effect is estimated from the marginal association between Y_1 and Y_2, and it may be a biased estimate if the relationship between Y_1 and Y_2 is distorted by a third variable Y_3, which in our setting is the stratification variable. The stratification variable may be a confounder of the association between grouping and outcome, see, for instance Greenland and Rothman (2008). The marginal association will not necessarily be identical to the conditional associations, even if the associations between Y_1 and Y_2 are homogenous over the K strata.

Effect modification occurs when the association between Y_1 and Y_2 varies over the levels of the stratification variable Y_3. In that case, we say that the association is heterogeneous, and when we define a statistical model for the association, we use an interaction term between Y_1 and Y_3 to model the effect modification. Our main focus will be on so-called *fixed effects models*. In a fixed effects model, the heterogeneity over the strata is described by stratum specific parameters, and the remaining variability is fully described by the binomial distribution. Alternatively, the heterogeneity can be described by an inter-stratum specific random variable, with just one common effect parameter for the association between Y_1 and Y_2. This is the *random effects model*, which we will consider in more detail in Section 10.15.

In a study that is designed with pre-stratification based on Y_3—such as in meta-analysis, wherein the included studies form the stratification variable—the first step in the analysis is to estimate the effect in each stratum. Then, we proceed to test whether there is homogeneity of the effect measures over the strata. Provided we can assume homogeneity, we then estimate the common effect measure.

In epidemiological studies, post-stratification on Y_3 is often relevant. First, we estimate the crude association between Y_1 and Y_2. Then, we estimate the effect measures in each strata and carry out a test of homogeneity. If homogeneity can be assumed, we estimate the common effect measure, which we refer to as the adjusted effect measure. If it differs substantially from the crude estimate, *confounding* may be present. If, on the other hand, there is heterogeneity of the effect measure, Y_3 is an effect modifier of the association between Y_1 and Y_2.

Whether the analysis is based on pre- or post-stratification, if there is heterogeneity of the associations over the strata, we report the estimates of the stratified associations. If homogeneity is assumed, we report the estimate of the common effect measure. In the upcoming Section 10.8, we describe four estimates of the common effect: the Mantel-Haenszel estimate, the inverse variance estimate, the Peto odds ratio estimate, and the maximum likelihood estimate. Homogeneity of effects over the strata is not a necessary assumption for the Mantel-Haenszel estimate nor the inverse variance estimate; an overall grouping effect can be estimated regardless of an underlying assumption of homogeneity over the strata. Landis et al. (1978) use the term *average partial association* for the overall effect. Note, however, that if there is heterogeneity and we assume homogeneity, the standard error of the overall effect is underestimated, and the resulting P-values will be downwardly biased. But more importantly, the interpretation of the average partial association is not straightforward in the presence of heterogeneity.

The hypothesis testing is defined as follows. The general link function is given in Equation 10.6; however, we repeat it here for convenience:

$$\text{link}(\pi_{ik}) = \alpha + \beta_{ik} + \gamma_k.$$

There are three null hypotheses that can be tested:

$$H_0 : \beta_{11} = \beta_{12} = \ldots = \beta_{1K} = 0,$$

$$H_1 : \beta_{11} = \beta_{12} = \ldots = \beta_{1K} = \beta,$$

or

$$H_2 = (H_0 \mid H_1) : \beta = 0.$$

Here, H_0 means independence within each stratum, H_1 means homogeneity conditioned on the strata, i.e., no interaction terms in the link function (see Section 10.4). Further, H_2 means conditional independence, under the assumption of no interaction. The differences between testing H_0 and H_2 has caused some confusion, see Cheng et al. (2010). Landis et al. (1978) use the term testing for *partial association* when testing H_2, to distinguish it from conditional independence under H_0. If interaction is present, a test of H_2 may be considered a test for average partial association, see also Section 10.13.3 and Sections 11.4–11.6.

The relationship between the hypotheses has led to a two-step procedure.

First, test the null hypothesis H_1, i.e., test for homogeneity of effects over the strata. If H_1 is rejected, report the stratum specific effect estimates. If H_1 is not rejected, estimate the common effect and perform a test for the null hypothesis H_2.

10.7 Confidence Intervals for the Stratum Specific Effects

Whether or not we can regard the effect as homogenous over the strata, we can always estimate the effect in each stratum with a confidence interval. In many situations, for instance in meta-analysis, this will be a natural first step in the analysis of stratified 2×2 tables. Chapter 4 presented confidence intervals for the difference between probabilities (Section 4.5), the ratio of probabilities (Section 4.7), and the odds ratio (Section 4.8). In principle, any of the intervals in those sections can be used; however, if the main purpose is to estimate a common effect measure over the K strata, as is often the case with meta-analysis, we usually use a simple Wald interval for the stratum specific estimates. For the difference between probabilities, the Wald interval is

$$\hat{\Delta}_k \pm z_{\alpha/2} \widehat{\mathrm{SE}}(\hat{\Delta}_k);$$

for the ratio of probabilities, the Wald interval (on the log scale) is

$$\log \left(\hat{\phi}_k \right) \pm z_{\alpha/2} \widehat{\mathrm{SE}}(\hat{\phi}_k);$$

and for the odds ratio, The Wald interval (on the log scale) is

$$\log \left(\hat{\theta}_k \right) \pm z_{\alpha/2} \widehat{\mathrm{SE}}(\hat{\theta}_k).$$

Here, $z_{\alpha/2}$ is the upper $\alpha/2$ percentile of the standard normal distribution. The effect estimates and standard errors are shown in Table 10.8. The resulting confidence intervals correspond to the Wald interval in Section 4.5.2 for the difference between probabilities, the Katz log interval in Section 4.7.2 for the ratio of probabilities, and the Woolf logit interval in Section 4.8.2 for the odds ratio.

If one or more of the observed cell counts in a stratum are zero, we may not be able to estimate the effect or its standard error in that stratum. For the log link, this happens when $n_{11k} = 0$ or $n_{21k} = 0$. For the logit link, this happens when either one of the four cell counts is zero. In such cases, we follow the standard convention of adding $1/2$ observations to all cells in that stratum. This adjustment is due to Haldane (1956), who used this approach to estimate odds ratios.

TABLE 10.8

Stratum specific effect estimates and standard errors for three effect
measures

Effect measure	Estimate	Standard error
Difference between probabilities (Δ_k)	$\dfrac{n_{11k}}{n_{1+k}} - \dfrac{n_{21k}}{n_{2+k}}$	$\sqrt{\dfrac{n_{11k} \cdot n_{12k}}{n_{1+k}^3} + \dfrac{n_{21k} \cdot n_{22k}}{n_{2+k}^3}}$
Ratio of probabilities (ϕ_k)	$\dfrac{n_{11k}/n_{1+k}}{n_{21k}/n_{2+k}}$	$\sqrt{\dfrac{1}{n_{11k}} - \dfrac{1}{n_{1+k}} + \dfrac{1}{n_{21k}} - \dfrac{1}{n_{2+k}}}$
Odds ratio (θ_k)	$\dfrac{n_{11k}/n_{12k}}{n_{21k}/n_{22k}}$	$\sqrt{\dfrac{1}{n_{11k}} + \dfrac{1}{n_{12k}} + \dfrac{1}{n_{21k}} + \dfrac{1}{n_{22k}}}$

10.8   Estimating a Common Effect

When we can assume homogeneity of the association between the grouping and
the outcome over the strata, the objective of the analysis will be to estimate
the common association. In a randomized clinical trial, for example—or in a
systematic review of several trials on the same subject—the objective is to
estimate the overall treatment effect.

The information from each 2×2 table can be summarized in several ways.
First, we have to decide on the link function and the related effect measure
(see Section 10.4). Second, we have to decide on which estimate to use. In
the following subsections, we consider four estimates: the Mantel-Haenszel estimate, the inverse variance estimate, the Peto odds ratio estimate, and the
maximum likelihood estimate. The maximum likelihood estimate is model-
based, and the estimate is calculated under the assumption of homogeneity of
effects over the strata, which means no interaction effect of the grouping variable and the stratification variable on the outcome. The other estimates are
heuristic, in the sense that they are linear combinations of the effect estimates
in the strata. The Mantel-Haenszel estimate is based on weights given by the
cell numbers and the total numbers in the marginals in the stratified 2×2
tables. The inverse variance estimate uses weights that are inverses of the estimated variances of the stratum specific effect estimates. The inverse variance
estimate is the sample variant of the minimum variance linear estimate, see
Cochran and Carroll (1953). The Peto odds ratio estimate is included because
of its widespread use in meta-analysis, see Palmer and Sterne (2016); however, it tends to perform poorly, and some have recommended that it should
be avoided (Hirji, 2006, p. 271).

10.8.1   Mantel-Haenszel and Inverse Variance Estimates

Three link functions, the linear, log, and logit links, were presented in Section 10.4. These three link functions are directly related to the difference between probabilities (linear link), the ratio of probabilities (log link), and the odds ratio (logit link); see Table 10.7, which includes the maximum likelihood estimates of the three effect measures and their notation.

Now, let $\hat{\psi}_k$ denote the effect estimate in stratum k for an arbitrary link function, where $\hat{\psi}_k = \hat{\Delta}_k$ for the linear link, $\hat{\psi}_k = \hat{\phi}_k$ for the log link, and $\hat{\psi}_k = \hat{\theta}_k$ for the logit link. To be able to estimate the effect in a stratum that has one or more zero cell counts, we use the Haldane (1956) adjustment of adding $1/2$ observations to all cell counts in that stratum. For the ratio of probabilities, we make this adjustment when $n_{11k} = 0$ or $n_{21k} = 0$. For the odds ratio, we make the adjustment when either one of the four cell counts is zero, see also Section 10.7.

The Mantel-Haenszel estimate of the overall effect is given by

$$\hat{\psi}_{\mathrm{MH}} = \frac{\sum_{k=1}^{K} w_k \hat{\psi}_k}{\sum_{k=1}^{K} w_k}. \tag{10.14}$$

The inverse variance estimate (Cochran, 1954a) has weights equal to the inverse of the variance of the stratum specific effect estimates. It has the general form:

$$\hat{\psi}_{\mathrm{IV}} = \frac{\sum_{k=1}^{K} v_k \hat{\psi}_k}{\sum_{k=1}^{K} v_k}. \tag{10.15}$$

Note that in Equation 10.15, $\hat{\psi}_k$ is defined differently than in Equation 10.14. The relative effect measures are combined on the log scale, so that $\hat{\psi}_k = \log\left(\hat{\phi}_k\right)$ for the log link and $\hat{\psi}_k = \log\left(\hat{\theta}_k\right)$ for the logit link. For the linear link, the effect measures are still combined on their natural scale: $\hat{\psi}_k = \hat{\Delta}_k$. Table 10.9 gives the formulas for the weights used in the Mantel-Haenszel and inverse variance estimates.

The inverse variance estimate is the sample variant of the minimum variance linear estimate (MVLE). The inverse variance estimate of the odds ratio is also called the Woolf estimate, see Woolf (1955) and Palmer and Sterne (2016), and as such, it is a MVLE. The Mantel-Haenszel estimate, on the other hand, is derived heuristically, and it is neither a MVLE, a maximum likelihood estimate, nor a conditional maximum likelihood estimate.

The inverse variance estimate and the maximum likelihood estimate (Section 10.8.3) are consistent for large stratum samples; however, they can be

TABLE 10.9

Weights for the Mantel-Haenszel (MH) and inverse variance estimates

Effect measure	MH (w_k)	Inverse variance (v_k)
Difference between probabilities	$\dfrac{n_{1+k}\cdot n_{2+k}}{n_{++k}}$	$1\Big/\left(\dfrac{n_{11k}\cdot n_{12k}}{n_{1+k}^3}+\dfrac{n_{21k}\cdot n_{22k}}{n_{2+k}^3}\right)$
Ratio of probabilities	$\dfrac{n_{1+k}\cdot n_{21k}}{n_{++k}}$	$1\Big/\left(\dfrac{1}{n_{11k}}-\dfrac{1}{n_{1+k}}+\dfrac{1}{n_{21k}}-\dfrac{1}{n_{2+k}}\right)$
Odds ratio	$\dfrac{n_{12k}\cdot n_{21k}}{n_{++k}}$	$1\Big/\left(\dfrac{1}{n_{11k}}+\dfrac{1}{n_{12k}}+\dfrac{1}{n_{21k}}+\dfrac{1}{n_{22k}}\right)$

biased in small samples. The bias of the maximum likelihood estimate is due to the many nuisance parameters, which again is a result of a large number of strata.

For the odds ratio, the Mantel-Haenszel estimate is favorable to both the MVLE and the maximum likelihood estimate in terms of mean square error, which is the sum of the bias and the variance (Breslow, 1981; Hauck, 1984). The Mantel-Haenszel estimate is the solution to the first iteration when solving the maximum likelihood equations, see Breslow (1981), and it may seem counterintuitive that the solution to the first iteration is better than the maximum likelihood estimate itself.

The conditional maximum likelihood estimate of the odds ratio may be of interest as an alternative to the Mantel-Haenszel estimate. It is obtained by conditioning on the marginals, such that $T(\mathbf{n}) = \sum_{k=1}^{K} n_{11k}$ is a sufficient statistic for θ, the common odds ratio (see Section 10.5). The Mantel-Haenszel estimate and the conditional maximum likelihood estimate are both consistent. The conditional likelihood estimate is generally more efficient than the Mantel-Haenszel estimate, in the sense that it has less asymptotic variance. For $\theta = 1$, they are equally efficient; however, the Mantel-Haenszel estimate performs very well because it is, in general, more than 90% efficient for any value of θ and more than 95% efficient when θ is in the range 0.2 to 5.0 (Breslow, 1981; Hauck, 1984; Greenland, 1989).

10.8.2 The Peto Odds Ratio Estimate

Peto and colleagues proposed an alternative estimate of the common odds ratio, based on the cell counts n_{11k}, for $k = 1, 2, \ldots, K$ (Yusuf et al., 1985; Fleiss, 1993). The distribution of n_{11k}, conditional on the marginals, is given in Equation 10.11. When $\theta_k = 1$, the conditional expectation and the variance in the hypergeometric distribution are

$$E(n_{11k} \mid \mathbf{n}_{++}) = \frac{n_{1+k}\cdot n_{+1k}}{n_{++k}} \qquad (10.16)$$

and
$$\text{Var}(n_{11k} \,|\, \mathbf{n}_{++}) = \frac{n_{1+k} \cdot n_{2+k} \cdot n_{+1k} \cdot n_{+2k}}{n_{++k}^2(n_{++k} - 1)}. \tag{10.17}$$

The Peto estimate of the odds ratio in stratum k is given by

$$\log\left(\hat{\theta}_{\text{Peto},k}\right) = \frac{\left[n_{11k} - \text{E}(n_{11k} \,|\, \mathbf{n}_{++})\right]}{\text{Var}(n_{11k} \,|\, \mathbf{n}_{++})},$$

and the inverse variance weighted estimate of the common odds ratio takes the form

$$\log\left(\hat{\theta}_{\text{Peto}}\right) = \frac{\displaystyle\sum_{k=1}^{K}\left[n_{11k} - \text{E}(n_{11k} \,|\, \mathbf{n}_{++})\right]}{\displaystyle\sum_{k=1}^{K}\text{Var}(n_{11k} \,|\, \mathbf{n}_{++})}.$$

For details on the derivations, see for instance, Fleiss (1993) or Deeks et al. (2001, p. 289). Fleiss (1993) also reports results about the close agreement between the logarithm of the odds ratio and the Peto estimate.

The Peto estimate of the odds ratio can work well in studies where the group sizes (n_{1+k} and n_{2+k} for $k = 1, 2, \ldots, K$) are of the same order of magnitude (Fleiss, 1993). Otherwise, the overall estimate may be biased, see Greenland and Salvan (1990). For effects close to $\theta = 1$, there will be only minor bias; however, for effects far from $\theta = 1$, the bias may be considerable. As already mentioned, Hirji (2006) warns that the Peto estimate performs poorly for small and sparse data and adds that it should be avoided.

10.8.3 Maximum Likelihood Estimates

In Section 10.4 (page 432), we noted that there are two roads to follow to estimate the overall effect of the grouping. In Sections 10.8.1 and 10.8.2, we estimated the effect as a weighted average of the effect in each stratum. An alternative approach is to produce a model-based estimate with the principle of maximum likelihood.

We assume that the group sizes (n_{1+k}, n_{2+k}) in each stratum, $k = 1, 2, \ldots, K$, are given, and that the sampling distribution is the one in Equation 10.10. Calculating the maximum likelihood estimate of the common effect measure relies on the assumption of a link function without interaction. Tests of no interaction will be described in Section 10.9. If we insert any of the link functions without interaction into Equation 10.4, we obtain maximum likelihood estimates of the grouping effect and the stratum effects: $\{\hat{\alpha}, \hat{\beta}, \hat{\gamma}_2, \hat{\gamma}_3, \ldots, \hat{\gamma}_K\}$, where $\hat{\beta}$ is the estimate of the common grouping effect.

10.9   Tests of Homogeneity of Effects over the K Strata

In Section 10.6, we recommended testing for the homogeneity of the effect over the strata, before estimating the common effect, and if heterogeneity is present, to report the stratum specific effect estimates. We have three effect measures of interest, related to three link functions. The tests for homogeneity depend on the effect measure; however, they all have the same null hypothesis:

$$H_1 : \beta_{11} = \beta_{12} = \ldots = \beta_{1K} = \beta. \qquad (10.18)$$

The reason why we label this null hypothesis as "H_1" is given in Section 10.6.

For the exponential family, the log odds are the canonical parameters, which make them appealing as an effect measure. Many tests have been proposed for testing homogeneity of the odds ratios over the strata, and several of these will be presented in Section 10.10. But first, we present the Cochran Q, likelihood ratio, and Pearson chi-squared tests, which are versatile tests that can be used regardless of the effect measure.

10.9.1   The Cochran Q Test

Let $\hat{\psi}_k$ denote the effect estimate in stratum k for an arbitrary link function, where $\hat{\psi}_k = \hat{\Delta}_k$ for the linear link, $\hat{\psi}_k = \log\left(\hat{\phi}_k\right)$ for the log link, and $\hat{\psi}_k = \log\left(\hat{\theta}_k\right)$ for the logit link. The Cochran Q test (Cochran, 1954a) can be used with either the Mantel-Haenszel estimate or the inverse variance estimate. If the log or logit link is used, $\hat{\psi}_{\mathrm{MH}}$ and $\hat{\psi}_{\mathrm{IV}}$ are defined on the log scale. Note that for the Mantel-Haenszel estimate, this is different from Section 10.8.1, wherein the Mantel-Haenszel estimate was defined on the natural scale.

The Cochran Q test statistic for the Mantel-Haenszel estimate is

$$Q_{\mathrm{MH}}(\mathbf{n}) = \sum_{k=1}^{K} v_k \left(\hat{\psi}_k - \hat{\psi}_{\mathrm{MH}}\right)^2, \qquad (10.19)$$

and the Cochran Q test statistic for the inverse variance estimate is

$$Q_{\mathrm{IV}}(\mathbf{n}) = \sum_{k=1}^{K} v_k \left(\hat{\psi}_k - \hat{\psi}_{\mathrm{IV}}\right)^2, \qquad (10.20)$$

where the weights v_k depend on the effect measure and are given in Table 10.9. Note that the inverse variance weights (v_k) are used in both Equations 10.19 and 10.20; the Mantel-Haenszel weights (w_k) are only used to combine the stratum specific estimates in Equation 10.14. The test statistics Q_{MH} and Q_{IV} can be approximated by a chi-squared distribution with $K - 1$ degrees of freedom, such that P-values are obtained as

$$P\text{-value} = \Pr\left[\chi^2_{K-1} \geq Q_{\mathrm{MH}}(\mathbf{n})\right],$$

and similarly for Q_{IV}.

For the logit link (the odds ratio) with the inverse variance estimate, the Cochran Q test statistic takes the form

$$Q_{IV}(\mathbf{n}) = \sum_{k=1}^{K} v_k \left[\log\left(\hat{\theta}_k\right) - \log\left(\hat{\theta}_{IV}\right) \right]^2, \tag{10.21}$$

which is also called the Woolf homogeneity statistic (Woolf, 1955).

10.9.2 The Likelihood Ratio Test

Note that Equation 10.6 is without restrictions on the parameters and thereby is a saturated model. The expected cell counts under a saturated model is simply n_{ijk}. Under the null hypothesis $H_1 : \beta_{11} = \ldots = \beta_{1K} = \beta$, the maximum likelihood estimates of the parameters in the general model with only main effects (Equation 10.6) were given in Section 10.8.3, and the expected cell counts can be expressed as

$$m_{i1k} = n_{i+k}\hat{\pi}_{ik}, \quad \text{and} \quad m_{i2k} = n_{i+k}(1 - \hat{\pi}_{ik}), \tag{10.22}$$

for $i = 1, 2$ and $k = 1, 2, \ldots, K$. The $\hat{\pi}_{ik}$ in Equation 10.22 are the estimated conditional probabilities of success obtained by inserting the maximum likelihood parameter estimates derived in Section 10.8.3—under the null hypothesis H_1—into Equation 10.6. The likelihood ratio statistic is then given by

$$T_{LR}(\mathbf{n}) = 2 \sum_{k=1}^{K} \sum_{j=1}^{2} \sum_{i=1}^{2} n_{ijk} \log \frac{n_{ijk}}{m_{ijk}},$$

where we set all terms for which $n_{ijk} = 0$ to zero. For large samples, T_{LR} is chi-squared distributed with $K - 1$ degrees of freedom.

10.9.3 The Pearson Chi-Squared Test

The Pearson chi-squared statistic is given by

$$T_{Pearson}(\mathbf{n}) = \sum_{k=1}^{K} \sum_{j=1}^{2} \sum_{i=1}^{2} \frac{(n_{ijk} - m_{ijk})^2}{m_{ijk}}, \tag{10.23}$$

where the m_{ijk} are given in Equation 10.22. Like the previous tests statistics, $T_{Pearson}$ is approximately chi-squared distributed with $K - 1$ degrees of freedom.

10.9.4 Evaluations of Tests

In this section, we briefly evaluate the tests of homogeneity presented so far. This will not be a thorough evaluation but rather an illustration of some of the properties of the tests. In previous evaluations in this book, we have for the most part used exact calculations of actual significance levels and power for the tests, and exact calculations of coverage probabilities, interval width, and location for the confidence intervals. In this chapter, we consider tests and intervals for stratified 2×2 tables, which are three-way tables in contrast with the two-way tables in earlier chapters. Exact calculations for stratified 2×2 tables are very computer intensive and beyond the scope of this book. Throughout this chapter, we therefore use simulations (with $10\,000$ replications) instead of exact calculations to evaluate the methods. For the tests, we only consider simulated actual significance levels, and for the intervals, we only consider simulated coverage probabilities.

Figures 10.1 and 10.2 show the simulated actual significance levels of the four tests of homogeneity of the difference between probabilities over $K = 3$ strata. In both figures, the effect is set to $\Delta_k = -0.2$ for $k = 1, 2, 3$, thus the null hypothesis of homogeneity is satisfied. The only difference between the two situations is the sample size. In Figure 10.1, we have a sample size of 40 observations in each stratum, evenly distributed between the two groups. In Figure 10.2, the sample size is 100 observations in each stratum. As we can see in both figures, the Cochran Q test, regardless of whether it is based on the Mantel-Haenszel or inverse variance estimate, has actual significance levels that are very liberal at about 7% (Figure 10.1) and 6% (Figure 10.2), whereas the likelihood ratio and Pearson chi-squared tests have actual significance levels much closer to the nominal level of 5%. The Pearson chi-squared test performs a little bit better than the likelihood ratio test.

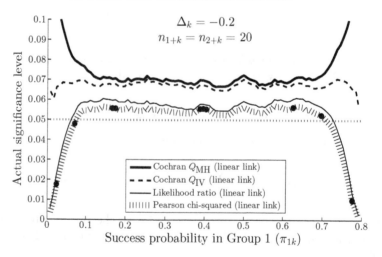

FIGURE 10.1
Simulated actual significance levels of four tests for homogeneity of the difference between probabilities over $K = 3$ strata

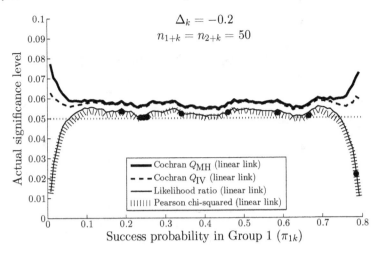

FIGURE 10.2
Simulated actual significance levels of four tests for homogeneity of the difference between probabilities over $K = 3$ strata

If we have an unbalanced sample size across the strata, for instance with $n_{i+1} = 10$, $n_{i+2} = 20$, and $n_{i+3} = 60$ (for $i = 1, 2$), all tests increase their actual significance level, and in particular the two Cochran Q tests, which display very undesirable performance, see Figure 10.3. Only the Pearson chi-squared test performs satisfactorily in this situation.

Figures 10.4–10.9 show plots of the actual significance levels of the same four tests, only now for the ratio of probabilities (Figures 10.4–10.6) and the odds ratio (Figures 10.7–10.9). The sample sizes and the effect sizes are similar to those for the difference between probabilities in Figures 10.1–10.3. The one thing that is common across all nine figures is the excellent performance of the Pearson chi-squared test. The Cochran Q tests, which had actual significance levels quite a lot above the nominal level for the difference between probabilities, have actual significance levels most often below the nominal level for the two relative effect measures. Sometimes, this conservatism is quite severe, and it is only when $n_{i+k} = 50$ that the performance of the Cochran Q tests can be regarded as acceptable, and then only for parts of the parameter space. The likelihood ratio test performs generally a little worse for the ratio of probabilities and the odds ratio than it did for the difference between probabilities. It sometimes has actual significance levels close to the nominal level; however, most often its actual significance levels are quite a lot higher than the nominal level. As a rule that applies to all the tests, performance deteriorates when the sample sizes in the strata are not equal.

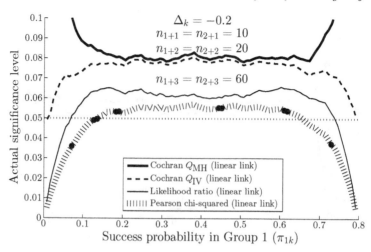

FIGURE 10.3
Simulated actual significance levels of four tests for homogeneity of the difference between probabilities over $K = 3$ strata

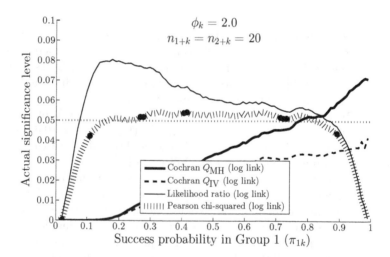

FIGURE 10.4
Simulated actual significance levels of four tests for homogeneity of the ratio of probabilities over $K = 3$ strata

10.10 Additional Tests of Homogeneity of Odds Ratios over the K Strata

As for 2×2 tables, the odds ratio is an effect measure of great interest in $2 \times 2 \times K$ tables. It is applicable for all the sampling designs in Table 10.5,

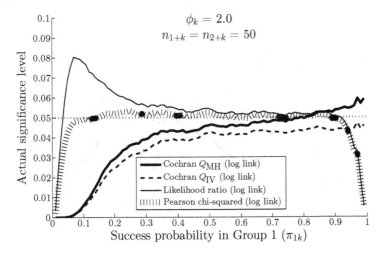

FIGURE 10.5
Simulated actual significance levels of four tests for homogeneity of the ratio
of probabilities over $K = 3$ strata

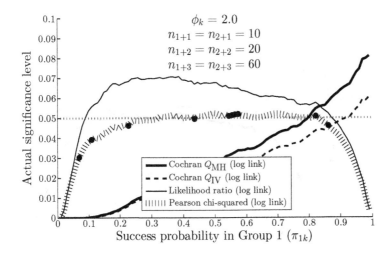

FIGURE 10.6
Simulated actual significance levels of four tests for homogeneity of the ratio
of probabilities over $K = 3$ strata

and it has agreeable mathematical properties. For the exponential family, the
log odds is the canonical parameter, and a variety of tests can be derived
based on conditioning on the marginals.

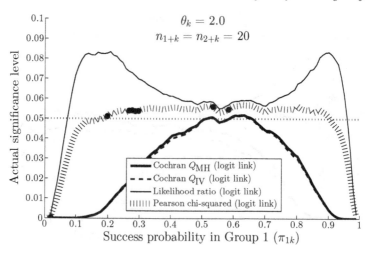

FIGURE 10.7
Simulated actual significance levels of four tests for homogeneity of the odds
ratio over $K = 3$ strata

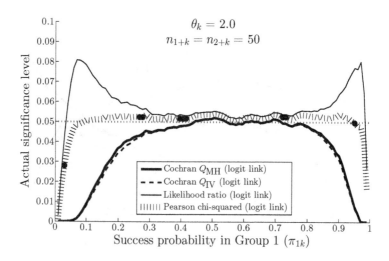

FIGURE 10.8
Simulated actual significance levels of four tests for homogeneity of the odds
ratio over $K = 3$ strata

We denote the odds ratio in stratum k by θ_k, and the null hypothesis of
interest—here labeled "H_1" for reasons laid out in Section 10.6—is

$$H_1 : \theta_1 = \theta_2 = \ldots = \theta_K.$$

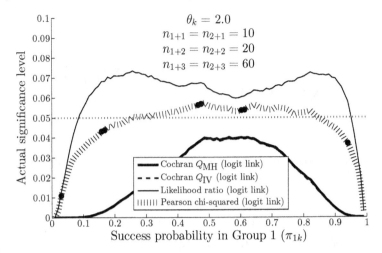

FIGURE 10.9
Simulated actual significance levels of four tests for homogeneity of the odds ratio over $K = 3$ strata

In the following subsections, we first present tests that are expected to have agreeable properties for large samples, then we present exact and mid-P tests, which are suitable for small samples.

10.10.1 The Breslow-Day Test

Breslow and Day (1980, p. 142) proposed a test for homogeneity of odds ratios as an alternative to the Woolf (homogeneity) test in Equation 10.21. The Woolf test and the Breslow-Day test usually produce similar results for large cell frequencies; however, for sparse data, the performance of the Woolf test is poor, and it is thus not recommended for general use (Breslow and Day, 1980, p. 143).

The Breslow-Day test was corrected by Tarone (1985), a correction that was wholeheartedly embraced by Breslow in his seminal paper on the case-control study (Breslow, 1996). Both versions—the original test from Breslow and Day (1980) and the corrected one from Tarone (1985) and Breslow (1996)—are commonly referred to as the *Breslow-Day test*, although the corrected test is sometimes referred to as the *Tarone test*. Because the uncorrected test is not recommended for small and sparse data, make sure that when using statistical software packages, you get the version with the Tarone correction.

Let $\hat{\theta}_{\text{MH}}$ denote the Mantel-Haenszel estimate of the odds ratio, which is assumed to be constant over the strata. Let $m_{11k} = \text{E}(n_{11k}; \hat{\theta}_{\text{MH}})$ be the expected cell count in cell $(1,1)$ in the kth table, given the Mantel-Haenszel

estimate of the common odds ratio. The (corrected) Breslow-Day statistic has
the form

$$T_{BD}(\mathbf{n}) = \sum_{k=1}^{K} \frac{(n_{11k} - m_{11k})^2}{v_{11k}} - \frac{\left(\sum_{k=1}^{K} n_{11k} - \sum_{k=1}^{K} m_{11k}\right)^2}{\sum_{k=1}^{K} v_{11k}}, \qquad (10.24)$$

where the second part of Equation 10.24 is the Tarone correction. Here, m_{11k}
is the solution to the quadratic equation

$$\hat{\theta}_{MH} = \frac{m_{11k}(n_{2+k} - n_{+1k} + m_{11k})}{(n_{+1k} - m_{11k})(n_{1+k} - m_{11k})} \qquad (10.25)$$

that is positive and less than both n_{1+k} and n_{+1k}. The two solutions to Equation 10.25 are

$$\frac{-s_k + \sqrt{s_k^2 - 4rt_k}}{2r} \quad \text{and} \quad \frac{-s_k - \sqrt{s_k^2 - 4rt_k}}{2r},$$

where $s_k = n_{2+k} - n_{+1k} + \hat{\theta}_{MH}(n_{+1k} + n_{1+k})$, $r = 1 - \hat{\theta}_{MH}$, and $t_k = -\hat{\theta}_{MH} \cdot n_{1+k} \cdot n_{+1k}$. Further, v_{11k} in Equation 10.24 is an estimate of the variance of n_{11k} under the assumption of a common odds ratio:

$$v_{11k} = \left(\frac{1}{m_{11k}} + \frac{1}{n_{+1k} - m_{11k}} + \frac{1}{n_{1+k} - m_{11k}} + \frac{1}{n_{2+k} - n_{+1k} + m_{11k}}\right)^{-1}.$$

For details, see Breslow and Day (1980, p. 142) or Hosmer et al. (2013, p. 85).
For large sample sizes, T_{BD} is well approximated by the chi-squared distribution with $K - 1$ degrees of freedom.

As an alternative to the Breslow-Day test in Equation 10.24, we can use
the unconditional maximum likelihood estimate $\hat{\theta}_{ML}$ instead of the Mantel-Haenszel estimate $\hat{\theta}_{MH}$. We then need to calculate $m_{11k} = E(n_{11k}; \hat{\theta}_{ML})$ from
Equation 10.25 with $\hat{\theta}_{ML}$ on the left-hand side. The resulting test is an unconditional score test, see, for instance, Reis et al. (1999).

As pointed out by, for instance, Breslow (1981), the approximation to the
chi-squared distribution may be poor when the data are sparse. An alternative
test for sparse data, based on the maximum conditional likelihood estimate of
the common odds ratio, was considered by Liang and Self (1985), and which
we outline in Section 10.10.3. Another alternative is to calculate an exact or
mid-P test, which we will do in the next section.

10.10.2   Exact and Mid-P Tests

When the data are sparse, the asymptotic tests in Section 10.10.1 are inappropriate because the chi-squared approximation may not hold. We can derive an exact test based on the Breslow-Day statistic by conditioning on $T(\mathbf{n}) = \sum_{k=1}^{K} n_{11k}$, which is a sufficient statistic for θ. Let $\mathbf{n}$ denote the observed table, and let $\mathbf{x}$ denote any possible table. The P-value for the *Breslow-Day exact test* is

$$P\text{-value} = \sum_{\Omega(\mathbf{x}|\mathbf{n}_{++},t)} I\big[T_{\mathrm{BD}}(\mathbf{x}) \geq T_{\mathrm{BD}}(\mathbf{n})\big] f(x_{111}, x_{112}, \ldots, x_{11K} \mid \theta, \mathbf{n}_{++}, t),$$

$$(10.26)$$

where $\Omega(\mathbf{x}|\mathbf{n}_{++}, t)$ is the set of possible tables with all marginal fixed and $t = \sum_{k=1}^{K} n_{11k}$, $I()$ is the indicator function, $T_{\mathrm{BD}}()$ is the Breslow-Day statistic in Equation 10.24, and $f()$ is the conditional sampling distribution in Equation 10.13.

Exact conditional tests can be conservative; however, the mid-P approach (see Section 1.10 for a general description), which involves a small modification to the exact conditional test, can reduce this conservatism. The mid-P value of the *Breslow-Day mid-P test* is calculated by

$$\text{mid-}P\text{ value} = \sum_{\Omega(\mathbf{x}|\mathbf{n}_{++},t)} I\big[T_{\mathrm{BD}}(\mathbf{x}) > T_{\mathrm{BD}}(\mathbf{n})\big] f(\mathbf{x} \mid \theta, \mathbf{n}_{++}, t)$$

$$+\ 0.5 \cdot \sum_{\Omega(\mathbf{x}|\mathbf{n}_{++},t)} I\big[T_{\mathrm{BD}}(\mathbf{x}) = T_{\mathrm{BD}}(\mathbf{n})\big] f(\mathbf{x} \mid \theta, \mathbf{n}_{++}, t), \quad (10.27)$$

where $f(\mathbf{x} \mid \theta, \mathbf{n}_{++}, t)$ is a shorthand for $f(x_{111}, x_{112}, \ldots, x_{11K} \mid \theta, \mathbf{n}_{++}, t)$.

Another exact test is obtained with the point probability $f()$ as test statistic. This is the *Zelen exact test*, first suggested by Zelen (1971), see also Gart (1970, 1971). The P-value for the Zelen exact test is thus

$$P\text{-value} = \sum_{\Omega(\mathbf{x}|\mathbf{n}_{++},t)} I\big[f(\mathbf{x} \mid \theta, \mathbf{n}_{++}, t) \leq f(\mathbf{n} \mid \theta, \mathbf{n}_{++}, t)\big] f(\mathbf{x} \mid \theta, \mathbf{n}_{++}, t),$$

and the Zelen mid-P test is

$$\text{mid-}P\text{ value} = \sum_{\Omega(\mathbf{x}|\mathbf{n}_{++},t)} I\big[f(\mathbf{x} \mid \theta, \mathbf{n}_{++}, t) > f(\mathbf{n} \mid \theta, \mathbf{n}_{++}, t)\big] f(\mathbf{x} \mid \theta, \mathbf{n}_{++}, t)$$

$$+\ 0.5 \cdot \sum_{\Omega(\mathbf{x}|\mathbf{n}_{++},t)} I\big[f(\mathbf{x} \mid \theta, \mathbf{n}_{++}, t) = f(\mathbf{n} \mid \theta, \mathbf{n}_{++}, t)\big] f(\mathbf{x} \mid \theta, \mathbf{n}_{++}, t).$$

Zelen (1971) also gave an asymptotic approximation to the exact test; however, this was later shown to be incorrect by Halperin et al. (1977).

Yet another exact and mid-P test can be calculated with the Pearson chi-squared statistic in Equation 10.23. Simply substitute T_{Pearson} for T_{BD} in Equation 10.26 to obtain the *Pearson exact test*, and substitute T_{Pearson} for T_{BD} in Equation 10.27 to obtain the *Pearson mid-P test*.

The mid-P tests in this section are, as far as we are aware (per April 2017), not available in any major software package, and because they are quite difficult to calculate, we will not consider them further.

10.10.3 The Conditional Score Test

Let $\hat{\theta}_{\mathrm{CML}}$ denote the conditional maximum likelihood estimate of the common odds ratio under the assumption of homogeneity of the odds ratios across strata. With standard theory for the one-parameter exponential family (Lehmann and Casella, 1998, p. 450), we find that the conditional maximum likelihood estimate is the solution to the likelihood equation

$$\sum_{k=1}^{K} n_{11k} = \sum_{k=1}^{K} \mathrm{E}(n_{11k} \,|\, \hat{\theta}_{\mathrm{CML}}),$$

which must be solved iteratively. Mehta et al. (1985), Vollset and Hirji (1991), and Martin and Austin (1991) have proposed methods for computing the conditional maximum likelihood estimate.

The variance of the conditional maximum likelihood estimate is (Liang and Self, 1985)

$$1 \left/ \sum_{k=1}^{K} \mathrm{Var}(n_{11k} \,|\, \hat{\theta}_{\mathrm{CML}}), \right.$$

where $\mathrm{E}(n_{11k} \,|\, \hat{\theta}_{\mathrm{CML}})$ and $\mathrm{Var}(n_{11k} \,|\, \hat{\theta}_{\mathrm{CML}})$ in the above equation are given by Satten and Kupper (1990) as

$$\mathrm{E}(n_{11k} \,|\, \hat{\theta}_{\mathrm{CML}}) =$$

$$\frac{n_{1+k} n_{+1k}}{n_{2+k} - n_{+1k} - 1} \frac{F(1 - n_{1+k}; 1 - n_{+1k}; n_{2+k} - n_{+1k} + 2; \hat{\theta}_{\mathrm{CML}})}{F(-n_{1+k}; -n_{+1k}; n_{2+k} - n_{+1k} + 1; \hat{\theta}_{\mathrm{CML}})},$$

where $F(a; b; c; t)$ is the Fisher non-central hypergeometric distribution, see also Section 4.3, and

$$\mathrm{Var}(n_{11k} \,|\, \hat{\theta}_{\mathrm{CML}}) = \frac{\mathrm{E}(n_{11k} \,|\, \hat{\theta}_{\mathrm{CML}}) \left[n_{2+k} - n_{+1k} + \mathrm{E}(n_{11k} \,|\, \hat{\theta}_{\mathrm{CML}}) \right]}{\hat{\theta}_{\mathrm{CML}} - 1}$$

$$- \frac{\hat{\theta}_{\mathrm{CML}} \left[n_{1+k} - \mathrm{E}(n_{11k} \,|\, \hat{\theta}_{\mathrm{CML}}) \right] \left[n_{+1k} - \mathrm{E}(n_{11k} \,|\, \hat{\theta}_{\mathrm{CML}}) \right]}{\hat{\theta}_{\mathrm{CML}} - 1}.$$

If $n_{2+k} - n_{+1k} - 1 \leq 0$, the columns and the rows in stratum k can be interchanged, see Satten and Kupper (1990).

The conditional score test statistic has the form

$$T_{\mathrm{CST}}(\mathbf{n}) = \sum_{k=1}^{K} \frac{\left[n_{11k} - \mathrm{E}(n_{11k} \,|\, \hat{\theta}_{\mathrm{CML}}) \right]^{2}}{\mathrm{Var}(n_{11k} \,|\, \hat{\theta}_{\mathrm{CML}})}.$$

The reference distribution for T_{CST} is the chi-squared distribution with $K - 1$ degrees of freedom; see Liang and Self (1985) for more information.

An *exact conditional score test* is obtained by substituting T_{CST} for T_{BD} in Equation 10.26. In a similar manner, we obtain the *mid-P conditional score test* by substituting T_{CST} for T_{BD} in Equation 10.27.

10.10.4 The Peto Test

The Peto test for homogeneity is defined by the test statistic:

$$T_{\text{Peto}}(\mathbf{n} \mid \mathbf{n}_{++}) = \sum_{k=1}^{K} \text{Var}(n_{11k} \mid \mathbf{n}_{++}) \left[\log \left(\hat{\theta}_{\text{Peto},k} \right) - \log \left(\hat{\theta}_{\text{Peto}} \right) \right]^2,$$

where the terms in the equation are defined in Section 10.8.2. The Peto test statistic is asymptotically chi-squared distributed with $K - 1$ degrees of freedom.

10.10.5 Evaluations of Tests

Figures 10.10–10.12 show simulated actual significance levels of the Breslow-Day and Peto tests of homogeneity of odds ratios, with the same parameter values as in Section 10.9.4. Compared with the results of the Cochran Q, likelihood ratio, and Pearson chi-squared tests in Figures 10.7–10.9, the Breslow-Day test performs similarly to the Cochran Q tests, only somewhat less conservatively. It performs quite well for the case with $n_{i+k} = 50$; however, it does not perform as well as the Pearson chi-squared test, and it can be very conservative for small or large π_{1k}. The Peto test is more stable across the range of π_{1k}; however, it is often more conservative than the Breslow-Day test in the parts of the parameter space that are likely to be of most relevance.

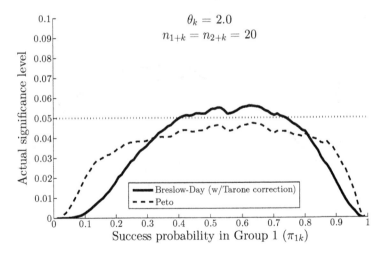

FIGURE 10.10
Simulated actual significance levels of the Breslow-Day and Peto tests for homogeneity of the odds ratio over $K = 3$ strata

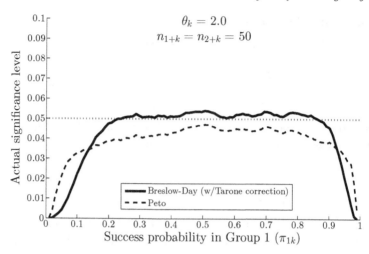

FIGURE 10.11
Simulated actual significance levels of the Breslow-Day and Peto tests for homogeneity of the odds ratio over $K = 3$ strata

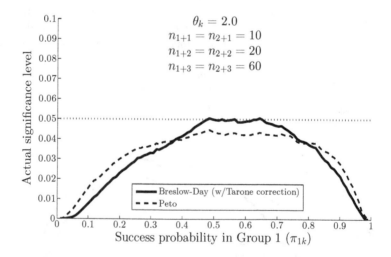

FIGURE 10.12
Simulated actual significance levels of the Breslow-Day and Peto tests for homogeneity of the odds ratio over $K = 3$ strata

10.11 Tests of a Common Difference between Probabilities

Provided that we have reasons to assume homogeneity of effects over the strata, we are interested in estimating the common effect, and testing the null hypothesis of no effect. In this section, we consider the difference between probabilities, which is the effect measure of the model with linear link. The difference between probabilities is of particular interest in meta-analysis, see, for instance, Palmer and Sterne (2016) or Hirji et al. (2012). In Sections 10.12 and 10.13, we consider the ratio of probabilities and the odds ratio, respectively.

The null hypothesis of interest is

$$H_2 : \Delta = 0, \tag{10.28}$$

where $\Delta = \Delta_1 = \Delta_2 = \ldots = \Delta_K$ and $\Delta_k = \beta_{1k}$, $k = 1, 2, \ldots, K$, see Equation 10.7. This null hypothesis states that there is conditional independence between the grouping and the outcome. The rationale for referring to the null hypothesis as "H_2" in this case is provided in Section 10.6.

10.11.1 The Likelihood Ratio Test

The likelihood ratio test is a versatile test, it works regardless of the link function. We present it here for the difference between probabilities; however, the expression is the same also for the ratio of probabilities and the odds ratio.

Under the null hypothesis in Equation 10.28—or the general null hypothesis $\beta = 0$, which also apply to the log and logit links—the expected cell counts are

$$m_{ijk}^* = \frac{n_{i+k} \cdot n_{+jk}}{n_{++k}}.$$

In Section 10.9.2, the expected cell counts under the null hypothesis H_1 : $\beta_{11} = \beta_{12} = \ldots = \beta_{1K} = \beta$ were given as m_{ijk}, see Equation 10.22. The likelihood ratio test for testing H_2 is

$$T_{\mathrm{LR}}(\mathbf{n}) = 2 \sum_{k=1}^{K} \sum_{j=1}^{2} \sum_{i=1}^{2} m_{ijk} \log \frac{m_{ijk}}{m_{ijk}^*},$$

where we set all terms for which $m_{ijk} = 0$ to zero. For large samples, T_{LR} is chi-squared distributed with one degree of freedom.

10.11.2 The Pearson Chi-Squared Test

Rao (1965, p. 329) proposed a Pearson-type statistic for testing nested hypotheses, see also Agresti and Ryu (2010). This test can be applied to stratified 2 × 2 tables. With m_{ijk} and m_{ijk}^* from the previous section, we have that the Pearson chi-squared statistic is

$$T_{\mathrm{Pearson}}(\mathbf{n}) = \sum_{k=1}^{K} \sum_{j=1}^{2} \sum_{i=1}^{2} \frac{(m_{ijk} - m_{ijk}^*)^2}{m_{ijk}^*}, \tag{10.29}$$

which is approximately chi-squared distributed with one degree of freedom.

Like the likelihood ratio test in the previous section, the Pearson chi-squared test is applicable also for the ratio of probabilities and the odds ratio with the same notation as in Equation 10.29.

10.11.3   The Wald Test Based on the Mantel-Haenszel Estimate

The Mantel-Haenszel estimate of the common difference between probabilities is

$$\hat{\Delta}_{\text{MH}} = \frac{\sum_{k=1}^{K} w_k \hat{\Delta}_k}{\sum_{k=1}^{K} w_k},$$

where the weights w_k are given in Table 10.9. Greenland and Robins (1985)—see also Deeks et al. (2001) or Palmer and Sterne (2016, p. 24)—showed that the estimated standard error of the Mantel-Haenszel estimate is

$$\widehat{\text{SE}}(\hat{\Delta}_{\text{MH}}) = \sqrt{A/B^2},$$

where

$$A = \sum_{k=1}^{K} \left(n_{11k}n_{12k}n_{2+k}^3 + n_{21k}n_{22k}n_{1+k}^3\right)/n_{1+k}n_{2+k}n_{++k}^2$$

and

$$B = \sum_{k=1}^{K} n_{1+k}n_{2+k}/n_{++k}.$$

The Wald test statistic for the Mantel-Haenszel estimate is then

$$Z_{\text{Wald}}^{\text{MH}}(\mathbf{n}) = \frac{\hat{\Delta}_{\text{MH}}}{\widehat{\text{SE}}(\hat{\Delta}_{\text{MH}})},$$

with asymptotic P-value equal to

$$P\text{-value} = \Pr\left[Z \geq \left|Z_{\text{Wald}}^{\text{MH}}(\mathbf{n})\right|\right],$$

where Z is a standard normal variable.

10.11.4 The Wald Test Based on the Inverse Variance Estimate

The inverse variance estimate of the common difference between probabilities is

$$\hat{\Delta}_{\text{IV}} = \frac{\sum_{k=1}^{K} v_k \hat{\Delta}_k}{\sum_{k=1}^{K} v_k},$$

where the weights v_k are given in Table 10.9. The estimated standard error of the inverse variance estimate is

$$\widehat{\text{SE}}(\hat{\Delta}_{\text{IV}}) = 1 \Big/ \sqrt{\sum_{k=1}^{K} v_k}.$$

The Wald test statistic for the inverse variance estimate is then

$$Z_{\text{Wald}}^{\text{IV}}(\mathbf{n}) = \frac{\hat{\Delta}_{\text{IV}}}{\widehat{\text{SE}}(\hat{\Delta}_{\text{IV}})},$$

and the reference distribution of $Z_{\text{Wald}}^{\text{IV}}$ is the standard normal distribution.

10.11.5 Evaluations of Tests

The Pearson chi-squared test of a common difference between probabilities does not perform as well as the Pearson chi-squared test for homogeneity, see Figures 10.13–10.15. All the tests in these figures have actual significance levels above the nominal level for almost the entire parameter space, and sometimes by quite a bit. The Wald test based on the Mantel-Haenszel estimate is by far the best test; however, it too can violate the nominal level by a noticeable margin.

10.12 Tests of a Common Ratio of Probabilities

The ratio of probabilities is the effect measure of the model with log link. Like the difference between probabilities, the ratio of probabilities is an effect measure of particular interest in meta-analysis (Palmer and Sterne, 2016; Hirji et al., 2012). The null hypothesis of conditional independence between the grouping and the outcome is now

$$H_2 : \phi = 1,$$

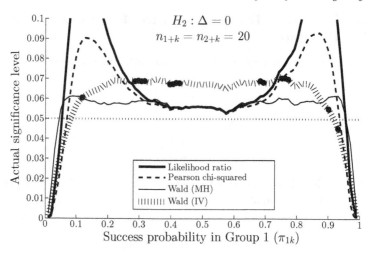

FIGURE 10.13
Simulated actual significance levels of four tests of a common difference between probabilities $(K = 3)$

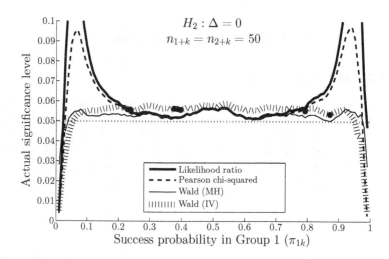

FIGURE 10.14
Simulated actual significance levels of four tests of a common difference between probabilities $(K = 3)$

where $\phi = \phi_1 = \phi_2 = \ldots = \phi_K$ and $\phi_k = \exp(\beta_{1k})$, $k = 1, 2, \ldots, K$, see Equation 10.8.

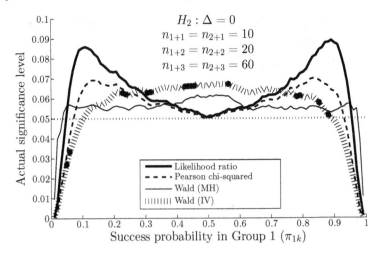

FIGURE 10.15
Simulated actual significance levels of four tests of a common difference between probabilities $(K = 3)$

10.12.1 The Likelihood Ratio Test

The likelihood ratio test for testing H_2 for the common ratio of probabilities is the same as that for testing H_2 for the common difference between probabilities in Section 10.11.1. For convenience, we repeat the expression for the likelihood ratio test statistic here:

$$T_{\text{LR}}(\mathbf{n}) = 2 \sum_{k=1}^{K} \sum_{j=1}^{2} \sum_{i=1}^{2} m_{ijk} \log \frac{m_{ijk}}{m_{ijk}^*},$$

where we set all terms for which $m_{ijk} = 0$ to zero. We refer to Section 10.11.1 for the definitions of m_{ijk} and m_{ijk}^*. The approximate large-sample distribution of T_{LR} is the chi-squared distribution with one degree of freedom.

10.12.2 The Pearson Chi-Squared Test

The Pearson chi-squared statistic for the common ratio of probabilities is equal to its corresponding statistic for the difference between probabilities in Section 10.11.2. It is defined by

$$T_{\text{Pearson}}(\mathbf{n}) = \sum_{k=1}^{K} \sum_{j=1}^{2} \sum_{i=1}^{2} \frac{(m_{ijk} - m_{ijk}^*)^2}{m_{ijk}^*},$$

and it is approximately chi-squared distributed with one degree of freedom.

10.12.3 The Wald Test Based on the Mantel-Haenszel Estimate

The Mantel-Haenszel estimate of the common ratio of probabilities is

$$\hat{\phi}_{\text{MH}} = \frac{\sum_{k=1}^{K} w_k \hat{\phi}_k}{\sum_{k=1}^{K} w_k},$$

where the weights w_k are given in Table 10.9. The estimated standard error of the logarithm of the Mantel-Haenszel estimate is

$$\widehat{\text{SE}}\left[\log\left(\hat{\phi}_{\text{MH}}\right)\right] = \sqrt{C/DE},$$

where

$$C = \sum_{k=1}^{K} \left(n_{1+k} n_{2+k} n_{+1k} - n_{11k} n_{21k} n_{++k}\right)/n_{++k}^2$$

and

$$D = \sum_{k=1}^{K} n_{11k} n_{2+k}/n_{++k} \quad \text{and} \quad E = \sum_{k=1}^{K} n_{21k} n_{1+k}/n_{++k},$$

see Greenland and Robins (1985), Deeks et al. (2001), and Palmer and Sterne (2016, p. 24). The Wald test statistic for the Mantel-Haenszel estimate is then

$$Z_{\text{Wald}}^{\text{MH}}(\mathbf{n}) = \frac{\log\left(\hat{\phi}_{\text{MH}}\right)}{\widehat{\text{SE}}\left[\log\left(\hat{\phi}_{\text{MH}}\right)\right]}.$$

We obtain an asymptotic P-value from a comparison of $Z_{\text{Wald}}^{\text{MH}}$ with the standard normal distribution.

10.12.4 The Wald Test Based on the Inverse Variance Estimate

The inverse variance estimate of the ratio of probabilities is

$$\hat{\phi}_{\text{IV}} = \frac{\sum_{k=1}^{K} v_k \hat{\phi}_k}{\sum_{k=1}^{K} v_k},$$

where the weights v_k are given in Table 10.9. The estimated standard error of the logarithm of the inverse variance estimate is

$$\widehat{SE}\left[\log\left(\hat{\phi}_{IV}\right)\right] = 1 \Big/ \sqrt{\sum_{k=1}^{K} v_k},$$

and the Wald test statistic for the inverse variance estimate is

$$Z_{Wald}^{IV}(\mathbf{n}) = \frac{\log\left(\hat{\phi}_{IV}\right)}{\widehat{SE}\left[\log\left(\hat{\phi}_{IV}\right)\right]}.$$

The reference distribution of Z_{Wald}^{IV} is the standard normal distribution.

10.12.5 Evaluations of Tests

Figures 10.16–10.18 show simulated actual significance levels of the likelihood ratio test, the Pearson chi-squared test, the Wald test based on the Mantel-Haenszel estimate, and the Wald test based on the inverse variance estimate for the null hypothesis of no common effect, with the ratio of probabilities as the effect measure. These are the same tests as those that were evaluated for the difference between probabilities in Section 10.11.5. For the difference between probabilities, all tests were liberal, with actual significance levels above the nominal level, and only the Wald test based on the Mantel-Haenszel estimate performed acceptably. The situation for the ratio of probabilities is similar in the sense that the Wald test based on the Mantel-Haenszel estimate is still the best-performing test. The situation is different in that the two Wald tests are now conservative instead of liberal, and the likelihood ratio and Pearson chi-squared tests have actual significance levels closer to the nominal level than they had for the difference between probabilities, although not always at an acceptable level.

10.13 Tests of a Common Odds Ratio

The odds ratio plays an important part in the analysis of stratified 2 × 2 tables, and much theory has been developed for the logit link. The Mantel-Haenszel estimate for the common odds ratio was first considered by Mantel and Haenszel (1959), and later extended to the difference between probabilities and the ratio of probabilities by Nurminen (1981), Tarone (1985), and Greenland and Robins (1985). The odds ratio is the effect measure of interest in case-control studies (see Section 4.8); however, because of its agreeable mathematical properties and role in logistic regression, it is also much used in

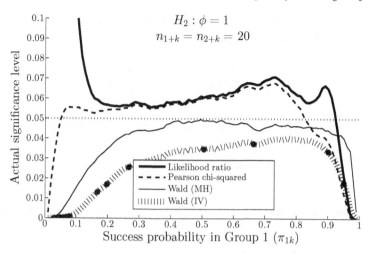

FIGURE 10.16
Simulated actual significance levels of four tests of a common ratio of probabilities $(K = 3)$

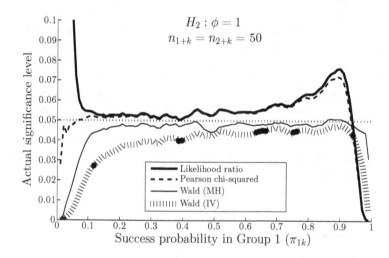

FIGURE 10.17
Simulated actual significance levels of four tests of a common ratio of probabilities $(K = 3)$

almost any other study design and setting, including meta-analysis. The null hypothesis for testing the common odds ratio effect is

$$H_2 : \theta = 1,$$

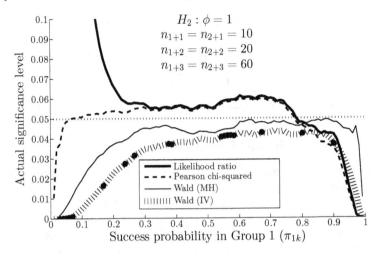

FIGURE 10.18
Simulated actual significance levels of four tests of a common ratio of probabilities ($K = 3$)

where $\theta = \theta_1 = \theta_2 = \ldots = \theta_K$ and $\theta_k = \exp(\beta_{1k})$, $k = 1, 2, \ldots, K$, see Equation 10.9.

10.13.1 The Likelihood Ratio Test

The likelihood ratio test for testing H_2 for the common odds ratio is the same as those for testing H_2 for the common ratio of probabilities (Section 10.11.1) and the difference between probabilities (Section 10.12.1). For convenience, we repeat the expression for the likelihood ratio test statistic here:

$$T_{\mathrm{LR}}(\mathbf{n}) = 2 \sum_{k=1}^{K} \sum_{j=1}^{2} \sum_{i=1}^{2} m_{ijk} \log \frac{m_{ijk}}{m_{ijk}^*},$$

where we set all terms for which $m_{ijk} = 0$ to zero. We refer to Section 10.11.1 for the definitions of m_{ijk} and m_{ijk}^*. The approximate large-sample distribution of T_{LR} is the chi-squared distribution with one degree of freedom.

10.13.2 The Pearson Chi-Squared Test

The Pearson chi-squared statistic for the common odds ratio is equal to its corresponding statistic for the difference between probabilities (Section 10.11.2) and the ratio of probabilities (Section 10.12.2). It is defined by

$$T_{\text{Pearson}}(\mathbf{n}) = \sum_{k=1}^{K}\sum_{j=1}^{2}\sum_{i=1}^{2} \frac{(m_{ijk} - m_{ijk}^{*})^2}{m_{ijk}^{*}},$$

and it is approximately chi-squared distributed with one degree of freedom.

10.13.3   The Cochran-Mantel-Haenszel Test

The Cochran-Mantel-Haenszel test was proposed in the path breaking article by Mantel and Haenszel (1959), an article that is cited more than 11 000 times. Cochran (1954b) had suggested a similar test that used the binomial distribution instead of the hypergeometric distribution used in Mantel and Haenszel (1959). Today, the Mantel-Haenszel form of the test statistic is used; however, the test and its generalizations are usually referred to as the Cochran-Mantel-Haenszel test. The test statistic is

$$T_{\text{CMH}}(\mathbf{n}) = \left[\sum_{k=1}^{K} \frac{n_{11k}n_{22k} - n_{12k}n_{21k}}{n_{++k}}\right]^2 \bigg/ \sum_{k=1}^{K} \frac{n_{1+k}n_{2+k}n_{+1k}n_{+2k}}{n_{++k}^2(n_{++k}-1)}.$$

(10.30)

The Cochran-Mantel-Haenszel test statistic is approximately chi-squared distributed with one degree of freedom.

The expectation and variance of n_{11k}, conditional on the marginals, are given in Equations 10.16 and 10.17, and with the derivations in Section 10.8.2, we have that the Wald test statistic based on the Peto estimate of the odds ratio is

$$Z_{\text{Peto}}(\mathbf{n}) = \frac{\log\left(\hat{\theta}_{\text{Peto}}\right)}{\widehat{\text{SE}}\left[\log\left(\hat{\theta}_{\text{Peto}}\right)\right]} = \frac{\sum_{k=1}^{K}\left[n_{11k} - \text{E}(n_{11k}\,|\,\mathbf{n}_{++})\right]}{\sqrt{\sum_{k=1}^{K}\text{Var}(n_{11k}\,|\,\mathbf{n}_{++})}}.$$

(10.31)

We now have that $T_{\text{CMH}} = Z_{\text{Peto}}^2$. The test statistic Z_{Peto} is also described in Deeks et al. (2001, p. 298).

The Cochran-Mantel-Haenszel test is a score test in a logistic regression model. Day and Byar (1979) have shown that it is the score test for the association between the grouping and the outcome, when the stratification variable is already included in the model, see also Hosmer et al. (2013, p. 86).

The Cochran-Mantel-Haenszel test performs best when the associations over the strata are similar, which can be expected when there is homogeneity of associations. When there are stratum differences in opposite directions, on the other hand, and they are of similar size, the Cochran-Mantel-Haenszel test may fail to reveal the association. When there is heterogeneity over the strata, the Cochran-Mantel-Haenszel test can also be used; however, it then must be considered as a test of average partial association (see Section 10.6). As noted by Landis et al. (2005), the Cochran-Mantel-Haenszel test is concerned with the broader hypothesis of no partial association, which includes no interaction and no average partial association.

For 2 × 2 and $r \times c$ tables, the Cochran's criterion states that the Pearson chi-squared test should be used only if at least 80% of the expected cell counts are greater than five and all expected cell counts are greater than one. For the 2 × 2 table, this amounts to demanding that all expected cell counts are above five, and Cochran's criterion is sometimes referred to as the *rule of five*. The approximation of the Cochran-Mantel-Haenszel test statistic to the chi-squared distribution also has a "rule of five"; however, the criterion is now that for a particular cell, the minimum of the sum over the strata of the difference between the expected counts and the minimum possible sum and the sum over the strata of the difference between the maximum possible sum and the expected counts should not be less than five. This criterion was proposed by Mantel and Fleiss (1980) and has been given the name of the *Mantel-Fleiss criterion*. With a similar notation as in Mantel and Fleiss (1980), we let

$$(n_{11k})_L = \max(0, n_{+1k} - n_{2+k}) \quad \text{and} \quad (n_{11k})_U = \min(n_{1+k}, n_{+1k}),$$

and we denote the expected cell counts in cell $(1, 1)$ under the null hypothesis by $m_{11k} = n_{1+k} \cdot n_{+1k}/n_{++k}$. Then, the Mantel-Fleiss criterion can be expressed as

$$\min \left[\sum_{k=1}^{K} m_{11k} - \sum_{k=1}^{K} (n_{11k})_L, \sum_{k=1}^{K} (n_{11k})_U - \sum_{k=1}^{K} m_{11k} \right] > 5.$$

One of the (many) advantages of the Cochran-Mantel-Haenszel test is that the criterion for the chi-squared approximation relates to the total sum over the strata, and not the individual cell counts themselves.

10.13.4   The RBG Test

Several Wald-type tests for the null hypothesis $\theta = 1$ exist, and in this and the next section, we present two of them. Robins et al. (1986) proposed a variance estimate of $\hat{\theta}_{MH}$ that performs well in both small and large samples. The corresponding estimated standard error is

$$\widehat{SE}\left[\log\left(\hat{\theta}_{MH}\right)\right]_{RBG} = \sqrt{\frac{C}{2A^2} + \frac{D + E}{2AB} + \frac{F}{2B^2}}, \tag{10.32}$$

where

$$A = \sum_{k=1}^{K} n_{11k} \cdot n_{22k}/n_{++k},$$

$$B = \sum_{k=1}^{K} n_{12k} \cdot n_{21k}/n_{++k},$$

$$C = \sum_{k=1}^{K} (n_{11k} + n_{22k}) \cdot n_{11k} \cdot n_{22k}/n_{++k}^2,$$

$$D = \sum_{k=1}^{K} (n_{11k} + n_{22k}) \cdot n_{12k} \cdot n_{21k}/n_{++k}^2,$$

$$E = \sum_{k=1}^{K} (n_{12k} + n_{21k}) \cdot n_{11k} \cdot n_{22k}/n_{++k}^2,$$

and

$$F = \sum_{k=1}^{K} (n_{12k} + n_{22k}) \cdot n_{12k} \cdot n_{21k}/n_{++k}^2.$$

See also Deeks et al. (2001) or Palmer and Sterne (2016, p. 24). The variance estimate in Equation 10.32 is called the *RBG variance of the Mantel-Haenszel estimate,* named after Robins, Breslow, and Greenland (Robins et al., 1986).

The Wald test statistic based on the RBG variance is

$$Z_{\text{RBG}}(\mathbf{n}) = \frac{\log\left(\hat{\theta}_{\text{MH}}\right)}{\widehat{\text{SE}}\left[\log\left(\hat{\theta}_{\text{MH}}\right)\right]_{\text{RBG}}},$$

and we obtain a *P*-value by comparing Z_{RBG} to the standard normal distribution.

10.13.5   The Woolf Test

The Woolf estimate (Woolf, 1955) may also give rise to a Wald-type test for the common odds ratio. From Equation 10.15, we know that the inverse variance estimate of the odds ratio is

$$\hat{\theta}_{\text{IV}} = \frac{\sum_{k=1}^{K} v_k \hat{\theta}_k}{\sum_{k=1}^{K} v_k},$$

with weights v_k given in Table 10.9. The estimated standard error of this estimate is

$$\widehat{\text{SE}}\left[\log\left(\hat{\theta}_{\text{IV}}\right)\right] = 1 \bigg/ \sqrt{\sum_{k=1}^{K} v_k},$$

and the corresponding Wald test is

$$Z_{\text{Woolf}}(\mathbf{n}) = \frac{\log\left(\hat{\theta}_{\text{IV}}\right)}{\widehat{\text{SE}}\left[\log\left(\hat{\theta}_{\text{IV}}\right)\right]}.$$

P-values can be obtained from the standard normal distribution.

10.13.6 The Gart Exact, Mid-*P*, and Asymptotic Tests

An exact test for the common odds ratio, often referred to as the *Gart exact test*, is based on the sufficient statistic $T(\mathbf{n}) = \sum_{k=1}^{K} n_{11k}$, see Gart (1970, 1971). Note that $T(\mathbf{n}) = \sum_{k=1}^{K} n_{11k}$ is a special case of $T_{\text{linrank}}(\mathbf{n}) = \sum_{k=1}^{K} a_k n_{11k}$, given in Section 5.7.4 as the Cochran-Armitage exact conditional test, with $a_k = 1$ for $k = 1, 2, \ldots, K$.

The one-sided Gart exact (conditional) *P*-value for an observed value $t_0 = T(\mathbf{n})$ is given by

$$\text{one-sided } P\text{-value} = \sum_{t \geq t_0} f(t \mid \mathbf{n}_{++}), \qquad (10.33)$$

where the sum is over all tables with marginals equal to $\mathbf{n}_{++}$ and with equal or greater value of the sufficient statistic T than the observed table, and $f(t \mid \mathbf{n}_{++})$ is the conditional sampling distribution, given the marginals $\mathbf{n}_{++}$, in Equation 10.13. Note that $f(t \mid \mathbf{n}_{++})$ is different from the $f(\mathbf{x} \mid \mathbf{n}_+)$ used in the Cochran-Armitage exact conditional test, see Equation 5.23.

To compute two-sided *P*-values, we use the principle of twice the smallest tail. Provided Equation 10.33 gives the smallest of the two tail probabilities, by summation over all tables for which $\{t \geq t_0\}$ and $\{t \leq t_0\}$, respectively, the two-sided *P*-value is twice the one in Equation 10.33.

A mid-*P* version of the Gart exact test can be obtained in the usual manner, with the following modification to Equation 10.33:

$$\text{one-sided mid-}P\text{ value} = \sum_{t > t_0} f(t \mid \mathbf{n}_{++}) + 0.5 \cdot \sum_{t = t_0} f(t \mid \mathbf{n}_{++}).$$

Gart (1971) also suggested an asymptotic version of the exact test above. The Gart asymptotic test statistic is derived in the same way as the Wald statistic based on the Peto estimate of the odds ratio, see Equation 10.31, and is thereby equal to the Cochran-Mantel-Haenszel test statistic in Equation 10.30.

10.13.7 An Exact Unconditional Test

Under the null hypothesis of a common odds ratio equal to one, $\theta = 1$, there is a common success probability for each stratum: $\pi_{1k} = \pi_{2k} = \pi_k$, for $k =$

$1, 2, \ldots, K$. For a suitable test statistic T, the one-sided exact unconditional P-value is

$$\text{one-sided } P = \max_{0 \le \pi_1, \ldots, \pi_K \le 1} \sum_{\Omega(\mathbf{x}|\mathbf{n}_+)} I\big[T(\mathbf{x}) \ge T(\mathbf{n})\big] \cdot f(\mathbf{x} \,|\, \pi_1, \ldots, \pi_K; \mathbf{n}_+),$$

where $\Omega(\mathbf{x}|\mathbf{n}_+)$ denotes the set of all tables with fixed row sums $\mathbf{n}_+$ (see Section 10.5), $I()$ is the indicator function, and $f()$ is the sampling distribution in Equation 10.10 with the additional restriction of a common success probability for each stratum.

The exact unconditional test is of potential interest; however, because of the need to maximize over the nuisance parameters $\pi_1, \ldots, \pi_K$, it is extremely time consuming to calculate. We would expect the exact unconditional test to be superior to the exact conditional test. We also expect the mid-P test to perform similarly to the exact unconditional test, as we have seen for the 2×2 table (see Section 4.4.9).

10.13.8   Evaluations of Tests

An evaluation of the performances of the likelihood ratio, Pearson chi-squared, Cochran-Mantel-Haenszel, RBG, and Woolf tests for the null hypothesis of a common odds ratio equal to one can be seen in Figures 10.19–10.21. The Cochran-Mantel-Haenszel test performs excellently in all the three figures. The Pearson chi-squared test also performs very well, although it seems to be slightly liberal for small sample sizes. Interestingly, both tests perform very well also when the sample size is quite unequally balanced, as in Figure 10.21. The likelihood ratio test has acceptable actual significance levels for parts of the range of π_{1k}; however, it is much too liberal for small or large π_{1k}. The two Wald tests (RBG and Woolf), are conservative for most of the parameter space and only perform well in a narrow range of π_{1k} values. Increasing the sample size seems to remedy this, at least to some extent.

10.14   Confidence Intervals for the Common Effect

If we can assume that there is no heterogeneity of the effect measure over the strata, we estimate the common stratum effect (Section 10.8) and test the null hypothesis of no effect (Section 10.11–10.13). In this section, we present confidence intervals for the common effect measure, and we consider the difference between probabilities, the ratio of probabilities, and the odds ratio.

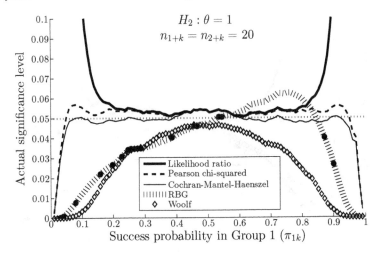

FIGURE 10.19
Simulated actual significance levels of five tests of a common odds
ratio $(K = 3)$

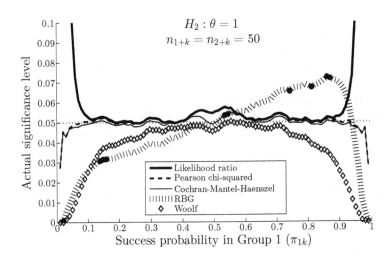

FIGURE 10.20
Simulated actual significance levels of five tests of a common odds
ratio $(K = 3)$

10.14.1 Asymptotic Confidence Intervals

Table 10.10 lists Wald and Wald-type confidence intervals for the difference
between probabilities, the ratio of probabilities, and the odds ratio, based on

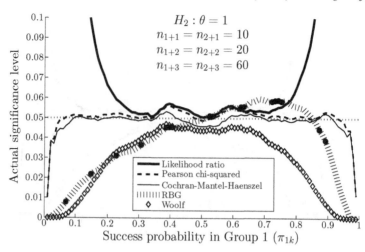

FIGURE 10.21
Simulated actual significance levels of five tests of a common odds ratio ($K = 3$)

three estimates: the Mantel-Haenszel estimate, the inverse variance estimate, and the maximum likelihood estimate. Expressions for the Mantel-Haenszel and inverse variance estimates are given in Section 10.8.1 with weights in Table 10.9. The estimated standard errors for the difference between probabilities are described in Sections 10.11.3 and 10.11.4, and the estimated standard errors for the ratio of probabilities are shown in Sections 10.12.3 and 10.12.4. The estimated standard errors for the odds ratio are given in Sections 10.13.4 and 10.13.5.

10.14.2   The Gart Exact and Mid-P Intervals

In Section 10.13.6, we derived the Gart exact and mid-P tests for the null hypothesis $\theta = 1$. The exact conditional P-value was derived from the exact distribution of $T(\mathbf{n}) = \sum_{k=1}^{K} n_{11k}$, the sufficient statistic for θ, given the marginals $\mathbf{n}_{++}$, see Equation 10.13. More generally, the null hypothesis $\theta = \theta_0$ can be tested with the exact distribution of T inserted θ_0. This is a uniformly most powerful unbiased test, see also Section 1.6.2. We derive an exact confidence interval by inverting the test, as was proposed by Cornfield (1956) for the 2×2 table (see Sections 4.8.6 and 5.8.4); see also Gart (1970) and Cox (1970, p. 48).

With the notation and definitions surrounding Equation 10.33 (the Gart exact test, p. 467), we obtain the Gart exact $100(1 - \alpha)\%$ confidence interval

TABLE 10.10
Asymptotic confidence intervals for the difference between
probabilities, the ratio of probabilities, and the odds ratio

Name	Estimate	Confidence interval
Difference between probabilities (Δ)		
Wald	Mantel-Haenszel	$\hat{\Delta}_{\mathrm{MH}} \pm z_{\alpha/2}\widehat{\mathrm{SE}}(\hat{\Delta}_{\mathrm{MH}})$
Wald	Inverse variance	$\hat{\Delta}_{\mathrm{IV}} \pm z_{\alpha/2}\widehat{\mathrm{SE}}(\hat{\Delta}_{\mathrm{IV}})$
Wald	Maximum likelihood	$\hat{\beta}_{\mathrm{ML,linear}} \pm z_{\alpha/2}\widehat{\mathrm{SE}}(\hat{\beta}_{\mathrm{ML,linear}})$
Ratio of probabilities (ϕ)		
Wald	Mantel-Haenszel	$\hat{\phi}_{\mathrm{MH}} \exp\left\{\pm z_{\alpha/2}\widehat{\mathrm{SE}}\left[\log\left(\hat{\phi}_{\mathrm{MH}}\right)\right]\right\}$
Wald	Inverse variance	$\hat{\phi}_{\mathrm{IV}} \exp\left\{\pm z_{\alpha/2}\widehat{\mathrm{SE}}\left[\log\left(\hat{\phi}_{\mathrm{IV}}\right)\right]\right\}$
Wald	Maximum likelihood	$\exp\left[\hat{\beta}_{\mathrm{ML,log}} \pm z_{\alpha/2}\widehat{\mathrm{SE}}(\hat{\beta}_{\mathrm{ML,log}})\right]$
Odds ratio (θ)		
RBG	Mantel-Haenszel	$\hat{\theta}_{\mathrm{MH}} \exp\left\{\pm z_{\alpha/2}\widehat{\mathrm{SE}}\left[\log\left(\hat{\theta}_{\mathrm{MH}}\right)\right]\right\}$
Woolf	Inverse variance	$\hat{\theta}_{\mathrm{IV}} \exp\left\{\pm z_{\alpha/2}\widehat{\mathrm{SE}}\left[\log\left(\hat{\theta}_{\mathrm{IV}}\right)\right]\right\}$
Wald	Maximum likelihood	$\exp\left[\hat{\beta}_{\mathrm{ML,logit}} \pm z_{\alpha/2}\widehat{\mathrm{SE}}(\hat{\beta}_{\mathrm{ML,logit}})\right]$

$z_{\alpha/2}$ is the upper $\alpha/2$ percentile of the standard normal distribution

(L, U) by—iteratively—solving the two equations

$$\sum_{t \geq t_0} f(t, L \mid \mathbf{n}_{++}) = \alpha/2 \qquad (10.34)$$

and

$$\sum_{t \leq t_0} f(t, U \mid \mathbf{n}_{++}) = \alpha/2, \qquad (10.35)$$

where L and U take the place of θ in Equation 10.13. If $t_0 = t_{\min}$, set $L = 0$.
If $t_0 = t_{\max}$, set $U = \infty$. The quantities $t_{\min}$ and $t_{\max}$ are defined at the very
end of Section 10.5 (page 435).

Efficient algorithms to find L and U have been proposed by Mehta et al.
(1985), Vollset and Hirji (1991), and Martin and Austin (1991).

The Gart mid-P confidence interval can be calculated with the following
modifications to Equations 10.34 and 10.35:

$$\sum_{t > t_0} f(t, L \mid \mathbf{n}_{++}) + 0.5 \cdot \sum_{t = t_0} f(t, L \mid \mathbf{n}_{++}) = \alpha/2$$

and

$$\sum_{t<t_0} f(t, U \mid \mathbf{n}_{++}) \ + \ 0.5 \cdot \sum_{t=t_0} f(t, U \mid \mathbf{n}_{++}) = \alpha/2.$$

Note that these confidence intervals are for the conditional maximum likelihood estimate ($\hat{\theta}_{\mathrm{CML}}$) of the common odds ratio, see also Section 10.10.3.

10.14.3   Evaluations of Intervals

In this section, we evaluate the coverage probability (see Section 1.4 for a general description) of the asymptotic confidence intervals. We will use simulations instead of exact calculations for reasons laid out in Section 10.9.4, which, briefly summarized, has to do with the huge computational effort required to do exact calculations.

Figures 10.22–10.24 show the simulated coverage probabilities of the three Wald intervals for the common difference between probabilities. The common effect measure is fixed at $\Delta = -0.25$ and the number of strata is $K = 3$ in all three figures. All three Wald intervals are too narrow such that their coverage probabilities are below the nominal level. The Wald interval based on the Mantel-Haenszel estimate is the interval with coverage probabilities closest to the nominal level, whereas the Wald interval based on the inverse variance estimate is the most liberal. The inverse variance interval performs particularly poorly for unbalanced sample sizes (Figure 10.24). The Wald intervals perform quite similarly to the Wald interval for the 2×2 table, which is evaluated in a similar setting in Figure 4.17 (page 125).

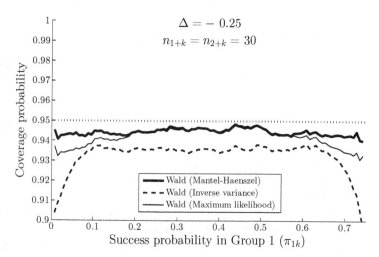

FIGURE 10.22
Simulated coverage probabilities of three confidence intervals for the common difference between probabilities ($K = 3$)

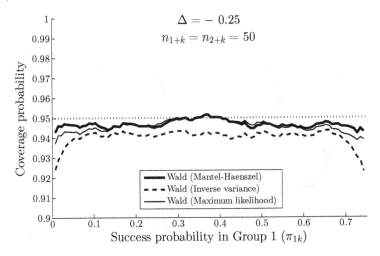

FIGURE 10.23
Simulated coverage probabilities of three confidence intervals for the common difference between probabilities ($K = 3$)

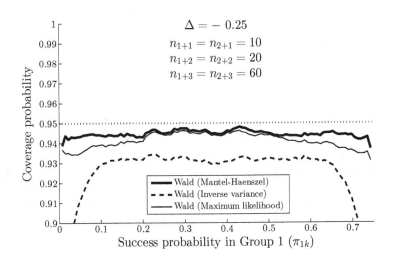

FIGURE 10.24
Simulated coverage probabilities of three confidence intervals for the common difference between probabilities ($K = 3$)

The coverage probabilities of the three Wald intervals for the common ratio of probabilities are illustrated in Figures 10.25–10.27. The three intervals perform quite similarly, with coverage probabilities above the nominal level, except when $\pi_{1k} > 0.6$, for which the Wald interval based on the inverse variance estimate has coverage probabilities below the nominal level. There does not seem to be a large effect of unbalanced sample sizes (Figure 10.27). The Mantel-Haenszel and maximum likelihood intervals perform similarly to the corresponding Wald intervals for the 2×2 table (the Katz log and adjusted log intervals), see Figure 4.27 (page 147).

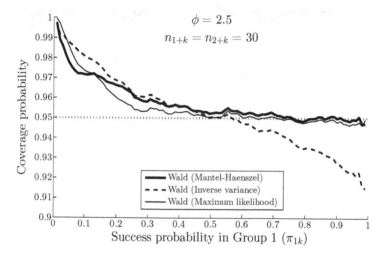

FIGURE 10.25
Simulated coverage probabilities of three confidence intervals for the common ratio of probabilities ($K = 3$)

Figures 10.28–10.30 show simulated coverage probabilities of the confidence intervals for the common odds ratio. The effect measure is fixed at $\theta = 5.0$, which makes the results comparable to those of the Woolf and Gart adjusted logit intervals for the 2×2 table in Figure 4.36 (page 167). The Wald interval based on the maximum likelihood estimate, and partly the Woolf interval based on the inverse variance estimate, performs well; whereas the RBG interval based on the Mantel-Haenszel estimate is quite poor, with usually very conservative—that is, wide—intervals.

10.15 Random Effects Models for Stratified 2 × 2 Tables

10.15.1 Introduction

So far in this chapter, we have considered so-called fixed effects models, for which the general model is given in Equation 10.6. The variability is described

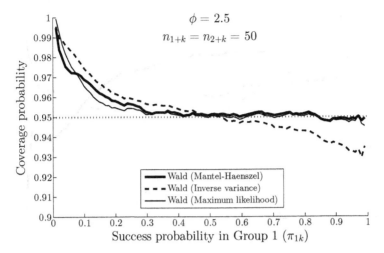

FIGURE 10.26
Simulated coverage probabilities of three confidence intervals for the common ratio of probabilities ($K = 3$)

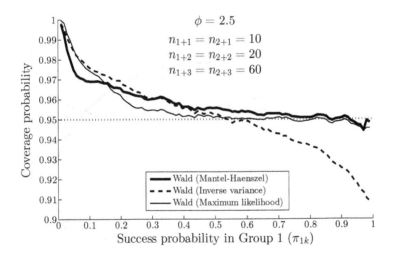

FIGURE 10.27
Simulated coverage probabilities of three confidence intervals for the common ratio of probabilities ($K = 3$)

by the binomial distribution, which fully describes the sampling variability, regardless of the link function.

A priori, there are K parameters that describe the association between

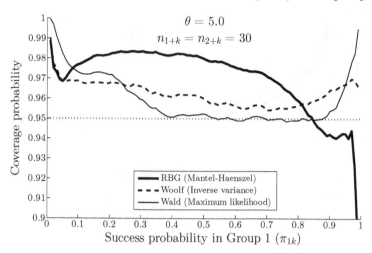

FIGURE 10.28
Simulated coverage probabilities of three confidence intervals for the common odds ratio ($K = 3$)

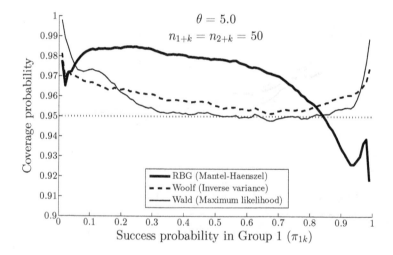

FIGURE 10.29
Simulated coverage probabilities of three confidence intervals for the common odds ratio ($K = 3$)

the grouping and the outcome. Any such model is called a *fixed effects model*. When there is no heterogeneity in the parameters, we have just one fixed parameter to estimate, and we say we have a common effect model.

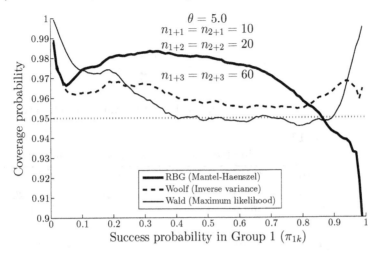

FIGURE 10.30
Simulated coverage probabilities of three confidence intervals for the common
odds ratio $(K = 3)$

A fixed effects model is appropriate when the strata comprise a—loosely
defined—closed set, for instance, when we stratify by a given variable with predefined categories, like sex. Our interest lies specifically in the K parameters
that described the association, conditioned on the strata.

Alternatively, we may consider the K strata as a sample from a population
of strata. In a systematic review, for example, we include K studies (the
strata), and it may be natural to think of these studies as a sample from a
larger pool of possible studies. Then, the association between the grouping
and the outcome is a random variable, and the effect estimate based on the
data from the K strata is just one observation from a distribution of effect
estimates. This is a random effects model. For an introduction to fixed effects
and random effects models, see Borenstein et al. (2010).

Previously in this chapter, we have considered three effect measures (and
their corresponding link functions): the difference between probabilities (linear link), the ratio of probabilities (log link), and the odds ratio (logit link).
Regardless of the link function, we denote the effect parameter in stratum k
by ψ_k, for $k = 1, 2, \ldots, K$, and we let ψ denote the common effect parameter over the strata. Moreover, let U_k and s_k^2 denote the effect estimates and
the variances, see Table 10.11. Following Normand (1999) and Higgins et al.
(2009), we assume that for the fixed effects model

$$U_k \sim N\left(\psi, s_k^2\right),$$

where $N(\mu, \sigma^2)$ denotes the normal distribution. If we assume that the s_k^2 are

478

Statistical Analysis of Contingency Tables

known, the maximum likelihood estimate of ψ is

$$\hat{\psi}_{\mathrm{ML}} = \sum_{k=1}^{K} w_k U_k \Big/ \sum_{k=1}^{K} w_k,$$

where $w_k = 1/s_k^2$. We recognize $\hat{\psi}_{\mathrm{ML}}$ as the inverse variance estimate.

TABLE 10.11

Effect estimates and estimated variances for three effect measures

Effect measure	Estimate (U_k)	Variance (s_k^2)
Difference between probabilities	$\dfrac{n_{11k}}{n_{1+k}} - \dfrac{n_{21k}}{n_{2+k}}$	$\dfrac{n_{11k} \cdot n_{12k}}{n_{1+k}^3} + \dfrac{n_{21k} \cdot n_{22k}}{n_{2+k}^3}$
Ratio of probabilities	$\dfrac{n_{11k}/n_{1+k}}{n_{21k}/n_{2+k}}$	$\dfrac{1}{n_{11k}} - \dfrac{1}{n_{1+k}} + \dfrac{1}{n_{21k}} - \dfrac{1}{n_{2+k}}$
Odds ratio	$\dfrac{n_{11k}/n_{12k}}{n_{21k}/n_{22k}}$	$\dfrac{1}{n_{11k}} + \dfrac{1}{n_{12k}} + \dfrac{1}{n_{21k}} + \dfrac{1}{n_{22k}}$

In a random effects model, we assume that the ψ_k are random variables, such that

$$U_k \mid \psi_k, s_k \sim N\big(\psi_k, s_k^2\big)$$

and

$$\psi_k \mid \psi, \tau^2 \sim N\big(\psi, \tau^2\big).$$

Here, τ^2 is the inter-stratum variation. When τ^2 is known, the maximum likelihood estimate of ψ is

$$\hat{\psi}_{\mathrm{ML}}(\tau) = \sum_{k=1}^{K} w_k(\tau) U_k \Big/ \sum_{k=1}^{K} w_k(\tau), \qquad (10.36)$$

where $w_k(\tau) = 1/(s_k^2 + \tau^2)$. In the following two sections, we will consider two alternative estimates of the inter-stratum variance τ^2.

10.15.2 The DerSimonian-Laird Estimate and Confidence Interval

In Section 10.9.1, we presented the Cochran Q test for testing homogeneity of the stratum specific parameters ψ_k. With the notation from the preceding section, the Cochran Q test statistic is

$$Q = \sum_{k=1}^{K} w_k \big(U_k - \hat{\psi}_{\mathrm{ML}}\big).$$

DerSimonian and Laird (1986) derived a method of moments estimate of τ given as

$$\hat{\tau}_{\mathrm{DL}}^2 = \left[Q - (K-1)\right] \Big/ \left(\sum_{k=1}^{K} w_k - \frac{\sum_{k=1}^{K} w_k^2}{\sum_{k=1}^{K} w_k}\right).$$

If $\hat{\tau}_{\mathrm{DL}} < 0$, set $\hat{\tau}_{\mathrm{DL}} = 0$. If we insert $\hat{\tau}_{\mathrm{DL}}$ into the maximum likelihood estimate of the common effect (Equation 10.36), we get

$$\hat{\psi}_{\mathrm{DL}} = \sum_{k=1}^{K} w_k(\hat{\tau}_{\mathrm{DL}}) U_k \Big/ \sum_{k=1}^{K} w_k(\hat{\tau}_{\mathrm{DL}}),$$

where $w_k(\hat{\tau}_{\mathrm{DL}}) = 1/(s_k^2 + \hat{\tau}_{\mathrm{DL}}^2)$. A Wald interval for ψ is then

$$\hat{\psi}_{\mathrm{DL}} \pm z_{\alpha/2} \sqrt{1 \Big/ \sum_{k=1}^{K} w_k(\hat{\tau}_{\mathrm{DL}})}.$$

10.15.3 The Restricted Maximum Likelihood Estimate and Confidence Interval

Restricted maximum likelihood is often preferred to maximum likelihood for estimation of variance components, see, for instance, Harville (1977). Restricted maximum likelihood gives less bias in the estimated variance components than maximum likelihood. The restricted maximum likelihood estimate of τ^2 is given by

$$\hat{\tau}_{\mathrm{REML}}^2 = \frac{\displaystyle\sum_{k=1}^{K} w_k^2(\hat{\tau}_{\mathrm{REML}}) \left[\frac{K}{K-1}(U_k - \hat{\psi}_{\mathrm{REML}})^2 - s_k^2\right]}{\displaystyle\sum_{k=1}^{K} w_k^2(\hat{\tau}_{\mathrm{REML}})}, \qquad (10.37)$$

where

$$\hat{\psi}_{\mathrm{REML}} = \sum_{k=1}^{K} w_k(\hat{\tau}_{\mathrm{REML}}) U_k \Big/ \sum_{k=1}^{K} w_k(\hat{\tau}_{\mathrm{REML}}),$$

and $w_k(\hat{\tau}_{\mathrm{REML}}) = 1/(s_k^2 + \hat{\tau}_{\mathrm{REML}}^2)$. Equation 10.37 must be solved iteratively. Statistical software packages that include mixed models support the estimation. An algorithm to solve Equation 10.37 is given in Appendix A of DerSimonian and Kacker (2007).

Because $\hat{\psi}_{\mathrm{REML}}$ is approximately normally distributed with mean ψ and variance $1 / \sum_{k=1}^{K} w_k(\hat{\tau}_{\mathrm{REML}})$, a Wald interval for ψ is

$$\hat{\psi}_{\mathrm{REML}} \pm z_{\alpha/2} \sqrt{1 \Big/ \sum_{k=1}^{K} w_k(\hat{\tau}_{\mathrm{REML}})}.$$

10.16   Examples Analyzed

10.16.1   Smoking and Lung Cancer

We start by considering the data from Doll and Hill (1950) on the association between smoking and lung cancer, see Table 10.2. We have two strata, one for males and one for females. Note that these data come from a case-control study, and only the odds ratio is relevant as effect measure, see Section 10.4. We use the Wald confidence interval from Section 10.7 to estimate the separate associations between smoking and lung cancer for males and females (Table 10.12). The estimated associations for both sexes are significant; however, the association seems to be much stronger for males than females. Can we combine the effect estimates for males and females and provide a common effect estimate of the association between smoking and lung cancer? We turn to the tests for homogeneity to find the answer.

TABLE 10.12
Estimated odds ratios for the
association between smoking and lung
cancer for males and females, based on
the data in Table 10.2

Sex	Estimate	95% Wald CI
Males	14.0	3.33 to 59.3
Females	2.47	1.17 to 5.19

Table 10.13 shows the results of several tests for homogeneity of the odds ratio over the two strata. All tests are significant at the 5% level, regardless of whether the tests are model-based or based on the Mantel-Haenszel or inverse variance estimates.

Because we have evidence of heterogeneity, we do not go on to estimate and test for a common odds ratio with these data. In the next example, though, we will go through all the steps in the process of analyzing stratified 2×2 tables.

10.16.2   Prophylactic Use of Lidocaine in Myocardial Infarction (Meta-Analysis)

We now turn to the data in Table 10.3, originally published by Hine et al. (1989), and later reanalyzed by Normand (1999). The data comprise the results of six randomized clinical trials of the effect of prophylactic use of lidocaine in patients with myocardial infarction. The outcome of the trials was mortality. As a first step in the analysis of the data in Table 10.3, we estimate the treatment effect for each trial individually.

With the methods in Section 10.7, we obtain the results in Table 10.14.

TABLE 10.13
Results of tests of homogeneity of the odds ratio over strata
based on the data on smoking and lung cancer in Table 10.2

Test	Statistic	*P*-value
Cochran Q_{MH}	4.93 (df = 1)	0.0264
Cochran Q_{IV}	4.42 (df = 1)	0.0355
Likelihood ratio	5.74 (df = 1)	0.0166
Pearson chi-squared	5.04 (df = 1)	0.0247
Breslow-Day (w/Tarone correction)	5.18 (df = 1)	0.0229
Breslow-Day exact		0.0322
Zelen exact		0.0322
Peto	2.82 (df = 1)	0.0930

There are some apparent differences between the trial-specific estimates; however, they all point in the same direction—toward an increased probability of mortality in the lidocaine group—and, if we take the wide confidence intervals into account, the results are not that different. By themselves, none of the trials show a significant harmful effect of lidocaine.

Before we start combining the trial-specific results, we test for homogeneity. Table 10.15 gives the results, which are very consistent across effect measures and estimation method: there are no indications of heterogeneity.

We may thus proceed with estimating a common treatment effect. We will do this with the Mantel-Haenszel estimate, the inverse variance estimate, and the maximum likelihood estimate, and for the three effect measures: difference between probabilities, ratio of probabilities, and odds ratio. We also provide the corresponding confidence intervals (see Section 10.14) along with the *P*-values for the tests of no effect (see Table 10.16).

If the difference between probabilities is our effect measure, we can see that the estimated effect of the lidocaine treatment is to increase the probability of mortality by about 3 percentage points. The 95% confidence intervals range from about 0.5 percentage points to about 5.5 percentage points. The tests of no difference indicate that the treatment effect is significant, although not at a strong level.

The results for the ratio of probabilities and the odds ratio follow the same pattern as the results for the difference between probabilities: the harmful effect of lidocaine is significant, although only slightly so. The estimated ratio of probabilities is about 1.7, which means that the risk of mortality is 1.7 times higher with lidocaine treatment compared with the control treatment. The estimated odds ratio is slightly higher than the estimated ratio of probabilities, which is as expected, because the odds ratio is, in general, further from the null effect than the ratio of probabilities. Note that the confidence interval for the odds ratio that uses the RBG variance is the only interval that includes the null effect ($\theta = 1$), and the corresponding RBG test is the only test that has a *P*-value above 0.05.

TABLE 10.14

The estimated treatment effects of lidocaine
on mortality for six trials based on the data
in Table 10.3

Trial	Estimate	95% Wald CI
The difference between probabilities		
Chopra et al.	0.028	-0.055 to 0.111
Mogensen	0.000	-0.120 to 0.120
Pitt et al.	0.020	-0.036 to 0.076
Darby et al.	0.018	-0.047 to 0.083
Bennett et al.	0.035	-0.020 to 0.091
O'Brian et al.	0.044	-0.005 to 0.093
The ratio of probabilities		
Chopra et al.	2.21	0.21 to 23.4
Mogensen	1.00	0.27 to 3.75
Pitt et al.	1.54	0.45 to 5.31
Darby et al.	1.36	0.45 to 4.14
Bennett et al.	2.25	0.60 to 8.47
O'Brian et al.	2.61	0.85 to 8.00
The odds ratio		
Chopra et al.	2.27	0.20 to 26.1
Mogensen	1.00	0.23 to 4.28
Pitt et al.	1.57	0.43 to 5.74
Darby et al.	1.39	0.42 to 4.52
Bennett et al.	2.33	0.59 to 9.27
O'Brian et al.	2.73	0.85 to 8.78

In addition to the analyses of the common effect that we have reported in Table 10.16, we can also use a random effects model, as explained in Section 10.15. Here, we calculate the DerSimonian-Laird estimate and confidence interval and obtain

$$\hat{\Delta}_{DL} = 0.020 \ (95\% \ \text{CI: -0.009 to 0.049}),$$

$$\hat{\phi}_{DL} = 1.31 \ (95\% \ \text{CI: 0.77 to 2.21}),$$

and

$$\hat{\theta}_{DL} = 1.34 \ (95\% \ \text{CI: 0.77 to 2.35}).$$

The random effects estimates are smaller than the fixed effects estimates in Table 10.16, and the random effects confidence intervals are slightly wider than the fixed effects confidence intervals.

In systematic reviews, as part of the meta-analysis, the estimates for the individual study and the common effect are usually summarized in a forest plot. Figure 10.31 shows the forest plot for the data in Table 10.3, with the ratio of probabilities (relative risk) as the effect measure. The forest plot

TABLE 10.15
Results of tests of homogeneity over strata based on the data
on treatment of lidocaine and mortality in Table 10.3

Test	Statistic	P-value
The difference between probabilities		
Cochran Q_{MH}	0.87 (df = 5)	0.972
Cochran Q_{IV}	0.86 (df = 5)	0.973
Likelihood ratio	0.86 (df = 5)	0.973
Pearson chi-squared	0.86 (df = 5)	0.973
The ratio of probabilities		
Cochran Q_{MH}	1.58 (df = 5)	0.904
Cochran Q_{IV}	1.57 (df = 5)	0.904
Likelihood ratio	1.61 (df = 5)	0.901
Pearson chi-squared	1.61 (df = 5)	0.901
The odds ratio		
Cochran Q_{MH}	1.52 (df = 5)	0.911
Cochran Q_{IV}	1.51 (df = 5)	0.912
Likelihood ratio	1.54 (df = 5)	0.908
Pearson chi-squared	1.54 (df = 5)	0.909
Breslow-Day (w/Tarone correction)	1.54 (df = 5)	0.909
Breslow-Day exact		0.928
Zelen exact		0.924
Peto	1.41 (df = 5)	0.923

includes two estimates of the common effect: the Mantel-Haenszel estimate
and the DerSimonian-Laird estimate.

10.17   Additional Methods for Meta-Analysis

There are many statistical methods available for systematic reviews and meta-
analysis. We do not aim to provide a complete list here, but we will briefly
mention some of the most common ones. For a general introduction to systematic reviews, with emphasis on medical and health-care research, see Sutton
et al. (2000), Egger et al. (2001), and Fagerland (2015).

Assessing Between-Study Heterogeneity with L'Abbe Plots

The L'Abbe plot is a visual tool to examine heterogeneity between the included
studies in a meta-analysis. It is obtained by plotting, for each study, the
observed proportion of success for Group 1 on the *y*-axis versus the observed
proportion of success for Group 2 on the *x*-axis. If the odds ratio is the effect

TABLE 10.16
Estimates, confidence intervals, and tests of the common treatment effect of lidocaine on mortality, based on the data in Table 10.3

Method	Estimate (95% CI)	P-value
The difference between probabilities		
Mantel-Haenszel (Wald test)	0.028 (0.002 to 0.054)	0.0349
Inverse variance (Wald test)	0.029 (0.004 to 0.055)	0.0243
Maximum likelihood	0.030 (0.004 to 0.055)	
Likelihood ratio test		0.0216
Pearson chi-squared test		0.0230
The ratio of probabilities		
Mantel-Haenszel (Wald test)	1.73 (1.03 to 2.92)	0.0388
Inverse variance (Wald test)	1.70 (1.00 to 2.88)	0.0495
Maximum likelihood	1.73 (1.03 to 2.91)	
Likelihood ratio test		0.0361
Pearson chi-squared test		0.0373
The odds ratio		
Mantel-Haenszel (RBG test)	1.79 (0.96 to 3.34)	0.0682
Inverse variance (Woolf test)	1.76 (1.01 to 3.08)	0.0461
Maximum likelihood	1.79 (1.03 to 3.10)	
Conditional ML (Gart exact interval)	1.79 (1.00 to 3.25)	
Conditional ML (Gart mid-P interval)	1.79 (1.03 to 3.13)	
Likelihood ratio test		0.0347
Pearson chi-squared test		0.0359
Cochran-Mantel-Haenszel test		0.0365

measure, a logit scale is used on both axes. Under study homogeneity, the points (one for each study) form a straight line. A strong violation of this condition indicates study heterogeneity.

Figure 10.32 shows a L'Abbe plot of the six randomized clinical trials of prophylactic use of lidocaine in myocardial infarction. The slope of the straight line is defined by the estimated common effect, which we here take to be the Mantel-Haenszel estimate. Except for one trial—the Mogensen trial—the points adhere well to the straight line assumption.

We refer to Fagerland (2015, p. 452) for an illustration (with hypothetical data) of the ideal situations with virtually no study heterogeneity for the difference between probabilities, ratio of probabilities, and the odds ratio as effect measures.

Assessing Between-Study Heterogeneity with Meta-Regression

Meta-regression refers to a statistical model that includes one or more study features as explanatory variables. Let x denote such an explanatory variable, for instance, $x = 0$ for studies published before the year 2000 and $x = 1$ for

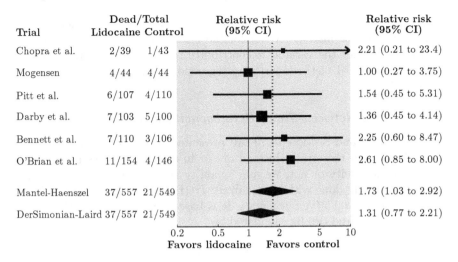

FIGURE 10.31
Forest plot of the relative risk (ratio of probabilities) of mortality with lidocaine versus control in six randomized clinical trials

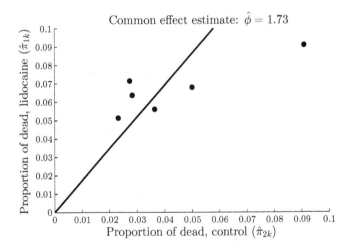

FIGURE 10.32
L'Abbe plot of six randomized clinical trials of prophylactic use of lidocaine in myocardial infarction

studies published after the year 2000. A (fixed effects) meta-regression model that investigates the effects of x can be written as:

$$\text{link}(\pi_{ik}) = \alpha + \beta_{ik} + \gamma x$$

where γ is the effect of x. A random effects meta-regression model may be defined in a similar manner. A test for the null hypothesis $\gamma = 0$ is a test to determine whether the variable x contributes to the study heterogeneity. A review of statistical methods for meta-regression is given by Thompson and Sharp (1999).

Summarizing Between-Study Heterogeneity with I^2

In Section 10.9 and Section 10.10, we presented several tests of homogeneity of effects over the strata. Instead of—or in addition to—testing whether there is homogeneity or not, we may quantify the amount of heterogeneity. This is commonly done with the quantity I^2 (Higgins et al., 2003), which is also called the *index of heterogeneity*. It is based on the Cochran Q statistic (Section 10.9.1), and can be calculated by

$$I^2 = 100\%(Q - \mathrm{df})/Q,$$

where $\mathrm{df} = K - 1$ are the degrees of freedom for the Cochran Q test. The I^2 index measures the percentage of total variation in study effects that is due to heterogeneity (as opposed to random variation). A low I^2 value thus indicates low heterogeneity, except that which can be explained by chance, and larger I^2 values indicate greater amounts of heterogeneity. The I^2 value should be accompanied by an uncertainty interval to indicate its level of precision, see Higgins and Thompson (2002, Appendix).

Assessing Risk of Bias in the Included Studies

Systematic reviews have traditionally included an assessment of the quality of the included studies, usually carried out with a quality scale or checklist. In recent years, however, the difference between methodological quality and risk of bias has been given increase attention (Higgins et al., 2011). Bias refers to the extent to which the results of a study should be believed, whereas study quality refers to the extent to which a study adheres to the highest *possible* standards for answering the research question. Hence, the methodological quality of a study could be of the highest possible standard, but the risk of bias may still be high. Because bias is ultimately more important than study quality, it is no longer recommended that systematic reviews include traditional tools, such as scales or checklists, to assess study quality (Higgins et al., 2011). Instead, an assessment of the risk of bias should be performed.

One way to assess bias is with the *Cochrane Risk of Bias Tool* (Higgins et al., 2011), which consists of six domains: random sequence generation, allocation concealment, blinding of participants and personnel, blinding of outcome assessment, incomplete outcome data, selective outcome reporting, and other potential threats to validity. The final element may or may not be included, depending on the research question or the included studies, or both. Each included study is assessed for each domain and given a label of either *low risk*, *high risk*, or *unclear risk*. *Unclear risk* is used if information is lacking or if the potential for bias is uncertain. An overall judgement of *low risk of bias* may be given to a study if it is awarded a *low risk* on all domains.

Assessing Publication Bias with Funnel Plots and Tests for Funnel Plot Asymmetry

Publication bias refers to absence of information caused by either non-publication of entire studies (missing studies), or selective outcome reporting in published studies based on their results (missing outcomes). Studies that report a statistically significant result ($P \leq 0.05$) are more likely to be published than studies that do not show a statistically significant result ($P > 0.05$). Selective outcome reporting frequently occurs for the same reason: outcomes that show a statistically significant result are more likely to be reported than outcomes that do not show a statistically significant result. Missing studies and missing outcomes may be identified through rigorous searching of protocol registries, comparisons of protocols with published reports, and establishing contact with study authors; however, this will only reveal part of the missing information.

One tool to identify missing studies is the funnel plot (Sterne et al., 2011). Funnel plots target small study bias, in which small studies tend to show larger estimates of effects and greater variability than larger studies. The funnel plot is a scatter plot with effect estimates on the x-axis and study precision—usually the standard error—on the y-axis The y-axis is reversed, such that studies with low precision are placed at the bottom and studies with greater precision are placed at the top. If the effect measure is a ratio measure (such as the ratio of probability or the odds ratio), the x-axis is log transformed. In the absence of missing studies, the shape of the scatter plot resembles a symmetrical inverted funnel with a wide base (consisting of small studies with large variability of effect estimates) and a narrow top (consisting of large studies with small variability of effect estimates). Publication bias is indicated if large "holes" can be seen in the plot, most often in the bottom part where the precision is low, or if asymmetry is present. Note that study heterogeneity—or other determinants unrelated to publication bias—may also cause holes in the plot (Sterne et al., 2011).

Figure 10.33 shows a funnel plot of the six clinical trials of prophylactic use of lidocaine in myocardial infarction. The dotted vertical line represents the estimated common effect, here the logarithm of the Mantel-Haenszel estimate: $\log(\hat{\phi}) = \log(1.73) = 0.55$. Also shown in the figure is a triangular region defined by the solid lines and the x-axis, within which 95% of the trials are expected to lie in the absence of publication bias and heterogeneity. With only six trials, it is impossible to give this figure a proper interpretation. We need several more trials to determine whether the funnel plot is asymmetric or contains holes. Note that the trial represented by the point at the bottom of the plot is the Chopra et al. trial, which only has a total of three deaths and a small total sample size, and thereby has a considerably higher standard error than the other trials. This can also be seen in the forest plot (Figure 10.31), wherein the Chopra et al. trial has a much wider confidence intervals than the other trials.

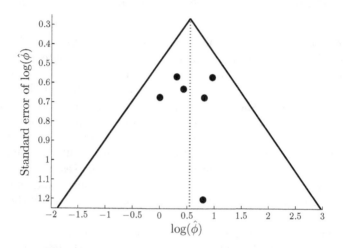

FIGURE 10.33
Funnel plot of six randomized clinical trials of prophylactic use of lidocaine in myocardial infarction

We may also test the null hypothesis of a symmetric funnel plot. One of the most commonly used methods—among many available—is the Egger test for asymmetry (Egger et al., 1997b). It is obtained by regressing the standardized effect size on the inverse of the standard error (the precision):

$$\text{Effect}/\text{SE} = \alpha + \beta(1/\text{SE}) + \epsilon,$$

where Effect is the estimated effect, SE is the standard error of Effect, and ϵ is random noise. Under the null hypothesis of a symmetric plot, the regression line is expected to go through the origin ($\alpha = 0$), with the slope (β) indicating the size and direction of the common effect. A test of $H_0 : \alpha = 0$ is a test for a symmetric funnel plot, and the size of α indicates the extent of asymmetry. Unfortunately, the test has low power, especially when the number of studies is smaller than 10 and the asymmetry is less than considerable.

Cumulative Meta-Analysis

Cumulative meta-analysis, first proposed by Lau et al. (1992), starts by sorting the included studies according to publication date. A meta-analysis may then be performed on the studies available at any chosen point in time. In this way, it is possible to identify the point in time at which the evidence for a treatment or exposure effect first reached statistical significance. A useful visual aid is to draw up a cumulative forest plot, where the first row represents the first published study and subsequent rows represent the common effect after adding the next published study.

Sensitivity Analysis

A sensitivity analysis examines to what extent the inclusion or exclusion of one or more studies changes the results of the meta-analysis. For instance, high risk of bias studies, studies below a certain size, or studies published before a particular year may be excluded, and the meta-analysis is then repeated on the remaining subset of studies. Sensitivity analysis may be regarded as a tool to explore study heterogeneity, see Egger et al. (1997a).

Individual Subject Meta-Analysis

When the data from all included studies in a systematic review are available at the subject level, we can carry out an individual subject meta-analysis. This form of meta-analysis greatly increases power and facilitates the investigation of study heterogeneity. The statistical methods required to perform individual subject meta-analysis include multilevel and hierarchical models, see Turner et al. (2000) and Higgins et al. (2001).

10.18   Recommendations

Table 10.17 summarizes our recommended methods for the analysis of stratified 2×2 tables. Note that the recommendations for the difference between probabilities and the ratio of probabilities are the same, whereas the recommendations for the odds ratio are slightly different and more nuanced. A general result worth mentioning is the excellent performance of the Pearson chi-squared test for homogeneity, which we recommend for all three effect measures. We also specifically note that our recommended estimate of the common effect, regardless of the link function, is the Mantel-Haenszel estimate. The following subsections discuss our recommendations in more detail.

10.18.1   Testing Homogeneity of Effects over the K Strata

For the difference between probabilities and the ratio of probabilities, the Pearson chi-squared test is the only test that performs consistently well. The likelihood ratio test is often too liberal, and the Cochran Q tests with Mantel-Haenszel estimate and inverse variance estimate are either too liberal (difference between probabilities) or too conservative (ratio of probabilities) to be recommended.

For the logit link, we agree with Reis et al. (1999), who recommend the

TABLE 10.17

Recommended tests, estimates, and confidence intervals (CIs) for stratified 2×2 tables

Analysis	Recommended methods	Sample sizes
The difference between probabilities		
Tests of homogeneity of Δ_k	Pearson chi-squared	all
Tests of a common Δ	Wald (Mantel-Haenszel)	all
Estimate of Δ	Mantel-Haenszel	all
	Maximum likelihood	all
CIs for the common Δ	Wald (Mantel-Haenszel)	all
	Wald (ML*)	all
The ratio of probabilities		
Tests of homogeneity of ϕ_k	Pearson chi-squared	all
Tests of a common ϕ	Wald (Mantel-Haenszel)	all
Estimate of ϕ	Mantel-Haenszel	all
	Maximum likelihood	all
CIs for the common ϕ	Wald (Mantel-Haenszel)	all
	Wald (ML*)	all
The odds ratio		
Tests of homogeneity of θ_k	Pearson chi-squared	all
	Zelen exact	small/medium
Tests of a common θ	Cochran-Mantel-Haenszel	all
	Pearson chi-squared	medium/large
Estimate of θ	CML[†]	small/medium
	Mantel-Haenszel	medium/large
CIs for the common θ	Gart mid-P	small/medium
	Wald (ML*)	large

*Maximum likelihood
[†]Conditional maximum likelihood

Pearson chi-squared test for testing homogeneity of odds ratios, both for small and large sample sizes. The Breslow-Day test for large samples and the Zelen test for small samples seem to be much used in practice, see for instance, Agresti (2013, pp. 242, 608). Studies by Gavaghan et al. (2000) and Reis et al. (1999) support the use of these tests. Among the asymptotic tests, Reis et al. (1999) recommend the Breslow-Day test and the Pearson chi-squared test. Among the exact tests, the Zelen, Pearson chi-squared, and conditional score tests are recommended. Hosmer et al. (2013, p. 86) advocate the use of logistic regression and the likelihood ratio test to assess homogeneity of odds ratios. Reis et al. (1999), however, find that the likelihood ratio test is too liberal, with actual significance levels above the nominal level. This is similar to our findings in Section 10.10.5. We further find that the Cochran Q test is too conservative to be of any interest, except for very large sample sizes.

10.18.2 Tests of a Common Effect

For the difference between probabilities and the ratio of probabilities, the Wald test based on the Mantel-Haenszel estimate performs acceptably, although far from excellent. For the difference between probabilities, all the tests that we have considered have actual significance levels above the nominal level. That includes the Wald test based on the Mantel-Haenszel estimate, although this test violates the nominal level to a lesser degree than the others. For the ratio of probabilities, the Wald test based on the Mantel-Haenszel estimate is somewhat conservative, but not much. We refer to Sections 10.11.5 and 10.12.5 for more details.

For the odds ratio, the Cochran-Mantel-Haenszel test is, by many, considered the standard test for a common effect in stratified 2 × 2 tables, and this is with good reason. The Cochran-Mantel-Haenszel test has a simple form; it compares the observed and expected cell counts, summed over the strata. The required sample size is related to total cell counts, summed over the strata, and not to the individual stratum cell counts, see the Mantel-Fleiss criterion in Section 10.13.3.

The Cochran-Mantel-Haenszel test can be considered as a "quasi-exact" test, which may explain its excellent performance. Mantel and Haenszel (1959) derived the test statistic from the exact hypergeometric distribution. When large sample theory is applied, the Cochran-Mantel-Haenszel test statistic is asymptotically chi-squared distributed with one degree of freedom. The exact conditional tests are known to be conservative. For the Cochran-Mantel-Haenszel test, the conservatism of an exact test is compensated by the approximation to the chi-squared distribution. Thus the best of two principles may be obtained—a test statistic derived as an exact test, with the distribution approximated to its asymptotic distribution.

The Cochran-Mantel-Haenszel test performs excellently for all sample sizes. The Pearson chi-squared test also performs well, especially for large sample sizes.

10.18.3 Estimates and Confidence Intervals

For the difference between probabilities, the ratio of probabilities, and the odds ratio, the Mantel-Haenszel estimate, the inverse variance estimate, and the maximum likelihood estimate can serve as effect estimates for the common effect. For the odds ratio, the conditional maximum likelihood estimate can be used in addition.

The Mantel-Haenszel and inverse variance estimates are weighted estimates of the stratum specific effect estimates, with weights given in Table 10.9.

The inverse variance and maximum likelihood estimates are consistent for large samples, but can be seriously biased when the sample size is small and the number of strata is large. For the inverse variance estimate, the bias is due to how the cell counts relate to the weights. For the maximum likelihood estimate, too many strata will give too many nuisance parameters. Although the Mantel-Haenszel estimate is derived heuristically, and is neither a minimum variance linear estimate nor a maximum likelihood estimate, it has many preferable properties. It can be calculated regardless of the sample size, it is easy to calculate, and it is found in most statistical software packages.

For the odds ratio, the conditional maximum likelihood estimate may be an attractive alternative. It is based on conditioning, and the likelihood is free of nuisance parameters. Maximum likelihood and conditional maximum likelihood are asymptotically equivalent in the so-called fixed strata case, i.e., when the number of strata is low. But when the number of strata increases, maximum likelihood no longer produces a consistent estimate, and conditional maximum likelihood performs better. Both for small and large sample sizes, the Mantel-Haenszel estimate is an alternative. The Mantel-Haenszel estimate has very good efficiency, it is equal to or a little inferior to the conditional maximum likelihood, see Breslow (1981), Hauck (1984), and Emerson (1994). Emerson proposed to calculate the conditional maximum likelihood in cases when the total sample size is less than 1000, or when there is severe unbalance in the marginal cell count, and compared the conditional maximum likelihood with the Mantel-Haenszel estimate. Conditional maximum likelihood is generally not included in statistical software packages; however, it is found in StatXact 11 (Cytel Inc., Cambridge, MA).

For the difference of probabilities and the ratio of probabilities, we can recommend both Mantel-Haenszel and maximum likelihood for confidence interval estimation. For the odds ratio, the Mantel-Haenszel estimate performs well, but the RBG variance seems to overestimate the variance and provide intervals that are too wide. We recommend the maximum likelihood confidence interval for large sample sizes, with not too many strata. For small sample sizes, or if there are many strata, the conditional maximum likelihood can be recommended. The related confidence interval is the Gart exact interval, preferably with a mid-P correction. Calculation of the conditional maximum likelihood, with the Gart exact and mid-P intervals, can be carried out in StatXact.

11

Other Stratified Tables

11.1 Introduction

Stratification was discussed in the introduction to Chapter 10, where stratified unpaired 2×2 tables were introduced. Stratification is, of course, relevant for any two-way contingency table, and this chapter considers stratified paired 2×2 tables (Section 11.3), stratified $r \times 2$ tables (Section 11.4), stratified $2 \times c$ tables (Section 11.5), and stratified $r \times c$ tables (Section 11.6).

The approach to the analysis of the stratified tables in this chapter follows the outline in Section 10.6, wherein a guide to the analysis of stratified 2×2 tables was presented. First, we perform a test of homogeneity over the strata. If there is heterogeneity, that is, if the association between the row and column variables varies with the strata, stratum specific effect estimates are reported. If the null hypothesis of homogeneity is not rejected, we estimate the common effect and test the null hypothesis of no common effect.

But first, we introduce four examples that will be used to illustrate the methods later in the chapter.

11.2 Examples

11.2.1 Los Angeles Retirement Community Study of Endometrial Cancer (Stratified Paired 2×2 Tables)

The results of a study of the effect of exogenous estrogens on the risk of endometrial cancer were reported by Mack et al. (1976) and later analyzed by Breslow and Day (1980, p. 162). Each case of endometrial cancer was matched with a control who was alive, at the age of plus minus one year of the case, and living in the community at the time when the case was diagnosed. Breslow and Day (1980, p. 168) report results from the analysis of the association between gallbladder disease and endometrial cancer. A total of 63 matched pairs were analyzed both unstratified and stratified by age under and over 70 years. Table 11.1 presents the data stratified by age, which make up two paired 2×2 tables.

TABLE 11.1

Matched paired data on endometrial cancer and gallbladder (GB)
disease, stratified by age (Mack et al., 1976)

	Controls (no cancer)			
	GB	No GB		
Cases (cancer)	disease	disease	Total	Age
GB disease	2	7	9 (32%)	
No GB disease	1	18	19 (68%)	< 70 years
Total	3 (11%)	25 (89%)	28 (100%)	
GB disease	2	6	8 (23%)	
No GB disease	4	23	27 (73%)	≥ 70 years
Total	6 (17%)	29 (83%)	35 (100%)	

The total number of observations: $N = 63$

11.2.2 Ille-et-Vilaine Study of Oesophageal Cancer (Stratified $r \times 2$ Tables)

Tuyns et al. (1977) collected data in the French department of Ille-et-Vilaine
on 200 male cases of oesophageal cancer. A control sample of 775 males were
selected, see Breslow and Day (1980, p. 122). Breslow and Day also present
a contingency table for the association between alcohol consumption and oesophageal cancer, stratified by age (Breslow and Day, 1980, p. 151). Alcohol
consumption is categorized in four categories and age is categorized in six
categories, so we have six 4×2 tables, which we also can regard as a $4 \times 2 \times 6$
table. In Table 11.2, we have re-arranged the data by swapping the rows and
columns. We have done this solely because a $4 \times 2 \times 6$ table takes up considerably more space than a $2 \times 4 \times 6$ table. This is also the format in which
Breslow and Day (1980) present these data.

11.2.3 Adolescent Placement Study (Stratified $2 \times c$ Tables)

In Section 6.2.1, we presented data from a study by Fontanella et al. (2008)
of the association between danger to others as an ordered variable with four
outcome categories and sex as the group variable. Here, we present the same
data stratified by ethnicity (Table 11.3). Ethnicity has two categories: white
and non-white. We thus have stratified ordered 2×4 tables with two strata.

TABLE 11.2
Oesophageal cancer and alcohol consumption, stratified by age (Tuyns et al., 1977)

Cancer	Alcohol consumption (g/day)				Total	Age*
	0–39	40–79	80-119	120+		
Cases	0	0	0	1 (20%)	1 (0.8%)	
Controls	61 (100%)	45 (100%)	5 (100%)	4 (80%)	115 (99%)	25–34
Total	61	45	5	5	116	
Cases	1 (1.1%)	4 (5.0%)	0	4 (40%)	9 (4.5%)	
Controls	88 (99%)	76 (95%)	20 (100%)	6 (60%)	190 (95%)	35–44
Total	89	80	20	10	199	
Cases	1 (1.3%)	20 (25%)	12 (31%)	13 (87%)	46 (22%)	
Controls	77 (99%)	61 (75%)	27 (69%)	2 (13%)	167 (78%)	45–54
Total	78	81	39	15	213	
Cases	12 (13%)	22 (26%)	24 (56%)	18 (69%)	76 (31%)	
Controls	77 (87%)	62 (74%)	19 (44%)	8 (31%)	166 (69%)	55–64
Total	89	84	43	26	242	
Cases	11 (15%)	25 (47%)	13 (45%)	6 (75%)	55 (34%)	
Controls	60 (85%)	28 (53%)	16 (55%)	2 (25%)	106 (66%)	65–74
Total	71	53	29	8	161	
Cases	4 (15%)	4 (33%)	2 (100%)	3 (100%)	13 (30%)	
Controls	23 (85%)	8 (67%)	0	0	31 (70%)	75+
Total	27	12	2	3	44	

The total number of observations: $N = 975$
*Age in years

11.2.4 Psychiatric Diagnoses and BMI (Stratified $r \times c$ Tables)

We return to the study by Mangerud et al. (2014) on the association between psychiatric diagnoses and body mass index (BMI), which we first encountered in Section 7.2.3. Under the assumption that both psychiatric diagnoses and BMI are unordered, the overall association was found to be statistically significant ($P < 0.01$, see Table 7.15). The participants in this study were

TABLE 11.3

Danger to others for males and females, stratified by
ethnicity (Fontanella et al., 2008)

	Unlikely	Possible	Probable	Likely	Total	Ethnicity
		Danger to others				
Male	3 (2.4%)	13 (10%)	42 (34%)	66 (53%)	124	
Female	28 (18%)	38 (25%)	37 (24%)	50 (32%)	154	Non-white
Total	31 (11%)	53 (19%)	79 (28%)	116 (42%)	278	
Male	5 (4.6%)	15 (14%)	30 (27%)	60 (55%)	110	
Female	17 (14%)	35 (29%)	32 (27%)	36 (30%)	120	White
Total	22 (9.6%)	50 (22%)	62 (27%)	96 (42%)	230	

The total number of observations: $N = 508$

both males and females, and we suspect that sex may influence the association between diagnoses and BMI. Table 11.4 displays the cross-classification of psychiatric diagnoses and BMI, stratified by sex, in a $6 \times 3 \times 2$ table.

TABLE 11.4

Psychiatric diagnoses and weight categories based on age- and sex-adjusted
BMI, stratified by sex (Mangerud et al., 2014)

Main diagnosis	Thin	Normal	Overweight	Total	Sex
Mood (affec.) dis.	1 (5.9%)	12 (71%)	4 (24%)	17	
Anxiety disorders	3 (5.7%)	34 (64%)	16 (30%)	53	
Eating disorders	1 (50%)	1 (50%)	0	2	
Autism spec. dis.	3 (9.1%)	20 (61%)	10 (30%)	33	Males
Hyperkinetic dis.	12 (9.9%)	82 (68%)	27 (22%)	121	
Other disorders	6 (19%)	15 (48%)	10 (32%)	31	
Total	26 (10%)	164 (64%)	67 (26%)	257	
Mood (affect.) dis.	2 (3.1%)	43 (67%)	19 (27%)	64	
Anxiety disorders	5 (5.4%)	68 (73%)	20 (22%)	93	
Eating disorders	5 (26%)	13 (68%)	1 (5.3%)	19	
Autism spec. dis.	2 (40%)	1 (20%)	2 (40%)	5	Females
Hyperkinetic dis.	7 (7.6%)	48 (52%)	37 (40%)	92	
Other disorders	1 (5.0%)	11 (55%)	8 (40%)	20	
Total	22 (7.5%)	184 (63%)	87 (30%)	293	

11.3   Stratified Paired 2 × 2 Tables

11.3.1   Notation and Statistical Model

Stratified paired 2×2 tables are a generalization of the paired 2×2 table that was considered in detail in Chapter 8. The notation that we will use for stratified paired 2×2 tables is thus a generalization of the notation in Section 8.3. We assume that we have K strata, with the parameters shown in Table 11.5. For stratum k, the observed counts are given in Table 11.6.

TABLE 11.5
The joint probabilities of stratum k in paired 2×2 tables

	Event B		
Event A	Success	Failure	Total
Success	π_{11k}	π_{12k}	π_{1+k}
Failure	π_{21k}	π_{22k}	π_{2+k}
Total	π_{+1k}	π_{+2k}	1

TABLE 11.6
The observed counts of stratum k in paired 2×2 tables

	Event B		
Event A	Success	Failure	Total
Success	n_{11k}	n_{12k}	n_{1+k}
Failure	n_{21k}	n_{22k}	n_{2+k}
Total	n_{+1k}	n_{+2k}	n_{++k}

For the difference between marginal probabilities, see Section 8.6, we consider the effect measure $\Delta_k = \pi_{1+k} - \pi_{+1k}$, with effect estimate $\hat{\Delta}_k = (n_{12k} - n_{21k})/n_{++k}$. Homogeneity over the strata means

$$\Delta_1 = \Delta_2 = \ldots = \Delta_K.$$

For the ratio of marginal probabilities, see Section 8.8, we consider the effect measure $\phi_k = \pi_{1+k}/\pi_{+1k}$. Then, homogeneity over the strata means

$$\phi_1 = \phi_2 = \ldots = \phi_K.$$

For the subject specific model, we consider the effect measure $\theta_{k,\text{cond}} = \pi_{12k}/\pi_{21k}$, with effect estimate $\hat{\theta}_{k,\text{cond}} = n_{12k}/n_{21k}$, see Section 8.9.1. Define

$n_{d,k}$ as the sum of the discordant pairs in stratum k: $n_{d,k} = n_{12k} + n_{21k}$. Conditional on n_{11k} and the number of discordant pairs, n_{12k} is binomially distributed with parameters μ_k and $n_{d,k}$, where

$$\mu_k = \frac{\pi_{1+k}}{\pi_{1+k} + \pi_{+1k}},$$

for $k = 1, 2, \ldots, K$. Conditional homogeneity over the strata means

$$\mu_1 = \mu_2 = \ldots = \mu_K. \tag{11.1}$$

11.3.2 Testing Homogeneity over the Strata

A test for marginal homogeneity of the differences between probabilities was presented by Zhao et al. (2014), which also included R code for computing the test statistic and a P-value. A similar test for marginal homogeneity of the ratios of probabilities is given in Lachin (2011, p. 230). If we assume homogeneity of the differences or the ratios, inverse variance estimates can be calculated. Here, we do not go into details of these tests and estimates, but instead concentrate on the tests for conditional homogeneity.

Testing for conditional homogeneity (Equation 11.1) is simply a test for equal probabilities in K binomial series, see Breslow and Day (1980, p.168) or Lachin (2011, p. 229). The data can be rearranged in a $2 \times K$ table, with one column for each stratum, see Table 11.7.

TABLE 11.7
Cell counts for K binomial series in stratified paired 2×2 tables

Counts	Stratum 1	Stratum 2	...	Stratum K	Total
Cell (1,2)	n_{121}	n_{122}	$\cdots$	n_{12K}	n_{12+}
Cell (2,1)	n_{211}	n_{212}	$\cdots$	n_{21K}	n_{21+}
Total	$n_{d,1}$	$n_{d,2}$	$\cdots$	$n_{d,K}$	$n_{d,+}$

Conditional on the number of discordant pairs, the problem of testing homogeneity is transformed to the problem of testing independence in an unordered $2 \times K$ table. The recommended tests are the Pearson chi-squared test for large sample sizes, and the Fisher-Freeman-Halton exact test—with or without a mid-P correction—in small samples, see Section 7.11. Lui and Lin (2014) performed a Monte Carlo evaluation of the Pearson chi-squared and likelihood ratio tests and found that they performed well with respect to type I error. The Fisher-Freeman-Halton test, however, was not included in that study.

The likelihood ratio statistic can also be calculated from the data in Table 11.7; however, it is not among the recommended tests for such a table. Note that matched data can be analyzed with conditional logistic regression.

The likelihood ratio test for conditional homogeneity can then be derived by taking the difference in the minus two log likelihoods for the conditional logistic regression models with and without an interaction term between the exposure variable and the stratum variable.

If there is heterogeneity, the effect estimate for each stratum should be presented. Then, the stratum specific estimate of the conditional odds ratio in stratum k is

$$\hat{\theta}_{k,\mathrm{cond}} = \frac{n_{12k}}{n_{21k}},$$

see Section 8.9. For the recommended confidence interval, see Table 8.15.

If, on the other hand, we can assume homogeneity, we estimate a common effect over the strata, and this is the topic of the next section.

11.3.3 Estimation and Testing of a Common Effect over the Strata

If we assume marginal homogeneity over the strata, the estimate of the common effect is given as a weighted stratum average:

$$\hat{\Delta} = \frac{\displaystyle\sum_{k=1}^{K} w_k \hat{\Delta}_k}{\displaystyle\sum_{k=1}^{K} w_k},$$

with inverse variance weights $w_k = 1/\mathrm{Var}(\hat{\Delta}_k)$, and

$$\mathrm{Var}(\hat{\Delta}_k) = n_{12k} + n_{21k} - \frac{(n_{12k} - n_{21k})^2}{n_{++k}},$$

see Section 8.6.2 and Zhao and Rahardja (2013). A similar derivation for the ratio of probabilities can be found in Lachin (2011, p. 230).

When we have conditional homogeneity, the common effect can be estimated by

$$\hat{\theta}_{\mathrm{cond}} = \frac{n_{12+}}{n_{21+}},$$

which is also the conditional maximum likelihood estimate. The recommended confidence intervals for the common effect is the same as those for the stratum specific conditional estimates and include the transformed Wilson score interval for all sample sizes and the Wald interval for large sample sizes; for details, see Table 8.15.

11.3.4   Example

Los Angeles Retirement Community Study of Endometrial Cancer (Table 11.1)

For the data in Table 11.1, the estimated conditional odds ratio for age under 70 years is $\hat{\theta}_{1,\text{cond}} = 7.0$, and the transformed Wilson score interval for the odds ratio is (1.12 to 43.6). The Wald interval is (0.86 to 56.9), which is considerably wider. For age above 70 years, the estimated conditional odds ratio is $\hat{\theta}_{2,\text{cond}} = 1.5$ with transformed Wilson score interval equal to (0.46 to 4.95), and with Wald interval equal to (0.42 to 5.32). Again the Wald interval is wider than the transformed Wilson score. There seems to be a sizeable difference between the two stratum specific estimates, although the confidence intervals are wide and overlap by a considerable margin. Next, we test for homogeneity to find out whether it is sensible to combine the data and calculate a common effect.

To test for homogeneity, we rearrange the data into Table 11.8. Because there are only two strata, testing homogeneity is equivalent to testing independence—or association—in a 2×2 table. Our recommendations for testing association in 2×2 tables are given in Table 4.24 (page 174). For small samples, such as that in Table 11.8, we recommend the Fisher mid-P test, the Suissa and Shuster exact unconditional test, or the Fisher-Boschloo exact unconditional test. The results for these tests are given in Table 11.9. Also included in that table is the Pearson chi-squared test, which, due to the small sample size, is not a recommended test for these data; we have included it for reference.

TABLE 11.8
Cell counts for the $K = 2$ binomial series based on the data in Table 11.1

Counts	Stratum 1	Stratum 2	Total
Cell (1,2)	7	6	13
Cell (2,1)	1	4	5
Total	8	10	18

None of the tests indicate heterogeneity of the strata specific effects, and we proceed to estimate the common effect. With the conditional odds ratio as the effect measure, we get that $\hat{\theta}_{\text{cond}} = 2.6$ with a 95% transformed Wilson score interval of (0.966 to 7.000). The Wald interval is (0.927 to 7.293). We can also calculate a test for no association ($H_0 : \theta_{\text{cond}} = 1$). The recommendations for this test include the McNemar asymptotic test (see Table 8.15, page 384), which gives $P = 0.0593$.

TABLE 11.9
Results of tests of homogeneity over strata based on the data in
Table 11.1

Test	Statistic	*P*-value
Pearson chi-squared*	1.67 (df = 1)	0.196
Fisher mid-*P*		0.216
Suissa-Shuster exact unconditional†		0.246
Fisher-Boschloo exact unconditional†		0.246

*Not a recommended test for these data
†With Berger and Boos procedure ($\gamma = 0.0001$)

11.4 Stratified $r \times 2$ Tables

11.4.1 Introduction

In Section 5.6, we described tests for unspecific ordering in the $r \times 2$ table, which in the context of stratified $r \times 2$ tables, may be considered a case
where $K = 1$ (one stratum). A Pearson chi-squared and a likelihood ratio
test were presented, both with an asymptotic chi-bar-squared distribution.
A generalization to the stratified case with $K \geq 2$ was suggested by Agresti
and Coull (1996, 2002), which also include procedures for estimating and testing for order restrictions of the grouping effects and a method for calculating
order restricted maximum likelihood estimates of the grouping effect. The proposed tests use the likelihood ratio test statistic, which asymptotically follow
a chi-bar-squared distribution. The weights of the distribution are difficult to
determine. As pointed out by Agresti and Coull (1996, 2002), P-values can
be found by simulating from the exact distribution, instead of calculating the
exact P-values, which are very computer intensive. To our knowledge, little
is known about the properties of these methods, and few analyses of actual
data sets can be found in the literature.

Our recommendations for the $r \times 2$ table are summarized in Table 5.12.
Although the Cochran-Armitage test was derived as a test for trend in the
linear probability model, it is also appropriate for the logit model. In general,
we recommend use of logit models; however, we recommend the Cochran-
Armitage test or the Mantel-Haenszel test for testing trend. The Wald test
and the Wald confidence interval also perform well, and we recommend both
for a consistent trend analysis.

11.4.2 Notation and Statistical Model

Following the notation in Section 5.3, we let π_{ik} for $i = 1, 2, \ldots, r$ and $k = 1, 2, \ldots, K$ denote binomial probabilities in the distribution of n_{i1k} given the marginal n_{i+k}. The conditional probability of success or failure is given by a simple generalization of Equations 5.6 and 5.7 for the K strata.

When the grouping variable (Y_1) is ordinal, with attached scores $\{a_1, a_2, \ldots, a_r\}$, and there is no interaction between the outcome variable (Y_2) and the stratification variable (Y_3), the logit is given as

$$\text{logit}(\pi_{ik}) = \alpha_k + \beta_k a_i,$$

for $i = 1, 2, \ldots, r$ and $k = 1, 2, \ldots, K$. In general, we recommend the use of equally spaced scores, see Section 5.5.

The maximum likelihood estimates of the β_k in the logit model above are simple generalizations of the maximum likelihood estimate in Section 5.8.2. We recommend maximum likelihood estimates for the logit link when estimating a trend, see Section 5.9.4.

11.4.3 The Likelihood Ratio Test of Homogeneity of Trend over the K Strata

Initially, we are interested in testing the null hypothesis of homogeneity of the trend over the strata. The null hypothesis is

$$H_0 : \beta_1 = \beta_2 = \ldots = \beta_K, \tag{11.2}$$

and we can test it with the likelihood ratio statistic. Let L_0 be the maximum likelihood function under the null hypothesis (Equation 11.2), and let L_1 be the maximum likelihood function under the alternative model in Section 11.4.2. Then, the likelihood ratio statistic is

$$T_{\text{LR}}(\mathbf{n}) = -2(L_0 - L_1).$$

The likelihood ratio statistic is asymptotically chi-squared distributed with $K - 1$ degrees of freedom.

11.4.4 The Wald Test of a Common Trend

If we can assume homogeneity of the trend over the strata, i.e., $\beta_1 = \beta_2 = \ldots = \beta_K = \beta$, we can find the maximum likelihood estimate of the common trend (β). Let $\hat{\beta}_{\text{ML}}$ be the maximum likelihood estimate, and let $\widehat{\text{SE}}(\hat{\beta}_{\text{ML}})$ be its estimated standard error. The null hypothesis of interest is now

$$H_0 : \beta = 0. \tag{11.3}$$

The Wald test statistic for this null hypothesis is

$$Z_{\text{Wald}}(\mathbf{n}) = \frac{\hat{\beta}_{\text{ML}}}{\widehat{\text{SE}}(\hat{\beta}_{\text{ML}})}.$$

The Wald statistic is approximately standard normally distributed.

11.4.5 The Cochran-Mantel-Haenszel Asymptotic, Exact, and Mid-P Tests of a Common Trend

The null hypothesis of a common trend in Equation 11.3 can also be tested with an extension of the Mantel-Haenszel test for trend in Section 5.7.6. A straightforward generalization of Equation 5.24 to K strata gives

$$T_{\text{CMH}}(\mathbf{n}) = \frac{\sum_{k=1}^{K} n_{++k}\left[\bar{d}_k - \text{E}(\bar{d}_k)\right]^2}{\sum_{k=1}^{K} n_{++k}^2 \text{Var}(\bar{d}_k)} = \frac{\sum_{k=1}^{K} n_{++k} r_{ab,k}\sqrt{v_{ak}v_{bk}}}{\sum_{k=1}^{K} n_{++k}^2 v_{ak}v_{bk}/(n_{++k}-1)}, \quad (11.4)$$

where d_k, $\text{E}(\bar{d}_k)$, and $\text{Var}(\bar{d}_k)$ are given in Section 5.7.6, $r_{ab,k}$ is the correlation between Y_1 and Y_2 in stratum k, and v_{ak} and v_{bk} are the variances of Y_1 and Y_2 in stratum k. The test statistic in Equation 11.4 is approximately chi-squared distributed with one degree of freedom. Stokes et al. (2012, p. 94) have suggested that the chi-squared approximation to T_{CMH} is appropriate when $\sum_{k=1}^{K} n_{++k} \geq 40$.

An exact conditional test based on T_{CMH} can be expressed as

$$\text{one sided exact } P\text{-value} = \sum_{\Omega(\mathbf{x}|\mathbf{n}_{++})} I\left[T_{\text{CMH}}(\mathbf{x}) \geq T_{\text{CMH}}(\mathbf{n})\right] f(\mathbf{x}\,|\,\mathbf{n}_{++}),$$

$$(11.5)$$

where the summation is over all possible (stratified) tables with the same row and column sums $(\mathbf{n}_{++})$ as the observed (stratified) table, $I()$ is the indicator function, and $f(\mathbf{x}\,|\,\mathbf{n}_{++})$ is the product of the probability distribution in Equation 5.9 over the K strata. The twice the smallest tail two-sided P-value is obtained in the usual manner by doubling the one-sided P-value in Equation 11.5, provided this is the smallest of the two possible one-sided P-values defined by $T_{\text{CMH}}(\mathbf{x}) \geq T_{\text{CMH}}(\mathbf{n})$ and $T_{\text{CMH}}(\mathbf{x}) \leq T_{\text{CMH}}(\mathbf{n})$.

The corresponding mid-P is defined by

$$\text{one sided mid-}P \text{ value} = \sum_{\Omega(\mathbf{x}|\mathbf{n}_{++})} I\left[T_{\text{CMH}}(\mathbf{x}) > T_{\text{CMH}}(\mathbf{n})\right] f(\mathbf{x}\,|\,\mathbf{n}_{++})$$

$$+ \; 0.5 \cdot \sum_{\Omega(\mathbf{x}|\mathbf{n}_{++})} I\left[T_{\text{CMH}}(\mathbf{x}) = T_{\text{CMH}}(\mathbf{n})\right] f(\mathbf{x}\,|\,\mathbf{n}_{++}).$$

11.4.6 Example

Ille-et-Vilaine Study of Oesophageal Cancer (Table 11.2)

The first stratum contains only one case and is for that reason deleted from the analysis. We start with the likelihood ratio test for homogeneity over the strata. The observed value of the test statistic is $L_{\text{LR}}(\mathbf{n}) = 5.788$, and with

$K - 1 = 4$ degrees of freedom, we get $P = 0.216$. We thus have no evidence of heterogeneity of the trend over the strata.

We can now test the null hypothesis of no common trend (Equation 11.3) with the Wald test and the Cochran-Mantel-Haenszel test. Both tests give a P-value < 0.001.

The estimate of the common trend is calculated as the maximum likelihood estimate of the common parameter in Section 11.4.4. For the data in Table 11.2, the estimate is $\hat{\beta}_{\mathrm{ML}} = 1.094$ with a 95% confidence interval equal to (0.891 to 1.296).

11.5   Stratified $2 \times c$ Tables

The topic of this section is stratified $2 \times c$ tables, where the c columns can be regarded as either unordered or ordered. The methods in this section are thereby related to—and sometimes generalizations of—the methods in Chapter 6. In Section 11.5.1, we consider unordered categories, and in Section 11.5.2, we consider ordered categories.

11.5.1   Unordered Categories

We now generalize the notation we introduced in Section 6.3 to the case of stratified $2 \times c$ tables. The probability of an outcome equal to j given group i in stratum k is denoted by

$$\Pr(Y_2 = j \mid Y_1 = i, Y_3 = k) = \pi_{j|ik},$$

for $i = 1, 2$, $j = 1, 2, \ldots, c$, and $k = 1, 2, \ldots, K$. A generalization of the logit model for stratified $2 \times c$ tables can be expressed as

$$\log\left(\frac{\pi_{j|ik}}{\pi_{1|ik}}\right) = \alpha_{jk} + \beta_{jk}x_i,$$

for $i = 1, 2$, $j = 2, 3, \ldots, c$, and $k = 1, 2, \ldots, K$, where $x_1 = 1$ and $x_2 = 0$. Then,

$$\beta_{jk} = \log\left(\frac{\pi_{j|1k}/\pi_{1|1k}}{\pi_{j|2k}/\pi_{j|2k}}\right),$$

or, equivalently, that

$$\exp\left(\beta_{jk}\right) = \frac{\pi_{j|1k}/\pi_{1|1k}}{\pi_{j|2k}/\pi_{j|2k}},$$

which is the odds ratio for an outcome in category j compared with category 1 for Group 1 relative to Group 2.

The null hypothesis of no interaction is given by

$$H_0 : \beta_{j1} = \beta_{j2} = \ldots = \beta_{jK} = \beta_j,$$

for $j = 2, 3, \ldots, k$. The principle of maximum likelihood will now provide us with maximum likelihood estimates of the parameters in the model with no interaction effect. Let m_{ijk} denote the estimated expected cell counts. The likelihood ratio test statistic for testing the null hypothesis of no interaction is given by

$$T_{\text{LR}}(\mathbf{n}) = 2 \sum_{k=1}^{K} \sum_{j=1}^{c} \sum_{i=1}^{2} n_{ijk} \log \frac{n_{ijk}}{m_{ijk}}.$$

T_{LR} follows, approximately, a chi-squared distribution with $(c-1)(K-1)$ degrees of freedom.

Testing for conditional independence means testing the null hypothesis

$$H_0 : \beta_j = 0,$$

for $j = 2, 3, \ldots, c$. Let m_{ijk}^* denote the expected cell counts under the null hypothesis. The likelihood ratio test statistic for testing conditional independence can be expressed by

$$T_{\text{LR}}(\mathbf{n}) = 2 \sum_{k=1}^{K} \sum_{j=1}^{c} \sum_{i=1}^{2} m_{ijk} \log \frac{m_{ijk}}{m_{ijk}^*}.$$

This likelihood ratio statistic is approximately chi-squared distributed with $c-1$ degrees of freedom.

The Mantel-Haenszel approach is an appealing alternative, particularly when the number of strata increases and the data are sparse. Yanagawa and Fujii (1990) proposed a generalized Breslow-Day test of homogeneity, and provided an algorithm for computing the test statistic.

The Mantel-Haenszel estimate of a common odds ratio in stratified 2×2 tables has been extended to stratified $2 \times c$ tables by, among others, Greenland (1989) and Yanagawa and Fujii (1990, 1995); see, in particular, Yanagawa and Fujii (1990) for an explicit expression. Greenland (1989) also gave an expression of the variance of the generalized Mantel-Haenszel estimate. With the estimated standard error from this variance, we may construct a Wald test and confidence interval for the common effect. For the example in Greenland (1989), it is noteworthy that the confidence intervals based on the maximum likelihood method and the generalized Mantel-Haenszel estimate are almost identical.

The Cochran-Mantel-Haenszel test statistic for conditional independence is

$$T_{\text{CMH}}(\mathbf{n}) = \sum_{j=1}^{c} \frac{N}{n_{+j+}} \left(n_{1j+} - \frac{n_{1++} \cdot n_{+j+}}{N} \right)^2 \Bigg/ \sum_{k=1}^{K} \frac{n_{1+k} \cdot n_{2+k}}{n_{++k} - 1},$$

which for large samples has a chi-squared distribution with $c - 1$ degrees of freedom. For the derivation of the generalized Cochran-Mantel-Haenszel test, see Birch (1965).

11.5.2 Ordered Categories

As is the case for stratified $r \times 2$ tables, Agresti and Coull (1998b, 2002) have generalized estimates and tests for order restriction to the situation of stratified $2 \times c$ tables, including maximum likelihood estimates and likelihood ratio tests. Exact P-values can be derived; however, they are very computer-intensive to calculate, and Agresti and Coull (1998b, 2002) recommend to estimate them by Monte Carlo simulations. We note, as we did for stratified $r \times 2$ tables, that little is known about the properties of tests for stratified $2 \times c$ tables with order restrictions, and few examples of actual data analyses can be found in the literature.

Let the outcome variable Y_2 be ordinal with scores $\{b_1, b_2, \ldots, b_c\}$ attached to the categories.

The Mantel-Haenszel test for conditional independence in an ordered $2 \times c$ table was described in Section 6.9, and we extend it here to stratified $2 \times c$ tables. Let $\bar{b}_{1k}$ and $\bar{b}_{2k}$ be the mean scores in Group 1 and Group 2 in stratum k:

$$\bar{b}_{1k} = \sum_{j=1}^{c} b_j n_{1jk}/n_{1+k} \quad \text{and} \quad \bar{b}_{2k} = \sum_{j=1}^{c} b_j n_{2jk}/n_{2+k}.$$

Further, let $\bar{b}_k = \sum_{j=1}^{c} b_j n_{+jk}/n_{++k}$. The Cochran-Mantel-Haenszel test statistic for testing conditional independence in stratified $2 \times c$ tables is

$$T_{\text{CMH}}(\mathbf{n}) = \frac{\sum_{j=1}^{c} \dfrac{n_{1+k} \cdot n_{2+k}}{n_{++k}} \left(\bar{b}_{1k} - \bar{b}_{2k}\right)^2}{\sum_{k=1}^{K} \dfrac{n_{++k}}{n_{++k} - 1} \sum_{j=1}^{c} \left(b_j - \bar{b}_k\right)^2 \dfrac{n_{+jk}}{n_{++k}}}.$$

The asymptotic distribution of T_{CMH} is chi-squared with one degree of freedom.

11.5.3 The Proportional Odds Model for Stratified $2 \times c$ Tables

The proportional odds model presented in Section 6.11 is the most popular model for analyzing ordinal data. For stratified $2 \times c$ tables with ordered outcomes and without interaction between the grouping and the strata, the proportional odds model can be expressed as

$$\text{logit}\left[\Pr(Y_{2k} \leq j)\right] = \alpha_{jk} - \beta x_i,$$

for $i = 1, 2$, $j = 1, 2, \ldots, c - 1$ and $k = 1, 2, \ldots, K$, where $x_1 = 1$ and $x_2 = 0$.
Let γ_k be a stratum specific parameter, with $\gamma_1 = 0$. The simplified model

$$\text{logit}\left[\Pr(Y_{2k} \leq j)\right] = \alpha_j + \gamma_k - \beta x_1 \qquad (11.6)$$

has proportional odds also for the stratum variable. It is appropriate in many
applications (Liu and Agresti, 1996), and it is the model that is used in statistical software packages.

For testing the fit of the proportional odds model, the Pearson goodness-
of-fit test works well for ordered $2 \times c$ tables, see Section 6.13.1. The Pearson
test can readily be generalized to stratified $2 \times c$ tables, and we expect it to
perform well also in this situation.

For the proportional odds model in Equation 11.6, we can calculate a
maximum likelihood estimate $\hat{\beta}_{\text{ML}}$ and a standard error estimate $\widehat{\text{SE}}(\hat{\beta}_{\text{ML}})$. A
Wald test for the null hypothesis $\beta = 0$ and a confidence interval for β can
then be produced in the usual manner.

The maximum likelihood estimates for a proportional odds model usually
perform well if the number of strata is small; however, when the number of
strata is large, these estimates are known to be biased. Liu and Agresti (1996)
proposed an alternative estimate, based on a Mantel-Haenszel type estimate
for an ordinal response in two groups. It has a rather simple form, and with
a generalization of the method by Robins et al. (1986), a variance estimate
can be produced. Because the Mantel-Haenszel estimate performs well for
stratified 2×2 table with many strata, we expect the Mantel-Haenszel type
estimate for stratified $2 \times c$ tables to perform well too.

Conditional maximum likelihood estimates, which eliminate the stratum
specific parameters, are an established alternative when the number of strata
is large. Mukherjee et al. (2008) proposed a method based on the conditional
likelihoods obtained from all binary collapses of the ordinal scale. It turns out
that the method by Mukherjee et al. (2008) coincides with the method by Liu
and Agresti (1996) for the simpler case of ordinal data for two groups.

11.5.4 Example

Adolescent Placement Study (Table 11.3)

We now test the proportional odds model in Equation 11.6 on the data in
Table 11.3. The Pearson goodness-of-fit test gives a P-value of < 0.001, which
means that the assumption of proportional odds over the strata can be rejected. We then go on to fit a separate proportional odds model for each
stratum. When we test the fit of those two models, we obtain $P = 0.011$ for
the stratum "white" and $P = 0.763$ for the stratum "non-white" (with the
Pearson goodness-of-fit test). This indicates that the proportional odds model
fits well for non-white but not for white.

To examine the fit of the proportional odds model further, we calculate the
cumulative odds ratios for white and non-white separately. With the notation

from Section 6.10.4, see in particular Equations 6.21–6.23, we get, for non-white:

$$OR(1,2\text{-}4) = \frac{5/(15+30+60)}{17/(35+32+36)} = 0.28,$$

$$OR(1\text{-}2,3\text{-}4) = \frac{(5+15)/(30+60)}{(17+15)/(32+36)} = 0.29,$$

$$OR(1\text{-}3,4) = \frac{(5+15+30)/60}{(17+32+36)/17} = 0.35,$$

and for white:

$$OR(1,2\text{-}4) = \frac{3/(13+42+66)}{28/(38+37+50)} = 0.11,$$

$$OR(1\text{-}2,3\text{-}4) = \frac{(3+13)/(42+66)}{(28+38)/(37+50)} = 0.20,$$

$$OR(1\text{-}3,4) = \frac{(3+13+42)/66}{(28+38+37)/50} = 0.43.$$

We observe that the cumulative odds ratios are very similar for non-white, but increase by a factor of about two for white. This likely explains the P-values of the Pearson goodness-of-fit test for the separate proportional odds models.

If we disregard the fact that there is heterogeneity for white, we could estimate a common parameter for males relative to females over the strata by ethnicity with a simple proportional odds model, and test the common effect with a Wald test or a Cochran-Mantel-Haenszel test. The results are shown in Table 11.10.

TABLE 11.10
Results for the proportional odds model, the Wald test, and the
Cochran-Mantel-Haenszel test for the data in Table 11.3

Method	Estimate (95% CI)	Statistic	*P*-value
Proportional odds model	3.12 (2.32 to 4.36)		
Wald		44.42	< 0.001
Cochran-Mantel-Haenszel		48.55	< 0.001

Because of its simplicity, this model may seem tempting; however, it might cover up the interesting findings we obtained with separate models for white and non-white. Note that we could have analyzed the data for non-white separately with a proportional odds model.

11.6   Stratified $r \times c$ Tables

In this section, we present methods for two multicategory variables, stratified on a third multicategory variable. The data are summarized in K $r \times c$ tables, and an example is provided in Table 11.4. The stratified $r \times c$ table generalizes the $r \times c$ table, which was the topic of Chapter 7. Here, as in Chapter 7, we consider three situations: (i) the unordered case, where the grouping variable and the outcome variable are both unordered (Section 11.6.1); (ii) the singly ordered case, where the grouping variable is nominal and the outcome variable is ordinal (Section 11.6.2); and (iii) the doubly ordered case, where both the grouping variable and the outcome variable are ordered (Section 11.6.3).

In all the three following sections, we derive generalizations of the Cochran-Mantel-Haenszel test of conditional independence (Landis et al., 1978, 2005; Iannario and Lang, 2016), each tailored to the specific situation outlined above. We start by presenting some general formulas.

Let $\mathbf{n}_k = \{n_{11k}, n_{12k}, \ldots, n_{1ck}, \ldots, n_{r1k}, n_{r2k}, \ldots, n_{rck}\}^{\mathrm{T}}$, and let $\mathbf{m}_k$ denote the estimated expectation of $\mathbf{n}_k$. Under the null hypothesis of conditional independence,

$$\mathbf{m}_k = \{n_{1+k} \cdot n_{+1k}, n_{1+k} \cdot n_{+2k}, \ldots, n_{r+k} \cdot n_{+ck}\}^{\mathrm{T}} / n_{++k}.$$

Conditional on the marginals and under the null hypothesis, let $\mathbf{V}_k$ denote the estimated covariance matrix of $\mathbf{n}_k$. The elements of $\mathbf{V}_k$ are

$$\widehat{\mathrm{Cov}}(n_{ijk}, n_{i'j'k'}) = \frac{n_{i+k}(\delta_{ii'} n_{++k} - n_{i'+k}) n_{+jk}(\delta_{jj'} n_{++k} - n_{+j'k})}{n_{++k}^2 (n_{++k} - 1)},$$

where $\delta_{ab} = 1$ if $a = b$, and $\delta_{ab} = 0$ if $a \neq b$.

The Cochran-Mantel-Haenszel test statistics derived in the following sections are approximately chi-squared distributed. Note that this approximation will, in general, only be satisfactory if the total number of observations, summed over the strata, is sufficiently large.

11.6.1 Stratified Unordered Tables

The saturated logit model for stratified $r \times c$ tables can be written as

$$\log \left(\frac{\pi_{j|ik}}{\pi_{1|ik}} \right) = \alpha_{jk} + \beta_{ij} + \gamma_{jk} + \delta_{ijk},$$

for $i = 1, 2, \ldots, r$, $j = 2, 3, \ldots, c$, and $k = 1, 2, \ldots, K$. Here, β_{ij} is the effect of the grouping, γ_{jk} is the stratum effect, and δ_{ijk} is the interaction effect between the grouping and the strata. We have that $\beta_{1j} = 0$, $\gamma_{j1} = 0$, and $\delta_{ijk} = 0$ for $i = 2$ and $k = 2, 3, \ldots, K$. When $\delta_{ijk} = 0$ for all i, j, and k, there is no interaction effect. Let $\hat{\alpha}_j$ and $\hat{\beta}_{ij}$ be the maximum likelihood estimates of the main effects in the logit model under the assumption of no interaction effect. We may then test the null hypothesis of conditional independence:

$$H_0 : \beta_{1j} = \beta_{2j} = \ldots = \beta_{rj} = 0,$$

for $j = 2, 3, \ldots, c - 1$, with a likelihood ratio test, which has $(r - 1)(c - 1)$ degrees of freedom.

To calculate the Cochran-Mantel-Haenszel test, we let $\mathbf{A}_{1k}$ be the matrix of dimension $rc \times rc$ defined by

$$\mathbf{A}_{1k} = \left[(\mathbf{I}_{r-1}, \mathbf{0}_{r-1}) \otimes (\mathbf{I}_{c-1}, \mathbf{0}_{c-1}) \right],$$

where $\otimes$ is the Kronecker product (Harville, 1997, p. 333) of the two matrices. The generalized Cochran-Mantel-Haenszel test statistic for stratified unordered $r \times c$ tables is

$$T_{\mathrm{CMH}}(\mathbf{n}) = \left[\sum_{k=1}^{K} \mathbf{A}_{1k}(\mathbf{n}_k - \mathbf{m}_k) \right]^{\mathrm{T}} \left[\sum_{k=1}^{K} \mathbf{A}_{1k} \mathbf{V}_k \mathbf{A}_{1k}^{\mathrm{T}} \right]^{-1} \left[\sum_{k=1}^{K} \mathbf{A}_{1k}(\mathbf{n}_k - \mathbf{m}_k) \right],$$

which is approximately chi-squared distributed with $(r-1)(c-1)$ degrees of freedom. The generalized Cochran-Mantel-Haenszel test is the score test for the null hypothesis of conditional independence in the logit model for stratified $r \times c$ tables (Day and Byar, 1979).

When $K = 1$, T_{CMH} is equal to the Pearson chi-squared statistic for unordered $r \times c$ tables, see Section 7.5.1.

11.6.2   Stratified Singly Ordered Tables

For stratified $r \times c$ tables, we can express the proportional odds model as

$$\mathrm{logit}\left[\Pr(Y_{2k} \leq j) \right] = \alpha_{jk} - \beta_i,$$

for $i = 1, 2, \ldots, r$, $j = 1, 2, \ldots, c-1$, and $k = 1, 2, \ldots, K$, where $\beta_1 = 0$. As was the case for stratified $2 \times c$ tables, the model

$$\mathrm{logit}\left[\Pr(Y_{2k} \leq j) \right] = \alpha_j + \gamma_k - \beta x_{1k}$$

has proportional odds also for the stratum variable and may be appropriate (Liu and Agresti, 1996). If we calculate the maximum likelihood estimates $\hat{\beta}_{i,\mathrm{ML}}$ and the standard error estimates $\widehat{\mathrm{SE}}(\hat{\beta}_{i,\mathrm{ML}})$, we can use a Wald test for the null hypothesis $H_0 : \beta_2 = \beta_3 = \ldots = \beta_r = 0$ in the proportional odds model.

Mantel and Byar (1978) and Landis et al. (1978, 2005), among others, have generalized the Cochran-Mantel-Haenszel test to the case with vectors of scores for the grouping variable and the outcome variable. Let $\mathbf{a}_k$ and $\mathbf{b}_k$ be vectors of scores with $\mathbf{a}_k = \{a_1, a_2, \ldots, a_r\}^{\mathrm{T}}$ and $\mathbf{b}_k = \{b_1, b_2, \ldots, b_c\}^{\mathrm{T}}$ for all strata. Assume that the outcome categories are ordered and the grouping categories are unordered. Further, let the matrix $\mathbf{A}_{2k}$ of dimension $(r-1) \times rc$ be given by

$$\mathbf{A}_{2k} = \mathbf{b}_k^{\mathrm{T}} \otimes \left[\mathbf{I}_{r-1}, \mathbf{0}_{r-1} \right].$$

Then, the generalized Cochran-Mantel-Haenszel test statistic for stratified singly ordered $r \times c$ tables is expressed by

$$T_{\mathrm{CMH}}(\mathbf{n}) = \left[\sum_{k=1}^{K} \mathbf{A}_{2k}(\mathbf{n}_k - \mathbf{m}_k) \right]^{\mathrm{T}} \left[\sum_{k=1}^{K} \mathbf{A}_{2k} \mathbf{V}_k \mathbf{A}_{2k}^{\mathrm{T}} \right]^{-1} \left[\sum_{k=1}^{K} \mathbf{A}_{2k}(\mathbf{n}_k - \mathbf{m}_k) \right],$$

which is approximately chi-squared distributed with $r - 1$ degrees of freedom.

The test statistic compares the mean row scores over the strata, and thereby generalizes the Mantel-Haenszel statistic for stratified $2 \times c$ tables. If midranks are assigned as scores, the generalized Cochran-Mantel-Haenszel statistic also generalizes the Kruskal-Wallis statistic for stratified tables, see Section 7.6.2. As noted in that section, the Kruskal-Wallis test is equivalent to a score test for the proportional odds model with an unordered grouping variable.

11.6.3 Stratified Doubly Ordered Tables

For stratified $r \times c$ tables with scores assigned to the grouping variable, the proportional odds model can be written as

$$\text{logit}\big[\Pr(Y_2 \leq j \mid Y_1 = i, Y_3 = k)\big] = \alpha_{jk} - \beta a_i,$$

or, more simply, as

$$\text{logit}\big[\Pr(Y_2 \leq j \mid Y_1 = i, Y_3 = k)\big] = \alpha_j + \gamma_k - \beta a_i,$$

for $i = 1, 2, \ldots, r$, $j = 1, 2, \ldots, c-1$, and $k = 1, 2, \ldots, K$. When $\beta = 0$, there is conditional independence between Y_1 and Y_2. A Wald test or a one-degree-of-freedom likelihood ratio test can be derived.

When the grouping variable and the outcome variable both are ordered, with scores $\mathbf{a}_k = \{a_1, a_2, \ldots, a_r\}$ and $\mathbf{b}_k = \{b_1, b_2, \ldots, b_c\}$ assigned to the grouping and the outcome, respectively, the Cochran-Mantel-Haenszel statistic takes the form

$$T_{\text{CMH}}(\mathbf{n}) = \frac{\left\{ \sum\limits_{k=1}^{K} \left[\sum\limits_{i=1}^{r} \sum\limits_{j=1}^{c} a_i b_j n_{ijk} - \text{E}\left(\sum\limits_{i=1}^{r} \sum\limits_{j=1}^{c} a_i b_j n_{ijk} \right) \right] \right\}^2}{\sum\limits_{k=1}^{K} \text{Var}\left(\sum\limits_{i=1}^{r} \sum\limits_{j=1}^{c} a_i b_j n_{ijk} \right)}.$$

This test statistic is approximately chi-squared distributed with one degree of freedom. The expectation and variance under the null hypothesis are

$$\text{E}\left(\sum_{i=1}^{r} \sum_{j=1}^{c} a_i b_j n_{ijk} \right) = \left(\sum_{i=1}^{r} a_i n_{i+k} \right) \left(\sum_{j=1}^{c} b_j n_{+jk} \right) \bigg/ n_{++k}$$

and

$$\text{Var}\left(\sum_{i=1}^{r} \sum_{j=1}^{c} a_i b_j n_{ijk} \right) = \frac{A \cdot B}{n_{++k} - 1},$$

where

$$A = \sum_{i=1}^{r} a_i^2 n_{i+k} - \left(\sum_{i=1}^{r} a_i n_{i+k} \right)^2 \bigg/ n_{++k}$$

and

$$B = \sum_{j=1}^{c} b_j^2 n_{+jk} - \left(\sum_{j=1}^{c} b_j n_{+jk} \right)^2 \Big/ n_{++k}.$$

This test is also the score test for testing $\beta = 0$ in the proportional odds model above, see also Agresti (2013, pp. 318–319).

11.6.4 Example

Psychiatric Diagnoses and BMI (Table 11.4)

We return to the example of psychiatric diagnoses and weight categories, stratified by sex (Table 11.4). The rows (diagnoses) in this table are unordered; however, the weight categories in the columns (thin, normal, overweight) can be regarded as both unordered or ordered (see discussion in Section 7.2.3). We thus have either a stratified unordered 6×3 table or a stratified singly ordered 6×3 table. In either case, $K = 2$. Table 11.11 shows the results of applying the Cochran-Mantel-Haenszel test to the data.

TABLE 11.11

Results of the generalized Cochran-Mantel-Haenszel test on the data of psychiatric diagnoses and weight categories, stratified by sex (Table 11.4)

Ordering	Scores	Test statistic	df*	P-value
Unordered	Equally spaced	27.91	10	0.0019
Singly ordered	Equally spaced	14.66	5	0.0119
Singly ordered	Mid-ranks	13.92	5	0.0162

*Degrees of freedom for the chi-squared distribution

Under the assumption of ordered weight categories, the two test statistics—one with equally spaced scores and one with mid-ranks as scores—give similar results. If we assume that the weight categories are unordered, a stronger association is indicated. In both cases, partial association, given sex is found. The next step in the analyses would be to consider the association between psychiatric diagnoses and weight categories for females and males separately. This can be done with the Pearson chi-squared test (Section 7.5.1) under the assumption of unordered tables, and with the Kruskal-Wallis test (Section 7.6.2) under the assumption of singly ordered tables. We refer to Section 7.5.9 for examples of the analysis of unordered $r \times c$ tables, and to Section 7.6.4 for examples of the analysis of singly ordered $r \times c$ tables.

11.7 Recommendations

11.7.1 Stratified Paired 2 × 2 Tables

Tests for marginal homogeneity of the differences between probabilities and the ratios of probabilities are based on test statistics that are weighted sums of the stratum specific effect estimates, with weights given by the variances of the effect estimates in the strata. For large sample sizes such tests may perform well, but they can generally not be recommended for small stratum sample sizes.

Homogeneity of the conditional odds ratios is done by testing for homogeneity in a $2 \times K$ table. Then, we recommend the Pearson chi-squared test for large sample sizes, and the Fisher-Freeman-Halton mid-P test.

Common effects for differences and ratios of probabilities are inverse variance weighted estimates, with Wald confidence intervals, which can be recommended only for large stratum sample sizes. For the conditional odds ratio, the estimated odds ratio is simply the ratio of the total number of discordant pairs. The recommended confidence intervals are the transformed Wilson score interval for all samples and the Wald interval for large samples.

11.7.2 Stratified $r \times 2$, $2 \times c$, and $r \times c$ Tables

For stratified $r \times 2$, $2 \times c$, and $r \times c$ tables, one can use the likelihood ratio test or the generalized Mantel-Haenszel tests. For large samples and few strata, the tests will perform equally well. If the number of strata is large, the generalized Cochran-Mantel-Haenszel tests perform better. Exact tests are, in general, too computer intensive to be a practical choice; however, Monte Carlo simulations can be used to estimate exact P-values.

A model-based approach to the analysis of stratified tables has major advantages. The parameters of the logit model can be estimated, and likelihood ratio tests and confidence intervals can be calculated. The proportional odds model is an obvious choice, because of its widespread availability in general-purpose statistical software packages.

Following the discussion in Section 10.8.1 (pp. 439–440), we do not recommend the minimum variance linear estimate for the linear trend. We prefer the logit link and the likelihood ratio test for testing heterogeneity over the strata and no trend. For small samples, we recommend the Cochran-Mantel-Haenszel test. To estimate the overall trend, we recommend the maximum likelihood estimate.

12

Sample Size Calculations

12.1 Introduction

In all research, planning and conduct of a study is crucial. A poor study design may ruin a whole research project, regardless of the statistical analysis. Planning and conduct of a study is not within the scope of this book; however, sample size calculation and power analysis are tightly connected to statistical issues which are covered throughout this book.

Sample size calculation is particularly important in the biomedical and health sciences. It is an integral part of the design of any study, and it is mandatory in a protocol for a clinical trial. Ethical aspects are particularly relevant in this field. A too small and a too large study are both unethical. A too small study is unethical because the likeliness of rejecting the statistical null hypothesis may be too low. A too large study means bringing unnecessary many individuals into the study.

In this chapter, we will review the statistical concepts underlying sample size calculation. In most settings, the null hypothesis is that there is "no effect", such as no difference between the groups, or no association between the variables. In clinical trials, this is the case for a superiority study, where the objective is to show that one treatment is better than another, and the null hypothesis is that there is no difference. Some clinical studies are equivalence studies or non-inferiority studies, with other null hypotheses, which we introduce in Section 12.2.2.

Our intention with this chapter is not to give a full review of sample size calculations. Several textbooks are written on this topic, for instance, Chow et al. (2008), Cohen (1988), Machin et al. (2009), and Ryan (2013). Other textbooks contain sections on sample size calculations, see Pocock (1983), Fleiss et al. (2003), or Lachin (2011). We will refer to these books throughout this chapter.

The methods described in this chapter are available in commercial software packages, for instance, NCSS PASS (https://www.ncss.com/software/pass/), Power and Precision (http://www.power-analysis.com/), and StatXact 11 (http://www.cytel.com/software/statxact). NCSS PASS and StatXact 11 also compute sample sizes based on exact tests. Most general purpose statistical software packages, such as Stata, SAS, and R, include some options for sample size calculations. We recommend that a software package is used for sample size calculations.

The calculations of sample size and power in this chapter were done with NCSS PASS 15, except for the calculations of the Suissa-Shuster exact unconditional test in Table 12.4 and the McNemar exact unconditional test in Table 12.6, which were done with StatXact 11.

12.2 General Principles

Sample size calculations are most often based on a hypothesis test, although some prefer confidence interval estimation, see Section 12.2.3. When testing a statistical hypothesis, there are four possible outcomes, as summarized in Table 12.1.

TABLE 12.1

Possible outcomes of a hypothesis test

	H_0 is true	H_0 is false
H_0 not rejected	A (correct)	B (Type II error)
H_0 rejected	C (Type I error)	D (correct)

H_0 is the null hypothesis

The events C and B lead to incorrect decisions. The event C is called a *Type I error*. The probability of C is the probability of rejecting a true null hypothesis, which is usually required not to exceed a prespecified significance level denoted by α. The event B is called a *Type II error*. The probability of B is denoted by β. The event D is correct, and the probability of D is $1 - \beta$, which is denoted by the *power of the test*. Note that some authors denote the power by β and the probability of Type II error by $1 - \beta$.

The value of α should be small and is often set to $\alpha = 0.05$. If it is particularly undesirable to reject a null hypothesis that might be true, it is appropriate to lower the significance level to, say, $\alpha = 0.01$. In addition, a high power is desirable. For example, in clinical trials, it is common to require a power of at least 0.80 or 0.90. Usually, one tolerates a higher probability of a Type II error than a Type I error.

For a given test procedure, these four quantities are connected:

- Significance level
- Specified effect size (under the alternative hypotheses); actually, in many cases a full set of parameters under the alternative hypothesis are specified
- Power (for the specified effect size)

- Sample size

In principle, if three of these are specified, the fourth can be calculated. One can calculate the power given values of the three other quantities. Or, as will be the main focus in this chapter, calculate sample size, given the first three quantities. As already noted, a common choice of significance level is 0.05, and common choices of power are 0.80 or 0.90. Specification of the effect size is crucial. Usually, the sample size depends most heavily on this choice. One criterion suggested in the literature is to choose the smallest effect size that would be of practical interest, for example, the smallest effect size regarded as clinically important. This may, however, lead to an unrealistically large sample size. Alternatively, one could use what is regarded as a realistic effect size, for example based on what is found in the literature, or based on a pilot study.

12.2.1 Testing for No Effect

Usually, the null hypothesis is that there is no effect or no association. In clinical trials, this is the case in *superiority studies*. With the same notation as in Section 1.3.4, let Δ be the parameter of interest, and let the null hypothesis be

$$H_0 : \Delta = 0 \quad \text{versus} \quad H_A : \Delta \neq 0.$$

There is an asymmetry in the conclusions that can be drawn from a statistical test procedure. The probability of accepting the alternative when the null hypothesis is true shall be below the specified significance level α. The symmetric conclusion of accepting the null hypothesis when it is not rejected is an unacceptable procedure. Such a decision may be wrong with a probability as high as $1 - \alpha$. The asymmetry in testing hypotheses is very eloquently expressed in the title of an article by Altman and Bland (1995): *"Absence of evidence is not evidence of absence"*.

Sample Size Calculations for Superiority Studies

Let $T(\mathbf{x})$ be any test statistic for which large values provide evidence against the null hypothesis, and let $T(\mathbf{n})$ be the observed value of the test statistic. The null hypothesis with two-sided alternatives should be rejected if

$$\Pr[T(\mathbf{x}) \geq T(\mathbf{n}) \mid \Delta = 0] \leq \alpha/2. \tag{12.1}$$

The power of the test for a specified alternative Δ_1 is

$$1 - \beta = \Pr[T(\mathbf{x}) \geq T(\mathbf{n}) \mid \Delta_1]. \tag{12.2}$$

For a given test statistic and any given Δ_1, Equations 12.1 and 12.2 can be solved to meet the required level and power. Let $\hat{\Delta}$ be an estimate of Δ, and let $\widehat{\text{SE}}(\hat{\Delta})$ be an estimate of the standard error of $T(\mathbf{x})$. Alternatively, we can use the standard error under the null hypothesis, denoted by $\widehat{SE}_0$, as we will see in Sections 12.3.1 and 12.4.1. Note that the standard error is a function of the sample size N.

Assume that $T(\mathbf{x})$ is asymptotically standard normally distributed. Then, the null hypothesis is rejected if

$$|\hat{\Delta}| \geq z_{\alpha/2}\widehat{SE}(\hat{\Delta}), \qquad (12.3)$$

where $z_{\alpha/2}$ is the upper $\alpha/2$ percentile of the standard normal distribution. Because the required power should be $1 - \beta$, we have to solve the equation

$$1 - \beta = \Pr\left[\hat{\Delta} \geq z_{\alpha/2}\widehat{SE}(\hat{\Delta}) \,|\, \Delta_1\right],$$

by inserting an approximation to $\widehat{SE}(\hat{\Delta})$.

In the preceding chapters, we have derived asymptotic and exact tests. Algebraic expressions of the sample size can be found for the asymptotic tests, as in Equation 12.3. Often, small studies must be designed and exact tests must be used. Sample size calculations based on power calculations for exact tests should then be calculated from the exact probability distribution, instead of with an asymptotic approximation.

12.2.2   Testing for Equivalence and Non-inferiority

In biomedicine and the health sciences, and, in particular, in clinical trials, testing for superiority, with the alternative of "better than", "worse than", or "not equal to", is not always the relevant issue. Now, assume a new drug is developed which is cheaper, has less side effects, or is easier to administrate, than the currently used drug. Then, the new drug will be preferred if it is equally efficient, or at least not less efficient, than the current drug. Then, an equivalence study or a non-inferiority study will be relevant, rather than a superiority study. An excellent introduction to non-inferiority and equivalence studies is given by Piaggio et al. (2012), and an in-depth coverage is found in the book by Wellek (2010).

Consider a randomized controlled trial that compares two treatments. Let Group 1 consist of patients on the new treatment, and let Group 2 be the control group. Let Δ be the effect measure for the new treatment, relative to the control. If $\Delta = 0$, there is no difference in effects of the new treatment. For $\Delta > 0$, the new treatment has a positive effect, for $\Delta < 0$, the new treatment has a negative effect. The treatments are considered equivalent if the effect in absolute value is less than a prespecified value Δ_0, that is $|\Delta| < \Delta_0$. If $\Delta > -\Delta_0$, the new treatment is said to be non-inferior to the old one. Here, Δ_0 is set to a value that is too small to be regarded as clinically important. The size of Δ_0 is, of course, crucial for the sample size calculation.

In an equivalence study, the null hypothesis is

$$H_0 : \Delta \leq -\Delta_0 \quad \text{or} \quad \Delta \geq \Delta_0,$$

versus the alternative of equivalence:

$$H_A : -\Delta_0 < \Delta < \Delta_0.$$

In a sense, the null hypothesis and the alternative hypothesis are interchanged compared with a superiority study. For details, see Wellek (2010, p. 11) and Lachin (2011, Section 2.6.8). Note that asymmetric bounds can be defined, see Wellek (2010), but here it will suffice with symmetric bounds.

Non-inferiority is tested by

$$H_0 : \Delta \leq -\Delta_0$$

versus the alternative

$$H_A : \Delta > -\Delta_0.$$

Let $\hat{\Delta}$ be an estimate of Δ, and let $(\hat{\Delta}_L, \hat{\Delta}_U)$ be a $100(1-2\alpha)\%$ confidence interval such that the non-coverage in each tail is at most α. Equivalence is claimed if the confidence interval falls within the upper and lower bounds set by Δ_0, that is

$$-\Delta_0 < \hat{\Delta}_L \quad \text{and} \quad \hat{\Delta}_U < \Delta_0.$$

It may seem counterintuitive that the equivalence test is based on a $100(1-2\alpha)\%$ confidence interval. Wellek (2010, p. 33) explains why such a confidence level gives a probability of making a type I error less than or equal to α.

Non-inferiority is claimed if

$$\hat{\Delta}_L > -\Delta_0.$$

The null and alternative hypotheses for testing superiority, equivalence, and non-inferiority are illustrated in Figure 12.1.

Our applications of equivalence and non-inferiority will be for sample size calculations when comparing two probabilities in a 2×2 table and for comparing two probabilities in a paired 2×2 table. For comparing two probabilities in a 2×2 table, the effect measure Δ will be the difference between probabilities, see Section 12.4.4. Sample size calculations are also available for the ratio of two probabilities and the odds ratio. In these cases, calculations are based on the log ratio and the logit of the probabilities, and symmetric intervals $(-\Delta_0, \Delta_0)$ are adequate. For a paired 2×2 table, the effect measure is the difference between the marginal probabilities for effect for new treatment and standard treatment, see Section 12.9.5.

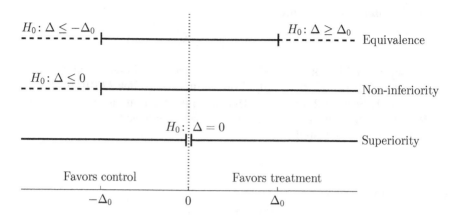

FIGURE 12.1
Null hypotheses (dashed lines) and alternative hypotheses (solid lines) for
testing superiority, equivalence, and non-inferiority

Sample Size Calculations for Equivalence and Non-Inferiority Studies

Equivalence is claimed if $\hat{\Delta} \pm z_\alpha \widehat{SE}(\hat{\Delta}) \in (-\Delta_0, \Delta_0)$. The probability that the confidence interval falls within $(-\Delta_0, \Delta_0)$ when $\Delta = 0$ is

$$\begin{aligned}
1 - \beta &= \Pr\left[-\Delta_0 + z_\alpha \widehat{SE}(\hat{\Delta}) \le \hat{\Delta} \le \Delta_0 - z_\alpha \widehat{SE}(\hat{\Delta})\right] \\
&= \Phi\left[\Delta_0/\widehat{SE}(\hat{\Delta}) - z_\alpha\right] - \Phi\left[-\Delta_0/\widehat{SE}(\hat{\Delta}) + z_\alpha\right] \\
&= 2\Phi\left[\Delta_0/\widehat{SE}(\hat{\Delta}) - z_\alpha\right] - 1,
\end{aligned}$$

where the standard error is calculated under the null hypothesis. Thus,

$$\Delta_0/\widehat{SE}(\hat{\Delta}) - z_\alpha = z_{\beta/2}$$

and

$$\widehat{SE}(\hat{\Delta}) = \Delta_0/(z_\alpha + z_{\beta/2}). \tag{12.4}$$

Because $\widehat{SE}(\hat{\Delta})$ is a function of N, and the required sample size can be calculated by solving Equation 12.4 with respect to N.

Non-inferiority means

$$\hat{\Delta} - \widehat{SE}(\hat{\Delta})z_\alpha > -\Delta_0.$$

The sample size should be such that

$$\Pr\left[\hat{\Delta} - z_\alpha \widehat{SE}(\hat{\Delta}) > -\Delta_0\right] = 1 - \beta$$

and

$$\widehat{SE}(\hat{\Delta}) = \Delta_0/(z_\alpha + z_\beta), \tag{12.5}$$

which, for an approximation to $\widehat{SE}(\hat{\Delta})$, can be solved for N, to give the required sample size for a non-inferiority study.

We return to examples of sample size calculations for non-equivalence and non-inferiority studies for 2×2 tables in Section 12.4.4.

12.2.3 Sample Size Calculations based on Confidence Interval Estimation

As an alternative to sample size calculations based on statistical power, one may calculate sample sizes based on confidence interval estimation. Hypothesis testing and confidence interval estimation are related; a confidence interval can be derived by inversion of the hypothesis test. The equation for the test statistic can be inverted to give the confidence limits, and the confidence interval is made up by all values between these limits. The equation for the sample size is usually simple to solve. But specification of the expected confidence interval length may be difficult.

Sample size calculations based on confidence intervals seem most relevant in cases when the objective is the estimation of a parameter with a certain precision (i.e., the width of a confidence interval for the parameter). Such parameters may be a proportion in a 1×2 table, such as the prevalence of a disease, or an association in a 2×2 table, such as the association between an exposure and a disease. We refer to Machin et al. (2009, Chapter 10) for details on the derivations of sample sizes based on confidence intervals.

Sample size calculations based on power analysis is by far the most commonly used method, and we concentrate on that. We will, however, give examples of sample size calculations based on confidence intervals for a 1×2 table (Section 12.3.3) and a 2×2 table (Section 12.4.5).

12.3 The 1×2 Table

In a 1×2 table, a binomially distributed outcome variable X with parameters n and π is observed. The null hypothesis of interest is

$$H_0 : \pi = \pi_0 \quad \text{versus} \quad H_A : \pi \neq \pi_0.$$

12.3.1   The Score Test

In Chapter 2, the score test is given as

$$Z_{\text{score}} = \frac{X/n - \pi_0}{\sqrt{\dfrac{\pi_0(1 - \pi_0)}{n}}},$$

which has an asymptotic standard normal distribution. The null hypothesis is rejected if

$$|X - n\pi_0| \geq z_{\alpha/2}\sqrt{n\pi_0(1 - \pi_0)},$$

where $z_{\alpha/2}$ is the upper $\alpha/2$ percentile of the standard normal distribution.

Let now π_1 be the parameter value where a power of $1 - \beta$ should be obtained. Then

$$
\begin{aligned}
1 - \beta &= \Pr\left[X - n\pi_0 \leq z_{\alpha/2}\sqrt{n\pi_0(1 - \pi_0)}\,\big|\,\pi_1\right] \\
&= \Pr\left[\frac{X - n\pi_1}{\sqrt{n\pi_1(1 - \pi_1)}} \leq \frac{z_{\alpha/2}\sqrt{n\pi_0(1 - \pi_0)} + n(\pi_0 - \pi_1)}{\sqrt{n\pi_1(1 - \pi_1)}}\,\bigg|\,\pi_1\right]
\end{aligned}
$$

and

$$z_\beta = \frac{z_{\alpha/2}\sqrt{n\pi_0(1 - \pi_0)} + n(\pi_0 - \pi_1)}{\sqrt{n\pi_1(1 - \pi_1)}},$$

where z_β is the upper β percentile of the standard normal distribution. The required sample size is

$$n = \left[\frac{z_{\alpha/2}\sqrt{\pi_0(1 - \pi_0)} + z_\beta\sqrt{\pi_1(1 - \pi_1)}}{\pi_0 - \pi_1}\right]^2. \tag{12.6}$$

For an alternative derivation, see Fleiss et al. (2003, p. 31).

For probabilities close to 0 or 1, use of the arcsine transformation may be preferable (Ryan, 2013, p. 104). Sample size calculations based on the arcsine transformation are rarely found in statistical software packages. The reason may be that exact binomial calculations are now easy to perform, also for probabilities close to 0 or 1. Casagrande et al. (1978) found that the arcsine transformation may give too small sample sizes.

12.3.2   Exact Binomial Test

In this section, we derive a method for sample size calculations based on the exact binomial distribution. Let the rejection region be defined by the number x_0, such that

$$\Pr(X \geq x_0 \,|\, \pi_0) \leq \alpha/2.$$

The power function is

$$\Pr(X \geq x_0 \,|\, \pi_1) + \Pr(X \leq n - x_0 \,|\, \pi_1).$$

For a given power $1 - \beta$, the required sample size can be found as the solution of

$$\Pr(X \geq x_0 \,|\, \pi_1) + \Pr(X \leq n - x_0 \,|\, \pi_1) \geq 1 - \beta. \qquad (12.7)$$

See Chow et al. (2008, p. 118) for a similar derivation of sample size for the one-sided null hypothesis. Chow et al. (2008, p. 119) present a table of sample sizes for the binomial test with one-sided alternatives. Calculations for one-sided and two-sided alternatives can be done in many statistical software packages.

12.3.3 Example

We use the example of Chow et al. (2008, p. 13), which was also analyzed by Ryan (2013, p. 109). The null value is $\pi_0 = 0.3$, and the objective is to detect an alternative value of $\pi_1 = 0.5$. A two-sided test is used, with a level of $\alpha = 0.05$ and a power of $1 - \beta = 0.80$. Table 12.2 shows calculated sample size and associated exact power for the score test and the exact binomial test. The exact power is calculated from the binomial distribution. The calculations were performed with NCSS PASS.

TABLE 12.2

Sample sizes and exact power for a power analysis in a 1 × 2 table

Test	Sample size	Exact power
Score test	44	77.43%
Exact binomial test	47	80.91%

The score test is known to perform excellently. The exact binomial test is more conservative than the score test, and it has lower power. This explains why the required sample size for the score test is lower than for the exact binomial test. The power of the score test meets the requirement of an asymptotic power of 80%, but due to the lower sample size, the exact power is lower than for the exact test. In line with the recommendations in Table 2.7, we recommend sample size calculations based on the score test, but for small sample sizes, exact enumerations based on the binomial distribution can be used.

To illustrate sample size calculations based on confidence intervals, we set the confidence width to 0.2, and calculate the sample sizes for $\pi = 0.3, 0.4$, and 0.5. A width of 0.2 means that we allow for a 10 percentage points reduction or a 10 percentage points increase in the probability π. The confidence level is set to 95%. The resulting sample sizes are given in Table 12.3.

We observe that the sample sizes are similar, although the sample sizes for the Wilson score interval is somewhat smaller. This is as expected, based on Figure 2.11, which illustrates that the Wilson score interval is narrower than the Wald interval.

TABLE 12.3
Sample sizes for confidence interval
calculations in a 1×2 table, with
interval width set to 0.2

Interval method	π		
	0.3	**0.4**	**0.5**
Wald interval	81	93	97
Wilson score interval	78	89	93

12.4   The 2×2 Table

Analysis of 2×2 tables is covered in Chapter 4. We use the notation in
Table 4.7, and let Y_1 and Y_2 be the row and column variables, with

$$\Pr(Y_2 = 1 \mid Y_1 = i) = \pi_i$$

and

$$\Pr(Y_1 = 1 \mid Y_2 = j) = \gamma_j,$$

for $i = 1, 2$ and $j = 1, 2$. The null hypothesis to be tested in a 2×2 table
can be written in several equivalent forms, see Table 4.10. The Pearson chi-
squared test, which is a score test, works for any design. Thus, the sample size
formulas that we will present in Section 12.4.1 also give the required sample
size for a case-control study with specified values of γ_1 and γ_2 rather than π_1
and π_2.

12.4.1   The Score Test (Pearson Chi-Squared Test)

The score test statistic is given in Section 4.4.4 as

$$Z = \frac{\dfrac{n_{11}}{n_{1+}} - \dfrac{n_{21}}{n_{2+}}}{\sqrt{\dfrac{n_{+1}}{N}\dfrac{n_{+2}}{N}\left(\dfrac{1}{n_{1+}} + \dfrac{1}{n_{2+}}\right)}}.$$

The Pearson chi-squared statistic equals the square of the score test statistic,
see also Section 4.4.4. Now, consider a design with equal sample sizes, that is,
$n_{1+} = n_{2+} = n$. We approximate the test statistic with

$$Z = \frac{\hat{\pi}_1 - \hat{\pi}_2}{\sqrt{\bar{\pi}(1 - \bar{\pi})/2n}},$$

where $\hat{\pi}_1 = n_{11}/n_{1+}$, $\hat{\pi}_2 = n_{21}/n_{2+}$, and $\bar{\pi} = (\pi_1 + \pi_2)/2$. Let $\Delta = \pi_1 - \pi_2$
denote the difference between the probabilities to be detected. With the same

derivations as in Section 12.3, we get that the sample size in each group should be

$$n = \left[\frac{z_{\alpha/2}\sqrt{2\bar{\pi}(1-\bar{\pi})} + z_\beta\sqrt{\pi_1(1-\pi_1) + \pi_2(1-\pi_2)}}{\pi_1 - \pi_2} \right]^2. \tag{12.8}$$

The sample size calculations can also be based on the Z-unpooled test in Equation 4.9. The required sample size then is

$$n = \frac{(z_{\alpha/2} + z_\beta)^2 \left[\pi_1(1-\pi_1) + \pi_2(1-\pi_2)\right]}{(\pi_1 - \pi_2)^2}, \tag{12.9}$$

see Chow et al. (2008, p. 89). Campbell et al. (1995) consider Equation 12.9 as an approximation to Equation 12.8, valid unless π_1 and π_2 are less than 0.05.

12.4.2 Exact Conditional Test

When the samples are small, the simple approximate tests do not work well, see Section 4.4.9. By conditioning on the marginal totals, we can obtain an exact conditional test, with exact P-values and exact conditional power calculated from the hypergeometric distribution. The unconditional power function can be obtained by summing over all outcomes with P-values less than or equal to α. Our recommended methods are the Fisher mid-P test and the exact unconditional test with either the Pearson test statistic (Suissa-Shuster exact unconditional test) or the Fisher exact test statistic (Fisher-Boschloo exact unconditional test), see Table 4.24.

Sample size calculations are not possible for the Fisher mid-P test; however, it can be done for the Fisher exact test. The calculations require an iterative procedure. Tables for sample sizes for the Fisher exact test are given in Chow et al. (2008, p. 123). The sample size can also be calculated in software packages like NCSS PASS and StatXact 11.

The unconditional power function takes a simple form, see Section 4.4.7. Sample size calculations for the Suissa-Shuster exact unconditional test can be performed in StatXact 11.

12.4.3 Confidence Interval Method

In Chapter 4, we considered the difference between probabilities, the ratio of probabilities, and the odds ratio as effect measures for the 2×2 table. The recommended confidence intervals are given in Table 4.24. For the difference between probabilities, our recommendations include the Wald interval and the Newcombe hybrid score interval; for the ratio of probabilities, we recommend (among others) the Katz log interval and the adjusted inverse sinh interval; and for the odds ratio, the recommended confidence intervals include the Woolf logit interval and the adjusted inverse sinh interval. All these intervals have closed-form expressions, which makes them attractive for sample size calculations.

12.4.4 Sample Size for Equivalence and Non-Inferiority

Equation 12.4 gave the general formula to be solved for the sample size of an equivalence study. For the 2×2 table, we get

$$\sqrt{\bar{\pi}(1 - \bar{\pi})/N} = \Delta_0/(z_\alpha + z_{\beta/2}),$$

where Δ_0 is the upper bound for equivalence, and $\bar{\pi} = (\pi_1 + \pi_2)/2$. Then,

$$N = \left(\frac{z_\alpha + z_{\beta/2}}{\Delta_0}\right)^2 \bar{\pi}(1 - \bar{\pi}).$$

For the non-inferiority study, we get from Equation 12.5 that

$$\sqrt{\bar{\pi}(1 - \bar{\pi})/N} = \Delta_0/(z_\alpha + z_\beta),$$

and the required sample size is

$$N = \left(\frac{z_\alpha + z_\beta}{\Delta_0}\right)^2 \bar{\pi}(1 - \bar{\pi}).$$

Note that in Chow et al. (2008, p. 90) these formulas are calculated under the alternative, with specified π_1 and π_2. The above calculations are performed under the null hypothesis $\pi_1 = \pi_2$, and with a probability of $1 - \beta$ that the confidence interval falls within $(-\Delta_0, \Delta_0)$ for equivalence and above $-\Delta_0$ for non-inferiority.

12.4.5 Example

Table 12.4 shows the required sample size and exact power for a few cases of π_1 and π_2, for two-sided tests with power 80% and significance level 5%. The picture is the same for all three choices of parameters. The Fisher exact test is conservative and this has lower power than the Pearson chi-squared test. The required sample size is higher. The Fisher exact test is not among the recommended tests for 2×2 tables, see Table 4.24 (page 174). The Suissa-Shuster exact unconditional test is among the recommended tests for small to medium sample sizes, and the Pearson chi-squared test is among the recommended tests for large sample sizes.

For confidence interval estimation, we consider the example with $\pi_1 = 0.20$ and $\pi_2 = 0.30$, and we only consider the confidence intervals recommended for large samples. The difference in sample size with the other intervals is small. Table 12.5 shows the results. First, note that the confidence interval widths are such that the confidence intervals do not cover the null values in the hypothesis tests for the effect measures. Second, note that the sample sizes

TABLE 12.4

Required sample size for testing the hypothesis of independence for
selected cases of π_1 and π_2 in a 2×2 table

Test	Sample size per group	Exact power
$\pi_1 = 0.05, \pi_2 = 0.10$		
Pearson chi-squared	435	81.11%
Fisher exact	465	80.09%
Suissa-Shuster exact uncond.	448	80.04%
$\pi_1 = 0.20, \pi_2 = 0.30$		
Pearson chi-squared	294	80.28%
Fisher exact	311	80.05%
Suissa-Shuster exact uncond.	299	80.12%
$\pi_1 = 0.20, \pi_2 = 0.40$		
Pearson chi-squared	82	80.74%
Fisher exact	90	80.17%
Suissa-Shuster exact uncond.	84	80.23%

based on the three effect measures are of similar size. Third, the sample sizes
calculated from the confidence interval method is somewhat higher than the
numbers calculated from the power analyses (Table 12.4).

TABLE 12.5

Required sample size in a 2×2 table based on the confidence interval
method, with $\pi_1 = 0.20$ and $\pi_2 = 0.30$

Effect measure	Interval method	Interval width	Sample size per group	95% CI
Difference between probabilities	Wald	0.10	385	0.05 to 0.15
Ratio of probabilities	Katz log	0.8	351	1.15 to 1.95
Odds ratio	Woolf logit	1.4	346	1.11 to 2.31

12.5 The Ordered $r \times 2$ Table

In an ordered $r \times 2$ table, we are interested in testing for trend, see Section 5.7.
As in Chapter 5, we let $\Pr(Y_2 = 1 \mid Y_1 = i) = \pi_i$ for $i = 1, 2, \ldots, r$. Then, the
null hypothesis is

$$H_0 : \pi_1 = \pi_2 = \ldots = \pi_r,$$

which we test against one of the two alternatives

$$H_A : \pi_1 \leq \pi_2 \leq \ldots \leq \pi_r \quad \text{or} \quad H_A : \pi_1 \geq \pi_2 \geq \ldots \geq \pi_r,$$

with at least one strict inequality.

Suppose that we have assigned the scores $\{a_1, a_2, \ldots, a_r\}$ to the rows in the $r \times 2$ table. With a linear link function, we assume that

$$\pi_i = \alpha + \beta a_i,$$

for $i = 1, 2, \ldots, r$, see Equation 5.4. The null hypothesis is now

$$H_0 : \beta = 0 \quad \text{versus} \quad H_A : \beta > 0 \quad \text{or} \quad H_A : \beta < 0.$$

12.5.1 Sample Size Requirements for Ordered $r \times 2$ Tables

The Cochran-Armitage test is based on the estimated slope $\hat{\beta}_{CA}$, see Equation 5.16. The test statistic is then given by

$$Z_{CA} = \frac{U}{s_0} = \frac{\hat{\beta}_{CA}}{\text{Var}(U \mid H_0)},$$

where

$$s_0^2 = \pi(1 - \pi) \sum_{i=1}^{r} n_{i+}(a_i - \bar{a})^2,$$

see also Equation 5.19.

Let now $r_i = n_{i+}/n_{1+}$ be the ratio of the sample size in Group i relative to Group 1, which must be specified. With $A = \sum_{i=1}^{r} r_i \pi_i (a_i - \bar{a})$, the sample size in the first group is given by

$$n_{1+} = \frac{1}{A^2} \left\{ z_\alpha \sqrt{\pi(1 - \pi) \sum_{i=1}^{r} r_i (a_i - \bar{a})^2} + z_\beta \sqrt{\sum_{i=1}^{r} r_i \pi_i (1 - \pi_i)(a_i - \bar{a})^2} \right\},$$

(12.10)

where $\pi = \sum_{i=1}^{r} r_i \pi_i / \sum_{i=1}^{r} r_i$. The sample sizes in the remaining groups are given by $n_{i+} = r_i \cdot n_{1+}$. For references, see Nam (1987) and Chow et al. (2008, p. 294).

For the calculation of the sample sizes, we have to specify the scores, the probabilities π_i, and the relative sample sizes r_i, for $i = 1, 2, \ldots, r$. Since $\text{Var}(U \mid H_A)$ is increasing with π_i, there may be reasons to place more observations for π_i close to 0.5. The gain, however, may be small, and when there is little knowledge available, equal sample sizes may be used.

The sample size in Equation 12.10 was derived for a two-sided alternative. For one-sided alternatives, substitute $z_{\alpha/2}$ for z_α in Equation 12.10.

12.5.2 Example

In our analysis of Table 5.3 (elevated troponin T levels in stroke patients), we found a P-value of 0.074 for the Cochran-Armitage test for trend, see Section 5.7.14. Suppose that we want to carry out a study on the same problem. At the planning stage, we have to decide on the trend itself, the assignment of scores, and the relative sample sizes of the groups. Assume that the probabilities $\{\pi_1, \pi_2, \pi_3, \pi_4, \pi_5\}$ are set to $\{0.05, 0.08, 0.12, 0.15, 0.17\}$. We assign equally spaced scores, and for the moment, we assume equal number of observations in the five groups. With a two-sided test, with significance level 0.05 and a power of 0.80, the required sample size is 82 persons in each group, with a total sample size of 410. We observe that the sample sizes in Table 5.3 for the groups with the highest degree of severity are too small compared with the sample size calculations we have just performed.

Suppose now that we require the sample sizes to be based on the variances, such that $\{r_1, r_2, r_3, r_4, r_5\} = \{1, 1.5, 2.2, 2.7, 3.0\}$. The calculated sample sizes now are $\{53, 79, 116, 143, 159\}$, with a total sample size of 550. These numbers are similar to those in Table 5.3.

12.6 The Ordered $2 \times c$ Table

The ordered $2 \times c$ table is studied in Chapter 6. The proportional odds model is given in Equation 6.24 as

$$\log\left[\frac{(\pi_{1|1} + \pi_{2|1} + \ldots + \pi_{j|1})/(\pi_{j+1|1} + \pi_{j+2|1} + \ldots + \pi_{c|1})}{(\pi_{1|2} + \pi_{2|2} + \ldots + \pi_{j|2})/(\pi_{j+1|2} + \pi_{j+2|2} + \ldots + \pi_{c|2})}\right] = -\beta,$$

for $j = 1, 2, \ldots, c-1$. The common odds ratio in the proportional odds model is $\theta = \exp(-\beta)$. The null hypothesis of no effect is

$$H_0 : \theta = 1.$$

The sample size requirements for the proportional odds model were derived by Whitehead (1993). Let θ_1 denote the value of the common odds ratio for which the specified power of $1 - \beta$ should be attained, when a test with significance level α is performed. Let further $\bar{\pi}_1, \bar{\pi}_2, \ldots, \bar{\pi}_c$ be the mean of the probabilities for the categories. With the notation in Chapter 6, $\bar{\pi}_j = (\pi_{1j} + \pi_{2j})/2$. As in Whitehead (1993), we let the allocation ratio A be the number of observations in Group 2 (control) relative to Group 1 (experimental), i.e. $A = n_{2+}/n_{1+}$. Whitehead (1993) gives the following formula for the sample size

$$N = \frac{3(A+1)^2(z_{\alpha/2} + z_\beta)^2}{A \cdot \left[\log(\theta_1)\right]^2 \left(1 - \sum_{j=1}^{c} \bar{\pi}_j^3\right)}. \tag{12.11}$$

Campbell et al. (1995) give the formula for equal allocation of observations, that is $A = 1$, which simplifies Equation 12.11 to

$$N = \frac{6(z_{\alpha/2} + z_{1-\beta})^2}{\log(\theta_1)^2 \left(1 - \sum_{j=1}^{c} \bar{\pi}_j^3\right)}.$$

As noted in Section 6.14.3, the score test for the null hypothesis $H_0 : \theta = 1$ is also the Wilcoxon-Mann-Whitney test. Thus, a sample size calculation may alternatively be done for the Wilcoxon-Mann-Whitney test. Chow et al. (2008, p. 361) have made these derivations. The resulting sample size formula is of closed form; however, it is very complicated, and it would require a pilot study to obtain all the quantities needed for the calculations. It seems to be less useful than the sample size calculation based on the proportional odds model.

The sample size calculations above rely on the assumption of the proportional odds model. Rabbee et al. (2003) proposed an asymptotic method for any linear rank test under general assumptions of ordered alternatives. Methods based on the proportional odds model may not be appropriate for large deviations from the proportional odds assumption or for small sample sizes. Rabbee et al. (2003) derived formulas for exact power and an asymptotic approximation to the exact power. The calculations are complicated; however, power calculations under general assumptions can be done in StatXact 11.

12.7   The Unordered $r \times c$ Table

The parameters of an unordered $r \times c$ table with fixed total sum is given in Table 7.10. The null hypothesis of independence between the row and column variable is

$$H_0 : \pi_{ij} = \pi_{i+}\pi_{j+},$$

for $i = 1, 2, \ldots, r$ and $j = 1, 2, \ldots, c$. When the row sums are fixed, the conditional probabilities are given in Table 7.11. Then, the null hypothesis of independence is

$$H_0 : \pi_{j|1} = \pi_{j|2} = \ldots = \pi_{j|r},$$

for $j = 1, 2, \ldots, c$. Regardless of the design, the null hypothesis can be tested with the Pearson chi-squared test. This test is one of the recommended tests for large samples, and it has the form

$$T_{\text{Pearson}} = \sum_{i,j} \frac{(n_{ij} - m_{ij})^2}{m_{ij}},$$

where $m_{ij} = n_{i+}n_{+j}/N$. With $\hat{\pi}_{ij} = n_{ij}/N$, $\hat{\pi}_{i+} = n_{i+}/N$, and $\hat{\pi}_{+j} = n_{+j}/N$, the test statistic can be written as

$$T_{\text{Pearson}} = \sum_{i,j} \frac{N(\hat{\pi}_{ij} - \hat{\pi}_{i+}\hat{\pi}_{+j})^2}{\hat{\pi}_{i+}\hat{\pi}_{+j}}.$$

12.7.1 Sample Size Requirements for Unordered $r \times c$ Tables

Under the null hypothesis, T_{Pearson} follows a chi-squared distribution with $(r-1)(c-1)$ degrees of freedom. Under the alternative hypothesis, T_{Pearson} follows a non-central chi-squared distribution with non-centrality parameter

$$\gamma = N \sum_{i,j} \frac{(\pi_{ij} - \pi_{i+}\pi_{+j})^2}{\pi_{i+}\pi_{+j}}.$$

Let the cumulative distribution of the chi-squared distribution with $(r-1)(c-1)$ degrees of freedom and non-centrality parameter γ be denoted by $F_{(r-1)(c-1),\gamma}$. Then, the null hypothesis of independence is rejected if

$$T_{\text{Pearson}} \geq F^{-1}_{(r-1)(c-1),\gamma}(1-\alpha).$$

The power function is given by

$$F_{(r-1)(c-1),\gamma}\left[F^{-1}_{(r-1)(c-1),\gamma}(1-\alpha)\right].$$

For a given power of $1-\beta$, the non-centrality parameter γ is found by solving the equation

$$F_{(r-1)(c-1),\gamma}\left[F^{-1}_{(r-1)(c-1),\gamma}(1-\alpha)\right] = \beta.$$

This equation is solved in two steps:

1. Calculate $F_{(r-1)(c-1),\gamma}(1-\alpha) = \gamma$. This can be done with the function `invnchi2tail(df,`α`)` with df $= (r-1)(c-1)$ in Stata (StataCorp LP, College Station, TX), or with a similar command in any other software package that supports the inverse cumulative non-central chi-squared distribution. The function returns the upper α percentile of the chi-squared distribution with $(r-1)(c-1)$ degrees of freedom.
2. Find the non-centrality parameter $\lambda_{\alpha,\beta}$ in the equation $F_{(r-1)(c-1),\gamma}(y) = \beta$. This can be done in Stata with the command `npnchi2(df,` y`,` β`)`.

The required sample size can now be calculated as

$$N = \lambda_{(r-1)(c-1),\alpha,\beta}\left[\sum_{i,j} \frac{(\pi_{ij} - \pi_{i+}\pi_{+j})^2}{\pi_{i+}\pi_{+j}}\right]^{-1}, \qquad (12.12)$$

see also Chow et al. (2008, p. 151). When designing a study, the parameters π_{ij} must be specified before the calculations in Equation 12.12 can be performed.

12.7.2   Example

In the analysis of Table 7.3 (page 277), we found a strong association between treatment of ear infection and acute otitis externa. The study seems to be a little oversized. For planning another study, we used the results from that study as a pilot study for our next study. What would the required sample size be for the next study, given a significance level of 0.05 and a power of 0.80? With the data in Table 7.3, we find that

$$\sum_{i,j} \frac{(\pi_{ij} - \pi_{i+}\pi_{+j})^2}{\pi_{i+}\pi_{+j}} = 0.0883.$$

Since $\lambda_{2,0.05,0.20} = 9.634$, the required sample size is $N = 109$. As we see, the original study was probably oversized, and for the next study, it would suffice with half the sample size of the original study.

12.8   Singly and Doubly Ordered $r \times c$ Tables

In a singly ordered $r \times c$ table, we assume that the column categories are ordered, with assigned scores $\{b_1, b_2, \ldots, b_c\}$. The row variable is an unordered grouping variable. The Wilcoxon-Mann-Whitney test is the score test for the proportional odds model with two groups. Similarly, the Kruskal-Wallis test is equivalent to the score test for the proportional odds model with group as a categorical covariate, see Section 7.6.2. To our knowledge, not much research is done on sample size calculations for the Kruskal-Wallis test. One exception is the article by Fan et al. (2011).

Fan et al. (2011) generalize the results in Chow et al. (2008, p. 361) for the Wilcoxon-Mann-Whitney test to the Kruskal-Wallis test. The results are similar, the sample size is given in closed form; however, it is complicated, and detailed information from a pilot study is needed to perform the calculations. To obtain precise information for sample size calculations, a moderately large number of observations would be needed in the pilot study. As pointed out by Fan et al. (2011), insufficient data from a pilot data may lead to unsatisfactory sample size calculations.

In a doubly ordered $r \times c$ table, we also assign scores $\{a_1, a_2, \ldots, a_r\}$ to the rows. Then, the linear-by-linear test for association is given by

$$T_{\text{linear}} = (N-1)\hat{r}_{\text{P}}^2,$$

where $\hat{r}_{\text{P}}^2$ is the estimated Pearson correlation coefficient between the scores in the rows and the scores in the columns, see Section 7.9.1. The power function will be studied with the help of the Fisher Z transformation:

$$z(\hat{r}_{\text{P}}) = \frac{1}{2} \log\left(\frac{1+\hat{r}_{\text{P}}}{1-\hat{r}_{\text{P}}}\right), \tag{12.13}$$

which is asymptotically normally distributed with expectation equal to the true correlation coefficient and standard error equal to $\sqrt{1/(N-3)}$.

Let r_P be the population correlation that we want to detect. A given correlation, r_P, is inserted into Equation 12.13 to give $z(r_\text{P})$. With calculations similar to those in Section 12.3.1, we get that the required sample size is

$$N = \left[\frac{z_{\alpha/2} + z_\beta}{z(r_\text{P})} \right]^2 + 3. \qquad (12.14)$$

Machin et al. (2009, p.154) make similar calculations for a one-sided alternative, that is, replace $z_{\alpha/2}$ by z_α in Equation 12.14. They argue that one-sided tests are more usual in this situation. For testing linear-by-linear association in doubly ordered $r \times c$ tables, we regard a two-sided alternative to be more appropriate.

12.9 The Paired 2×2 Table

12.9.1 Sample Size Estimation when Testing for Marginal Homogeneity

The observed counts for a paired 2×2 table is shown in Table 8.1, and the probabilities are defined in Table 8.5. The null hypothesis of marginal homogeneity is

$$H_0 : \pi_{1+} = \pi_{+1},$$

which is equivalent to the null hypothesis of conditional independence. The McNemar test statistic has the form

$$T_\text{McNemar} = \frac{n_{12} - n_{21}}{\sqrt{n_{12} + n_{21}}},$$

which under the null hypothesis is approximately distributed $N(0,1)$. If we condition on n_{11} and $n_{12} + n_{21}$, the McNemar test statistic reduces to

$$T_\text{McNemar} = n_{12}.$$

The cell count n_{12} is binomially distributed with parameter $\mu = \pi_{12}/(\pi_{12} + \pi_{21})$, which under the null hypothesis, equals $1/2$.

12.9.2 The McNemar Asymptotic Test

Under the alternative, the approximate power of the McNemar test is

$$\Phi\left[\frac{\sqrt{N}(\pi_{12} - \pi_{21}) - z_{\alpha/2}\sqrt{\pi_{12} + \pi_{21}}}{\sqrt{\pi_{12} + \pi_{21} - (\pi_{12} - \pi_{21})^2}} \right],$$

and the required sample size is

$$N = \frac{\left[z_{\alpha/2}\sqrt{\pi_{12} + \pi_{21}} + z_\beta \sqrt{\pi_{12} + \pi_{21} - (\pi_{12} - \pi_{21})^2} \right]^2}{(\pi_{12} - \pi_{21})^2}.$$

12.9.3   The McNemar Exact Conditional Test

Let $n_d = n_{12} + n_{21}$ be the number of discordant pairs. Conditioned on n_d, the cell count x_{12} is binomially distributed with parameter $\mu = \pi_{12}/(\pi_{12} + \pi_{21})$. Then, the problem can be defined as a 1×2 table (see Section 12.3), with $\mu = 1/2$ under the null hypothesis. The power and the required sample size for the McNemar exact conditional test can be calculated by enumeration of the binomial distribution. Let the required number of discordant pairs be N_d. Then, the required sample size for the total number of pairs is $N = N_d/(\pi_{12} + \pi_{21})$.

12.9.4   The McNemar Exact Unconditional Test

The McNemar exact unconditional test is derived in Section 8.5.5. The required sample size can be calculated with the P-value and power calculations in Section 8.5.5.

12.9.5   Sample Size Calculations for Equivalence and Non-Inferiority

Let, as usual, Δ_0 be the predefined equivalence bound. The null hypothesis of non-equivalence is

$$H_0 : \pi_{1+} - \pi_{+1} \le -\Delta_0 \quad \text{or} \quad \pi_{1+} - \pi_{+1} \ge \Delta_0,$$

versus the alternative of equivalence

$$H_A : -\Delta_0 < \pi_{1+} - \pi_{+1} < \Delta_0.$$

Similarly, the null hypothesis of non-inferiority is given by

$$H_0 : \pi_{1+} - \pi_{+1} \le -\Delta_0,$$

versus the alternative

$$H_A : \pi_{1+} - \pi_{+1} > -\Delta_0.$$

We recommend the sample size calculations proposed by Liu et al. (2002). We refer to their article for detailed derivations, and only give the formula for the required sample size for an equivalence study and for a non-inferiority study.

With the notation in the Appendix of Liu et al. (2002), we define

$$\bar{\pi}_{2,12} = \frac{1}{2} \left(\pi_{12} + \Delta_0 + \sqrt{\pi_{12}^2 + \Delta_0^2 - 2\pi_{12}\Delta_0^2} \right)$$

and

$$\bar{w} = \frac{\sqrt{2\pi_{12}}}{\sqrt{2\bar{\pi}_{2,12} - \Delta_0 - \Delta_0^2}}.$$

The required sample size for an attained power of $1 - \beta$ at the null hypothesis of equal treatment effect, and with significance level α, is

$$n = 2\pi_{12}\left(\frac{z_\alpha/\bar{w} + z_{\beta/2}}{\Delta_0}\right)^2. \tag{12.15}$$

The required sample size for a non-inferiority study is obtained by inserting $z_{\alpha/2}$ and z_β instead of z_α and $z_{\beta/2}$ in Equation 12.15.

A table of required sample sizes for combinations of π_{12} and Δ_0 is given in Liu et al. (2002).

12.9.6 Example

In the following example, we calculate required sample size by the three methods above. The calculations are based on a significance level of 0.05 and required power of 80%. To calculate the sample size, we need information on $\pi_{12} + \pi_{21}$ and $\pi_{12} - \pi_{21}$, or, equivalently, π_{12} and π_{21}. Table 12.6 shows the calculated sample size for three different scenarios.

TABLE 12.6

Required sample size or selected cases of π_{12} and π_{21} in a paired 2×2 table

Test	Sample size	Exact power
$\pi_{12} = 0.10, \pi_{21} = 0.05$		
McNemar asymptotic test	469	77.56%
McNemar exact conditional test	498	80.07%
McNemar exact unconditional test	463	80.00%
$\pi_{12} = 0.40, \pi_{21} = 0.30$		
McNemar asymptotic test	547	78.61%
McNemar exact conditional test	565	80.00%
McNemar exact unconditional test	551	80.00%
$\pi_{12} = 0.40, \pi_{21} = 0.20$		
McNemar asymptotic test	115	77.03%
McNemar exact conditional test	124	80.00%
McNemar exact unconditional test	116	80.00%

The results are similar to those for the 2×2 table (Table 12.4). Due to the conservatism of the McNemar exact conditional test, the sample size for that test is higher than that for the McNemar asymptotic test and the McNemar exact unconditional test, which are two of our three recommended tests for the paired 2×2 table, see Table 8.15 (page 384). For the two recommended tests, the sample sizes are quite similar. We observe that the exact tests have power equal to 80%, whereas the power of the asymptotic test is less than 80%, which can be ascribed to a somewhat higher actual significance level for the McNemar asymptotic test than the two exact tests. As noted in Chapter 8, however, the McNemar asymptotic test hardly ever violates the nominal significance level.

12.10   The Paired $c \times c$ Table

Methods for paired $c \times c$ tables were considered in Chapter 9. The notation for the observed cell counts of a paired $c \times c$ table is given in Table 9.1, and the notation for the probabilities follow the same format. We have two null hypotheses of interest. The hypothesis of marginal homogeneity is

$$H_0 : \pi_{i+} = \pi_{+i},$$

for $i = 1, 2, \ldots, c$, and the hypothesis of symmetry is

$$H_0 : \pi_{ij} = \pi_{ji},$$

for $i, j = 1, 2, \ldots, c$.

The Stuart test is recommended for testing marginal homogeneity; however, it has a complicated form, and there is no obvious way to make sample size calculations for the test statistic. Symmetry can be tested with the McNemar-Bowker test, which is a generalization of the McNemar test for the paired 2×2 table. Note that symmetry implies marginal homogeneity.

12.10.1   Sample Size Calculations for Testing Symmetry

The McNemar-Bowker test has the form

$$T_{\text{McNemar-Bowker}} = \sum_{i>j} \frac{(n_{ij} - n_{ji})^2}{n_{ij} + n_{ji}}.$$

Under the null hypothesis, $T_{\text{McNemar-Bowker}}$ is chi-squared distributed with $c(c-1)/2$ degrees of freedom. Under the alternative, the test statistic follows a non-central chi-squared distribution with non-centrality parameter

$$\lambda = N \sum_{i>j} \frac{(\pi_{ij} - \pi_{ji})^2}{\pi_{ij} + \pi_{ji}}.$$

Now, we proceed as did for the unordered $r \times c$ table and calculate $\lambda_{c(c-1)/2,\alpha,\beta}$ by the two-step procedure in Section 12.7.1. The required sample size then becomes

$$N = \lambda_{c(c-1)/2,\alpha,\beta} \left[\sum_{i>j} \frac{(\pi_{ij} - \pi_{ji})^2}{\pi_{ij} + \pi_{ji}} \right]^{-1}.$$

12.11 Stratified 2 × 2 Tables

Estimation and testing in stratified 2×2 tables are covered in Chapter 10, with a description of the different steps in the analysis given in Section 10.6. If there is heterogeneity over the strata, we calculate and report the effect estimates in each stratum. If the objective is to study heterogeneity over strata, we calculate the required sample size within each stratum, see Section 12.4 for the calculations of sample size in a 2 × 2 table. In this case, sample size calculations based on confidence intervals seem most relevant.

The main objective of the analysis of stratified 2 × 2 tables is usually to estimate and test the common effect over strata. For the odds ratio, we recommend the Cochran-Mantel-Haenszel test for the common odds ratio. The sample size for the Cochran-Mantel-Haenszel test is derived in Section 12.11.1.

12.11.1 Sample Size Requirements for Testing for Common Effect of the Odds Ratio

The Cochran-Mantel-Haenszel test is described in Section 10.13.3. It has the form

$$T_{\text{CMH}} = \left[\sum_{k=1}^{K} \frac{n_{11k}n_{22k} - n_{12k}n_{21k}}{n_{++k}} \right]^2 \Big/ \sum_{k=1}^{K} \frac{n_{1+k}n_{2+k}n_{+1k}n_{+2k}}{n_{++k}^2(n_{++k} - 1)},$$

which is approximately chi-squared distributed with one degree of freedom under the null hypothesis. Let $F_1^{-1}(\alpha)$ be the upper α percentile in the chi-squared distribution. The null hypothesis is rejected if

$$T_{\text{CMH}} \geq F_1^{-1}(\alpha).$$

Under the alternative, T_{CMH} follows non-central chi-squared distribution. Let $r_k = n_{++k}/N$ when the number of observations increases. Then, the non-centrality parameter is $N\lambda^2$, where

$$\lambda^2 = \frac{\left[\sum_{k=1}^{K} r_k(\pi_{11k}\pi_{22k} - \pi_{12k}\pi_{21k}) \right]^2}{\sum_{k=1}^{K} r_k\pi_{1+k}\pi_{2+k}\pi_{+1k}\pi_{+2k}}.$$

Since $T_{\text{CMH}} \geq F_1^{-1}(\alpha)$ is equivalent to $X \geq z_{\alpha/2}$, where $X \sim N(\sqrt{N}\lambda, 1)$, the probability of rejecting the null hypothesis under the alternative is at least $1 - \beta$, if the sample size is at least

$$N = \frac{(z_{\alpha/2} + z_\beta)^2}{\lambda^2}.$$

12.11.2   Example

A study of the association between coronary heart disease and the results of an electrocardiogram is planned. There is an association between sex and coronary heart disease, and stratification with respect to sex is relevant. The sample sizes for the two sexes are assumed to be equal, that is, we let $r_1 = r_2 = 1/2$. The probabilities in the strata are set as in Table 12.7, based on the example in Stokes et al. (2012, p. 64). With significance level 5% and 80% power, we get a total sample size of

$$N = \frac{(1.96 + 0.84)^2}{0.0229} = 342.$$

TABLE 12.7

Assumed association, in terms of probabilities, between coronary heart disease and electrocardiogram, stratified by sex

Electrocardiogram	Coronary heart disease		Sex
	Yes	No	
Positive	0.20	0.30	
Negative	0.15	0.35	Females
Total	0.35	0.65	
Positive	0.40	0.10	
Negative	0.25	0.25	Males
Total	0.65	0.35	

12.12   Recommendations

In general, we recommend sample size calculations based on power analysis, unless the research problem is formulated as the estimation of a parameter with a certain precision. In that case, sample size calculations based on confidence intervals may be the best choice.

Power analysis or sample size estimation is relevant at the planning stage. Post hoc power calculations are futile, although it has been recommended by some journals. Power is the probability of rejecting the null hypothesis in a (future) study. Once the study has been conducted, this probability is either 1 (if the null hypothesis was rejected) else 0. Post hoc power is fundamentally flawed, see Hoenig and Heisey (2001). After the study, meaningful quantifications of uncertainty are confidence intervals and P-values.

When the purpose of the study is to compare two groups, with a given total sample size, maximum power is generally obtained by allocating equal number of observations to the two groups. If for a given sample size, the allocation rate is 2:1 the loss in power is small. If, however, the ratio is 5:1, the loss in power is around 25%, see Campbell et al. (1995).

To safeguard against loss of power due to drop-outs from the study, it is generally recommended to increase the sample size obtained from the calculation. The amount to increase will depend on the setting and the expected drop-out rate, which may vary considerably from study to study. If no specific expectations exist, an increase of 10% is commonly used.

The calculation of the sample size, both for power analysis and confidence interval estimation, is in theory straightforward; however, it may be computationally demanding, and use of adequate statistical software is recommended. If an exact computation of sample size for a specific situation is not feasible, we recommend to carry out a sample size calculation for a simplified situation, to get an idea of the realism of the planned study. An approximate sample size calculation is better than no sample size calculation at all.

13

Miscellaneous Topics

13.1 Diagnostic Accuracy

13.1.1 Introduction

Diagnostic tests play a central role in medicine. Diagnostic tests may be given to patients with symptoms or signs of a disease, or, in screening programs, to persons with no such symptoms or signs. In this section, we will describe diagnostic tests that are binary: a positive test indicates that the person has the disease, and a negative test indicates that the person does not have the disease. Other diagnostic tests may have an ordinal outcome, or a continuous numeric outcome. Histology results in breast cancer, for example, can be classified as benign, uncertain, or malignant, which may be regarded as an ordinal outcome. Serum concentration of PSA, which is a marker for prostate cancer, is an example of a test with a continuous outcome.

The properties of a diagnostic test are related to the true state of the disease. In this section, we assume that there are only two states of the disease in question, namely diseased or non-diseased, and that the true disease status of each subject is known. How to compare the results of two diagnostic tests or two raters in the absence of the "true" status is the topic of the upcoming Section 13.2.

In this section, we use the notation a, b, c, d instead of $n_{11}, n_{12}, n_{21}, n_{22}$ for the observed counts of the 2×2 table, because this is a well-established notation in the literature on diagnostic tests.

This section is partly based on Lydersen (2012), with permission from Gyldendal Akademisk. For a recent and comprehensive textbook on diagnostic tests, we refer to Zhou et al. (2011).

13.1.2 Measures of Diagnostic Accuracy

For a binary test with two possible disease statuses, there are four possible combined outcomes: for a patient who has the disease, a positive test result is a *true positive*, and a negative result is a *false negative*. If the patient does not have the disease, a negative test result is a *true negative*, and a positive result is a *false positive*. We let the indicator variable D denote the disease status, with $D = 0$ for non-diseased and $D = 1$ for diseased. The indicator variable T denotes the test results, with $T = 0$ for a negative test and $T = 1$ for a positive test. The four possible combinations of D and T are illustrated in Table 13.1.

TABLE 13.1

The four possible combinations of disease status and diagnostic test result for a binary test

	Test result	
Disease status	**Positive**	**Negative**
Diseased	True positive	False negative
Non-diseased	False positive	True negative

The probabilities that the test result gives the true disease status for a diseased and a non-diseased individual are called *sensitivity* and *specificity*, respectively:

$$\text{Sensitivity} = \Pr(\text{Positive test} \mid \text{Diseased}) = \Pr(T = 1 \mid D = 1)$$

and

$$\text{Specificity} = \Pr(\text{Negative test} \mid \text{Non-diseased}) = \Pr(T = 0 \mid D = 0).$$

If we have data on disease status and test results for patients such as in Table 13.2, these probabilities can be estimated as follows:

$$\text{Sensitivity} = \frac{a}{a + b}$$

and

$$\text{Specificity} = \frac{d}{c + d}.$$

Confidence intervals and hypothesis tests for sensitivity and specificity can be calculated as described in Chapter 2.

TABLE 13.2

The observed counts of a binary diagnostic test

	Test result		
Disease status	**Positive**	**Negative**	**Total**
Diseased	a	b	$a + b$
Non-diseased	c	d	$c + d$
Total	$a + c$	$b + d$	$a + b + c + d$

Sometimes, other terms are used for sensitivity and specificity. The sensitivity can be called the *true-positive fraction* or *true-positive rate*, and a is the number of *true positives*. Similarly, the specificity can be called the *true-negative fraction* or *true-negative rate*, and d is the number of *true negatives*.

Further, b is the number of *false negatives*, and $b/(a+b)$ is the *false-negative fraction* or *false-negative rate*. The number of *false positives* is c, and $c/(c+d)$ is the *false-positive fraction* or *false-positive rate*.

In a clinical setting, the sensitivity and specificity are usually not the probabilities of primary interest. On the contrary, one is interested in the probabilities for disease status given the test result. These are called the *predictive values* and are defined as

$$
\begin{aligned}
\text{PPV} &= \text{Positive predictive value} \\
&= \Pr(\text{Diseased} \,|\, \text{Positive test}) \\
&= \Pr(D = 1 \,|\, T = 1)
\end{aligned}
$$

and

$$
\begin{aligned}
\text{NPV} &= \text{Negative predictive value} \\
&= \Pr(\text{Non-diseased} \,|\, \text{Negative test}) \\
&= \Pr(D = 0 \,|\, T = 0).
\end{aligned}
$$

With the notation in Table 13.2, these probabilities can be estimated as

$$
\widehat{\text{PPV}} = \frac{a}{a+c}
$$

and

$$
\widehat{\text{NPV}} = \frac{d}{b+d}.
$$

With the help of Bayes' theorem, the predictive values can be expressed in terms of the sensitivity, specificity, and $\Pr(D = 1)$, the prevalence of the disease, as

$$
\begin{aligned}
\text{PPV} &= \Pr(D = 1 \,|\, T = 1) & \text{(13.1)} \\
&= \frac{\Pr(T = 1 \,|\, D = 1) \cdot \Pr(D = 1)}{\Pr(T = 1 \,|\, D = 1) \cdot \Pr(D = 1) \; + \; \Pr(T = 1 \,|\, D = 0) \cdot \Pr(D = 0)} \\
&= \frac{\text{Sensitivity} \cdot \text{Prevalence}}{\text{Sensitivity} \cdot \text{Prevalence} \; + \; (1 - \text{Specificity}) \cdot (1 - \text{Prevalence})}
\end{aligned}
$$

and

$$
\begin{aligned}
\text{NPV} &= \Pr(D = 0 \,|\, T = 0) & \text{(13.2)} \\
&= \frac{\Pr(T = 0 \,|\, D = 0) \cdot \Pr(D = 0)}{\Pr(T = 0 \,|\, D = 1) \cdot \Pr(D = 1) \; + \; \Pr(T = 0 \,|\, D = 0) \cdot \Pr(D = 0)} \\
&= \frac{\text{Specificity} \cdot (1 - \text{Prevalence})}{(1 - \text{Sensitivity}) \cdot \text{Prevalence} \; + \; \text{Specificity} \cdot (1 - \text{Prevalence})}.
\end{aligned}
$$

The sensitivity and specificity are direct properties of the test, and in most settings, they do not depend on the disease prevalence. The predicted values, on the other hand, depend strongly on the prevalence. Consider a diagnostic test with sensitivity 0.95 and specificity 0.92, which in most settings would be regarded as a highly accurate test. In a population with prevalence 0.001, the above formulas give $\widehat{\text{PPV}} = 0.012$ and $\widehat{\text{NPV}} = 0.99995$, whereas in a population with prevalence 0.1, the predictive values become $\widehat{\text{PPV}} = 0.57$ and $\widehat{\text{NPV}} = 0.994$. The positive predictive values are typically very low in a low prevalence population. As another example, consider a diagnostic test with sensitivity 0.95 and a very high specificity of 0.999. In a population with disease prevalence 0.001, we obtain $\widehat{\text{PPV}} = 0.49$ and $\widehat{\text{NPV}} = 0.99995$. Compared with the $\widehat{\text{PPV}} = 0.012$ obtained with a specificity of 0.92, a positive predictive value of 0.49 may be acceptable in some situations. This illustrates that diagnostic tests for screening ought to have high specificity.

Likelihood ratios are another way of quantifying the properties of a diagnostic test, see for instance Lydersen (2012) or Zhou et al. (2011); however, they are considerably less used than sensitivity and specificity.

It has been proposed that the accuracy of a diagnostic test should be summed into one measure, such as the *Youden's index*, which is defined as

$$J = \text{Sensitivity} + \text{Specificity} - 1.$$

If a diagnostic test is perfect, then sensitivity and specificity both equal 1, and $J = 1$. In a *noninformative diagnostic test*, the probability of a positive test does not depend on the disease status. Then, Sensitivity $= 1 -$ Specificity, and $J = 0$. Values of J between -1 and 0 could arise if the test result is negatively associated with the disease, which is not likely to occur in practice. Youden's index implicitly implies that the sensitivity and specificity have equal importance, because they are equally weighted in the definition of the index. Thus, the practical usefulness of the index is highly questionable (Armitage et al., 2002, pp. 693–694).

Another proposed measure is the total fraction of correct results as a measure of accuracy, which can be estimated by $(a + d)/(a + b + c + d)$. With the law of total probability, we obtain

$$
\begin{aligned}
\text{Pr(Correct test result)} \;=\; &\text{Pr(Correct test result} \mid \text{Diseased)} \cdot \text{Pr(Diseased)} \\
&+ \text{Pr(Correct test results} \mid \text{Non-diseased)} \cdot \text{Pr(Non-diseased)} \\
=\; &\text{Sensitivity} \cdot \text{Prevalence} \;+\; \text{Specificity} \cdot (1 - \text{Prevalence}).
\end{aligned}
$$

We see that also this measure depends on the disease prevalence in addition to the sensitivity and specificity. This is an undesirable property of a measure for diagnostic accuracy.

It is not possible to sum up the properties of a diagnostic test in one measure without disregarding important information. Our general advice is that the accuracy of a binary diagnostic test should be given by two measures: sensitivity and specificity.

13.1.3 Confidence Intervals and Hypothesis Tests for a Single Diagnostic Test

It is straightforward to construct confidence intervals and tests for sensitivity and specificity with methods for the 1×2 table and the binomial distribution, as described in Chapter 2. Predictive values, on the other hand, depend on the disease prevalence in the target population. If the disease prevalence in the studied subjects equals the prevalence in the target population—as is the case in cohort or cross-sectional studies—we can use the same methods for confidence intervals and tests for the predictive values as for sensitivity and specificity. But, to obtain a sufficient number of diseased subjects to estimate sensitivity, many studies of diagnostic accuracy include a higher proportion of diseased subjects than that of the target population. Then, the scope of the analysis of predictive values may be limited to exploring populations with a known prevalence of disease.

In a target population with a known disease prevalence, we can estimate predictive values by inserting the estimated sensitivity, specificity, and prevalence into Equations 13.1 and 13.2. Confidence intervals for the predictive values can be obtained with the logit transformation of Bayes' formula, see Zhou et al. (2011, pp. 107–109) or Mercaldo et al. (2007):

$$\text{logit}\left(\widehat{\text{PPV}}\right) = \log\left[\frac{\text{Sensitivity} \cdot \text{Prevalence}}{(1 - \text{Specificity}) \cdot (1 - \text{Prevalence})}\right],$$

$$\text{logit}\left(\widehat{\text{NPV}}\right) = \log\left[\frac{\text{Specificity} \cdot (1 - \text{Prevalence})}{(1 - \text{Sensitivity}) \cdot \text{Prevalence}}\right],$$

$$\text{Var}\left[\text{logit}\left(\widehat{\text{PPV}}\right)\right] = \frac{1 - \text{Sensitivity}}{\text{Sensitivity}} \cdot \frac{1}{a + b} + \frac{\text{Specificity}}{1 - \text{Specificity}} \cdot \frac{1}{c + d},$$

and

$$\text{Var}\left[\text{logit}\left(\widehat{\text{NPV}}\right)\right] = \frac{\text{Sensitivity}}{1 - \text{Sensitivity}} \cdot \frac{1}{a + b} + \frac{1 - \text{Specificity}}{\text{Specificity}} \cdot \frac{1}{c + d}.$$

A $100(1 - \alpha)\%$ confidence interval for PPV can then be calculated as

$$\frac{\exp\left\{\text{logit}\left(\widehat{\text{PPV}}\right) \pm z_{\alpha/2}\sqrt{\widehat{\text{Var}}\left[\text{logit}\left(\widehat{\text{PPV}}\right)\right]}\right\}}{1 + \exp\left\{\text{logit}\left(\widehat{\text{PPV}}\right) \pm z_{\alpha/2}\sqrt{\widehat{\text{Var}}\left[\text{logit}\left(\widehat{\text{PPV}}\right)\right]}\right\}}, \tag{13.3}$$

where $z_{\alpha/2}$ is the upper $\alpha/2$ percentile of the standard normal distribution. The confidence interval for NPV has a similar expression as Equation 13.3. The confidence intervals for PPV and NPV work well unless PPV or NPV has estimate equal to 1. In that case, we can add $z_{\alpha/2}/2$ pseudo-frequencies to each of the four cell counts before computing the confidence interval (Zhou et al., 2011, pp. 107–109).

Note that the above method assumes that the disease prevalence in the target population is known. We refer to Zhou et al. (2011, pp. 107–109) for recommended methods to include uncertainty due to an estimated prevalence.

13.1.4   Example: Chlamydia Rapid Test versus Polymerase Chain Reaction

Chlamydia thractomatis is the most common sexually transmitted bacterial infection in many countries. In men, chlamydia infection in the lower urinary tract can result in ureteritis and sometimes epididymitis. Nadala et al. (2009) investigated the performance of a new rapid urine test for chlamydia in men in two cohorts: a young people's sexual health centre in Birmingham (site 1) and a genitourinary medicine clinic in London (site 2). Polymerase chain reaction (PCR) testing at a separate laboratory was chosen as the gold standard for testing chlamydia in this study. The results are summarized in Table 13.3.

TABLE 13.3

Performance of chlamydia rapid test (CRT) versus polymerase chain reaction (PCR), based on data from (Nadala et al., 2009)

Disease status (PCR)	Test result (CRT)		Total	
	Positive	Negative		
Diseased	18	2	20	
Non-diseased	8	426	434	Site 1
Total	26	428	454	
Diseased	72	17	89	
Non-diseased	9	659	668	Site 2
Total	81	676	757	

The estimated sensitivities are $18/20 = 0.900$ (site 1) and $72/89 = 0.809$ (site 2), and the estimated specificities are $426/434 = 0.982$ (site 1) and $659/668 = 0.987$ (site 2). Note that there are just a few diseased cases and a low prevalence at site 1. The sensitivity does not differ significantly between the two sites (Fisher mid-$P = 0.36$). The estimated positive predictive values are $\widehat{PPV} = 18/26 = 0.692$ (site 1) and $\widehat{PPV} = 72/81 = 0.889$ (site 2), and the negative predicted values are $\widehat{NPV} = 426/428 = 0.995$ (site 1) and $\widehat{NPV} = 659/676 = 0.975$ (site 2). Note that there is a substantial difference between the positive predictive values at site 1 and site 2 (Fisher mid-$P = 0.028$). This is as expected, because the prevalence differs substantially between the two sites, at $20/454 = 0.044$ (site 1) and $89/757 = 0.118$ (site 2).

Now, let us consider predictive values for site 2. The disease prevalence in this sample is $89/668 = 0.118$. If we assume that this is representative for the target population, the estimated predictive values with confidence intervals are $\widehat{PPV} = 72/81 = 0.889$ (95% CI 0.802 to 0940) and $\widehat{NPV} = 659/676 = 0.975$ (95% CI 0.960 to 0.984). Here, we have used the Wilson score interval, see Section 2.4.3.

Now, let us instead assume that we have a target population with known disease prevalence equal to 0.01. With Equation 13.3, we obtain $\widehat{PPV} = 0.378$ (95% CI 0.802 to 0940) and $\widehat{NPV} = 0.998$ (95% CI 0.997 to 0.999). As expected, the PPV is lower and the NPV is higher in a sample with lower prevalence of disease.

13.1.5 Confidence Intervals and Hypothesis Tests for Two Diagnostic Tests

Now, we consider the situation where we compare two diagnostic tests. The samples from the two tests may be independent, or they may be related through a matched pairs design where two different diagnostic tests are applied to each subject. In either case, the purpose is to find out if one of the tests is superior to the other.

In the independent samples case, one test is applied to one sample of subjects, and the other test is applied to another sample of subjects. The sensitivity and specificity of the two tests can be compared with methods for the independent 2×2 table, as described in Chapter 4, see in particular, the recommendations in Table 4.24 (page 174).

In the matched pairs design, we can compare the sensitivity and specificity of the two tests with methods for the paired 2×2 table, as described in Chapter 8, see in particular, the recommendations in Table 8.15 (page 384). Predictive values, as noted in the previous section, depend on the disease prevalence. Hence, a comparison of predictive values is best performed with a matched pairs design, in which the two tests are applied to the same subjects with the same disease prevalence (Zhou et al., 2011, p. 171). Then, confidence intervals and tests for the predicted values can use the same methods as the sensitivity and specificity.

Note that the matched pairs design is substantially more efficient than the independent sample design; however, in some situations, it may be unpractical or even unethical to use a matched pairs design (Zhou et al., 2011, p. 66).

13.2   Inter-Rater Agreement

13.2.1   Introduction

In Section 13.1, we considered agreement between a diagnostic test on one side, and a gold standard—representing the truth—on the other side. Now, we turn to measures of agreement between raters in the absence of a gold standard. For example, the raters may be radiologists assessing whether a tumor is present or not on given x-rays. These measures of agreement are quantified in the absence of a gold standard, and hence, do not quantify the degree to which the raters agree with the true status of the subjects. The assessment given by the raters may have a dichotomous outcome, more than two nominal categories, more than two ordinal outcomes, or a continuous outcome.

Table 13.4 gives an overview of relevant inter-rater reliability measures, depending on type of data and number of raters. The table is based on Gisev et al. (2013), who also provide an introductory description of these measures. Inter-rater agreement is most often measured in just one measure, such as one of those listed in Table 13.4. In a diagnostic setting, however, it may be easier for different raters to agree on one status, for example healthy, than on diseased. Then, it can be more useful to have two measures of agreement: *positive agreement* and *negative agreement*, analogous to sensitivity and specificity for diagnostic tests.

TABLE 13.4

Examples of inter-rater agreement measures by types of data (not exhaustive), based on Gisev et al. (2013)

	Binary or nominal categorical	Ordinal	Continuous
2 raters	Cohen's kappa	Cohen's weighted kappa	Bland-Altman plot
	ICC*	ICC	ICC
> 2 raters	Fleiss' kappa	Kendall coefficient of concordance	
	ICC	ICC	ICC

*ICC = inter-rater correlation coefficient

In this section, we will concentrate on agreement for categorical data. *Cohen's kappa* (Cohen, 1960)—which is a measure of agreement between two raters—is by far the most used single measure of agreement in this context. Cohen's kappa, and its generalization called *Cohen's weighted kappa* will be described along a few other measures of agreement. We will also describe and discuss measures of positive and negative agreement, measures that we believe should be applied more often in diagnostic medicine than is the case today.

This section is partly based on Lydersen (2012), with permission from Gyldendal Akademisk. For a recent and comprehensive textbook on inter-rater agreement, we refer to Gwet (2014). An introduction to positive and negative agreement is provided by De Vet et al. (2013).

13.2.2 Cohen's Kappa for Dichotomous or Nominal Categories

Consider a situation where two raters each classify subjects in $c \geq 2$ categories, numbered from 1 to c. These are paired data. If the question is whether the marginal distributions of the two raters are different, or to quantify how much they differ, the relevant methods to use can be found in Chapter 8 ($c = 2$) and Chapter 9 ($c > 2$). In this section, the question of interest is how much the two raters agree.

Let π_{ij} denote the joint probability that a subject is classified in Category i by Rater 1 and Category j by Rater 2. The observed counts are denoted as in Table 13.5, which has the same notation for the counts as Table 9.1 (page 388).

TABLE 13.5
The observed counts for paired classifications into $c \geq 2$ categories from two raters

	Rater 2				
Rater 1	Category 1	Category 2	...	Category c	Total
Category 1	n_{11}	n_{12}	...	n_{1c}	n_{1+}
Category 2	n_{21}	n_{22}	...	n_{2c}	n_{2+}
$\vdots$	$\vdots$	$\vdots$	$\ddots$	$\vdots$	$\vdots$
Category c	n_{c1}	n_{c2}	...	n_{cc}	n_{c+}
Total	n_{+1}	n_{+2}	...	n_{+c}	N

An intuitive measure of agreement is the probability that the two raters agree, which is

$$\pi_{\mathrm{a}} = \pi_{11} + \pi_{22} + \ldots + \pi_{cc}.$$

Part of this agreement, however, is due to chance. Suppose that Rater 1 assigns to Category i with probability $\pi_{i+} = \sum_{j=1}^{c} \pi_{ij}$, and Rater 2 assigns to Category j with probability $\pi_{+j} = \sum_{i=1}^{c} \pi_{ij}$, independently of Rater 1. Then, Cohen's probability of agreement by chance is given by

$$\pi_{\mathrm{e}} = \pi_{1+}\pi_{+1} + \pi_{2+}\pi_{+2} + \ldots + \pi_{c+}\pi_{+c}.$$

Cohen's kappa is defined as the relative proportion of agreement exceeding that by chance:

$$\kappa = \frac{\pi_a - \pi_e}{1 - \pi_e}.$$

Cohen's kappa takes the maximum value of 1 for perfect agreement, 0 for agreement no better than chance, and negative values for agreement worse than chance, which is unlikely to occur in practice.

Among a total number N of subjects, n_{ij} are categorized in Category i by Rater 1 and in Category j by Rater 2, see Table 13.5. We can estimate the probabilities defined above with the observed proportions:

$$\hat{\pi}_{ij} = \frac{n_{ij}}{N}, \quad \hat{\pi}_{i+} = \frac{n_{i+}}{N}, \quad \text{and} \quad \hat{\pi}_{+j} = \frac{n_{+j}}{N}.$$

How Do We Interpret Kappa?

How do we interpret an estimated value of κ, say $\hat{\kappa} = 0.60$? While no absolute rules or guidelines are possible, the rough guideline in Table 13.6—based on Landis and Koch (1977) and slightly modified by Altman (1991)—is much used. We may thus say that $\hat{\kappa} = 0.60$ indicates a moderate agreement between two raters. While the guideline in Table 13.6 adds some interpretation to the estimated value of kappa, we clearly recommend that the data are displayed as a contingency table, and not only as a summary measure of agreement.

TABLE 13.6
Rough guideline for interpreting kappa, based on Altman (1991)

Value of κ	Strength of agreement
< 0.20	Poor
0.21–0.40	Fair
0.41–0.60	Moderate
0.61–0.80	Good
0.81–1.00	Very good

Confidence Intervals for Cohen's Kappa

The approximate standard error of Cohen's kappa for dichotomous as well as nominal categories is given by Altman et al. (2000) as

$$\widehat{\text{SE}}(\hat{\kappa}) = \sqrt{\frac{\hat{\pi}_a(1 - \hat{\pi}_a)}{N(1 - \hat{\pi}_e)^2}}, \tag{13.4}$$

and hence, an approximate $100(1 - \alpha)\%$ confidence interval for κ is given by

$$\hat{\kappa} \pm z_{\alpha/2}\widehat{\text{SE}}(\hat{\kappa}),$$

where $z_{\alpha/2}$ is the upper $\alpha/2$ percentile of the standard normal distribution.

Note that some software packages may use other expressions for the standard error. We refer to Liebetrau (1983) and Fleiss et al. (2003, p. 610) for alternative expressions.

13.2.3   Cohen's Weighted Kappa

So far, we have not considered degree of disagreement, which is relevant when there are more than two categories. We can assign a weight of 1 to counts along the diagonal, which represent full agreement, and weights between 0 and 1 for off-diagonal counts. We may thus view the unweighted kappa as a special case of this general setting, where all the off-diagonal counts have a weight of 0. When the outcome categories can be considered ordinal, it is natural to assign larger weights to counts near the diagonal.

Let w_{ij} denote the weight for the cell in row i and column j. *Cohen's linear weighted kappa* uses the weights

$$w_{ij} = 1 - \frac{|i - j|}{c - 1}.$$

With $c = 5$ categories, the linear weights are 1 for a cell on the diagonal and $3/4$, $2/4$, $1/4$, and 0 for cells off the diagonal. Alternatively, the *quadratic weighted kappa* uses the weights

$$w_{ij} = 1 - \frac{(i - j)^2}{(c - 1)^2}.$$

With $c = 5$ categories, the quadratic weights are 1 for a cell on the diagonal and $15/16$, $12/16$, $7/16$, and 0 for cells off the diagonal.

The weighted observed and chance agreements are given by

$$\hat{\pi}_{a,w} = \frac{1}{N} \sum_{i=1}^{c} \sum_{j=1}^{c} w_{ij} n_{ij}$$

and

$$\hat{\pi}_{e,w} = \frac{1}{N^2} \sum_{i=1}^{c} \sum_{j=1}^{c} w_{ij} n_{i+} n_{+j},$$

respectively. The *weighted kappa* is then given by

$$\hat{\kappa}_w = \frac{\hat{\pi}_{a,w} - \hat{\pi}_{e,w}}{1 - \hat{\pi}_{e,w}}.$$

An estimate of the standard error of $\hat{\kappa}_w$ was given by Fleiss et al. (1969), which can be used to compute a confidence for κ_w. It is rather elaborate, and we do not repeat it here. We recommend that a software package is used to calculate confidence intervals for the weighted kappa.

In a situation where weighted kappa is appropriate, which weighting should be used? The literature provides no general advice; however, for the case of equal marginal distributions ($n_{i+} = n_{+i}$ for $i = 1, 2, \ldots, c$), $\hat{\kappa}_w$ with quadratic weights is equal to the *intraclass correlation coefficient* (ICC), except for a term involving the factor $1/N$ (Fleiss and Cohen, 1973). We regard this as a favorable property of the quadratic weighted kappa.

13.2.4   Other Measures of Agreement

Cohen's kappa is by far the most used measure of agreement; however, several other measures have been proposed. *Scott's pi* (Scott, 1955) is similar to Cohen's kappa, the only difference is the computation of the chance agreement. Cohen's kappa calculates the chance agreement probability in cell (i, j) as the product of the marginals ($\hat{\pi}_{i+}\hat{\pi}_{+i}$), whereas Scott's pi uses the squared average of the marginals: $\left[(\hat{\pi}_{i+} + \hat{\pi}_{+i})/2\right]^2$.

When calculating chance agreement, both Cohen's kappa and Scott's pi assume the raters to rate the subjects independently. In reality, some subjects may be easier to classify than others. Some x-rays, for example, show a clear picture where misclassification is improbable, whereas other x-rays are difficult to interpret. The two measures *Alpha* (Aickin, 1990) and *Gwet's AC_1* (Gwet, 2008) are based upon the assumption that only a portion of the observed ratings will potentially lead to agreement by chance. Their suggestions give measures that tend to 1 as the agreement tends to be perfect, irrespective of the prevalence. Unfortunately, the interpretation of the values between 0 and 1 is not straightforward. We refer to Gwet (2014) for a discussion of these measures.

13.2.5   More Than Two Raters

The notion of inter-rater agreement seems intuitive for two raters, but does not generalize to multiple raters in any obvious way. For instance, if four raters classify a subject into the categories 1, 1, 2, and 3, respectively, such that two raters agree on category 1 and the rest disagree, how do we quantify the agreement for this subject? This issue may be resolved in several ways, including the measure suggested by Fleiss (1971), which reduces to Scott's pi when the number of raters is two. Another measure was suggested by Conger (1980), which reduces to Cohen's kappa for the case of two raters. See also Gwet (2014) for descriptions of these measures. Note that Gwet (2014, p. 57) recommends the measure by Fleiss, which will, in general, be very close to that of Conger.

13.2.6   The Kendall Coefficient of Concordance

The Kendall coefficient of concordance can be used to quantify agreement between more than two raters when the ratings are on an ordinal scale. It is closely related to the Spearman correlation coefficient (see Section 7.9.2). If there are k raters, we can compute the Spearman correlation coefficient $\hat{\rho}_S$ for each of the $k(k-1)/2$ pairs of raters. Denote the average of the Spearman correlation coefficients by $\bar{\rho}_S$. The Kendall coefficient of concordance is given as

$$W = \bar{\rho}_S - (\bar{\rho}_S - 1)/k.$$

The larger the number of raters, the closer W is to the average Spearman correlation coefficient. The Kendall coefficient of concordance can vary between 0 (total absence of agreement) and 1 (perfect agreement). It can never take a negative value.

13.2.7 Positive and Negative Agreement for 2 × 2 Tables

So far in this section, we have described ways to quantify agreement in just one measure. We have noted that one should also report the actual contingency table from which the agreement is calculated. Still, particularly in diagnostic medicine, it may be useful to report two measures of agreement instead of just one. If the ratings are "healthy" and "diseased", it may be easier for the raters to agree on "healthy" than on "diseased". This is eloquently described by De Vet et al. (2013). Ideally, we would like to quantify the agreement for subjects who are actually healthy, and to quantify the agreement for subjects who are actually diseased, but we do not know the true disease status.

When the data are organized such that row 1 and column 1 correspond to a positive rating, positive agreement is defined as

$$\text{positive agreement} = \frac{n_{11}}{n_{11} + (n_{12} + n_{21})/2},$$

and negative agreement is defined as

$$\text{negative agreement} = \frac{n_{22}}{n_{22} + (n_{12} + n_{21})/2}.$$

These definitions of *specific agreement* were first suggested by Dice (1945), and later elaborated by Cichetti and Feinstein (1990) and Feinstein and Cichetti (1990). The measures of specific agreement do not suffer from poor agreement when there is a high prevalence and agreement in one category, as is the case with Cohen's kappa. More importantly, the measures of specific agreement convey more information than a single summary measure of agreement.

Confidence intervals for specific agreement have not been much studied. Graham and Bull (1998) report that an interval based on the delta method performs adequately with sample sizes of at least $N = 200$. They also describe a Bayesian interval that performs better in small samples.

13.2.8 Example: Children with Cerebral Palsy

Among children with cerebral palsy (CP), there is large variation in their ability to use their hands, ranging from difficulties only with in-hand manipulation, to more severe impairments that make it impossible to even grasp or hold. The Revised Bimanual Fine Motor Function (BFMF 2) is a classification system of the hand function in children with CP on a five-level scale, where level I is the best and level V the most limited function. Seventy-nine children aged 3 to 17 years were included in an inter-rater reliability study by Elvrum et al. (2017). Each child was guided to perform specific tasks with each hand separately, involving grasping and holding objects, as well as manipulation tasks of different degree of difficulty. The children's fine motor capacity was classified from video recordings using the BFMF 2. Table 13.7 shows the results for two raters, A and B.

TABLE 13.7
Revised Bimanual Fine Motor Function assessed by two raters on 79 children, based on data from Elvrum et al. (2017)

	Rater B					
Rater A	I	II	III	IV	V	Total
I	22	3	0	0	0	25
II	7	16	2	1	0	26
III	0	1	5	7	0	13
IV	0	0	1	8	1	10
V	0	0	0	1	4	5
Total	29	20	8	17	5	79

The estimated (overall) probability that the two raters agree is

$$\hat{\pi}_a = \frac{22 + 16 + 5 + 8 + 4}{79} = 0.70.$$

Cohen's probability of agreement by chance is estimated to be

$$\hat{\pi}_e = \frac{29 \cdot 25 + 20 \cdot 26 + 8 \cdot 13 + 17 \cdot 10 + 5 \cdot 5}{79^2} = 0.25,$$

and we estimate Cohen's kappa to be

$$\hat{\kappa} = \frac{0.70 - 0.25}{1 - 0.25} = 0.60.$$

The value of $\hat{\kappa}$ is somewhat less than the estimated overall agreement of 0.70, which means that at least some part of the overall agreement is due to chance. The estimated standard error (Equation 13.4) is 0.067, and a 95% confidence interval for κ is (0.46 to 0.73).

   The unweighted Cohen's kappa is generally not recommended for ordinal categories. If we calculate the weighted kappa for the data in Table 13.7, we obtain $\hat{\kappa}_w = 0.89$ with quadratic weights. A 95% confidence interval—calculated by StatXact 11—is (0.85 to 0.94). We observe a considerable difference between the unweighted and the weighted kappa.

Another limitation of the unweighted kappa is that it may depend strongly on the number of categories. Suppose now that the raters only distinguish between level I to III on one side versus level IV and V on the other side, such that we obtain the 2×2 table in Table 13.8. The estimated Cohen's kappa for this table is $\hat{\kappa} = 0.69$ (95% CI 0.49 to 0.88), which is quite a bit higher than for the table with five outcome categories.

TABLE 13.8

Dichotomized version of Table 13.7, based on data from Elvrum et al. (2017)

	Rater B		
Rater A	I–III	IV–V	Total
I–III	56	8	64
IV–V	1	14	15
Total	57	22	79

Two other undesirable properties of Cohen's kappa are worth mentioning. First, the kappa value depends on the marginal distribution (prevalence) of the categories, such that high prevalence and agreement in one category produces low values of kappa. Second, raters who disagree more on the marginal distribution may produce higher kappa values. These paradoxes can be resolved with other measures of agreement, such as positive and negative agreement.

Positive and negative agreement for Table 13.8 can now be calculated as

$$\text{positive agreement} = \frac{56}{56 + (8+1)/2} = 0.93,$$

and

$$\text{negative agreement} = \frac{14}{14 + (8+1)/2} = 0.76.$$

We note that these values are more in line with the quadratic weighted kappa than the unweighted kappa.

13.2.9 Recommendations

Cohen's kappa is probably the most used measure of agreement for categorical ratings. It is defined for dichotomous and nominal categories, and the weighted kappa is used for ordinal categories. Cohen's kappa is hampered by some undesirable properties related to the marginal distribution for the raters. In particular, if the raters predominantly agree on one of two dichotomous outcomes, the corresponding kappa will be low. Other measures of agreement exist that do not have this disadvantage; however, the interpretation of these measures for values between zero (no agreement) and one (perfect agreement) is less intuitive.

For ordinal ratings, we generally recommend a quadratic weighted kappa due to its close connection to the intraclass correlation coefficient.

In settings such as diagnostic medicine with dichotomous ratings, we recommend reporting the two measures of specific agreement: positive agreement and negative agreement, which together are more informative than a single measure of agreement. Moreover, specific agreement is not hampered by the problems of Cohen's kappa related to the marginal distributions.

Regardless of the measure of agreement that is used, we recommend that the contingency table is reported along the estimated measures and their confidence intervals.

13.3 Missing Data

13.3.1 Introduction

The methods in this book are generally described for contingency tables with complete data; that is, a subject is included in the analysis only if both the row and column variables, and, if relevant, the stratum variable, are observed. Subjects with missing values on at least one variable are not included in such an analysis. This is called *complete case analysis*. If there are subjects with partially missing data, disregarding these implies loss of power, and, more seriously, it can introduce bias. The examples shown elsewhere in this book either have complete data, or the table shows counts only for subjects with complete data. Table 7.9 (page 281), for example, shows the counts only for subjects with data on both time points. But some of the adolescents took part only at one of the time points, and thus have missing data on the row or column variable. A contingency table that includes the subjects with partially missing data is shown in Table 13.9. In some of the examples in this section, we will use a simplified version of this table, where the data have been dichotomized and only include subjects with data at the first time point (Young-HUNT 1), as shown in Table 13.10.

Today, an extensive literature on analysis with missing data exists. Some recent, comprehensive books are Little and Rubin (2002), Van Buuren (2012), and Molenberghs et al. (2015). These books contain sections on missing data in contingency tables, from different perspectives. Only a limited part of the missing data literature concerns contingency tables, and it is mainly from the last ten to twenty years.

In the present section, we will describe missing data mechanisms. Some principles for handling missing data are imputation, probability weighting,

TABLE 13.9
Self-rated health for 12- to 17-year-old adolescents in Young-HUNT 1 and four years later in Young-HUNT 2 (Breidablik et al., 2008), with missing data

Young-HUNT 1	Young-HUNT 2				Total	Missing
	Poor	Not very good	Good	Very good		
Poor	2	3	3	3	11	28
NVG*	2	58	98	14	172	407
Good	8	162	949	252	1371	2205
Very good	4	48	373	369	794	975
Total	16	271	1423	638	2348	3615
Missing	4	50	228	89	371	

*Not very good

TABLE 13.10
Dichotomized version of Table 13.9, for subjects with data at Young-HUNT 1

Young-HUNT 1	Young-HUNT 2			
	Poor	Good	Total	Missing
Poor	65	118	183	435
Good	222	1943	2165	3180
Total	287	2061	2348	3615

and full information maximum likelihood. *Imputation* can be done as single imputation, in which—usually—the estimated expectation of each missing value, conditional on observed data, is imputed in its place. In some specific settings, this can result in unbiased estimates. But the standard errors of the estimates will be underestimated. One way to remedy this is *multiple imputation*, where several complete data sets are created, and the imputed values vary between the data sets due to the random element in the imputation, see for example Sterne et al. (2009) or Van Buuren (2012). Multiple imputation is most relevant in situations where additional variables, sometimes denoted auxiliary variables, are available and can be used as predictors in regression models. Some other methods for variance estimation are also suggested in the literature. *Probability weighting*, on the other hand, means that some observations are "weighted up" since they represent more than one case, due to missingness. In Table 13.10, for example, only 183 of the 435 subjects reporting poor self-rated health at time 1 (Young-HUNT 1) have data at time 2 (Young-HUNT 2). Hence, these are regarded as sampled with probability 183/435 from the 435 candidates, so each of them "represents"

$435/183 = 2.38$ subjects, and are weighted with this factor in the analysis. *Full information maximum likelihood*, often abbreviated FIML, is particularly relevant if data are multivariate normally distributed. But judging by the literature today, FIML does not seem to have much merit yet in the context of contingency tables.

Different authors suggest different ways of handling missing data in contingency tables. For a general framework for handling missing data in contingency tables, we refer to Tian and Li (2017). There does not seem to exist a consensus about which handful of methods can be regarded as the most appropriate for contingency tables. We will thereby describe some main principles, and refer to the literature for details.

13.3.2 Missing Data Mechanisms: MCAR, MAR, and MNAR

The *missing data mechanism* describes the degree to which the probability that data are missing depends on observed data, unobserved data, or both. This is to be understood as a probability distribution, and not as a physical mechanism causing the missing data. The missing data mechanism can be thought of as a second stage of sampling. First, potentially, there exists a complete data set $\mathbf{Y}$, generated by a probability distribution for the sampling. At the second stage, not all variables for each subject are observed. Let $\mathbf{R}$ be a matrix of response (observation) indicators, sometimes called *missing data indicators*, with entries 0 or 1 for each variable on each subject, with 1 in positions for observed data, and 0 in positions for unobserved data. The probability distribution of $\mathbf{R}$ is called the *missing data mechanism*.

We usually distinguish between three types of data mechanisms, as first suggested in the seminal paper by Rubin (1976):

- MCAR: Missing completely at random
- MAR: Missing at random
- MNAR: Missing not at random

For example, consider a 2×2 table with unpaired data, as in Chapter 4. Assume now that the missingness pattern is monotone, such that the data can be missing only for the column variable Y_2. Let $R = 0$ if Y_2 is missing, and let $R = 1$ if Y_2 is observed. The probability that Y_2 is missing can, in general, be written

$$\Pr(R = 0 \mid Y_1 = i, Y_2 = j) = \phi_{ij},$$

for $i, j = 1, 2$, which, without further information on ϕ_{ij} is MNAR. Then, MCAR correspond to

$$\Pr(R = 0 \mid Y_1 = i, Y_2 = j) = \Pr(R = 0) \iff \phi_{ij} = \phi,$$

and MAR correspond to

$$\Pr(R = 0 \,|\, Y_1 = i, Y_2 = j) = \Pr(R = 0 \,|\, Y_1 = i) \iff \phi_{ij} = \phi_i.$$

Note that MCAR is a special case of MAR, which again is a special case of MNAR. Some authors use the term *ignorable missing* if the data are MAR or MCAR, and the parameters of the missing data mechanism are distinct from those of the sampling distribution.

The difference between MAR and MCAR can be illustrated for the self-rated health example in Table 13.10. The proportions missing at time 2 are $\hat{\phi}_1 = 435/(183 + 435) = 70.4\%$ if self-rated health at time 1 is poor, and $\hat{\phi}_2 = 3180/(3180 + 2165) = 59.5\%$ if self-rated health at time 1 is good. Thus, the proportion of missing at time 2 depends on the observed value at time 1. This implies that the data are not MCAR, but must be MAR or MNAR.

In attrition analyses, many authors report a P-value, which in this case would be $P < 0.001$, for the null hypothesis that the two proportions missing are equal. A P-value has little relevance in an attrition analysis, as also pointed out by Dumville et al. (2006); it is hardly ever relevant, or even realistic, to consider a null hypothesis that the data are MCAR. In practice, there will always be some degree of deviation from MCAR. The relevant issue is how much the cases with (partially) missing data deviate from the completely observed cases. Returning to the example, we see that the distribution of self-rated health at time 1 for subjects with complete data differs from those missing data at time 2. The proportions reporting poor health at time 1 are $183/2348 = 7.8\%$ and $435/3615 = 12.0\%$, respectively, depending on whether data at time 2 is available or missing. This may be regarded as a substantial difference.

One can verify from data inspection that data are not MCAR, such as in this example. It is, however, impossible to identify whether data are MNAR and MAR based on the data at hand. An assumption of MAR must be based on external knowledge; however, if data are MNAR, an analysis based on the MAR assumption will generally be less biased than an analysis based on the MCAR assumption. Sometimes, it is useful to carry out sensitivity analyses with one or a few MNAR models, see, for example, Molenberghs et al. (2015, Section 4.2.2) or Fleiss et al. (2003, Section 16.6).

13.3.3 Estimation in a 2 × 2 Table

Let the joint probability distribution of Variable 1 (Y_1) and Variable 2 (Y_2) be as shown in Table 13.11, where we have used the same notation as in Table 4.6.

In principle, estimation based on imputation of missing values can be rather straightforward. Let us return to the example in Table 13.10. A complete case analysis (disregarding subjects with partially missing data) results

TABLE 13.11

The joint probabilities of a 2 × 2 table

Variable 1	Variable 2		Total
	Category 1	Category 2	
Category 1	π_{11}	π_{12}	π_{1+}
Category 2	π_{21}	π_{22}	π_{2+}
Total	π_{+1}	π_{+2}	1

in the estimated cell probabilities

$$\{\hat{\pi}_{11}, \hat{\pi}_{12}, \hat{\pi}_{21}, \hat{\pi}_{22}\}_{\text{MCAR}} = \left\{ \frac{65}{2348}, \frac{118}{2348}, \frac{222}{2348}, \frac{1943}{2348} \right\}$$

$$= \{0.028, 0.050, 0.095, 0.828\}.$$

Based on complete case analysis, the estimated probability of poor self-rated health at time 2 is $\hat{\pi}_{+1} = \hat{\pi}_{11} + \hat{\pi}_{21} = 0.078$. The estimated conditional probability of poor self-rated health at time 1 given poor self-rated health at time 2 is $65/287 = 0.226$. The estimated difference between probabilities is 0.253, and the estimated odds ratio is 4.82.

Now we assume MAR, which implies that the conditional probabilities for self-rated health at time 2 are the same for those with missing data at time 2 as for those with complete data. The proportion reporting poor health at time 2 among those reporting poor health at time 1 is $65/183 = 0.355$, so we impute the count $435 \cdot 0.355 = 154.5$ in the upper left corner of the 2 × 2 table. Continuing like this, we obtain a table with imputed values as shown in Table 13.12, which we use to estimate the cell probabilities:

$$\{\hat{\pi}_{11}, \hat{\pi}_{12}, \hat{\pi}_{21}, \hat{\pi}_{22}\}_{\text{MAR}} = \left\{ \frac{219.5}{5963}, \frac{398.5}{5963}, \frac{548.1}{5963}, \frac{5195.4}{5963} \right\}$$

$$= \{0.037, 0.067, 0.092, 0.804\}.$$

Based on imputation assuming MAR, the estimated probability of poor self-rated health at time 2 is $\hat{\pi}_{+1} = \hat{\pi}_{11} + \hat{\pi}_{21} = 0.104$. This differs from the estimate assuming MCAR, as expected, since the data are not MCAR. The estimated conditional probability of poor self-rated health at time 2, given poor self-rated health at time 1, is $219.5/618 = 0.355$, and the estimated conditional probability of poor self-rated health at time 1 given poor self-rated health at time 2 is $219.5/767.6 = 0.286$. The latter estimate differs from the complete case estimate, as expected, since data are not MCAR. The estimated difference between probabilities is 0.253, and the estimated odds ratio is 4.82, exactly as in the complete case analysis, also as expected, since we assumed MAR and this is a 2 × 2 table (Fleiss et al., 2003, p. 497).

Variance estimation, needed to construct confidence intervals and hypothesis tests, is generally more challenging than estimation. Jackknife methods

TABLE 13.12
Dichotomized version of Table 13.9, for
subjects with data at Young-HUNT 1,
with imputed values under MAR

Young-	Young-HUNT 2		
HUNT 1	Poor	Good	Total
Poor	219.5	398.5	618
Good	548.1	5195.4	5345
Total	767.6	5195.4	5963

or multiple imputation can be used in some settings. For a 2 × 2 table with monotone missingness, as in the example above, Fleiss et al. (2003) present a closed-form expression based on the delta method.

13.3.4 Estimation in Stratified 2 × 2 Tables

Fleiss et al. (2003) also consider stratified 2 × 2 tables. The authors assume there is a common odds ratio across the K tables. They show that when either the explanatory or the outcome variable is MAR, the complete case Mantel-Haenszel estimate of the odds ratio is consistent. The authors also describe two methods for obtaining a variance estimate, one of which is based on the jackknife technique. When the stratifying variable is MAR, the complete case estimate is not consistent and one must turn to imputation or weighting methods.

13.3.5 Hypothesis Tests

Lipsitz and Fitzmaurice (1996) proposed a score test for testing independence in $r \times c$ tables. They also derived appropriate score statistics for the 2 × 2 table, unordered $r \times c$ table, and the doubly ordered $r \times c$ table. Under the null hypothesis of independence, these test statistics are asymptotically chi-squared distributed with 1, $(r-1)(c-1)$, and 1 degree of freedom, respectively. The tests are valid when data are MCAR or MAR. If data are complete, these test statistics coincide with the Pearson chi-squared statistic for unordered tables, and with the linear-by-linear test statistic for doubly ordered tables. The score statistics can be expressed in closed form (Lipsitz and Fitzmaurice, 1996); however, they are somewhat elaborate.

A score test for association in doubly ordered $r \times c$ tables, which is an extension of the test proposed by Lipsitz and Fitzmaurice (1996), was derived by Parzen et al. (2010), and it is unbiased when data are MAR. With complete data, the score test statistic coincides with the linear-by-linear test statistic (also called the *Mantel-Haenszel statistic*) for ordered tables.

Li et al. (2016) studied tests for unspecific ordering (see Section 5.6) in incomplete $r \times c$ tables. They described six test statistics: the likelihood ratio, score, global score, Hausman-Wald, Wald, and distance-based statistics. The score and the global score statistics have asymptotic chi-bar-squared distribution (see page 188). The authors used simulations to evaluate the power and actual significance levels of these tests. The likelihood ratio test based on bootstrapping performed satisfactory in small as well as large sample sizes, and the score test has almost as high power.

13.3.6   Methods for MNAR

Methods for MNAR are discussed in Fleiss et al. (2003, Section 16.6). Molenberghs et al. (2015, Section 4.2.2) show examples of analyses for a 2×2 table with monotone missingness, including MNAR models. They note that if missingness is not monotone, even this apparently simple situation becomes surprisingly complex for MNAR models.

13.4   Structural Zeros

13.4.1   Introduction

Zero counts can occur in contingency tables for two different reasons: sampling zeros or structural zeros. A *sampling zero*, also called *random zero*, is due to random variation, when zero is only one of the possible values under the assumed model. One example is the 2×2 table of CHRNA4 genotypes and presence of exfoliation syndrome (XFS) in the eyes (Table 4.5, page 88). For one of the genotypes, no subjects have presence of XFS in the eyes. When a sampling zero occurs, the maximum likelihood estimate does not always exist. Some methods for confidence intervals and hypothesis tests handle zero counts, and others do not. The methods we recommend for small samples usually handle sampling zeros without problems.

A *structural zero*, also called a *fixed zero*, on the other hand, occurs when a value or a combination of values are impossible to observe. Typically, this is due to design, or due to a physically impossible combination of values. Structural zeros are the topic of this section, and we will consider two different situations, representing two different types of research questions.

The first situation is a structural zero due to a two-way design in which the second variable is observed only if the first variable attains a certain value or certain values. This situation is primarily relevant for 2×2 tables, where the second variable is observed only if the outcome of the first variable is a success. Then, the question of interest will usually be to compare the probability of success for the first variable with the conditional probability of success for the second variable. Independence—which we consider in the 2×2 table without structural zeros—is not an issue for a single 2×2 table with a structural zero.

Even in larger tables, when there is a structural zero, the variables cannot be independent in the usual sense; however, a property called *quasi-independence* can be relevant. This is the second situation with structural zeros that we will consider. Quasi-independence can occur in $r \times c$ tables, where $r \geq 3$ and $c \geq 3$, but also in stratified tables, including stratified 2×2 tables.

The two separate situations are covered in Sections 13.4.2 and 13.4.3.

13.4.2 Structural Zero Due to Design

Table 13.13 shows an example of a 2×2 table with a structural zero due to design: it is possible for a subject to have a secondary infection only if it had a primary infection. Another example is when two consecutive treatments can be offered to patients, where treatment B is given after treatment A, and only if the patient was not cured by treatment A. In this context, it can be relevant to compare the probability of being cured from treatment A with the conditional probability of being cured by treatment B, given that treatment A failed. A third example is the use of two consecutive diagnostic tests (Macaskill et al., 2002). The "either positive" rule means that the individual is defined as test positive only if at least one test is positive. Then, test B is applied only if test A was negative, and the combined test achieves high sensitivity at the cost of lower specificity. In other settings, such as screening in a low prevalence population, an "either negative" rule can be used. Then, test B is applied only if test A was positive.

TABLE 13.13

Observed counts, with expected counts under the null hypothesis in parentheses, of primary and secondary pneumonia infection in calves, based on data from Agresti (2013, p. 20)

Primary infection	Secondary infection		Total
	Yes	No	
Yes	30 (38.1)	63 (39.0)	93
No	0 (-)	63 (78.9)	63
Total	30	126	156

Although structural zeros due to design can also occur in $r \times c$ tables or in stratified tables, we will limit our coverage to single 2×2 tables, where we will compare the probability of success for the first variable with the probability of success for the second variable, conditional on a success for the first variable.

Notation and Null Hypothesis

Let Y_1 denote Variable 1 and let Y_2 denote Variable 2. The two possible outcomes for both variables are success ("Yes" in Table 13.13), denoted by $Y_t = 1$, and failure ("No" in Table 13.13), denoted by $Y_t = 0$, for $t = 1, 2$. Let π_{ij} denote the cell probability for cell (i, j) in a 2×2 table, where cell $(2, 1)$ is a structural zero, as in Table 13.13. Then, we have that

$$\Pr(Y_1 = 1) = \pi_{11} + \pi_{12}$$

and

$$\Pr(Y_2 = 1 \mid Y_1 = 1) = \frac{\pi_{11}}{\pi_{11} + \pi_{12}}.$$

The null hypothesis of interest is that these two probabilities are equal:

$$H_0 : \pi_{11} + \pi_{12} = \frac{\pi_{11}}{\pi_{11} + \pi_{12}} \tag{13.5}$$

Under the null hypothesis, let $\Pr(Y_1 = 1) = \pi = \pi_{11} + \pi_{12}$. Then, it follows that the probability structure for the table is as shown in Table 13.14.

TABLE 13.14
Probability structure under the null hypothesis in Equation 13.5

	Variable 2		
Variable 1	Success	Failure	Total
Success	π^2	$\pi(1 - \pi)$	π
Failure	-	$1 - \pi$	$1 - \pi$
Total	π^2	$(\pi + 1)(1 - \pi)$	1

Straightforward derivation (see for example Agresti (2013, p. 21)) yields the maximum likelihood estimate:

$$\hat{\pi} = \frac{2n_{11} + n_{12}}{2n_{11} + 2n_{12} + n_{22}}. \tag{13.6}$$

Expected counts m_{ij} under the null hypothesis can be calculated in the usual manner, by multiplying the total sum N with the cell probabilities in Table 13.14, inserted $\hat{\pi}$.

The Pearson Chi-Squared Test

The Pearson chi-squared statistic is defined in the same way as for tables without structural zeros:

$$T_{\text{Pearson}}(\mathbf{n}) = \sum_{i,j} \frac{(n_{ij} - m_{ij})^2}{m_{ij}},$$

where the summation is over all cells except the one(s) with a structural zero. The degrees of freedom equals the number of terms in the sum minus one minus the number of free parameters under the null hypothesis, which, for a 2×2 table with one structural zero, amounts to df $= 3 - 1 - 1 = 1$. The counts $\{n_{11}, n_{12}, n_{22}\}$ are multinomially distributed, see also Chapter 3.

Tang and Tang (2003) compared the likelihood ratio test with a score test, and found that both tests performed well. The score based methods, however, have the advantage of being undefined in fewer scenarios than is the case for the likelihood ratio based methods.

Confidence Intervals

Lui (2000) compared three alternative confidence interval methods for the difference between $\pi_{11} + \pi_{12}$ and $\pi_{11}/(\pi_{11} + \pi_{12})$: an asymptotic interval based on the Wald statistic, and asymptotic interval based on the likelihood ratio test, and a third interval based on the principle of Feller's theorem. Lui concludes that the interval based on the likelihood ratio test performs well in all situations he studied. A Bayesian confidence interval was derived by Shi et al. (2009), who report it to perform as well as or better than the score based confidence interval. Confidence intervals for the ratio of probabilities are given in Bai et al. (2011), Lui (1998), and Tang et al. (2004).

Example: Primary and Secondary Infection in Calves (Table 13.13)

Based on the observed counts in Table 13.13, we can estimate the probabilities as

$$\Pr(Y_1 = 1) = \frac{30 + 63}{30 + 63 + 63} = 0.596$$

and

$$\Pr(Y_2 = 1 \mid Y_1 = 1) = \frac{30}{30 + 63} = 0.323.$$

The maximum likelihood estimate under the null hypothesis, calculated from Equation 13.6, is $\hat{\pi} = 0.494$. The expected counts in the upper left cell is $m_{11} = (30 + 63 + 63) \cdot 0.494^2 = 38.1$. Similar calculations yield the other values in parentheses in Table 13.13.

We can now calculate the Pearson chi-squared statistic, which gives $T_{\text{Pearson}} = 19.7$. Under the null hypothesis, T_{Pearson} is chi-squared distributed with one degree of freedom, and we find that there is strong evidence against the null hypothesis ($P < 0.0001$). This may indicate that the primary infection has some immunizing effect, which reduces the probability of a secondary infection.

The estimated difference between the probabilities is $0.596 - 0.323 = 0.273$. A 95% confidence interval for the difference, based on the likelihood ratio test, is (0.148 to 0.392), see Lui (2000).

13.4.3   Quasi-Independence

Table 13.15 shows a $4 \times 2 \times 2$ table with two structural zeros: menstrual problems are not possible for males. We regard this table as a stratified 4×2 table, stratified by sex.

TABLE 13.15
Teenagers' concern with health problems, stratified by sex, based on data from Brunswick (1971) as analyzed by Everitt (1992) and Fienberg (2007)

Health concern	Age group 12–15*	16–17*	Total	Sex
Sex, reproduction	4	2	6	
Menstrual problems	-	-	-	
How healthy am I	42	7	49	Males
Nothing	57	20	77	
Total	103	29	132	
Sex, reproduction	9	7	16	
Menstrual problems	4	8	12	
How healthy am I	19	10	29	Females
Nothing	71	31	102	
Total	103	56	159	

The total number of observations: $N = 291$
*Years

Quasi-Independence as Defined by Fienberg

Now, consider a two-dimensional $r \times c$ table with the total sum fixed. With the usual notation, let π_{ij} denote the probability of cell (i, j), and let π_{i+} and π_{+j} denote the row and column total probabilities, as in Section 7.3. In a table with structural zeros, rows and columns are defined as independent if $\pi_{ij} = \pi_{i+}\pi_{+j}$ for all (i, j). Consider a table with at least one structural zero, and let S denote the cells that are not structural zeros. Fienberg (1972) defines quasi-independence similarly to independence: the row and column variables are said to be quasi-independent if there exist a_i and b_j such that the cell probabilities can be written on the form

$$\pi_{ij} = \begin{cases} a_i b_j & \text{for } (i, j) \in S \\ 0 & \text{else} \end{cases}. \tag{13.7}$$

Actually, Fienberg (1972) defines quasi-independence in terms of expected counts; however, the above definition is equivalent. Note that for a 2×2 table with a structural zero, Equation 13.7 can always fit perfectly. In a $2 \times c$ table, quasi-independence is equivalent to independence in the table obtained by deleting the columns with a structural zero, and similarly for the $r \times 2$ table. Hence, for two-way tables, the concept of quasi-independence is of interest if $r \geq 3$ and $c \geq 3$. The above definition of quasi-independence can be generalized in a straightforward manner to stratified tables, such as stratified 2×2 tables (Johnson and May, 1995).

Fienberg (2007) analyzed the data in Table 13.15 with a loglinear model similar to the one in Equation 1.7 (page 17). The fully saturated model included all interactions up to three-way interaction between sex, age group, and health concern, and he compared models with all relevant subsets of these interactions. Fienberg found that the model without any of the interactions fit quite poorly, but that including only the interaction between sex and health concern gave an acceptable fit. If we inspect the data, we see that this interaction seems reasonable: among males, $49/132 = 37\%$ report "how healthy I am", compared with $29/159 = 18\%$ (or $29/147 = 20\%$ if we exclude menstrual problems) among females.

Quasi-Independence as Defined by Agresti

Agresti (2013, p. 430) defines quasi-independence for square $c \times c$ tables in a slightly different way:

$$
\pi_{ij} = \begin{cases} a_i b_j & \text{for } (i \neq j) \\ \delta_i & \text{for} (i = j) \end{cases} . \tag{13.8}
$$

Equation 13.8 defines independence to hold only for the off-diagonal cells in a symmetric table, without any restrictions on the diagonal cells. This definition applies to square tables with $c \geq 3$. Table 13.16 shows social class (lower, middle, upper) for pairs of British fathers and sons (Glass, 1954). The expected counts under the assumption of independence between fathers' and sons' class are also shown in the table. The observed counts on the diagonal are substantially higher than the expected counts, which indicates a clear tendency for the son to remain in the same social class as his father. This tendency is statistically highly significant; the Pearson chi-squared statistic, for instance, has observed value equal to 505.5, and with four degrees of freedom, we have $P = 2 \cdot 10^{-9}$. It is, however, possible that the conditional probability of the son's class, given that it is different from his father's class, does not depend on the father's class. If this is the case, there is quasi-independence as defined in Equation 13.8.

Equivalently, we could delete the diagonal cells of the table, define the deleted cells as structural zeros, and apply the definition of quasi-independence given in Equation 13.7. Table 13.17 shows the observed counts and the (new) expected counts under the assumption of quasi-independence. The observed and expected counts agree quite well, and considerably better than in Table 13.16. In fact, the Pearson chi-squared statistic for the data in Table 13.7 equals 0.61 (see Everitt (1992, p. 140)), and with one degree of freedom, we

TABLE 13.16

Observed counts, with expected counts under the
assumption of independence in parentheses, of social class
for British fathers and sons, based on data from Glass
(1954) as analyzed by Everitt (1992)

Father's status	Son's status		
	Lower	**Middle**	**Upper**
Lower	411 (245.7)	320 (345.3)	114 (254.0)
Middle	447 (439.1)	714 (617.0)	349 (453.8)
Upper	159 (322.1)	395 (466.7)	588 (343.2)

find no evidence of lack of fit for this model ($P = 0.44$). Conditional on the
son leaving the class of his father, the class in which he arrives is independent
of the departure class.

TABLE 13.17

The data from Table 13.16 with diagonal cells deleted
(treated as structural zeros), with expected counts under
the assumption of quasi-independence in parentheses,
from Everitt (1992)

Father's status	Son's status		
	Lower	**Middle**	**Upper**
Lower	-	320 (324.8)	114 (109.2)
Middle	447 (442.2)	-	349 (353.8)
Upper	159 (163.8)	395 (390.2)	-

Note that Table 13.16 is a (symmetric) paired $c \times c$ table (with $c = 3$).
Hence, an analysis of marginal homogeneity or an analysis of symmetry, as
described in Chapter 9, may also be relevant.

13.5   Categorization of Continuous Variables

In many cases, the variables in a contingency table are categorical in their nature, such as treatment group in a randomized controlled trial, or genotype. In
other cases, the categorical variable can originate from a continuous variable,
which has been split into categories. If we have an outcome variable and an
explanatory variable, and one or both of the variables are continuous, it is
usually a good idea to keep the variable as it is, and not categorize it prior
to the analysis. If the explanatory variable—representing, for instance, an

exposure—is continuous and the outcome is categorical, one can use binary, nominal, or ordinal logistic regression without categorizing the explanatory variable. If the outcome variable is continuous, one can use linear regression or a related method. We emphasize that there are substantial negative consequences of categorizing continuous variables in general and in dichotomizing in particular. Hence, categorization should not be carried out unless there are solid reasons for doing so. This is discussed at the end of this section.

Categorizing a continuous variable into a categorical variable is in a sense the opposite of polychoric correlation (see Section 7.9.5). Polychoric correlation is an estimated correlation for an unobserved bivariate continuous variable from which the observed contingency table could have been created. Categorizing a continuous variable, on the other hand, starts with an observed continuous variable and creates a categorical variable.

In the following, we describe some arguments used for categorizing continuous variables, as well as arguments for keeping variables on their original scale. We will also describe some different ways of carrying out categorizations. Assigning scores to the categories, which is relevant for several methods covered in this book, will be briefly discussed. Finally, we end this section with some recommendations and concluding remarks.

13.5.1 Arguments Used for Categorization

Categorization of continuous variables is frequently seen in applied research in many disciplines. MacCallum et al. (2002) report a survey of the common practice of categorizing continuous variables in leading psychological journals. In most cases, the authors do not justify why they categorize. Arguments used as justification are often among—or related to—the following:

- follow practices used in previous research

- simplifying analyses

- simplifying presentation of results

- modelling a possibly non-linear relationship

We discuss these arguments in more detail below.

Follow Practices Used in Previous Research

If previous publications on the research topic at hand have categorized with certain cut-off values, it may be useful to categorize in the same way, with the purpose of comparing the results. When a continuous scale is used in a diagnostic test, this can be a sensible strategy, in particular when the cut-off value is an established criterion for a positive test. The Hospital Anxiety and Depression Scale (HADS), for instance, contains 14 items (questions) to be answered by the patient, each with four response categories, which are scored from 0 to 3 (Zigmond and Snaith, 1983). Seven of the items constitute the anxiety subscale, and the seven other items constitute the depression scale. A sum score of 0 to 7 is regarded as a non-case (negative test), and a sum score of 8 to 21 is regarded as a case (positive test). Hence, many researchers who use the HADS dichotomize the responses with a cut-off value between 7 and 8.

Simplifying Analyses

There are situations for which categorizing a continuous variable can be done to make use of a simple analysis method. One example is a heavily skewed dependent variable in regression analysis: Skorpen et al. (2016) analyzed disease activity in pregnant women with systematic lupus erythematosus with the LAI-P (Lupus Activity Index in Pregnancy) scale as dependent variable. This LAI-P scale variable was heavily skewed, with 51.6% of the observations being 0 (no disease activity). The authors chose an ordinal logistic regression model, with disease activity in four categories as dependent variable. The categories were defined by the values 0, > 0 to 0.25, > 0.25 to 0.5, an > 0.5 on the LAI-P scale. An increase of 0.25 on this scale is regarded as clinically relevant, which is one of the arguments the authors used for this categorization.

Simplifying Presentation of Results

Sometimes authors choose to collapse continuous data into categories and present them in contingency tables rather than in, for example, scatterplots, to simplify the presentation. This is perfectly fine; however, we note that for such a purpose, the categorization should be carried out at the final—i.e., presentation—stage. The statistical analyses, on the other hand, should be performed with the data on the original scale.

Modelling a Possibly Nonlinear Relationship

If categorization is used to study a possibly nonlinear effect of a continuous explanatory variable, this variable must be categorized in at least three categories. For example, individuals can be categorized as underweight, normal weight, or overweight based on cut-off values for the body mass index (BMI), see Section 7.2.3. Being underweight, as well as being overweight, can be associated with increased risk for certain adverse health outcomes, so the effect of BMI can be U-shaped. Such a possibly non-linear relationship can be investigated with a categorized BMI variable. Alternatively, and much better, a more refined model can use the BMI variable on the original scale, and the nonlinear relationship can be modelled with non-linear regression techniques, such as fractional polynomials (Royston and Altman, 1994). A non-linear effect of a continuous explanatory variable can also be analyzed with spline regression, which usually gives similar results as fractional polynomial regression.

13.5.2 How to Categorize

As already noted, a continuous variable can be categorized according to established cut-off values from the literature, if such exist. Examples already
mentioned are diagnosis cut-off values for a continuous diagnostic test variable
and BMI categories. Sometimes, categorization is based on percentiles, such as
the median, quartiles, or quintiles. Newborn babies, for instance, are classified
as small for gestational age (SGA) if their birthweight is less than the 10th
percentile in the reference population, adjusted for gestational age and sex
(see Section 7.2.4). In some settings, researchers categorize data based on percentiles in their own sample. Such data driven percentiles have the additional
drawback that the cut-off values may differ between studies (Ravichandran
and Fitzmaurice, 2008).

13.5.3 How to Assign Scores

Methods for ordinal categorical data are described for the ordered $r \times 2$ table
in Chapter 5, for the ordered $2 \times c$ table in Chapter 6, for singly and doubly ordered $r \times c$ tables in Sections 7.6–7.10, for stratified ordered tables in
Sections 11.4–11.6, and for the paired $c \times c$ table with ordered categories in
Section 9.5. Some of these methods use category scores, and the strategy for
choosing the scores is briefly discussed in some of the aforementioned chapters
and sections.

Assigning scores to categories is an issue that is somewhat related to categorization of continuous variables, because the scores implicitly define the
underlying scale. It is common to use the row or column numbers as scores,
or equivalently, any equally spaced scores. Sometimes, there may be reasons
to use scores that are not equally spaced. If the categorical variable is based
on a continuous variable, mid-interval scores can be used. The duration of
symptoms for colorectal cancer in Table 7.7, for instance, were recorded as
< 1 week, 2–8 weeks, 2–6 months, and > 6 months. In Section 7.8.4, we analyzed these data with both equally spaced scores and the mid-interval scores
$\{0.5, 5, 17, 39\}$. The mid-interval scores represent, in some sense, the average
duration of symptoms (in weeks) for each of the row categories. Another example of the use of equally spaced scores and mid-interval scores is given for
severity of stroke in Section 5.7.14. Not surprisingly, the choice of scores may
affect the results, as illustrated in that example.

In Section 5.5, the issue of assigning scores is discussed in more detail. We
repeat the recommendation to use equally spaced scores, unless there is information about the categories that clearly advises us to do otherwise. Perhaps
needless to say, the choice of scores should be done prior to the analysis, to
limit the risk of spurious findings as a consequence of "shopping" for the set
of scores that provides the most significant—or otherwise interesting—results.

13.5.4 Recommendations

Many authors have demonstrated the negative consequences of categorizing a continuous variable, and their general advice is to perform the statistical analyses on the original (continuous) scale, see Altman and Royston (2006), Greenland (1995), MacCallum et al. (2002), Ravichandran and Fitzmaurice (2008), or Royston et al. (2006). Royston et al. (2006) further states that dichotomization of a continuous explanatory variable should never be done in regression analyses. One important drawback of categorization is the loss of statistical power. When alternative methods are relevant, researchers should strive to use the methods that provide the highest statistical power. Another problem is that dichotomization redefines the problem and may give effect measures that are difficult to interpret. Categorization of a variable implies that the effect is constant on each side of a cut-off value, and that it makes a jump at the cut-off value. Such a model is practically never realistic. Hence, our recommendation is to avoid categorization of continuous variables, unless there are solid reasons for doing so.

13.6 Ecological Inference

13.6.1 Introduction

Statistical analysis of contingency tables is concerned with associations between categorical variables, often between a grouping variable and an outcome variable, with data on the individual level. Sometimes, information on the individual level has not been obtained, and only aggregated information on the group level is available. Analysis of aggregated data may be a quick and easy way to study the relationship between the grouping and the outcome, and provide new insight into the research problem. JD Snow (1815–1858) became an epidemiological pioneer with his study of the cholera epidemics in London in 1848 and 1854. Based on data from the water supplies in London, Snow explained the correspondence between the cholera deaths and the contamination of a water pump on Broad Street (Snow, 1855). Without definite knowledge about the water intake and the case history of the incidences of the cholera epidemics, Snow approached the analysis of the association between water intake and the disease via an aggregated analysis. Snow produced detailed maps of cholera cases and the location of the water pumps, and he found clear evidence of a connection between the Broad Street water pump and the nearby deaths of cholera. This is an astonishing example of a statistical analysis of aggregated data, carried out before there was any specific knowledge of the transmission of cholera, and it had important implications for the public health in London.

Ecological inference, also called *aggregate inference*, is concerned with the association between grouping and outcome when only information about the marginals is obtained. With marginals from only one table, such as a 2×2 table, we do not have sufficient information to make inference on the cell counts. For each row in a 2×2 table, there are two parameters to estimate, but only one marginal. Different combinations of cell counts in a 2×2 table will result in the same marginals. Drawing conclusions about associations on the individual level based only on information from groups is called the *ecological fallacy*. To obtain valid ecological inference, specific assumptions must be made on the association between the two variables, see Sections 13.6.5 and 13.6.6. Research on ecological inference concentrates on the analysis of 2×2 tables, and in this section, we will mostly consider stratified 2×2 tables. Methods for ecological inference can be extended to $r \times c$ tables, see Brown and Payne (1986) and Wakefield (2004).

Duncan and Davis (1953) and Goodman (1953, 1959) were the earliest to describe methods for ecological inference. In the last 20 years, there has been an increasing interest in the topic, and a large number of articles and books have emerged. The book edited by King et al. (2004a) presents different aspects of ecological inference. The review articles by Wakefield (2004) and Glynn and Wakefield (2010) give overviews of the research area and can be recommended as introductions to the topic.

13.6.2 Ecological Inference for Stratified 2×2 Tables

We assume that we observe the marginals of K 2×2 tables, and we use the notation for stratified 2×2 tables in Chapter 10. Table 13.18 shows a summary of the data from table k. No cell counts are recorded, and the aim is to estimate the cell counts, given the marginals in the K tables. With the simplified notation in Chapter 10, see Equations 10.2 and 10.3, we let

$$\pi_{1k} = \Pr(Y_2 = 1 \mid Y_1 = 1, Y_3 = k)$$

and

$$\pi_{2k} = \Pr(Y_2 = 1 \mid Y_1 = 2, Y_3 = k).$$

The corresponding estimates are denoted by $\hat{\pi}_{1k}$ and $\hat{\pi}_{2k}$, and the estimated cell counts are $n_{1+k}\hat{\pi}_{1k}$ and $n_{2+k}\hat{\pi}_{2k}$.

TABLE 13.18
The observed marginals of table k

| | Variable 2 | | | |
Variable 1	$Y_2 = 1$	$Y_2 = 0$	Total	Variable 3
$Y_1 = 1$			n_{1+k}	
$Y_1 = 2$			n_{2+k}	$Y_3 = k$
Total	n_{+1k}	n_{+2k}	n_{++k}	

If we let $x_k = n_{1+k}/n_{++k}$ and $y_k = n_{+1k}/n_{++k}$, we obtain the simple relation

$$\pi_{2k} = \frac{y_k}{1 - x_k} - \frac{x_k}{1 - x_k}\pi_{1k}. \tag{13.9}$$

This relationship between π_{2k} and π_{1k} is defined as the *tomography lines*, which is important in ecological inference, see Section 13.6.5.

13.6.3   Method of Bounds

The method of bounds by Duncan and Davis (1953) was the first approach to making inference for ecological data. The method is deterministic, in the sense that it is not model-based, but uses the data directly to calculate lower and upper bounds for the unknown parameters π_{1k} and π_{2k}. The bounds provide deterministic information about the parameters of interest. For the observed marginals in Table 13.18, the bounds are given as

$$\pi_{1k} \in \left[\max\left(0, \frac{n_{+1k} - n_{2+k}}{n_{1+k}}\right), \min\left(\frac{n_{+1k}}{n_{1+k}}, 1\right)\right]$$

and

$$\pi_{2k} \in \left[\max\left(0, \frac{n_{+1k} - n_{1+k}}{n_{2+k}}\right), \min\left(\frac{n_{+1k}}{n_{2+k}}, 1\right)\right],$$

see King et al. (2004b). The overall bounds are then weighted averages over the K tables, with weights equal to n_{1+k}/n_{++k} and n_{2+k}/n_{++k} for π_{1k} and π_{2k}, respectively. The narrower the interval between these two bounds, the more information we have about the parameters. As noted by, for instance, King (1997), these intervals may be too wide to be useful.

13.6.4   Ecological Regression

Goodman (1953, 1959) proposed what used to be called *Goodman regression* but is now called *ecological regression*. Goodman (1953) used the relation

$$\frac{n_{+1k}}{n_{++k}} = \frac{n_{1+k}}{n_{++k}}\pi_{1k} + \frac{n_{2+k}}{n_{++k}}\pi_{2k},$$

and assumed that $\pi_{ik} = \pi_i$ for $i = 1, 2$ and $k = 1, 2, \ldots, K$. This is an assumption of homogeneity of the probabilities of success $(Y_2 = 1)$ for the two groups over the strata. Note that in this case, there is also homogeneity of

$$\pi_{1k} - \pi_{2k}, \quad \log\left(\frac{\pi_{1k}}{\pi_{2k}}\right), \quad \text{and} \quad \log\left[\frac{\pi_{1k}(1 - \pi_{1k})}{\pi_{2k}(1 - \pi_{2k})}\right]$$

over the strata. We recognize these expressions as the effect measures in stratified 2×2 tables, see Chapter 10.

If we rearrange the terms in Equation 13.9, we obtain

$$y_k = x_k \pi_1 + (1 - x_k)\pi_2. \tag{13.10}$$

The parameters π_1 and π_2 can be estimated by regressing the y_k on x_k and $1 - x_k$. Standard procedure to estimate π_1 and π_2 is by least squares, under the assumption of constant variance. Estimates outside the range $(0, 1)$ may be obtained. The maximum likelihood methods in the next section hold more promise. The assumption $\pi_{ik} = \pi_i$ for $i = 1, 2$ and $k = 1, 2, \ldots, K$ will hardly ever be met; however, as we will see in the next section, this assumption is necessary to obtain useful estimates.

13.6.5 Maximum Likelihood

With the notation in Section 13.6.2, and with

$$\theta_k = \frac{\pi_{1k}/(1 - \pi_{1k})}{\pi_{2k}/(1 - \pi_{2k})},$$

the likelihood function can be written as

$$(1 - \pi_{1k})^{n_{1+k}} \pi_{2k}^{n_{+1k}} (1 - \pi_{2k})^{n_{2+k} - n_{+1k}} \sum_{i=i_0}^{i_1} \binom{n_{1+k}}{i} \binom{n_{2+k}}{n_{+1k} - i} \theta_k^i,$$

where $i_0 = \max(0, n_{+1k} - n_{2+k})$ and $i_1 = \min(n_{1+k}, n_{+1k})$, see McCullagh and Nelder (1989, p. 353) or Wakefield (2004). The likelihood surface has a ridge along the tomography line in Equation 13.9, and the maximum likelihood estimates are obtained at one of the end points of the tomography line, see Steel et al. (2004). At the ends of the tomography line, the two parameters in Equation 13.9 are either 0 or 1, and again, we notice that little information is obtained from the marginals in only one table.

A more promising approach would be to assume homogeneity of the probabilities, see Section 13.6.4. Then the likelihood function is

$$\prod_{k=1}^{K} (1 - \pi_1)^{n_{1+k}} \pi_2^{n_{+1k}} (1 - \pi_2)^{n_{2+k} - n_{+1k}} \sum_{i=i_0}^{i_1} \binom{n_{1+k}}{i} \binom{n_{2+k}}{n_{+1k} - i} \theta^i,$$

see McCullagh and Nelder (1989, p. 353) or Steel et al. (2004). Parameter estimates $\hat{\pi}_1$ and $\hat{\pi}_2$ can be obtained with the Newton-Raphson iterative method.

Steel et al. (2004) also proposed a likelihood ratio test, a score test, and a Wald test for the null hypothesis $H_0 : \pi_{ik} = \pi_i$ for $i = 1, 2$ and $k = 1, 2, \ldots, K$.

13.6.6 Hierarchical Models

As mentioned above, when we remove the assumption that $\pi_{ik} = \pi_i$ for $i = 1, 2$ and $k = 1, 2, \ldots, K$, the parameters in Equation 13.9 are not identifiable. Instead of assuming heterogeneity, one may assume that the pairs (π_{1k}, π_{2k})

follow some continuous distribution function. King (1997) and King et al. (2004b) assumed that the (π_{1k}, π_{2k}) are independent realizations from a truncated bivariate distribution. Under the assumption of no autocorrelation of the pairs over the strata, and independence between the pairs and x_k, King (1997) derived the posterior distribution of (π_{1k}, π_{2k}) for each k. The assumption of independence between the pairs and x_k is controversial. King (1997) proposed diagnostic procedures for assessments of the assumptions; however, they too are controversial, see Wakefield (2004) and the references therein.

King et al. (1999) suggested a beta distribution for π_{1k} and π_{2k}. Then,

$$\pi_{ik} \mid \alpha_i, \beta_i \sim \text{Beta}(\alpha_i, \beta_i),$$

for $i = 1, 2$ and $k = 1, 2, \ldots, K$. The parameters α_1, β_1, α_2, and β_2 follow an exponential distribution with mean $1/\lambda$, see King et al. (1999), King et al. (2004b), and Wakefield (2004).

A simpler approach might be the one proposed by Wakefield (2004), which was later studied by Imai et al. (2008). For all k, we let $\gamma_k = \begin{bmatrix} \pi_{1k}/(1 - \pi_{1k}), \ \pi_{2k}/(1 - \pi_{2k}) \end{bmatrix}^{\mathrm{T}}$ be a 2×1 vector of the logits. Assume that

$$\gamma_k \mid \mu, \Sigma \sim N(\mu, \Sigma).$$

Here, the logits can be correlated. The maximum likelihood estimates can be calculated by the EM algorithm. As noted by Imai et al. (2008), this model can be extended to a Bayesian model by putting a conjugate prior on (μ, Σ). The R package `eco` by Imai et al. (2011) uses the EM algorithm for the likelihood model and the MCMC algorithm for the Bayesian model.

A

Appendix

A.1 List of Statistical Methods

(with recommendations in bold)

The 1×2 Table and the Binomial Distribution (Chapter 2)

The $1 \times c$ Table and the Multinomial Distribution (Chapter 3)

The 2 × 2 Table (Chapter 4)

The Ordered $r \times 2$ Table (Chapter 5)

The Ordered $2 \times c$ Table (Chapter 6)

The $r \times c$ Table (Chapter 7)

The Paired $c \times c$ Table (Chapter 9)

Stratified 2×2 Tables and Meta-Analysis (Chapter 10)

Other Stratified 2 × 2 Tables (Chapter 11)

A.2　List of Examples

The list is organized by chapter, and the first cited page for each example points to the introduction of the example and the table data.

Topic	Pages
The 1×2 Table (Chapter 2)	
The probability of male births to Indian immigrants in Norway	32, 40, 54
A randomized cross-over trial of probiotics versus placebo for patients with irritable bowel syndrome	33, 41, 55
The $1 \times c$ Table (Chapter 3)	
Distribution of genotype counts of SNP rs6498169 in patients with rheumatoid arthritis	68, 71, 77, 82
The 2×2 Table (Chapter 4)	
A lady tasting a cup of tea	86, 99, 119, 160
A randomized clinical trial of high versus standard dose of epinephrine	86, 100, 119, 131, 141, 162
A case-control study of the effect of GADA exposure on IPEX syndrome	87, 102, 122, 163
A cross-sectional study of the association between CHRNA4 genotypes and XFS	87, 103, 122, 132, 144, 165
The $r \times 2$ Table (Chapter 5)	
Alcohol consumption and malformations	180, 190, 204, 214
Elevated troponin T levels in stroke patients	181, 205, 214
The $2 \times c$ Table (Chapter 6)	
Adolescent placement study*	224, 242, 244, 259, 267
Postoperative nausea	225, 230, 234, 238, 261, 268
The $r \times c$ Table (Chapter 7)	
Treatment for ear infection	277, 292
Psychiatric diagnoses and physical activity	277, 293
Psychiatric diagnoses and BMI[†]	277, 295, 303, 309

*Also presented and analyzed for stratified $2 \times c$ tables in Chapter 11
[†]Also presented and analyzed for stratified $r \times c$ tables in Chapter 11

Table continues on next page

Bibliography

Aaberge R (2000) UMP unbiased tests for multiparameter testing problems with restricted alternatives. *Metrika*, **50**:179–193.

Agresti A (1983) Testing marginal homogeneity for ordinal categorical variables. *Biometrics*, **39**:505–510.

Agresti A (1992) A survey of exact inference for contingency tables (with discussion). *Statistical Science*, **7**:131–177.

Agresti A (1999) On logit confidence intervals for the odds ratio with small samples. *Biometrics*, **55**:597–602.

Agresti A (2003) Dealing with discreteness: Making 'exact' confidence intervals for proportions, difference of proportions, and odds ratios more exact. *Statistical Methods in Medical Research*, **12**:3–21.

Agresti A (2010) *Analysis of Ordinal Categorical Data*. 2nd edn. John Wiley & Sons, Inc., Hoboken, NJ.

Agresti A (2013) *Categorical Data Analysis*. 3rd edn. John Wiley & Sons, Inc., Hoboken, NJ.

Agresti A, Bini M, Bertaccini B, Ryu E (2008) Simultaneous confidence intervals for comparing binomial parameters. *Biometrics*, **64**:1270–1275.

Agresti A, Caffo B (2000) Simple and effective confidence intervals for proportions and differences of proportions result from adding two successes and two failures. *The American Statistician*, **54**:280–288.

Agresti A, Coull BA (1996) Order-restricted tests for stratified comparisons of binomial proportions. *Biometrics*, **52**:1103–1111.

Agresti A, Coull BA (1998a) Approximate is better than "exact" for interval estimation of binomial proportions. *The American Statistician*, **52**:119–126.

Agresti A, Coull BA (1998b) Order-restricted inference for monotone trend alternatives in contingency tables. *Computational Statistics & Data Analysis*, **28**:139–155.

Agresti A, Coull BA (2000) [Letter to the Editor] Approximate is better than "exact" for interval estimation of binomial proportions. *The American Statistician*, **54**:88.

Agresti A, Coull BA (2002) The analysis of contingency tables under inequality constraints. *Journal of Statistical Planning and Inference*, **107**:45–73.

Agresti A, Min Y (2001) On small-sample confidence intervals for parameters in discrete distributions. *Biometrics*, **57**:963–971.

Agresti A, Min Y (2002) Unconditional small-sample confidence intervals for the odds ratio. *Biostatistics*, **3**:379–386.

Agresti A, Min Y (2004) Effects and non-effects of paired identical observations in comparing proportions with binary matched-pairs data. *Statistics in Medicine*, **23**:65–75.

Agresti A, Min Y (2005a) Frequentist performance of Bayesian confidence intervals for comparing proportions in 2×2 contingency tables. *Biometrics*, **61**:515–523.

Agresti A, Min Y (2005b) Simple improved confidence intervals for comparing matched proportions. *Statistics in Medicine*, **24**:729–740.

Agresti A, Ryu E (2010) Pseudo-score confidence intervals for parameters in discrete statistical models. *Biometrika*, **97**:215–222.

Aickin M (1990) Maximum likelihood estimation of agreement in the constant predictive probability model, and its relation to Cohen's kappa. *Biometrics*, **46**:293–302.

Altman DG (1991) *Practical Statistics for Medical Research*. Chapman & Hall/CRC, Boca Raton, FL.

Altman DG (1998) Confidence intervals for the number needed to treat. *BMJ*, **317**:1309–1312.

Altman DG, Andersen PK (1999) Calculating the number needed to treat for trials where the outcome is time to an event. *BMJ*, **319**:1492–1495.

Altman DG, Bland JM (1995) Absence of evidence is not evidence of absence. *BMJ*, **311**:485.

Altman DG, Machin D, Bryant TN, Gardner MJ (eds.) (2000) *Statistics with Confidence*. 2nd edn. BMJ Publishing Group, London.

Altman DG, Royston P (2006) The cost of dichotomizing continuous variables. *BMJ*, **332**:1080.

Andersen EB (1970) Asymptotic properties of conditional maximum-likelihood estimators. *Journal of the Royal Statistical Society, Series B*, **32**:283–301.

Anderson JA (1984) Regression and ordered categorical variables. *Journal of the Royal Statistical Society, Series B*, **46**:1–30.

Angrist JA, Pische JS (2009) *Mostly Harmless Econometrics: An Empiricist's Companion.* Princeton University Press, Princeton, NJ.

Anscombe FJ (1948) The transformation of Poisson, binomial and negative-binomial data. *Biometrika,* **35**:246–254.

Armitage P (1955) Tests for linear trend in proportions and frequencies. *Biometrics,* **11**:375–386.

Armitage P, Berry G, Matthews JNS (2002) *Statistical Methods in Medical Research.* 4th edn. Blackwell Science, Malden, MA.

Bai P, Gan W, Shi L (2011) Bayesian confidence interval for the risk ratio in a correlated 2 × 2 table with structural zero. *Journal of Applied Statistics,* **12**:2805–2817.

Baptista J, Pike MC (1977) Exact two-sided confidence limits for the odds ratio in a 2 × 2 table. *Journal of the Royal Statistical Society, Series C,* **26**:214–220.

Barlow RE, Bartholomew DJ, Bremner JM, Brunk HD (1972) *Statistical Inference under Order Restriction: Theory and Application of Isotonic Regression.* John Wiley & Sons, Inc., New York.

Barnard GA (1945a) A new test for 2 × 2 tables. *Nature,* **156**:177.

Barnard GA (1945b) A new test for 2 × 2 tables. *Nature,* **156**:783–784.

Barnard GA (1947) Significance tests for 2 × 2 tables. *Biometrika,* **34**:123–138.

Bartholomew DJ (1959a) A test for homogeneity for ordered alternatives. *Biometrika,* **46**:36–48.

Bartholomew DJ (1959b) A test for homogeneity for ordered alternatives. II. *Biometrika,* **46**:328–335.

Bender R (2001) Calculating confidence intervals for the number needed to treat. *Controlled Clinical Trials,* **22**:102–110.

Bender R, Lange S (2001) Adjusting for multiple testing—when and how? *Journal of Clinical Epidemiology,* **54**:343–349.

Benjamini Y, Hochberg Y (1995) Controlling the false discovery rate: A practical and powerful approach to multiple testing. *Journal of the Royal Statistical Society, Series B,* **57**:289–300.

Bentur L, Lapidot M, Livnat G, Hakim F, Lidroneta-Katz C, Porat I, Vilozni D, Elhasid R (2009) Airway reactivity in children before and after stem cell transplantation. *Pediatric Pulmonology,* **44**:845–850.

Berger RL (1996) More powerful tests from confidence interval p values. *The American Statistician*, **50**:314–318.

Berger RL, Boos DD (1994) P-values maximized over a confidence set for the nuisance parameter. *Journal of the American Statistical Association*, **89**:1012–1016.

Berry G, Armitage P (1995) Mid-p confidence intervals: A brief review. *The Statistician*, **44**:417–423.

Bhapkar VP (1965) Categorical data analogs of some multivariate tests. Institute of Statistics Mimeo Series No. 450, University of North Carolina and University of Poona.

Bhapkar VP (1966) A note on the equivalence of two test criteria for hypotheses in categorical data. *Journal of the American Statistical Society*, **61**:228–235.

Birch MW (1965) The detection of partial association, II: the general case. *Journal of the Royal Statistical Society, Series B*, **27**:111–124.

Blaker H (2000) Confidence curves and improved exact confidence intervals for discrete distributions. *The Canadian Journal of Statistics*, **28**:783–798.

Blyth CR, Still HA (1983) Binomial confidence intervals. *Journal of the American Statistical Association*, **78**:108–116.

Bofin AM, Lydersen S, Hagmar BM (2004) Cytological criteria for the diagnosis of intraductal hyperplasia, ductal carcinoma in situ, and invasive carcinoma of the breast. *Diagnostic Cytopathology*, **31**:207–215.

Bonett DG, Price RM (2006) Confidence intervals for a ratio of binomial proportions based on paired data. *Statistics in Medicine*, **25**:3039–3047.

Bonett DG, Price RM (2012) Adjusted Wald confidence interval for a difference of binomial proportions based on paired data. *Journal of Educational and Behavioral Statistics*, **37**:479–488.

Bonett DG, Wright TA (2000) Sample size requirements for estimating Pearson, Kendall and Spearman correlations. *Biometrika*, **65**:23–28.

Borenstein M, Hedges LV, Higgins JPT, Rothstein HR (2010) A basic introduction to fixed effects and random effects models for meta-analysis. *Research Synthesis Methods*, **1**:97–111.

Boschloo RD (1970) Raised conditional level of significance for the 2×2-table when testing the equality of two probabilities. *Statistica Neerlandica*, **24**:1–9.

Bowker AH (1948) A test for symmetry in contingency tables. *Journal of the American Statistical Association*, **43**:572–574.

Brant R (1990) Assessing proportionality in the proportional odds model for ordinal logistic regression. *Biometrics*, **46**:1171–1178.

Breidablik HJ, Meland E, Lydersen S (2008) Self-rated health during adolescence: Stability and predictors of change (Young-HUNT study, Norway). *European Journal of Public Health*, **19**:73–78.

Breslow N (1981) Odds ratio estimators when the data are sparse. *Biometrika*, **68**:73–84.

Breslow NE (1996) Statistics in epidemiology: The case-control study. *Journal of the American Statistical Association*, **91**:14–28.

Breslow NE, Day NE (1980) *Statistical Methods in Cancer Research. Volume 1—The Analysis of Case-Control Studies*. International Agency for Research on Cancer, Lyon.

Bross IDJ (1958) How to use ridit analysis. *Biometrics*, **14**:18–38.

Brown LD, Cai T, DasGupta A (2001) Interval estimation for a binomial proportion. *Statistical Science*, **16**:101–133.

Brown PJ, Payne CD (1986) Aggregate data, ecological regression, and voting transitions. *Journal of the American Statistical Association*, **81**:452–460.

Brunk HD, Franck WE, Hanson DL, Hogg RV (1966) Maximum likelihood estimation of the distributions of two stochastically ordered random variables. *Journal of the American Statistical Association*, **61**:1067–1080.

Brunswick AF (1971) Adolescent health, sex and fertility. *American Journal of Public Health*, **61**:711–729.

Buonaccorsi JP, Laake P, Veierød MB (2014) On the power of the Cochran-Armitage test for trend in the presence of misclassification. *Statistical Methods in Medical Research*, **23**:218–243.

Campbell MJ, Julious SA, Altman DG (1995) Estimating sample size for binary, ordered categorical, and continuous outcomes in two group comparisons. *BMJ*, **311**:1145–1148.

Capuano AW, Dawson JD (2013) The trend odds model for ordinal data. *Statistics in Medicine*, **32**:2250–2261.

Casagrande JT, Pike MC, Smith PG (1978) An improved approximate formula for calculating sample size for comparing two binomial distributions. *Biometrics*, **34**:483–486.

Casella G (1986) Refining binomial confidence intervals. *The Canadian Journal of Statistics*, **14**:113–129.

Cavo M, Pantani L, Petrucci MT, Patriarca F, Zamagni E, Donnarumma D, Crippa C, Boccadoro M, Perrone G, Falcone A, Nozzoli C, Zambello R, Masini L, Furlan A, Brioli A, Derudas D, Ballanti S, Dessanti ML, De Stefano V, Carella AM, Marcatti M, Nozza A, Ferrara F, Callea V, Califano C, Pezzi A, Baraldi A, Grasso M, Musto P, Palumno A (2012) Bortezomib-Thalidomide-Dexamethasone is superior to Thalidomide-Dexamethasone as consolidation therapy after autologous hematopoietic stem cell transplantation in patients with newly diagnosed multiple myeloma. *Blood*, **120**:9–19.

Chacko VJ (1963) Testing homogeneity against ordered alternatives. *The Annals of Mathematical Statistics*, **34**:945–956.

Chacko VJ (1966) Modified chi-square test for ordered alternatives. *Sankhyā: The Indian Journal of Statistics, Series B*, **28**:185–190.

Chan ISF, Zhang Z (1999) Test-based exact confidence intervals for the difference of two binomial proportions. *Biometrics*, **55**:1202–1209.

Cheng PE, Liou M, Aston JAD (2010) Likelihood ratio test with three-way tables. *Journal of the American Statistical Association*, **105**:740–749.

Chow SH, Shao J, Wang H (2008) *Sample Size Calculations in Clinical Research*. 2nd edn. Chapman & Hall/CRC, Boca Raton, FL.

Cichetti DV, Feinstein AR (1990) High agreement but low kappa: II Resolving the paradoxes. *Journal of Clinical Epidemiology*, **43**:551–558.

Cochran WG (1954a) The combination of estimates from different experiments. *Biometrics*, **10**:101–129.

Cochran WG (1954b) Some methods for strengthening the common χ^2 tests. *Biometrics*, **10**:417–451.

Cochran WG, Carroll SP (1953) A sampling investigation on the efficiency of weighting inversely as the estimated variance. *Biometrics*, **9**:447–459.

Cohen A, Madigan D, Sackrowitz HB (2003) Effective directed tests for models with ordered categorical data. *Australian & New Zealand Journal of Statistics*, **45**:285–300.

Cohen J (1960) A coefficient of agreement for nominal scales. *Educational and Psychological Measurement*, **20**:37–46.

Cohen J (1988) *Statistical Power Analysis for the Behavioral Sciences*. 2nd edn. Lawrence Earlbaum, New Jersey.

Cole TJ, Bellizzi MC, Flegal KM, Dietz WH (2000) Establishing a standard definition of child overweight and obesity worldwide: International survey. *BMJ*, **320**:1240–1243.

Cole TJ, Flegal KM, Nicholls D, Jackson AA (2007) Body mass index cut offs to define thinness in children and adolescents: International survey. *BMJ*, **335**:194.

Colombi R, Forcina A (2016) Testing order restrictions in contingency tables. *Metrika*, **79**:73–90.

Conger AJ (1980) Integration and generalization of kappas for multiple raters. *Psychological Bulletin*, **88**:322–328.

Cook RJ, Sackett DL (1995) The number needed to treat: A clinically useful measure of treatment effect. *BMJ*, **310**:452–454.

Cornfield J (1956) A statistical problem arising from retrospective studies. *Proceedings of the Third Berkeley Symposium on Mathematical Statistics and Probability*, **4**:135–148.

Cox DR (1966) A simple example of a comparison involving quantal data. *Biometrika*, **53**:215–220.

Cox DR (1970) *The Analysis of Binary Data*. Chapman & Hall, London.

Cox DR, Reid N (1987) Parameter orthogonality and approximate conditional inference (with discussion). *Journal of the Royal Statistical Society, Series B*, **49**:1–39.

Cressie N, Read TRC (1989) Pearson's X^2 and the loglikelihood ratio statistic G^2: A comparative review. *International Statistical Review*, **57**:19–43.

Crow EL (1956) Confidence intervals for a proportion. *Biometrika*, **43**:423–435.

Dai S, Li X (2015) China's ratio of male to female babies remains high despite sixth annual fall. *BMJ*, **350**:h937.

Davis CS, Chung Y (1995) Randomization model methods for evaluating treatment efficacy in multicenter clinical trials. *Biometrics*, **51**:1163–1174.

Day NE, Byar DP (1979) Testing hypotheses in case-control studies— equivalence of Mantel-Haenszel statistic and logit score test. *Biometrics*, **35**:623–630.

De Vet HC, Mokkink LB, Terwee CB, Hoekstra OS, Knol DL (2013) Clinicians are right not to like Cohen's kappa. *BMJ*, **346**:f2125.

Deeks JJ, Altman DG, Bradburn MJ (2001) Statistical methods for examining heterogeneity and combining results from several studies in meta-analysis. In M Egger, G Davey Smith, DG Altman (eds.) *Systematic Reviews in Health Care: Meta-Analysis in Context*, 2nd edn. BMJ Books, London.

DerSimonian R, Kacker R (2007) Random-effects model for meta-analysis of clinical trials: An update. *Contemporary Clinical Trials*, **28**:105–114.

DerSimonian R, Laird N (1986) Meta-analysis in clinical trials. *Controlled Clinical Trials*, **7**:177–188.

Desu MM, Raghavarao D (2004) *Nonparametric Statistical Methods for Complete and Censored Data*. Chapman & Hall/CRC, Boca Raton, FL.

Dice LR (1945) Measures of the amount of ecologic association between species. *Ecology*, **26**:297–302.

Dmitrienko A, D'Agostino R (2013) Traditional multiplicity adjustment methods in clinical trials. *Statistics in Medicine*, **32**:5172–5218.

Doll R, Hill AB (1950) Smoking and carcinoma of the lung. *British Medical Journal*, **2**:739–748.

Donner A, Zou GY (2012) Closed-form confidence intervals for functions of the normal mean and standard deviation. *Statistical Methods in Medical Research*, **21**:347–359.

Dumville JC, Torgerson DJ, Hewitt CE (2006) Reporting attrition in randomised controlled trials. *BMJ*, **332**:969–971.

Duncan OD, Davis B (1953) An alternative to ecological correlation. *American Sociological Review*, **18**:665–666.

Edwards AL (1948) Note on the "correction for continuity" in testing the significance of the difference between correlated proportions. *Psychometrika*, **13**:185–187.

Efron B, Hinkley DV (1978) Assessing the accuracy of the maximum likelihood estimator: Observed versus expected Fisher information. *Biometrika*, **65**:457–482.

Egger M, Smith GD, Altman DG (eds.) (2001) *Systematic Reviews in Health Care: Meta-Analysis in Context*. 2nd edn. BMJ Books, London.

Egger M, Smith GD, Phillips AN (1997a) Meta-analysis: principles and procedures. *BMJ*, **315**:1533–1537.

Egger M, Smith GD, Schneider M, Minder C (1997b) Bias in meta-analysis detected by a simple, graphical test. *BMJ*, **315**:629–634.

Elvrum AKG, Beckung E, Sæther R, Lydersen S, Vik T, Himmelmann K (2017) Bimanual capacity of children with cerebral palsy: intra- and inter-rater reliability of a revised edition of the Bimanual Fine Motor Function classification. *Physical & Occupational Therapy in Pediatrics*, **37**:239–251.

Emerson JD (1994) Combining estimates of the odds ratio: the state of the art. *Statistical Methods in Medical Research*, **3**:157–178.

Everitt B (1992) *The Analysis of Contingency Tables*. 2nd edn. Chapman & Hall/CRC, London.

Ezra DG, Beaconsfield M, Sira M, Bunce C, Wormald R, Collin R (2010) The associations of floppy eyelid syndrome: A case control study. *Ophthalmology*, **117**:831–838.

Fagerland MW (2012) Exact and mid-p confidence intervals for the odds ratio. *The Stata Journal*, **12**:505–514.

Fagerland MW (2014) adjcatlogit, ccrlogit, and ucrlogit: Fitting ordinal logistic regression models. *The Stata Journal*, **14**:947–964.

Fagerland MW (2015) Evidence-Based Medicine and Systematic Reviews. In P Laake, H Benestad, BR Olsen (eds.) *Research in Medical and Biological Sciences*. Academic Press/Elsevier, London.

Fagerland MW, Hosmer DW (2013) A goodness-of-fit test for the proportional odds regression model. *Statistics in Medicine*, **32**:2235–2249.

Fagerland MW, Lydersen S, Laake P (2013) The McNemar test for binary matched-pairs data: Mid-p and asymptotic are better than exact conditional. *BMC Medical Research Methodology*, **13**:91.

Fagerland MW, Lydersen S, Laake P (2014) Recommended tests and confidence intervals for paired binomial proportions. *Statistics in Medicine*, **33**:2850–2875.

Fagerland MW, Lydersen S, Laake P (2015) Recommended confidence intervals for two independent binomial proportions. *Statistical Methods in Medical Research*, **24**:224–254.

Fagerland MW, Newcombe RG (2013) Confidence intervals for odds ratio and relative risk based on the inverse hyperbolic sine transformation. *Statistics in Medicine*, **32**:2823–2836.

Fan C, Zhang D, Zhang CH (2011) On sample size of the Kruskal-Wallis test with application to a mouse peritoneal cavity study. *Biometrics*, **67**:213–224.

Farrington CP, Manning G (1990) Test statistics and sample size formulae for comparative binomial trials with null hypothesis of non-zero risk difference or non-unity relative risk. *Statistics in Medicine*, **9**:1447–1454.

Feinstein AR, Cichetti DV (1990) High agreement but low kappa: I The problem of two paradoxes. *Journal of Clinical Epidemiology*, **43**:543–549.

Fieller EC, Hartley HO, Pearson ES (1957) Tests for rank correlation coefficients. *Biometrika,* **44**:470–481.

Fienberg SE (1972) Analysis of incomplete multi-way contingency tables. *Biometrics,* **28**:177–202.

Fienberg SE (2007) *The Analysis of Cross-Classified Categorical Data.* 2nd edn. Springer, New York.

Fischer D, Stewart AL, Bloch DA, Lorig K, Laurent D, Holman H (1999) Capturing the patients' view of change as a clinical outcome measure. *Journal of the American Medical Association,* **282**:1157–1162.

Fisher RA (1937) *The Design of Experiments.* 2nd edn. Oliver and Boyd, Edinburgh.

Fisher RA (1945) A new test for 2 × 2 tables. *Nature,* **156**:388.

Fleiss JL (1971) Measuring nominal scale agreement among many raters. *Psychological Bulletin,* **76**:378–382.

Fleiss JL (1993) The statistical basis of meta-analysis. *Statistical Methods in Medical Research,* **2**:121–145.

Fleiss JL, Cohen J (1973) The equivalence of weighted kappa and the intraclass correlation coefficient as measures of reliability. *Educational and Psychological Measurements,* **33**:613–619.

Fleiss JL, Cohen J, Everitt BS (1969) Large sample standard errors of kappa and weighted kappa. *Psychological Bulletin,* **72**:323–327.

Fleiss JL, Everitt BS (1971) Comparing the marginal totals of square contingency tables. *British Journal of Mathematical Statistics and Psychology,* **24**:117–123.

Fleiss JL, Levin B, Paik MC (2003) *Statistical Methods for Rates and Proportions.* 3rd edn. John Wiley & Sons, Inc., Hoboken, NJ.

Fontanella CA, Early TJ, Phillips G (2008) Need or availability? Modeling aftercare decisions for psychiatrically hospitalized adolescents. *Children and Youth Services Review,* **30**:758–773.

Gart JJ (1966) Alternative analyses of contingency tables. *Journal of the Royal Statistical Society, Series B,* **28**:164–179.

Gart JJ (1970) Point and interval estimation of the common odds ratio in combination of 2 × 2 tables with fixed marginal. *Biometrika,* **57**:471–475.

Gart JJ (1971) The comparison of proportions: A review of significance tests, confidence intervals and adjustment for stratification. *Review of the International Statistical Institute,* **39**:148–169.

Gart JJ, Nam J (1988) Approximate interval estimation of the ratio of binomial parameters: A review and correction for skewness. *Biometrics*, **44**:323–338.

Gavaghan DJ, Moore RA, McQuay HJ (2000) An evaluation of homogeneity tests in meta-analysis in pain using simulations of individual patient data. *Pain*, **85**:415–424.

Gisev N, Bell JS, Chen TF (2013) Interrater agreement and interrater reliability: key concepts, approaches, and applications. *Research in Social and Administrative Pharmacy*, **9**:330–338.

Glass DV (1954) *Social Mobility in Britain*. Free Press, Glencoe, IL.

Glynn A, Wakefield J (2010) Ecological inference in the social sciences. *Statistical Methodology*, **7**:307–322.

Gold ZR (1963) Tests auxiliary to χ^2 tests in a Markov chain. *The Annals of Mathematical Statistics*, **34**:57–74.

Good IJ (1956) On the estimation of small frequencies in contingency tables. *Journal of the Royal Statistical Society, Series B*, **18**:113–124.

Goodman LA (1953) Ecological regressions and behavior of individuals. *American Sociological Review*, **18**:663–664.

Goodman LA (1959) Some alternatives to ecological correlation. *American Journal of Sociology*, **64**:610–625.

Goodman LA (1964) Simultaneous confidence limits for contrasts among multinomial populations. *Annals of Mathematical Statistics*, **37**:716–725.

Goodman LA (1965) On simultaneous confidence intervals for multinomial parameters. *Technometrics*, **6**:191–195.

Goodman LA, Kruskal WH (1954) Measures of association for cross-classifications. *Journal of the American Statistical Association*, **49**:732–764.

Goodman LA, Kruskal WH (1979) *Measures of Association for Cross Classifications*. Springer, New York.

Graham P, Bull B (1998) Approximate standard errors and confidence intervals for indices of positive and negative agreement. *Journal of Clinical Epidemiology*, **51**:763–771.

Graubard BI, Korn EL (1987) Choice of column scores for testing independence in ordered $2 \times K$ contingency tables. *Biometrics*, **43**:471–476.

Greenland S (1989) Generalized Mantel-Haenszel estimators for K $2 \times J$ tables. *Biometrics*, **45**:183–191.

Greenland S (1995) Avoiding power loss associated with categorization and ordinal scores in dose-response and trend analysis. *Epidemiology*, **6**:450–454.

Greenland S (2000) Small-sample bias and corrections for conditional maximum-likelihood odds-ratio estimators. *Biostatistics*, **1**:113–122.

Greenland S, Robins JM (1985) Estimation of a common effect parameter from spares follow-up data. *Biometrics*, **41**:55–68.

Greenland S, Rothman KJ (2008) Introduction to stratified analysis. In KJ Rothman, S Greenland, TL Lash (eds.) *Modern Epidemiology*, 3rd edn. Lippincott Williams & Wilkins, Philadelphia.

Greenland S, Salvan A (1990) Bias in the one-step method for pooling study results. *Statistics in Medicine*, **9**:247–252.

Grove DM (1980) A test of independence against a class of ordered alternatives in a $2 \times c$ contingency table. *Journal of the American Statistical Association*, **75**:454–459.

Gwet KL (2008) Computing inter-rater reliability and its variance in the presence of high agreement. *British Journal of Mathematical and Statistical Psychology*, **61**:29–48.

Gwet KL (2014) *Handbook of Inter-Rater Reliability*. 4th edn. Advanced Analytics, LLC, Gaithersburg, Maryland.

Haberman SJ (1973) The analysis of residuals in cross-classified tables. *Biometrics*, **29**:205–220.

Haberman SJ (1982) Analysis of dispersion of multinomial responses. *Journal of the American Statistical Association*, **379**:568–580.

Haldane JBS (1954) An exact test for randomness of mating. *Journal of Genetics*, **52**:631–635.

Haldane JBS (1956) The estimation and significance of the logarithm of a ratio of frequencies. *Annals of Human Genetics*, **20**:309–311.

Halperin M, Waer JH, Byar DP, Mantel N, Brown CC, Koziol J, Gail M, Green SB (1977) Testing for interaction in an $I \times J \times K$ contingency table. *Biometrika*, **64**:271–275.

Harville DA (1977) Maximum likelihood approaches to variance component estimation and to related problems. *Journal of the American Statistical Association*, **72**:320–338.

Harville DA (1997) *Matrix Algebra from a Statistician's Perspective*. Springer, New York.

Hauck WW (1984) A comparative study of the conditional maximum likelihood estimation of a common odds ratio. *Biometrics*, **40**:1117–1123.

Hellevik O (2009) Linear versus logistic regression when the dependent variable is a dichotomy. *Quality & Quantity*, **43**:59–74.

Higgins JPT, Altman DG, Sterne JAC (2011) Assessing risk of bias in included studies. In JPT Higgins, S Green (eds.) *Cochrane Handbook for Systematic Reviews of Interventions*. The Cochrane Collaboration. Version 5.1.0 [Updated March 2011], Available from: `http://training.cochrane.org/handbook`.

Higgins JPT, Thompson SG (2002) Quantifying heterogeneity in a meta-analysis. *Statistics in Medicine*, **21**:1539–1558.

Higgins JPT, Thompson SG, Deeks JJ, Altman DG (2003) Measuring inconsistency in meta-analyses. *BMJ*, **327**:557–560.

Higgins JPT, Thompson SG, Spiegelhalter DJ (2009) A re-evaluation of random effects meta-analysis. *Journal of the Royal Statistical Society, Series A*, **172**:137–159.

Higgins JPT, Whitehead A, Turner RM, Omar RZ, Thompson SG (2001) Meta-analysis of continuous outcome data from individual patients. *Statistics in Medicine*, **20**:2219–2241.

Hine LK, Laird N, Hewitt P, Chalmers TC (1989) Meta-analytic evidence against prophylactic use of lidocaine in myocardial infarction. *Archives of Internal Medicine*, **149**:2694–2698.

Hirji KF (2006) *Exact Analysis of Discrete Data*. Chapman & Hall/CRC, Boca Raton, FL.

Hirji KM, Fagerland MW, Veierød M (2012) Meta-analysis. In M Veierød, S Lydersen, P Laake (eds.) *Medical Statistics in Clinical and Epidemiological Studies*. Gyldendal akademisk, Oslo.

Hitchcock DB (2009) Yates and contingency tables: 75 years later. *Electronic Journal for History of Probability and Statistics*, **5**:1–14.

Hjort NL (1988) On large-sample multiple comparison methods. *Scandinavian Journal of Statistics*, **15**:259–271.

Hoenig JM, Heisey DM (2001) The abuse of power: the pervasive fallacy of power calculations for data analysis. *The American Statistician*, **55**:19–24.

Holgado-Tello FP, Chacon-Moscoso S, Barbero-Garcia I, Vila-Abad E (2010) Polychoric versus Pearson correlations in exploratory and confirmatory factor analysis of ordinal variables. *Quality & Quantity*, **44**:153–166.

Hollander M, Wolfe DA, Chicken E (2014) *Nonparametric Statistical Methods.* 3rd edn. John Wiley & Sons, Inc., Hoboken, NJ.

Hosmer DW, Lemeshow S, Sturdivant RX (2013) *Applied Logistic Regression.* 3rd edn. John Wiley & Sons, Inc., Hoboken, NJ.

Hsueh HM, Liu JP, Chen JJ (2001) Unconditional exact tests for equivalence or noninferiority for paired binary endpoints. *Biometrics*, **57**:478–483.

Hwang JTG, Yang MC (2001) An optimality theory for mid p-values in 2×2 contingency tables. *Statistica Sinica*, **11**:807–826.

Iannario M, Lang JB (2016) Testing conditional independence in sets of $I \times J$ tables by means of moments and correlation score tests with application to HPV vaccine. *Statistics in Medicine*, **35**:4573–4587.

Imai K, Lu Y, Strauss A (2008) Bayesian and likelihood inference for 2×2 ecological tables: an incomplete-data approach. *Political Analysis*, **16**:41–69.

Imai K, Lu Y, Strauss A (2011) Eco: R package for ecological inference in 2×2 tables. *Journal of Statistical Software*, **42**.

Indredavik B, Rothweder G, Naalsund E, Lydersen S (2008) Medical complications in a comprehensive stroke unit and early supported discharge service. *Stroke*, **39**:414–420.

Ireland CT, Ku HH, Kullback S (1969) Symmetry and marginal homogeneity of an $r \times r$ contingency table. *Journal of the American Statistical Association*, **64**:1323–1341.

Irwin JO (1935) Tests of significance for differences between percentages based on small numbers. *Metron*, **12**:83–94.

Ivanova A, Berger VW (2001) Drawbacks to integer scoring for ordered categorical data. *Biometrics*, **57**:567–570.

Jha P, Kumar R, Vasa P, Dhingra N, Thiruchelvam D, Moineddin R (2006) Low male-to-female sex ratio of children born in India: National survey of 1.1 million households. *Lancet*, **367**:211–218.

Johnson WD, May WL (1995) Combining 2×2 tables that contain structural zeros. *Statistics in Medicine*, **14**:1901–1911.

Jonckheere AR (1954) A distribution-free k-sample test against ordered alternatives. *Biometrika*, **41**:133–145.

Jullumstrø E, Lydersen S, Møller B, Dahl O, Edna TH (2009) Duration of symptoms, stage at diagnosis and relative survival in colon and rectal cancer. *European Journal of Cancer*, **45**:2383–2390.

Kang SH, Ahn CW (2008) Tests for the homogeneity of two binomial proportions in extremely unbalanced 2 × 2 contingency tables. *Statistics in Medicine*, **27**:2524–2535.

Katz D, Baptista J, Azen SP, Pike MC (1978) Obtaining confidence intervals for the risk ratio in cohort studies. *Biometrics*, **34**:469–474.

Kendall MG (1945) The treatment of ties in rank problems. *Biometrika*, **33**:239–251.

King G (1997) *A Solution to the Ecological Inference Problem.* Princeton University Press, New Jersey.

King G, Rosen O, Tanner AM (1999) Binomial-beta hierarchical models for ecological inference. *Sociological Methods & Research*, **28**:61–90.

King G, Rosen O, Tanner AM (eds.) (2004a) *Ecological Inference: New Methodological Strategies.* Cambridge University Press, Cambridge.

King G, Rosen O, Tanner AM (2004b) Information in ecological inference: An introduction. In G King, O Rosen, AM Tanner (eds.) *Ecological Inference: New Methodological Strategies.* Cambridge University Press, Cambridge.

Kirk RE (2013) *Experimental Design. Procedures for the Behavioral Sciences.* 4th edn. Sage Publications, Thousand Oaks, CA.

Kleinbaum DG, Klein M (2010) *Logistic Regression: A Self-Learning Text.* 3rd edn. Springer, New York.

Kleinbaum DG, Kupper LL, Nizam A, Rosenberg ES (2014) *Applied Regression Analysis and Other Multivariable Methods.* 5th edn. Cengage Learning, Boston, MA.

Koehler KJ (1986) Goodness-of-fit tests for log-linear models in sparse contingency tables. *Journal of the American Statistical Association*, **81**:483–493.

Koopman PAR (1984) Confidence intervals for the ratio of two binomial proportions. *Biometrics*, **40**:513–517.

Kraemer HC (2006) Correlation coefficients in medical research: From product moment correlation to the odds ratio. *Statistical Methods in Medical Research*, **15**:525–545.

Lachin JM (2011) *Biostatistical Methods: The Assessment of Relative Risks.* 2nd edn. John Wiley & Sons, Inc., Hoboken, New Jersey.

Lampasona V, Passerini L, Barzaghi F, Lombardoni C, Bazzigaluppi E, Brigatti C, Bacchetta R, Bosi E (2013) Autoantibodies to harmonin and villin are diagnostic markers with IPEX syndrome. *PLOS ONE*, **8**:e78664.

Lancaster HO (1961) Significance tests in discrete distributions. *Journal of the American Statistical Association*, **56**:223–234.

Landis JR, Heyman ER, Koch GG (1978) Average partial association in three-way contingency tables: A review and discussion of alternative tests. *International Statistical Review*, **46**:237–254.

Landis JR, Koch GG (1977) The measurement of observer agreement for categorical data. *Biometrics*, **33**:159–174.

Landis JR, Sharp TJ, Kurits SJ, Koch GG (2005) Mantel-Haenszel methods. In P Armitage, T Colton (eds.) *Encyclopedia of Biostatistics*, 2nd edn. John Wiley & Sons, Inc., Chichester.

Lang JB (1996) Maximum likelihood methods for a generalized class of log-linear models. *Annals of Statistics*, **24**:726–752.

Larntz K (1978) Small-sample comparisons of exact levels for chi-squared goodness-of-fit statistics. *Journal of the American Statistical Association*, **73**:253–263.

Lau J, Antman EM, Jimenez-Silva J, Kupelnick B, Mosteller F, Chalmers TC (1992) Cumulative Meta-Analysis of Therapeutic Trials for Myocardial Infarction. *New England Journal of Medicine*, **327**:248–254.

Laupacis A, Sackett DL, Roberts RS (1988) An assessment of clinically useful measures of the consequences of treatment. *New England Journal of Medicine*, **318**:1728–1733.

Lehmann EL (1975) *Nonparametrics: Statistical Methods Based on Ranks*. Holden-Day, Inc., San Francisco, CA.

Lehmann EL, Casella G (1998) *Theory of Point Estimation*. Springer, New York.

Lehmann EL, Romano JP (2005) *Testing Statistical Hypothesis*. 3rd edn. Springer, New York.

Levin JR, Serlin RC, Seaman MA (1994) A controlled, powerful multiple-comparison strategy for several situations. *Psychological Bulletin*, **115**:153–159.

Li HQ, Tian GL, Jiang XJ, Tang NS (2016) Testing hypothesis for a simple ordering in incomplete contingency tables. *Computational Statistics and Data Analysis*, **99**:25–37.

Liang KY, Self SG (1985) Tests of homogeneity of odds ratios when the data are sparse. *Biometrika*, **72**:353–358.

Liebetrau AM (1983) *Measures of Association*. Sage Publications, Inc., Thousand Oaks, California.

Ligaarden SC, Axelsson L, Naterstad K, Lydersen S, Farup PG (2010) A candidate probiotic with unfavourable effects in subjects with irritable bowel syndrome: A randomised controlled trial. *BMC Gastroenterology*, **10**:16.

Lippert T, Skjærven R, Salvesen KÅ (2005) Why do some women only give birth to children of one sex (In Norwegian). *Tidsskrift for den norske legeforening*, **125**:3414–3417.

Lipsitz SR, Fitzmaurice GM (1996) The score test for independence in $R \times C$ contingency tables with missing data. *Biometrics*, **52**:751–762.

Little RJA, Rubin DB (2002) *Statistical Analysis with Missing Data*. 2nd edn. John Wiley & Sons, Inc., Hoboken, New Jersey.

Liu IM, Agresti A (1996) Mantel-Haenszel-type inference for cumulative odds ratios with a stratified ordinal response. *Biometrics*, **52**:1223–1234.

Liu JP, Hsueh MH, Hsieh E, Chen JJ (2002) Tests for equivalence or non-inferiority for paired binary data. *Statistics in Medicine*, **21**:231–245.

Lloyd CJ (2008) A new exact and more powerful unconditional test of no treatment effect from binary matched pairs. *Biometrics*, **64**:716–723.

Long JS (1997) *Regression Methods for Categorical and Limited Dependent Variables*. Sage Publications, Inc., Thousand Oaks, CA.

Lui KJ (1998) Interval estimation of the risk ratio between a secondary infection, given a primary infection, and the primary infection. *Biometrics*, **54**:706–711.

Lui KJ (2000) Confidence intervals of the simple difference between the proportions of a primary infection and a secondary infection, given the primary infection. *Biometrical Journal*, **42**:59–69.

Lui KJ (2004) *Statistical Estimation of Epidemiological Risk*. John Wiley & Sons, Inc., Hoboken, NJ.

Lui KJ, Lin CH (2014) Notes on test homogeneity of the odds ratio in matched pairs under stratified sampling: A Monte Carlo evaluation. *Communications in Statistics—Simulation and Computation*, **43**:2403–2414.

Lund LK, Vik T, Skranes J, Lydersen S, Brubakk AM, Indredavik MS (2012) Low birth weight and psychiatric morbidity; stability and change between adolescence and young adulthood. *Early Human Development*, **88**:623–629.

Lydersen S (2012) Diagnostic tests, ROC curves, and measures of agreement. In M Veierød, S Lydersen, P Laake (eds.) *Medical Statistics in Clinical and Epidemiological Studies*. Gyldendal akademisk, Oslo.

Lydersen S, Fagerland MW, Laake P (2009) Recommended tests for association in 2 × 2 tables. *Statistics in Medicine*, **28**:1159–1175.

Lydersen S, Fagerland MW, Laake P (2012a) Categorical data and contingency tables. In M Veierød, S Lydersen, P Laake (eds.) *Medical Statistics in Clinical and Epidemiological Studies*. Gyldendal akademisk, Oslo.

Lydersen S, Laake P (2003) Power comparison of two-sided exact tests for association in 2×2 contingency tables using standard, mid-p, and randomized test versions. *Statistics in Medicine*, **22**:3859–3871.

Lydersen S, Langaas M, Bakke Ø (2012b) The exact unconditional z-pooled test for equality of two binomial probabilities: Optimal choice of the Berger and Boos confidence coefficient. *Journal of Statistical Computation and Simulation*, **82**:1311–1316.

Lydersen S, Pradhan V, Senchaudhuri P, Laake P (2005) Comparison of exact tests for association in unordered contingency tables using standard, mid-p, and randomized test versions. *Journal of Statistical Computation and Simulation*, **75**:447–458.

Lydersen S, Pradhan V, Senchaudhuri P, Laake P (2007) Choice of test for association in small sample unordered $r \times c$ tables. *Statistics in Medicine*, **26**:4328–4343.

Macaskill P, Walter SD, Irwig L, Franco EL (2002) Assessing the gain in diagnostic performance when combining two diagnostic tests. *Statistics in Medicine*, **21**:2527–2546.

MacCallum RC, Zhang S, Preacher KJ, Rucker DD (2002) On the practice of dichotomization of quantitative variables. *Psychological Methods*, **7**:19–40.

Machin D, Campbell MJ, Tan SB, Tan SH (2009) *Sample Size Tables for Clinical Studies*. 3rd edn. John Wiley & Sons, Inc., Chichester, West Sussex.

Mack TM, Pike MC, Henderson BE, Pfeffer RI, Gerkins VR, Arthur M, Brown SE (1976) Estrogens and endometrial cancer in a retirement community. *New England Journal of Medicine*, **294**:1262–1267.

Mangerud WL, Bjerkeset O, Lydersen S, Indredavik MS (2014) Physical activity in adolescents with psychiatric disorders and in the general population. *Child and Adolescence Psychiatry and Mental Health*, **8**:2.

Mantel N (1963) Chi-square tests with one degree of freedom: Extensions of the Mantel-Haenszel procedure. *Journal of the American Statistical Association*, **58**:690–700.

Mantel N, Byar DP (1978) Marginal homogeneity, symmetry, and independence. *Communications in Statistics—Theory and Methods*, **7**:953–976.

Mantel N, Fleiss JL (1980) Minimum expected cell size requirements for the Mantel-Haenszel one-degree-of-freedom chi-squared test and a related rapid procedure. *American Journal of Epidemiology*, **112**:129–134.

Mantel N, Haenszel W (1959) Statistical aspects of the analysis of data from retrospective studies of disease. *Journal of the National Cancer Institute*, **22**:719–748.

Martin D, Austin H (1991) An efficient program for computing conditional maximum likelihood estimates and exact confidence limits for a common odds ratio. *Epidemiology*, **2**:359–362.

Maxwell AE (1970) Comparing the classification of subjects by two independent judges. *British Journal of Psychiatry*, **116**:651–655.

McCullagh P (1980) Regression models for ordinal data. *Journal of the Royal Statistical Society, Series B*, **42**:109–142.

McCullagh P, Nelder JA (1989) *Generalized Linear Models*. 2nd edn. Chapman & Hall/CRC, Boca Raton, FL.

McKelvey RD, Zavoina W (1975) A statistical model for the analysis of ordinal level dependent variables. *Journal of Mathematical Sociology*, **4**:103–120.

McNemar Q (1947) Note on the sampling error of the difference between correlated proportions or percentages. *Psychometrika*, **12**:153–157.

Mee RW (1984) Confidence bounds for the difference between two probabilities. *Biometrics*, **40**:1175–1176.

Mehrotra DV, Chan ISF, Berger RL (2003) A cautionary note on exact unconditional inference for a difference between two independent binomial proportions. *Biometrics*, **59**:441–450.

Mehta CR, Hillton JF (1993) Exact power of conditional and unconditional tests—going beyond the 2×2 contingency table. *The American Statistician*, **47**:91–98.

Mehta CR, Patel NR, Gray R (1985) Computing an exact confidence interval for the common odds ratio in several 2×2 tables. *Journal of the American Statistical Association*, **80**:969–973.

Mehta CR, Walsh SJ (1992) Comparison of exact, mid-p, and Mantel-Haenszel confidence intervals for the common odds ratio across several 2×2 contingency tables. *The American Statistician*, **46**:146–150.

Mercaldo ND, Lau KF, Zhou XH (2007) Confidence intervals for predictive values with an emphasis to case-control studies. *Statistics in Medicine*, **26**:2170–2183.

Meyer JP, Seaman MA (2013) A comparison of the exact Kruskal-Wallis distribution to asymptotic approximations for all sample sizes up to 105. *Journal of Experimental Education*, **81**:139–156.

Miettinen O, Nurminen M (1985) Comparative analysis of two rates. *Statistics in Medicine*, **4**:213–226.

Miller RG Jr (1981) *Simultaneous Statistical Inference.* 2nd edn. Springer, New York.

Mills JL, Graubard BI (1987) Is moderate drinking during pregnancy associated with an increased risk for malformations? *Pediatrics*, **80**:309–314.

Molenberghs G, Fitzmaurice G, Kenward MG, Tsiatis A, Verbeke G (2015) *Handbook of Missing Data Methodology.* Chapman & Hall/CRC, Boca Raton, FL.

Mukherjee B, Ahn J, Liu I, Rathouz PJ, Sánchez BN (2008) Fitting stratified proportional odds models by amalgamating conditional likelihoods. *Statistics in Medicine*, **27**:4950–4971.

Nadala EC, Goh BT, Magbanua JP, Barber P, Swain A, Alexander S, Laitila V, Michel CE, Mahilum-Tapay L, Ushiro-Lumb I, Ison C, Lee HH (2009) Performance evaluation of a new rapid urine test for chlamydia in men: prospective cohort study. *BMJ*, **339**:b2655.

Nam J (1987) A simple approximation for calculating sample sizes for detecting linear trend in proportions. *Biometrics*, **43**:701–705.

Nam J (1995) Confidence limits for the ratio of two binomial proportions based on likelihood scores: Non-iterative method. *Biometrical Journal*, **37**:375–379.

Nelder JA, Wedderburn RWN (1972) Generalized linear models. *Journal of the Royal Statistical Society, Series A*, **135**:370–384.

Neuhäuser M (2012) *Nonparametric Statistical Tests: A Computational Approach.* Chapman & Hall/CRC, Boca Raton, FL.

Newcombe RG (1998a) Improved confidence intervals for the difference between binomial proportions based on paired data. *Statistics in Medicine*, **17**:2635–2650.

Newcombe RG (1998b) Interval estimation for the difference between independent proportions: Comparison of eleven methods. *Statistics in Medicine*, **17**:873–890.

Newcombe RG (1998c) Two-sided confidence intervals for the single proportion: Comparison of seven methods. *Statistics in Medicine*, **17**:857–872.

Newcombe RG (1999) Confidence intervals for the number needed to treat—absolute risk reduction is less likely to be misunderstood. *BMJ*, **318**:1765–1767.

Newcombe RG (2001) Logit confidence intervals and the inverse sinh transformation. *The American Statistician*, **55**:200–202.

Newcombe RG (2011) Measures of location for confidence intervals for proportions. *Communications in Statistics—Theory and Methods*, **40**:1743–1767.

Newcombe RG (2013) *Confidence Intervals for Proportions and Related Measures of Effect Size*. Chapman & Hall/CRC, Boca Raton, FL.

Newcombe RG (2016) MOVER-R confidence intervals for ratios and products of two independently estimated quantities. *Statistical Methods in Medical Research*, **25**:1774–1778.

Newcombe RG, Nurminen MM (2011) In defence of score intervals for proportions and their differences. *Communications in Statistics—Theory and Methods*, **40**:1271–1282.

Normand SLT (1999) Meta-analysis: Formulating, evaluating, combining, and reporting. *Statistics in Medicine*, **18**:321–359.

Nurminen M (1981) Asymptotic efficiency of general noniterative estimators of common relative risk. *Biometrika*, **68**:525–530.

Oluyede BO (1993) A modified chi-square test for testing equality of two multinomial populations against an ordered restricted alternative. *Biometrical Journal*, **35**:997–1012.

Palmer TM, Sterne JAC (eds.) (2016) *Meta-Analysis in Stata: An Updated Collection from the Stata Journal*. 2nd edn. Stata Press, College Station, TX.

Parzen M, Lipsitz S, Metters R, Fitzmaurice GM (2010) Correlation when data are missing. *Journal of the Operational Research Society*, **61**:1049–1056.

Pearson K (1900) On the criterion that a given system of deviations from the probable in the case of a correlated system of variables is such that it can be reasonably supposed to have arisen from random sampling. *Philosophical Magazine Series 5*, **50**:157–175.

Perondi MBM, Reis AG, Paiva EF, Nadkarni VM, Berg RA (2004) A comparison of high-dose and standard-dose epinephrine in children with cardiac arrest. *New England Journal of Medicine*, **350**:1722–1730.

Peterson B, Harrell FE (1990) Partial proportional odds models for ordinal response variables. *Applied Statistics*, **39**:205–217.

Peterson J, Kaul S, Khashab M, Fisher A, Kahn JB (2007) Identification and pretherapy susceptibility of pathogens in patients with complicated urinary tract infection or acute pyelonephritis enrolled in a clinical study in the United States from November 2004 through April 2006. *Clinical Therapeutics*, **29**:2215–2221.

Pettigrew HM, Gart JJ, Thomas DG (1986) The bias and higher cumulants of the logarithm of a binomial variate. *Biometrika*, **73**:425–435.

Piaggio G, Elbourne DR, Pocock SJ, Evans SJ, Altman DG (2012) Reporting of noninferiority and equivalence randomized trials: extension of the CONSORT 2010 statement. *Journal of the American Medical Association*, **308**:2594–2604.

Pocock SJ (1983) *Clinical Trials. A Practical Approach*. John Wiley & Sons, Inc., New York.

Price RM, Bonett DG (2008) Confidence intervals for a ratio of two independent binomial proportions. *Statistics in Medicine*, **27**:5497–5508.

Pulkstenis E, Robinson TJ (2004) Goodness-of-fit tests for ordinal response regression models. *Statistics in Medicine*, **23**:999–1014.

Quesenberry CP, Hurst DC (1964) Large-sample simultaneous confidence intervals for multinomial parameters. *Technometrics*, **6**:191–195.

Rabbee N, Coull BA, Mehta C, Patel N, Senchaudhuri P (2003) Power and sample size for ordered categorical data. *Statistical Methods for Medical Research*, **12**:73–84.

Rabe-Hesketh S, Skrondal A (2012) *Multilevel and Longitudinal Modeling Using Stata*. 3rd edn. Stata Press, College Station, TX.

Rao CR (1965) *Linear Statistical Inference and its Applications*. John Wiley & Sons, Inc., New York.

Ravichandran C, Fitzmaurice GM (2008) To dichotomize or not to dichotomize? *Nutrition*, **24**:610–611.

Reis IM, Hirji KF, Afifi AA (1999) Exact and asymptotic tests for homogeneity in several 2×2 tables. *Statistics in Medicine*, **18**:893–906.

Ritland JS, Utheim TP, Utheim ØA, Espeseth T, Lydersen S, Semb SO, Rootwelt H, Elsås T (2007) Effects of APOE and CHRNA4 genotypes on retinal nerve fibre layer thickness at the optic disc and on risk for developing exfoliation syndrome. *Acta Ophthalmologica Scandinavica*, **17**:257–261.

Robertson T (1978) Testing for and against an order restriction on multinomial parameters. *Journal of the American Statistical Association*, **73**:197–202.

Robertson T, Wright FT, Dykstra RL (1988) *Order Restricted Statistical Inference*. John Wiley & Sons, Inc., Chichester.

Robins J, Greenland S, Breslow NE (1986) A general estimator for the variance of the Mantel-Haenszel odds ratio. *American Journal of Epidemiology*, **124**:719–723.

Rothman KJ, Greenland S, Lash TL (2008) *Modern Epidemiology.* 3rd edn. Lippincott Williams & Wilkins, Philadelphia, PA.

Royston P (2007) Profile likelihood for estimation and confidence intervals. *The Stata Journal*, **7**:376–387.

Royston P, Altman DG (1994) Regression using fractional polynomials of continuous covariates: Parsimonious parametric modelling. *Journal of the Royal Statistical Society, Series C*, **43**:429–467.

Royston P, Altman DG, Sauerbrei W (2006) Dichotomizing continuous predictors in multiple regression: a bad idea. *Statistics in Medicine*, **25**:127–141.

Rubin DB (1976) Inference and missing data. *Biometrika*, **63**:581–592.

Ryan TP (2013) *Sample Size Determination and Power.* John Wiley & Sons, Inc., Hoboken, New Jersey.

Ryu E, Agresti A (2008) Modeling and inference for an ordinal effect size measure. *Statistics in Medicine*, **27**:1703–1717.

Santner TJ, Snell MK (1980) Small-sample confidence intervals for $p_1 - p_2$ and p_1/p_2 in 2×2 contingency tables. *Journal of the American Statistical Association*, **75**:386–394.

Satten GA, Kupper LL (1990) Continued fraction representation for expected cell counts of a 2×2 table: A rapid and exact methods for conditional maximum likelihood estimation. *Biometrics*, **46**:217–223.

Scott WA (1955) Reliability of content analysis: the case of nominal scale coding. *The Public Opinion Quarterly*, **19**:321–325.

Senn S (2007) Drawbacks to noninteger scoring for ordered categorical data. *Biometrics*, **63**:296–298.

Shan G, Ma C (2016) Unconditional tests for comparing two ordered multinomials. *Statistical Methods in Medical Research*, **25**:241–254.

Shi L, Sun H, Bai P (2009) Bayesian confidence interval for the difference of the proportions in a 2×2 table with structural zero. *Journal of Applied Statistics*, **36**:483–494.

Sidik K (2003) Exact unconditional tests for testing non-inferiority in matched-pairs design. *Statistics in Medicine*, **22**:265–278.

Siegel S, Castellan NJ (1988) *Nonparametric statistics for the behavioral sciences.* 2nd edn. McGraw-Hill, New York.

Singh N, Pripp AH, Brekke T, Stray-Pedersen B (2010) Different sex ratios of children born to Indian and Pakistani immigrants in Norway. *BMC Pregnancy and Childbirth*, **10**:40.

Sison CP, Glaz J (1995) Simultaneous confidence intervals and sample size determination for multinomial proportions. *Journal of the American Statistical Association*, **90**:366–369.

Skinningsrud B, Lie BA, Husebye ES, Kvien TK, Førre Ø, Flatø B, Stormyr A, Joner G, Njølstad PR, Egeland T, Undlien DE (2010) A CLEC16A variant confers risk for juvenile idiopathic arthritis and anti-cyclic citrullinated peptide antibody negative rheumatoid arthritis. *Annals of the Rheumatic Diseases*, **69**:1471–1474.

Skorpen CG, Lydersen S, Gilboe IM, Skomsvoll JF, Salvesen KÅ, Palm Ø, Koksvik HSS, Jakobsen B, Wallenius M (2016) Disease activity during pregnancy and the first year postpartum in women with systematic lupus erythematosus. *Arthritis Care & Research*. Published online, doi: 10.1002/acr.23102.

Skrondal A, Rabe-Hesketh S (2004) *Generalized Latent Variable Modeling: Multilevel, Longitudinal, and Structural Equation Models*. Chapman & Hall/CRC, Boca Raton, FL.

Smyth GK (2003) Pearson's goodness of fit statistic as a score test statistic. In DR Goldstein (ed.) *Lecture Notes—Monograph Series Vol. 40, Statistics and Science: A Festschrift for Terry Speed*. Institute of Mathematical Statistics, pp. 115–126.

Snow JD (1855) *On the Mode of Communication*. New Churchill, London.

Stang A, Poole C, Bender R (2010) Common problems related to the use of number needed to treat. *Journal of Clinical Epidemiology*, **63**:820–825.

StatXact 11 (2015) *User Manual*. Cytel Software Corporation, Cambridge, MA.

Steel DG, Beh EJ, Chambers RL (2004) The information of aggregated data. In G King, O Rosen, AM Tanner (eds.) *Ecological Inference: New Methodological Strategies*. Cambridge University Press, Cambridge.

Sterne JAC, Sutton AJ, Ioannidis JPA, Terrin N, Jones DR, Lau J, Carpenter J, Rücker G, Harbord RM, Schmid CH, Tetzlaff J, Deeks JJ, Peters J, Macaskill P, Schwarzer G, Duval S, Altman DG, Moher D, Higgins JPT (2011) Recommendations for examining and interpreting funnel plot asymmetry in meta-analyses of randomised controlled trials. *BMJ*, **343**:d4002.

Sterne JAC, White IR, Carlin JB, Spratt M, Royston P, Kenward MG, Wood AM, Carpenter JR (2009) Multiple imputation for missing data in epidemiological and clinical research: potential and pitfalls. *BMJ*, **338**:b2393.

Sterne TE (1954) Some remarks on confidence or fiducial limits. *Biometrika,* **41**:275–278.

Stokes M, Davis CS, Koch GG (2012) *Categorical Data Analysis Using SAS.* 3rd edn. SAS Institute, Cary, NC.

Storvik G (2012) Bootstrapping. In M Veierød, S Lydersen, P Laake (eds.) *Medical Statistics in Clinical and Epidemiological Studies.* Gyldendal akademisk, Oslo.

Stuart A (1955) A test for homogeneity of the marginal distributions in a two-way classification. *Biometrika,* **42**:412–416.

Suissa S, Shuster JJ (1985) Exact unconditional sample sizes for the 2×2 binomial trial. *Journal of the Royal Statistical Society, Series A,* **148**:317–327.

Suissa S, Shuster JJ (1991) The 2×2 matched-pairs trial: Exact unconditional design and analysis. *Biometrics,* **47**:361–372.

Sutton AJ, Abrams KR, Jones DR, Sheldon TA, Song F (2000) *Methods for Meta-Analysis in Medical Research.* John Wiley & Sons, Inc., Chichester, West Sussex.

Sverdrup S (1990) The delta multiple comparison method. Performance and usefulness. *Scandinavian Journal of Statistics,* **17**:115–134.

Tang ML, Li HQ, Tang NS (2012) Confidence interval construction for proportion ratio in paired studies based on hybrid method. *Statistical Methods in Medical Research,* **21**:361–378.

Tang ML, Ling MH, Ling L, Tian G (2010) Confidence intervals for a difference between proportions based on paired data. *Statistics in Medicine,* **29**:86–96.

Tang ML, Tang NS, Carey VJ (2004) Confidence interval for rate ratio in a 2×2 table with structural zero: an application in assessing false negative ratio when combining two diagnostic tests. *Biometrics,* **60**:550–555.

Tang ML, Tang NS, Chan ISF (2005) Confidence interval construction for proportion difference in small-sample paired studies. *Statistics in Medicine,* **24**:3565–3579.

Tang NS, Tang ML (2003) Statistical inference for risk difference in an incomplete correlated 2×2 table. *Biometrical Journal,* **45**:34–46.

Tang NS, Tang ML, Chan ISF (2003) On tests of equivalence via non-unity relative risk for matched-pair design. *Statistics in Medicine,* **22**:1217–1233.

Tango T (1998) Equivalence test and confidence interval for the difference in proportions for the paired-sample design. *Statistics in Medicine,* **17**:891–908.

Tarone RE (1985) On heterogeneity tests based on efficient scores. *Biometrika*, **72**:91–95.

Terpstra TJ (1952) The asymptotic normality and consistency of Kendall's test against trend, when ties are present in one ranking. *Indagationes Mathematicae*, **14**:327–333.

Theil H (1970) On the estimation of relationships involving qualitative variables. *American Journal of Sociology*, **76**:103–154.

Thompson SG, Sharp SJ (1999) Explaining heterogeneity in meta-analysis: a comparison of methods. *Statistics in Medicine*, **18**:2693–2708.

Tian GL, Li HQ (2017) A new framework of statistical inference based on the valid joint sampling of the observed counts in an incomplete contingency table. *Statistical Methods in Medical Research*. Published online, doi: 10.1177/0962280215586591.

Turner RM, Omar RZ, Yang M, Goldstein H, Thompson SG (2000) A multilevel model framework for meta-analysis of clinical trials with binary outcomes. *Statistics in Medicine*, **19**:3417–3432.

Tuyns AJ, Péquignot G, Jensen OM (1977) Le cancer de l'oesophage en Ille-et-Vilaine en fonction des niveaux de consommation d'alcool et de tabac. *Bulletin du Cancer*, **64**:45–60.

Van Balen FAM, Smit WM, Zuithoff NPA, Verheij TJM (2003) Clinical efficacy of three common treatments in acute otitis externa in primary care: Randomised controlled trial. *BMJ*, **327**:1201.

Van Buuren S (2012) *Flexible Imputation of Missing Data*. Chapman & Hall/CRC, Boca Raton, FL.

VanderWeele TJ, Knol MJ (2014) A tutorial on interaction. *Epidemiological Methods*, **3**:33–72.

Venzon DJ, Moolgavkar SH (1988) A method for computing profile-likelihood-based confidence intervals. *Journal of the Royal Statistical Society, Series C*, **37**:87–94.

Vigderhous G (1979) Equivalence between ordinal measures of association and tests of significant differences between samples. *Quality & Quantity*, **13**:187–201.

Vollset SE, Hirji KF (1991) A micro computer program for exact and asymptotic analysis of several 2 × 2 tables. *Epidemiology*, **2**:217–220.

Wakefield J (2004) Ecological inference for 2 × 2 tables (with discussion). *Journal of the Royal Statistical Society, Series A*, **167**:385–445.

Walter SD (2001) Number needed to treat (NNT): Estimation of a measure of clinical benefit. *Statistics in Medicine*, **20**:3947–3962.

Wellek S (2010) *The Statistical Hypotheses of Equivalence and Noninferiority.* 2nd edn. Chapman & Hall/CRC, Boca Raton, FL.

Wellek S, Ziegler A (2012) Cochran-Armitage test verus logistic regression in the analysis of genetic assocation studies. *Human Heredity*, **73**:14–17.

Whitehead J (1993) Sample size calculations for ordered categorical data. *Statistics in Medicine*, **12**:2257–2271.

Wilson EB (1927) Probable inference, the law of succession, and statistical inference. *Journal of the American Statistical Association*, **22**:209–212.

Wooldridge JM (2013) *Introductory Econometrics: A Modern Approach.* 5th edn. South-Western Cengage Learning, Mason, OH.

Woolf B (1955) On estimating the relation between blood group and disease. *Annals of Human Genetics*, **19**:251–253.

Yanagawa T, Fujii Y (1990) Homogeneity tests with a generalized Mantel-Haenszel estimator for L $2 \times K$ tables. *Journal of the American Statistical Association*, **85**:744–748.

Yanagawa T, Fujii Y (1995) Projection-method Mantel-Haenszel estimator for K $2 \times J$ tables. *Journal of the American Statistical Association*, **90**:649–656.

Yates F (1934) Contingency tables involving small numbers and the χ^2 test. *Supplement to the Journal of the Royal Statistical Society*, **1**:217–235.

Yusuf S, Peto R, Lewis J, Collins R, Sleight P (1985) Beta blockage during and after a myocardial infarction: An overview of the randomized trials. *Progress in Cardiovascular Diseases*, **27**:335–371.

Zelen M (1971) The analysis of several 2×2 contingency tables. *Biometrika*, **58**:129–137.

Zhao YD, Rahardja D (2013) Estimation of the common risk difference in stratified paired binary data with homogeneous strata effect. *Journal of Biopharmaceutical Statistics*, **23**:848–855.

Zhao YD, Rahardja D, Wang DH, Shen H (2014) Testing homogeneity in stratified paired binary data. *Journal of Biopharmaceutical Statistics*, **24**:600–607.

Zheng G (2008) Analysis of ordered categorical data: Two score-independent approaches. *Biometrics*, **64**:1276–1279.

Zheng G, Gastwirth JL (2006) On the estimation of the variance in Cochran-Armitage trend test for genetic association using case-control studies. *Statistics in Medicine*, **25**:3150–3159.

Zheng G, Yang Y, Zhu X, Elston RC (2012) *Analysis of Genetic Association Studies*. Springer, New York.

Zhou XH, Obuchowski NA, McClish DK (2011) *Statistical Methods in Diagnostic Medicine*. 2nd edn. John Wiley & Sons, Inc., Hoboken, New Jersey.

Zigmond AS, Snaith RP (1983) The hospital anxiety and depression scale. *Acta Psychiatrica Scandinavica*, **67**:361–370.

Zou GY (2010) Confidence interval estimation under inverse sampling. *Computational Statistics and Data Analysis*, **54**:55–64.

Index